Environmental Toxicants and Lifestyle Diseases

Sheikh Raisuddin • Shahzad Ahmad
Mehjbeen Javed
Editors

Environmental Toxicants and Lifestyle Diseases

Springer

Editors
Sheikh Raisuddin
Department of Medical Elementology and Toxicology
Jamia Hamdard, School of Chemical and Life Sciences
New Delhi, Delhi, India

Shahzad Ahmad
Department of Medical Elementology and Toxicology
Jamia Hamdard, School of Chemical and Life Sciences
New Delhi, Delhi, India

Mehjbeen Javed
Department of Medical Elementology and Toxicology
Jamia Hamdard, School of Chemical and Life Sciences
New Delhi, Delhi, India

ISBN 978-981-95-0574-6 ISBN 978-981-95-0575-3 (eBook)
https://doi.org/10.1007/978-981-95-0575-3

This Springer imprint is published by the registered company Springer Nature Singapore Pte Ltd.
The registered company address is: 152 Beach Road, #21-01/04 Gateway East, Singapore 189721, Singapore

To our parents

Foreword

Globally, the presence of pollutants/contaminants in the air, water, soil, as well as in the food chain, is continuously on the rise. These pollutants are in the form of dust, gases, microplastics, electronic waste, industrial chemicals and effluents, agricultural chemicals, heavy metals, particulates, etc. This is essentially due to the increased anthropogenic activities as well as climate change and is adversely impacting the biodiversity. There is a global concern regarding the changing lifestyles in humans as well as an increased incidence of non-communicable diseases which are largely linked to environmental pollution and contaminated food. Although several lifestyle factors and environmental contaminants have been identified as causative agents for both physical and mental ailments, more work and in-depth understanding are needed. It is imperative to identify chemicals and pollutants that may lead to reproductive toxicity. Further, diagnosing a disease at an early stage with the help of biomarkers allows for better patient management as well as reduces morbidity and mortality.

This book provides comprehensive information with insights on environmental toxicology, and its impact on living organisms in general and human health in particular. The chapters address the current situation of toxicants like microplastics, endocrine-disruptive compounds (EDCs), polyaromatic hydrocarbons, nanoparticles, herbicides, etc. along with the prevalence of polycystic ovary syndrome (PCOS), infertility, obesity, cancer, cardiovascular disease, and chronic obstructive pulmonary disease. Also, the mitigation and management of lifestyle diseases have been discussed.

The book will indeed be very useful for policymakers, risk assessors, graduate and post-graduate students, researchers, environmental toxicologists, and others. I wish to compliment the editors for bringing out a comprehensive book on a very contemporary topic.

Director, Centre of BioMedical Research
Lucknow, India

Former Director, CSIR-Indian Institute of Toxicology Research
Lucknow, India

Alok Dhawan

Preface

The modern world is experiencing an unprecedented increase in lifestyle diseases, and many of them are closely associated with environmental factors including exposure to toxic chemicals. Human exposure to hazardous chemicals, pollutants, and synthetic compounds has multiplied manyfold in the last few decades through technological and industrial advancement requiring extensive use of chemicals. In recent times, many new chemical entities have been introduced in industrial processes for which appropriate risk assessment has yet to be undertaken. Such chemicals with potential human toxicity include microplastics, nanochemicals, chemicals leached from electronic waste, pharmaceuticals in environment, and chemical products from genetically modified organisms. Toxicants present in air, water, food, and consumer products have been implicated in a wide range of chronic diseases such as cardiovascular disorders, metabolic syndromes, reproductive dysfunctions, neurological disorders, and various forms of cancer. The interplay between environmental toxicants and lifestyle choices has made it imperative to understand their impact and devise effective prevention and management strategies.

This book, *Environmental Toxicants and Lifestyle Diseases*, presents a multidisciplinary and in-depth examination of the critical role of environmental toxicants in the etiology of lifestyle diseases. The book is structured into three major parts—emerging toxicants, their impact on human health, and strategies for disease prevention and management. A total of 26 chapters, divided into these three parts, provide a comprehensive insight into the understanding of the relationship of environmental toxicants with lifestyle diseases and are expected to address the various issues faced by the scientific community, health managers, and policymakers.

The first part theme is *Emerging Toxicants* and explores novel and persistent environmental toxicants that have gained scientific and public attention in recent years. This part contains ten chapters. Chapter "Microplastics: An Emerging Concern of Human Health," discusses the increasing presence of microplastics in the environment and their potential health risks. It covers sources of microplastics, their pathways into the human body, and their biological effects, including oxidative stress and inflammation. Chapter "Environmental Chemicals and Risk of Polycystic Ovary Syndrome (PCOS)," investigates the role of environmental chemicals in the pathogenesis of PCOS, a common reproductive disorder in women. The chapter reviews mechanisms by which these chemicals interfere with hormonal balance, leading to PCOS development. Chapter "Microplastics and Lung Diseases: An Overview," discusses the potential of airborne microplastics to exacerbate respiratory illnesses, including asthma and chronic obstructive pulmonary disease (COPD). Chapter "Occurrence, Fate, Effects, Risk Assessment, Biodegradation, and Bio-transformation of Polycyclic Aromatic Hydrocarbons (PAHs) and Health Hazards of Their Metabolites," covers the sources, environmental fate, and health effects of PAHs, a class of organic pollutants known for their carcinogenic potential. Chapter "Navigating Nanochemicals: Understanding the Landscape of Recent Emerging Contaminants," analyzes the applications, benefits, and potential risks of nanochemicals, particularly their unintended consequences on human health. Chapter "Human Health Dimension of Exposure to Copper Oxide Nanoparticles," highlights toxicity mechanisms, exposure pathways, and possible mitigation measures for nanoparticle-induced toxicity. Chapter "Glyphosate Herbicide: An Omnipresent Environmental Toxicant and Its Potential Human Health Impacts," discusses the controversial use of glyphosate, its presence in the food chain, and potential associations with diseases such as cancer. Chapter "Electronic Waste: An Emerging Pollutant of Human Health Concern and Biological Techniques for Their Treatment," examines the toxicity of e-waste, its impact on human health, and biological techniques for its treatment. Chapter "The Environmental Impact on Antimicrobial Resistance: A Threat to Biodiversity," discusses how environmental contamination with antibiotics promotes resistance, endangering biodiversity, and public health. Chapter "Role of Perfluoroalkyl Substances (PFAS) on the Reproductive System," evaluates the endocrine-disrupting effects of PFAS and their implications on the reproductive system and fertility.

The second part of this book contains 12 chapters broadly on the theme of *Environmental Toxicants and Lifestyle Diseases*. Chapters in this part delves into how environmental toxicants contribute to non-communicable diseases including lifestyle diseases. Chapter "Toxic Effect of Nanoparticles on the Gut Microbiota," analyzes the impact of engineered nanoparticles on gut microbiota composition and functions and their health implications. Chapter "The Nexus of Chemicals and Cancer: Environmental Toxicants and Lifestyle Effects Unveiled," discusses how carcinogens from environmental sources, combined with lifestyle habits, influence cancer risk. Furthermore, Chapter "Environmental Chemical-Induced Heat Shock Proteins in Toxicology: Past, Present, and Future Trends," explores the role of heat shock proteins in cellular stress responses triggered by environmental toxicants. Chapter "Role of Environmental Toxicants and Lifestyle Diseases in Infertility," examines how prolonged exposure to certain chemicals impairs reproductive function leading to, in certain cases, infertility. Chapter "Contribution of Environmental Toxicants and Lifestyle in the Development of Cardiovascular Diseases," discusses air pollution and chemically induced cardiovascular risks. Chapter "Association of Environmental Toxicants and Lifestyle in the Advancement of Infertility," analyzes data linking environmental stressors to declining fertility rates in the exposed population. Chapter "Elucidating the Molecular Pathophysiology and Mechanistic Insights of Chronic Obstructive Pulmonary Disease," explores the molecular mechanisms underlying COPD due to environmental pollutants, and

Chapter "Smoking as a Confounding Factor in Environmentally Induced Lifestyle Diseases," examines how smoking exacerbates the health effects of environmental toxicants. Chapter "Washed Covert Hair Dye Components are a Threat to Environmental Sustainability and a Potential Risk Factor for Increasing Cases of Breast Cancer," investigates the role of hair dye chemicals in carcinogenesis, while Chapter "Interactions Between Genetic and Environmental Risk Factors in Lifestyle Diseases," discusses gene-environment interactions contributing to chronic illnesses. Chapter "Effects of Dust Particles on Provoking the Risk of Lung Cancer," highlights the role of airborne particulate matter and other air-borne toxic chemicals in lung cancer development, while Chapter "Air Pollutants Aggravate Some Major Lifestyle Diseases and Beneficial Role of Medicinal Phytocompounds," explores medicinal phytocompounds as protective agents against air pollution-related diseases.

The third part of this book contains contributed chapters on the theme of *Management of Lifestyle Diseases Caused by Exposure to Toxicants* and focuses on strategies to counteract toxicant-induced diseases. There are several chapters on the potential role of natural products in mitigation of toxic effects and disease manifestation caused by exposure to toxic chemicals. This part comprises of four chapters. Chapter "Lifestyle Modification Strategies as the First Line of Chronic Disease Management," highlights how diet and physical activity mitigate toxicant-induced disorders, and Chapter "Natural Products for the Management of Oral Cancer: A Lifestyle Problem and its Solutions," examines bioactive compounds as alternative cancer therapies. Chapter "Natural Products as Promising Drug Candidates for Management of Lifestyle Diseases," explores phytochemicals with therapeutic potential. The last chapter of the book, "Diet and Nutrition in Disease Prevention and Maintenance of Healthy Lifestyle," analyzes dietary habits that reduce toxicant exposure risks.

This book brings together contributions from leading experts, researchers, and academicians from the fields of toxicology, environmental science, and biomedical research. It provides a detailed analysis of specific toxicants, their sources, mechanisms of toxicity, associated health risks, and potential remedial measures. We hope that *Environmental Toxicants and Lifestyle Diseases* serves as a critical reference for researchers, healthcare professionals, and policymakers to assess the health risk associated with exposure to environmental chemicals, especially with regard to their direct or indirect role in lifestyle diseases in humans. There are contributions on the strategies to mitigate environmental health risks. Overall, this volume brings together updated research and development regarding the role of chemicals in lifestyle diseases and strategies to mitigate them.

New Delhi, India

Sheikh Raisuddin
Shahzad Ahmad
Mehjbeen Javed

Acknowledgments

The editors are thankful to Prof. Basu Dev Banerjee, Adjunct Professor at the Department of Medical Elementology and Toxicology, Jamia Hamdard, New Delhi, and Santosh Deemed to be University, Ghaziabad, India, for critically reviewing each contributed chapter and offering valuable inputs for improvement of presentation aspects. The editors would also like to acknowledge Mr. Anish Cherian, Assistant, Internal Quality Assurance Cell (IQAC), Jamia Hamdard, New Delhi, who offered technical support in graphic designing of figures, and Mr. Sharif Khan, Assistant, IQAC, Jamia Hamdard, New Delhi, for secretarial assistance.

Contents

Emerging Toxicants

Microplastics: An Emerging Concern of Human Health 3
P. Said Hamid Thangal and P. A. Ahammed Shareef
Introduction 4
Microplastics 4
Classification of Microplastics 6
Classification of Microplastics Based on Types 6
Classification of Microplastics Based on Size 7
Classification of Microplastics Based on Shape 8
Classification of Microplastics Based on Chemical Nature 9
Sources of Microplastics 10
Formation of Microplastics 11
Biotic Degradation of Microplastics 11
Abiotic Degradation of Microplastics 11
Entry of Microplastics to the Human Body 12
Ingestion 12
Inhalation 13
Dermal contact 13
Potential Toxicity and Impact on Health 13
Impact on Animals 13
Impact on Humans 14
Digestive System 15
Respiratory System 16
Circulatory System 16
Reproductive System 16
Nervous System 16
Immune System 17
Interaction with Organic Pollutants, Heavy Metals, and Pathogens 17
Impact on Ecosystem and Environment 17
Control Measures 19
Decrease Plastic Manufacture and Use 19
Enhanced Waste Management Systems 20

Upgrading Wastewater Treatment Plants 20
Control of Microplastic Pollution from Textiles 20
Regulation on the Use of Microplastics in Cosmetics 21
Public Awareness and Education 21
Future Perspectives of Research 21
Conclusion 22
References 22

Environmental Chemicals and Risk of Polycystic Ovary Syndrome (PCOS) 25
Mehjbeen Javed, Suramya, Humaira Naaz Bhutto, Shaesta Shahid, Mohd. Anas Saifi, Shahzad Ahmad, and Sheikh Raisuddin
Introduction 26
PCOS Inducing Effects of EDCs 27
Heavy Metal-Induced PCOS Like Symptoms 30
Cadmium (Cd) 30
Lead (Pb) 31
Copper (Cu) 32
Iron (Fe) 32
Mercury (Hg) 33
Thallium (Tl) 33
Arsenic (As) 34
Heavy Metal Action Mechanism as EDCs 35
Pharmaceutical Agents as EDCs 37
Epidemiological Data 37
Mitigation of PCOS Symptoms 38
Conclusion 38
References 38

Microplastics and Lung Diseases: An Overview 43
Hina Javed and Mohd. Anas Saifi
Introduction 44
Classification of MPs Based on Origin 45
Sources and Features of Inhaled MPs 45
Effects of Industrial Emissions of MPs 46
MPs and Lungs 46
Chronic Obstructive Pulmonary Disease (COPD) 47
Asthma 48
Pneumoconiosis 48
Mechanisms of MP-Induced Lung Toxicity 48
Oxidative Stress and Inflammation 48
Cellular Damage and Apoptosis 49
Identification and Monitoring of MPs 50
Conclusion 50
References 51

Occurrence, Fate, Effects, Risk Assessment, Biodegradation, and Bio-transformation of Polycyclic Aromatic Hydrocarbons (PAHs) and Health Hazards of Their Metabolites 55
Annu Khatri, Krishan Kumar, and Indu Shekhar Thakur
Introduction 56
Occurrence and Types of Persistent Organic Pollutants 56
Intentionally POPs 57
Industrial Chemicals 58
Organochlorine Pesticide 58
Unintentionally POPs 58
Dioxins and Dibenzofurans 59
Polycyclic Aromatic Hydrocarbons 59
Sources of Polycyclic Aromatic Hydrocarbons 61
Transport and Fate of PAHs in the Environment 62
PAH Emission and Deposition in the Atmosphere 62
PAHs in Surface Soils 62
PAHs in Sediments 62
PAH in Food Occurrence 63
Degradation of Polycyclic Aromatic Hydrocarbons 64
Physical Degradation Method 64
Chemical Degradation Method 64
Biological Degradation Method 66
PAHs Degradation by Bacteria 66
PAHs Degradation by Fungi 66
PAHs Degradation by Algae 67
PAHs Degradation by Plant 67
Rhizoremediation 68
Vermiremediation 68
Omics Techniques for Investigating Sequence-Based and Function-Based PAH Bioremediation 69
Metabolites of Polycyclic Aromatic Hydrocarbon 72
Health Hazards and Toxicity of PAH Metabolites 73
Short-Term Health Effects (Acute) 74
Long-Term Health Effects (Chronic) 75
Extraction and Analysis of PAH 76
Fluorescence Spectroscopy 76
GC-MS (Gas Chromatography-Mass Spectrometry) 76
Risk Assessment of PAH 77
Future Perspective 77
Conclusion 78
References 80

Navigating Nanochemicals: Understanding the Landscape of Recent Emerging Contaminants . . . 83
Amrit Krishna Mitra and Santanu Chakrabarti
Introduction . . . 84
Applications of Nanochemicals: Current Trends and Emerging Uses . . . 85
Assessing the In Vitro Toxicity of Metal Oxide Nanoparticles: Insights and Implications . . . 86
Toxicity of Co_3O_4-Based NPs . . . 86
Toxicity of TiO_2-Based NPs . . . 89
Toxicity of CuO-Based NPs . . . 89
Toxicity of Iron Oxide-Based NPs . . . 89
Toxicity of MnO-Based NPs . . . 90
Toxicity of ZnO-Based NPs . . . 90
A Comparative Analysis of Nanoparticle Toxicity . . . 91
Assessing the In Vivo Toxicity of Metal Oxide Nanoparticles: Insights into Organ-Specific Damage and Systemic Effects . . . 91
Assessing the In Vitro Toxicological Impact of Metallic Nanoparticles . . . 92
In Vitro Toxicity Based on Silver Nanoparticles . . . 92
In Vitro Toxicity Based on Gold Nanoparticles . . . 93
In Vitro Toxicity Based on Copper Nanoparticles . . . 93
In Vitro Toxicity Based on Platinum Nanoparticles . . . 93
In Vitro Toxicity Based on Aluminum Nanoparticles . . . 94
In Vitro Toxicity Based on Palladium Nanoparticles . . . 94
Assessing the In Vivo Toxicological Impact of Metallic Nanoparticles . . . 95
In Vivo Toxicity of Silver-Based Nanoparticle . . . 95
In Vivo Toxicity of Gold-Based Nanoparticle . . . 95
In Vivo Toxicity of Copper-Based Nanoparticle . . . 95
In Vivo Toxicity of Nickel-Based Nanoparticle . . . 96
Toxicity of Non-metallic Nanoparticles . . . 96
Toxicity of Silica-Based Nanoparticles . . . 96
Toxicity of Quantum Dot-Based Nanoparticles . . . 96
Toxicity of Polymer-Based Nanoparticles . . . 97
Molecular Mechanisms of Nanoparticle-Induced Toxicity . . . 97
Evaluating the Safety of Nanochemicals: Risks and Mitigation Strategies . . . 98
Conclusion and Future Perspectives . . . 99
References . . . 99

Human Health Dimension of Exposure to Copper Oxide Nanoparticles . . . 103
Naveed Raza, Shabana Naz, Sajida Batool, and Rifat Ullah Khan
Nanotechnology . . . 104
Nanoparticles . . . 104
Nanoparticle Applications . . . 105
Metal Oxide Nanoparticles Toxicity . . . 106
Copper Oxide NPs . . . 106

Synthesis Methods of CuO NPs 108
Characteristics of CuO NPs 109
Toxicity of Copper Oxide NPs 110
Oxidative Stress 113
Dissolution 114
Human Exposure 114
CuO NPs-Induced Toxicity: What's Next? 116
Conclusion 116
References 118

Glyphosate Herbicide: An Omnipresent Environmental Toxicant and Its Potential Human Health Impacts 121
Reenu Singh, Sabreen Bashir, and Anand Mohan
Introduction 122
Glyphosate: Chemical Properties 123
Glyphosate: X-Ray Crystallography Structure 123
Glyphosate: Mode of Action 124
Glyphosate: Global Consumption 125
Glyphosate Resistance: Weeds and Plants 126
Glyphosate Resistance: Cropping Systems 127
Glyphosate Impact on Animals 127
Impact of Glyphosate on Humans 127
Carcinogenicity of Glyphosate 130
Epidemiological Studies 132
Conclusion 134
References 135

Electronic Waste: An Emerging Pollutant of Human Health Concern and Biological Techniques for Their Treatment 141
Manglam Bajpai, Mohit Nigam, Lalit Kumar Singh, and Garima Awasthi
Introduction 142
Distribution of E-waste 143
Electronic Waste Impact Related to Human Health 145
Electronic Waste Management 147
    Bioleaching 148
    Biomineralization 148
    Biotransformation 149
    Bioaccumulation 149
    Biosorption 150
Alternative Methods for Managing Electronic Waste 150
Conclusion 153
References 154

The Environmental Impact on Antimicrobial Resistance: A Threat to Biodiversity . . . 157
Farhan Ahmad Khan
Introduction . . . 158
The Scope of the AMR Problem . . . 158
Environmental Reservoirs and Pathways . . . 158
Human Activities and AMR Dissemination . . . 159
Methods of Antimicrobial Resistance (AMR) Transmission . . . 159
Environmental Impact of AMR . . . 160
Addressing the Environmental Dimensions of AMR . . . 160
Environmental Drivers of AMR . . . 160
Pharmaceutical Manufacturing and Waste . . . 161
Agricultural Practices . . . 161
Wastewater Treatment and Effluents . . . 161
Human Waste and Sewage . . . 161
Animal Husbandry . . . 162
Endocrine Disruptors . . . 162
Environmental Transmission Pathways . . . 162
Impact of Climate Change . . . 162
Mechanisms of AMR Spread in the Environment . . . 163
Horizontal Gene Transfer (HGT) . . . 163
Selective Pressure from Antimicrobial Pollutants . . . 164
Formation of Biofilms . . . 164
Role of Mobile Genetic Elements . . . 165
Environmental Transmission Pathways . . . 165
Environmental Impact of AMR . . . 166
Contamination of Natural Resources . . . 166
Disruption of Microbial Communities . . . 167
Implications for Wildlife . . . 167
Broader Ecological Consequences . . . 168
Key Challenges and Solutions . . . 168
Key Challenges . . . 169
Possible Resolutions . . . 170
Conclusion and future directions . . . 171
Summarizing the Challenges . . . 171
Comprehensive Strategies for Mitigation . . . 171
Strengthening Regulatory Frameworks . . . 171
Advanced Wastewater Treatment . . . 172
Implementing Comprehensive Global Surveillance and Monitoring Systems . . . 172
Research and Innovation . . . 172
Public Awareness and Education . . . 172
Promoting International Collaboration . . . 172
Prospects for the Future . . . 173
References . . . 173

Role of Perfluoroalkyl Substances (PFAS) on the Reproductive System 177
Suramya, Mehjbeen Javed, Shaesta Shahid, and Sheikh Raisuddin
Introduction 178
Physiochemical Properties 179
Source, Exposure, and Health Hazards 179
How Does PFAS Affect Male Reproductive Health? 180
How Does PFAS Affect Female Reproductive Health? 184
Conclusion 187
References 188

Environmental Toxicants and Lifestyle Diseases

Toxic Effect of Nanoparticles on the Gut Microbiota 195
Aryan Shukla, Abhishek Kumar Mishra, Nayanika Tiwari, and Garima Awasthi
Introduction 196
Application of Nanoparticles in Current Clinical Practices/Sources of Exposure 197
Overview of Human Microbiota 199
Emerging Concerns and Research Gaps 200
Properties of Metal Nanoparticles 200
    Physical and Chemical Properties 200
Structure and Chemical Formulae of NPs 202
Solubility and Aggregation Behavior 204
Negative Interaction Mechanisms with Biological Systems 204
Reactivity Inside Biological Systems 205
Mechanisms of Nanoparticle Toxicity 206
Oxidative Stress Induction 206
Alteration of Microbial Membrane Integrity 207
Impact of Metal Nanoparticles on Human Microbiota 208
Regulatory and Safety Considerations 210
Conclusion 212
References 214

The Nexus of Chemicals and Cancer: Environmental Toxicants and Lifestyle Effects Unveiled 217
Adela Perolla
Introduction 218
Definition and Sources of Environmental Toxicants 219
The Impact of Environmental Toxicants on the Development of Cancer 220
Lifestyle 223
Nutrition and Body Mass Index (BMI) 224
Drug Addiction 226
    Alcohol Use and the Risk of Cancer Development 227
The Correlation Between Tobacco Consumption and the Likelihood of Developing Cancer 228

Opioids and Illicit Drugs 230
Maternal Substance Use 231
Physical Activity 232
Combined Lifestyle Factors 233
Genetic Interactions 233
Conclusion 233
References 234

Environmental Chemical-Induced Heat Shock Proteins in Toxicology: Past, Present, and Future Trends 239
Ashutosh Pandey, Hitesh Singh Chouhan, Debapratim Kar Chowdhuri, and Anurag Sharma
Introduction 240
Heat Shock Proteins 240
Xenobiotic Stress: Impacts from Cellular to Organismal Level in Exposed Organisms 242
Implication of HSP in Xenobiotic Stress: A Perspective in Toxicology 244
  HSP as Sentinel Species Against Xenobiotic-Induced Stress 244
  Modulatory Role of HSP in Xenobiotic Stress 245
Protective Role of HSP in Xenobiotics-Induced Neurotoxicity 245
Modulation of Nephrotoxicity by HSP 247
Protective Role of HSP in Xenobiotics-Induced Genotoxicity 248
HSP Modulates Xenobiotics-Induced Hepatotoxicity 248
Protective Role of HSP in Xenobiotics-Induced Developmental and Reproductive Toxicity 249
HSP in Nanotoxicology 250
Prolonged/Long-Term Xenobiotic Stress: HSP and Hormesis 251
Limitations 254
Conclusions 254
References 255

Role of Environmental Toxicants and Lifestyle Diseases in Infertility 261
David Chukwuma Nwikwe
Introduction 262
Environmental Toxicants 264
  Endocrine-Disrupting Chemicals (EDCs) 264
  Smoking 265
  Heavy Metals 266
  Pollutants 267
  Radiations 268
  Pesticides 268
Lifestyle Diseases 268
  Excessive Alcohol Intake 269
  Obesity 269

Nutrition 270
Stress 270
Smoking 271
Age 271
Sexually Transmitted Diseases (STDs) 271
Interaction Between Environmental Toxicants and Lifestyle Diseases 272
Obesity and Endocrine Disruptive Compounds 272
Mitigation Strategies 273
Preventive Measures 273
Screening and Monitoring 273
Education and Awareness 273
Effective Policy and Regulation 274
Fostering Multidisciplinary Collaboration 274
Effective Research and Surveillance 274
Treatment Measures 274
Herbal Therapy 275
Traditional Practice 275
Counseling 275
Holistic Measure 276
Conclusion and Future Directions 276
References 277

Contribution of Environmental Toxicants and Lifestyle in the Development of Cardiovascular Diseases 281
Sarah Patel, Haroon Habib Beigh, and Kudaibergen Osmonaliev
Introduction 282
Environmental Toxicants and the Development of CVD 282
Lead 283
Cadmium 284
Arsenic 284
Pesticides 285
Lifestyle Factors That Play a Role in the Development of CVD 287
Smoking 287
Alcohol Consumption 289
Sedentary Lifestyle 290
Diet 291
Cardiovascular Disease 292
Conclusion 292
References 292

Association of Environmental Toxicants and Lifestyle in the Advancement of Infertility 295
Haroon Habib Beigh
Introduction 296
Ecological Toxicants and Their Contribution to Infertility 296

Harmful Chemicals and Its Association with Infertility 297
Dioxins 297
Plastic Additives (Bisphenol A) 298
Phthalates 299
Heavy Metals 300
Lead (Pb) 300
Cadmium (Cd) 301
Cobalt, Nickel, and Zinc 301
Air Pollution 303
Extreme Heat 304
Life Style Factors 305
Alcohol Consumption 305
Smoking 307
Eating Habits 308
Drug Abuse 308
BMI (Body Mass Index) 308
Caffeine Intake 309
Stress 309
Radiation and Cellular Phones 310
Exercise 310
Conclusion 310
References 311

Elucidating the Molecular Pathophysiology and Mechanistic Insights of Chronic Obstructive Pulmonary Disease 315
Saba Beigh
Introduction 316
Epidemiology and Risk Factors 318
Prevalence and Global Burden 318
Genetic Susceptibility 320
Environmental and Occupational Exposures 321
Pathophysiological Mechanisms 322
Inflammatory Mediators in COPD 323
Clinical Implications and Management 325
Pathophysiological Mechanisms and Outcomes 325
Non-pharmacological Approaches for Managing COPD 326
Nutrition 326
Pulmonary Rehabilitation 328
Non-invasive Positive Pressure Ventilation 329
Pharmacological Approaches for Refractory COPD Management 330
Conclusion 332
References 333

Smoking as a Confounding Factor in Environmentally Induced Lifestyle Diseases 337
Syed Amaan Ali, Neha Agrawal, Mohammad Atif, Mohd Mustafa, and Safia Habib
Introduction 338
Environmental Factors in Lifestyle Diseases 339
Understanding Confounding Factors 342
Mechanisms of Smoking-Related Health Effects 345
Pathophysiology and the Role of Smoking in Lifestyle Diseases 346
Impact of Smoking Cessation 347
Public Health Implications 348
Future Research Directions and Conclusion 350
References 351

Washed Covert Hair Dye Components are a Threat to Environmental Sustainability and a Potential Risk Factor for Increasing Cases of Breast Cancer 353
Hamda Khan, Safia Habib, Syed Amaan Ali, and Moinuddin
Introduction to Hair Dye 354
Hair Dye Ingredients 355
Hair Dye-Induced Adverse Health Effects 356
Hair Dye Use and Cancer 358
Hair Dye Contamination 359
  Hair Dye Usage Versus Hair Dye Contamination 359
  Hair Dye Contamination Is more Lethal than Hair Dye Usage 359
  Impact of Hair Dye Contamination on the Environment 360
  Industries and Sectors Contributing to Hair Dye Contamination 361
  Hair Dye Contamination-Induced Adverse Health Effects 363
  Hair Dye Contamination and Cancer 364
Breast Cancer and Its Global Incidence 364
  Hair Dye Contamination and Breast Cancer 365
Possible Mechanism of Breast Cancer Incidence Through Aquatic Pollution Via Hair Dye 366
  Drinking Water Contamination by Wastewater 367
  Environmental Chemicals Are Mammary Gland Carcinogens 368
  Breast Cancer Risk Near Polluted River 369
Treatment of Coloured Wastewater 369
  Coagulation 369
  Membrane Treatment of Aquatic Environment 370
Biological Therapy 371
  Enhanced Oxidation Process 372
Conclusion 373
References 373

Interactions Between Genetic and Environmental Risk Factors in Lifestyle Diseases 377
Faizyana Ali, M. Ashraf Ganie, Sami Jan, and Fouzia Rashid
Introduction 378
Causes of Lifestyle Diseases 379
Genetic and Environmental Risk Factors Influencing Main Lifestyle Disorders 379
Obesity 379
Genetic Factors Influencing Obesity 380
Environmental Factors Influencing Obesity 380
Diabetes Mellitus 382
Gene–Environment Interplay in DM 383
Genetic Factors 383
Other Genetic Associations 384
Environmental Factors 384
Viral Infections 385
Dietary Factors 385
Type 2 Diabetes Mellitus 385
Family-Based Linkage Analysis 386
The Candidate Gene Approach 386
Genome-Wide Association Studies (GWAS) 387
Cancer 387
Genetic and Environmental Risk Factors for Cancer 388
Genetic Factors 388
Environmental Factors 389
Interaction Between Genetic and Environmental Factors 389
Polycystic Ovary Syndrome (PCOS) 389
Genetic Basis of PCOS 390
Genome-Wide Association Studies 390
Environmental Risk Factors 391
Foetal Programming Hypothesis 391
Environmental Toxins 391
Dietary-Related Weight Gain 392
Lack of Exercise and Physical Activity 393
Advanced Glycation End Products 393
Interplay Between Genetic and Environmental Factors 393
Conclusion 394
References 394

Effects of Dust Particles on Provoking the Risk of Lung Cancer 397
Prasad Sherekar, Sanvidhan G. Suke, Pooja Patel, and Basu Dev Banerjee
Introduction 398
Carcinogens Provoking the Risk of Lung Cancer 399
Respirable Crystalline Silica Dust 399

Outdoor Air Pollutants 401
Cadmium, Vinyl Chloride Monomer and Welding Fumes 403
Radon and Benzo[a]pyrene 403
Hardwood Dust 404
Asbestos and Polycyclic Aromatic Hydrocarbons 404
Diesel Exhausts (DEs), Chromium-VI and Nickel 405
World Trade Center (WTC) Dust 405
Pathophysiological Mechanisms of Lung Cancer Development 406
Conclusion 408
References 408

Air Pollutants Aggravate Some Major Lifestyle Diseases and Beneficial Role of Medicinal Phytocompounds 411
Debosree Ghosh, Sourav Das, and Partha Sarathi Singha
Introduction 412
Impact of Air Pollutants on Lung Cancer 413
Impact of Air Pollutants on Cardiovascular Disorders (CVDs) 417
Impact of Air Pollutants on Asthma 419
Impact of Air Pollutants on Chronic Obstructive Pulmonary Disease 419
Impact of Air Pollutants on Emphysema 420
Impact of Air Pollutants on Retinopathy 421
Impact of Air Pollutants on Neurological Disorders 421
Impact of Air Pollutants on Skin Disorders 422
Impact of Air Pollutants on Diabetes 423
Impact of Air Pollutants on Other Lifestyle Diseases 423
Phytocompounds Against Air Pollutants-Mediated Lifestyle Diseases 423
Conclusion 424
References 429

Management of Lifestyle Diseases Caused by Exposure to Toxicants

Lifestyle Modification Strategies as the First Line of Chronic Disease Management 437
Saika Manzoor and Fouzia Rashid
Introduction 438
Non-communicable Disease Risk Factors 438
  Environmental Risk Factors 438
Social and Demographic Elements 439
Classification of Risk Factors for NCDs 440
Main NCD Diseases 441
Economic Burden of the Four Major Non-communicable Diseases 444
National-Level Economic Burden 445
Management of Risk Factors 445
Lifestyle Intervention for Prevention of Non-communicable Diseases 446
Dietary Interventions 446
Physical Activity 448

Stress Management 449
Support Groups to Prevent NCDs 450
Conclusion 451
References 451

Natural Products for the Management of Oral Cancer: A Lifestyle Problem and its Solution 455
Vikas Kumar, Komal Kumari, and Rizwanul Haque
Lifestyle Diseases 455
Oral Cancer 461
  Types of Oral Cancer 461
  Risk Factors of Oral Cancers 461
  Signaling Pathways Involved in Oral Cancer 462
Prevention of Lifestyle Diseases 463
Management of Lifestyle Diseases Using Natural Products 463
  Natural Products 463
Therapeutic Benefits of Natural Products Against Oral Cancer 465
Natural Inhibitor for Signaling Pathways Involved in Oral Cancer Progression 467
  Natural Inhibitor of PI3K/Akt/mTOR Pathway 467
Natural Inhibitor of Wnt/β-Catenin Signaling Pathway 469
Natural Inhibitor of MAPK/ERK Signaling Pathway 470
Natural Inhibitor of Notch Signaling Pathway 471
Conclusions 472
References 473

Natural Products as Promising Drug Candidates for Management of Lifestyle Diseases 477
Hussein Sakr and Srinivasa Rao Sirasanagandla
Cardiovascular Disorders and Natural Products 478
  Resveratrol 479
  Curcumin 481
  Ginseng 481
  Ginkgo biloba 482
  Ganoderma lucidum 483
  Gynostemma pentaphyllum 483
Liver Diseases and Natural Products 484
Environmental Toxins Induced Liver Diseases 485
  Flax Seeds (*Linum usitatissimum*) 485
  Resveratrol 486
  Curcumin 487
  Silymarin/Silybin 487
  Anthocyanins 488
  Garlic 488
Conclusion 489
References 489

Diet and Nutrition in Disease Prevention and Maintenance of Healthy Lifestyle 497
Srijit Das and Norsham Juliana
Introduction 498
Diseases Related to Faulty Diet and Nutrition 498
Type 2 Diabetes Mellitus 499
Obesity 500
Hypertension 501
Cardiovascular Diseases (CVD) 502
Osteological Disorders 502
Cancer 504
Immunodeficiency Conditions 504
Dental Problems 506
Mental Health Problems 506
Latest Concept of Food and Gut Microbiota 507
Environmental Pollutants and Nutritional Aspects 508
Conclusion 509
References 510

Editors and Contributors

About the Editors

Sheikh Raisuddin (ORCiD: http://orcid.org/0000-0003-0913-8906), a Commonwealth Fellow of the British Council (UK), FST (India), FAEB (India), Brain Pool Fellow (South Korea), and IUTOX Senior Fellow (USA), is currently serving as Dean (Academics) at Jamia Hamdard, New Delhi, India. Above all, he has been consistently ranked in the top 2% scientist of the world analysis conducted by Stanford University, USA, and also ranked 34th among the Indian Scientist in the field of Biology and Biochemistry by research.com. Prof. Raisuddin has obtained his Ph.D. degree from CSIR-Indian Institute of Toxicology Research awarded by Dr. R.-M. Lohia University (Avadh University), Lucknow. He has more than three decades of teaching and research experience at national and international levels and has guided more than 50 students for Ph.D. thesis in various fields of toxicology. Furthermore, there are over 180 research publications to his credit in reputed journals. Moreover, he has been honored with several awards for his contribution to teaching and research. He is an author and also an editor of several volumes of books and reputed journals and is currently serving as a Section Editor for *Drug and Chemical Toxicology* (Taylor & Francis).

Shahzad Ahmad (ORCiD: https://orcid.org/0000-0003-3764-3299) is an Assistant Professor in the Department of Medical Elementology and Toxicology at Jamia Hamdard, New Delhi, India. He also holds visiting faculty positions at Yesbud University, Zambia, and Lloyd Institute of Forensic Science. Dr. Ahmad's expertise lies in toxicology, with a focus on neuroendocrine, endocrine, environmental, computational, analytical, and forensic toxicology. He has a strong academic background, including a Ph.D. in Toxicology, and has published extensively in his field. His research interests include the effects of environmental toxicants on hormone biosynthetic pathways.

Mehjbeen Javed (ORCiD: https://orcid.org/0000-0003-0685-9046) has graduated from the Department of Zoology, Aligarh Muslim University, Aligarh, India. She has been a recipient of prestigious national fellowships and awards. Her areas of expertise include aquatic toxicology, heavy metal and pesticide toxicity, risk assessment, interaction studies, endocrine disruption, reproductive toxicity, and next-generation sequencing. Dr. Javed has published many studies on the toxicity of heavy metals and pesticides with special reference to biomarkers in international journals of repute. Her current research includes ovarian toxicity and transcriptome analysis. In addition to this, she is also an author and an editor of books published by Lambert Academic Publications and Springer, respectively.

Contributors

Neha Agrawal Department of Periodontics and Community Dentistry, Dr. Ziauddin Ahmad Dental College, Faculty of Medicine, Aligarh Muslim University, Aligarh, Uttar Pradesh, India

P. A. Ahammed Shareef Department of Zoology, PSMO College (affiliated to University of Calicut), Tirurangadi, Malappuram, Kerala, India

Shahzad Ahmad Department of Medical Elementology and Toxicology, Jamia Hamdard, School of Chemical and Life Sciences, New Delhi, Delhi, India

Faizyana Ali Department of Clinical Biochemistry, University of Kashmir, Srinagar, Jammu and Kashmir, India

Syed Amaan Ali Department of Periodontics and Community Dentistry, Dr. Ziauddin Ahmad Dental College, Faculty of Medicine, Aligarh Muslim University, Aligarh, Uttar Pradesh, India

Mohd. Anas Saifi Molecular Toxicology Laboratory, Department of Medical Elementology and Toxicology, School of Chemical and Life Sciences, Jamia Hamdard, New Delhi, India

M. Ashraf Ganie Department of Endocrinology, Sher I Kashmir Institute of Medical Sciences, Soura, Srinagar, Jammu and Kashmir, India

Mohammad Atif Department of Paediatric and Preventive Dentistry, Dr. Ziauddin Ahmad Dental College, Aligarh Muslim University, Aligarh, Uttar Pradesh, India

Garima Awasthi Amity Institute of Biotechnology, Amity University Uttar Pradesh, Lucknow Campus, Lucknow, Uttar Pradesh, India

Manglam Bajpai Amity Institute of Biotechnology, Amity University Uttar Pradesh, Lucknow Campus, Lucknow, Uttar Pradesh, India

Basu Dev Banerjee Department of Medical Elementology and Toxicology, School of Chemical and Life Sciences, Jamia Hamdard, New Delhi, India
Department of Biochemistry, Central Research Facility, Santosh Deemed to be University, Ghaziabad, Uttar Pradesh, India

Sabreen Bashir School of Bioengineering and Biosciences, Lovely Professional University, Phagwara, Punjab, India

Sajida Batool College of Veterinary Sciences, Faculty of Animal Husbandry & Veterinary Sciences, The University of Agriculture, Peshawar, Pakistan

Haroon Habib Beigh Faculty of Medicine, AlaToo International University, Bishkek, Kyrgyz Republic

Saba Beigh Department of Public Health, Faculty of Applied Medical Sciences, Al-Baha University, Al-Baha, Kingdom of Saudi Arabia

Humaira Naaz Bhutto Molecular Toxicology Laboratory, Department of Medical Elementology and Toxicology, School of Chemical and Life Sciences, Jamia Hamdard, New Delhi, India

Santanu Chakrabarti Department of Zoology, Government General Degree College, Singur, Hooghly, West Bengal, India

Hitesh Singh Chouhan National Institute of Neurological Disorders and Stroke, National Institutes of Health, Bethesda, MD, USA

Debapratim Kar Chowdhuri Embryotoxicology Laboratory, CSIR-Indian Institute of Toxicology Research (CSIR-IITR), Lucknow, Uttar Pradesh, India

Sourav Das Department of Zoology, Government General Degree College, Kharagpur II, Paschim Medinipur, West Bengal, India

Srijit Das Faculty of Medicine and Health Sciences, Sultan Qaboos University, Muscat, Oman

Debosree Ghosh Department of Physiology, Government General Degree College, Kharagpur II, Paschim Medinipur, West Bengal, India

Safia Habib Department of Biochemistry, Faculty of Medicine, Jawaharlal Nehru Medical College, Aligarh Muslim University, Aligarh, Uttar Pradesh, India

Rizwanul Haque Department of Biotechnology, School of Earth, Biological and Environmental Science, Central University of South Bihar, Gaya, Bihar, India

Seemi Jan Department of Gynaecology, Sher I Kashmir Institute of Medical Sciences Medical College Hospital, JVC, Srinagar, Jammu and Kashmir, India

Hina Javed Department of Chemistry, Aligarh Muslim University, Aligarh, Uttar Pradesh, India

Mehjbeen Javed Department of Medical Elementology and Toxicology, Jamia Hamdard, School of Chemical and Life Sciences, New Delhi, Delhi, India

Norsham Juliana Faculty of Medicine and Health Sciences, Universiti Sains Islam Malaysia, Nilai, Negeri Sembilan, Malaysia

Farhan Ahmad Khan Department of Pharmacology, Jawaharlal Nehru Medical College and Hospital, Aligarh Muslim University, Aligarh, Uttar Pradesh, India

Hamda Khan Department of Biochemistry, Jawahar Lal Nehru Medical College, Aligarh Muslim University, Aligarh, Uttar Pradesh, India

Rifat Ullah Khan College of Veterinary Sciences, Faculty of Animal Husbandry & Veterinary Sciences, The University of Agriculture, Peshawar, Pakistan

Annu Khatri Amity School of Earth and Environmental Sciences, Amity University Haryana, Manesar, Gurugram, Haryana, India

Komal Kumari Department of Biotechnology, School of Earth, Biological and Environmental Science, Central University of South Bihar, Gaya, Bihar, India

Vikas Kumar Department of Biotechnology, School of Earth, Biological and Environmental Science, Central University of South Bihar, Gaya, Bihar, India

Saika Manzoor Department of Clinical Biochemistry, University of Kashmir, Srinagar, Jammu and Kashmir, India

Abhishek Kumar Mishra Amity Institute of Biotechnology, Amity University Uttar Pradesh, Lucknow Campus, Lucknow, Uttar Pradesh, India

Amrit Krishna Mitra Department of Chemistry, Government General Degree College, Singur, Hooghly, West Bengal, India

Anand Mohan School of Bioengineering and Biosciences, Lovely Professional University, Phagwara, Punjab, India

Moinuddin Department of Biochemistry, Jawahar Lal Nehru Medical College, Aligarh Muslim University, Aligarh, Uttar Pradesh, India

Mohd Mustafa Department of Biochemistry, Faculty of Medicine, Jawaharlal Nehru Medical College, Aligarh Muslim University, Aligarh, Uttar Pradesh, India

Shabana Naz Department of Zoology, Government College University, Faisalabad, Pakistan

Mohit Nigam Department of Biochemical Engineering, School of Chemical Technology, Harcourt Butler Technical University, Kanpur, Uttar Pradesh, India

David Chukwuma Nwikwe Department of Chemical Sciences (Biochemistry Unit), Faculty of Science, Kings University, Odeomu, Osun State, Nigeria

Kudaibergen Osmonaliev Faculty of Medicine, AlaToo International University, Bishkek, Kyrgyz Republic

Ashutosh Pandey Department of Molecular & Human Genetics, Baylor College of Medicine, Houston, TX, USA

Pooja Patel Department of Biotechnology, Priyadarshini College of Engineering, Nagpur, Maharashtra, India

Sarah Patel Faculty of Medicine, AlaToo International University, Bishkek, Kyrgyz Republic

Adela Perolla Faculty of Medicine, University of Medicine, Tirana, Albania
Department of Internal Medicine, Service of Hematology, University Hospital Centre Mother Teresa, Tirana, Albania

Sheikh Raisuddin Department of Medical Elementology and Toxicology, Jamia Hamdard, School of Chemical and Life Sciences, New Delhi, Delhi, India

Fouzia Rashid Department of Clinical Biochemistry, University of Kashmir, Srinagar, Jammu and Kashmir, India

Naveed Raza Department of Zoology, Government College University, Faisalabad, Pakistan

Hussein Sakr Department of Physiology, College of Medicine and Health Sciences, Sultan Qaboos University, Muscat, Al-Khoudh, Oman

Shaesta Shahid Molecular Toxicology Laboratory, Department of Medical Elementology and Toxicology, School of Chemical and Life Sciences, Jamia Hamdard, New Delhi, India

Anurag Sharma Department of Environmental Health and Toxicology, Nitte (Deemed to be University), Nitte University Centre for Science Education and Research (NUCSER), Mangaluru, Karnataka, India

Prasad Sherekar Department of Biotechnology, Priyadarshini College of Engineering, Nagpur, Maharashtra, India

Aryan Shukla Amity Institute of Biotechnology, Amity University Uttar Pradesh, Lucknow Campus, Lucknow, Uttar Pradesh, India

Lalit Kumar Singh Department of Biochemical Engineering, School of Chemical Technology, Harcourt Butler Technical University, Kanpur, Uttar Pradesh, India

Reenu Singh School of Bioengineering and Biosciences, Lovely Professional University, Phagwara, Punjab, India

Partha Sarathi Singha Department of Chemistry, Government General Degree College, Kharagpur II, Paschim Medinipur, West Bengal, India

Srinivasa Rao Sirasanagandla Department of Human and Clinical Anatomy, College of Medicine and Health Sciences, Sultan Qaboos University, Muscat, Al-Khoudh, Oman

Sanvidhan G. Suke Department of Biotechnology, Priyadarshini College of Engineering, Nagpur, Maharashtra, India

Suramya Molecular Toxicology Laboratory, Department of Medical Elementology and Toxicology, School of Chemical and Life Sciences, Jamia Hamdard, New Delhi, India

Indu Shekhar Thakur Amity School of Earth and Environmental Sciences, Amity University Haryana, Manesar, Gurugram, Haryana, India

Said Hamid Thangal Department of Zoology, Korambayil Ahamad Haji Memorial Unity Women's College (affiliated to University of Calicut), Manjeri, Kerala, India

Nayanika Tiwari Amity Institute of Biotechnology, Amity University Uttar Pradesh, Lucknow Campus, Lucknow, Uttar Pradesh, India

Emerging Toxicants

Microplastics: An Emerging Concern of Human Health

P. Said Hamid Thangal and P. A. Ahammed Shareef

Abstract

Microplastics are small plastic particles, typically less than 5 mm in size, that pose serious environmental concern and human (and animal) health risks worldwide. They are classified based on type, size, shape and chemical composition. They are divided into primary microplastics, which are purposefully manufactured as small plastics (e.g., microbeads and nurdles), and secondary microplastics, formed by the biotic or abiotic degradation of larger plastics. They are further categorized based on their size into mesoplastics, microplastics, and nanoplastics. Classification based on shape include fragments, fibers, pellets, foams, films and beads. They are also classified as thermoplastics, thermosetting plastics, elastomers and biodegradable plastics based on their chemical composition. Microplastics are formed from a wide variety of sources and enter the living system via ingestion, inhalation or dermal contact, and distribute via circulatory system. They pose varying health risks through physical blockages, chemical toxicity, bioaccumulation and biomagnification. They have significant impact on various tissues and organ systems. They persist in ecosystem for longer period of time, accumulate in food chain, and pose risks to terrestrial and aquatics animals. They also effect the environment by disrupting soil structure and ecology, and interact with organic pollutants, heavy metals and pathogens. The control measures include reduce plastic manufacture/use, improved waste management,

P. S. H. Thangal
Department of Zoology, Korambayil Ahamad Haji Memorial Unity Women's College (affiliated to University of Calicut), Manjeri, Kerala, India

P. A. Ahammed Shareef (✉)
Department of Zoology, PSMO College (affiliated to University of Calicut), Tirurangadi, Malappuram, Kerala, India
e-mail: shareefkpr@psmocollege.ac.in

S. Raisuddin et al. (eds.), *Environmental Toxicants and Lifestyle Diseases*,
https://doi.org/10.1007/978-981-95-0575-3_1

regulative and legislative measures, public awareness and education, and innovative policies.

Keywords

Microplastics · Nanoplastics · Emerging contaminants · Environmental pollution · Human health · Toxicity

Introduction

Plastics are polymers synthesized by polymerization of monomers (the repeating units of long-chain polymers, composed of carbon, hydrogen, and oxygen held together by covalent bonds) extracted from petrochemicals and combined with other chemicals to improve the quality. They received wide acceptance due to their advantages over other materials, such as ease of manufacture, durability, resistance to degradation, tolerance to chemicals and higher temperatures, resistance to water, high strength, low cost, etc., made them an integral part of modern life, spanning numerous industries, including packaging, construction, healthcare, and electronics (Zhang et al. 2021).

The global production of plastics has increased from 2 million tons in 1950 to 400.3 million tons in 2022 and is expected to increase exponentially in the upcoming decades (Pilapitiya and Ratnayake 2024). Their applications range from single-use packaging to high-performance materials in industries, such as automotive and aerospace. Most of the plastics are manufactured for "single-use" with limited recyclability, resulting in increased production, leading to the generation of huge amount of waste. Their widespread use of plastics has also led to significant environmental challenges. The durability that makes plastics so useful in various applications also contributes to their persistence in the environment, resulting in pollution and the accumulation of plastic waste in landfills and oceans. This has led to growing concerns about the ecological impact of plastics, prompting initiatives for recycling, reduction, and the development of biodegradable alternatives (Zhang et al. 2021).

Microplastics

The degradation of synthetic plastics, which takes several years in nature, results in the formation of their fragments of different shapes and sizes, often called microplastics. Microplastics are small pieces of plastic, less than 5 mm in size, that occur in the environment as a consequence of the degradation of plastic. Originating from both primary sources, such as microbeads intentionally included in personal care products, and secondary sources, where larger plastic debris degrades into smaller fragments, microplastics now represent a universal pollutant (Wang et al. 2019). They have become a universal environmental concern, impacting terrestrial and aquatic ecosystems globally. This growing issue stems from their widespread use, slow degradation rates, and ability to traverse various environmental compartments, from the deepest oceans to the most remote terrestrial locations (Rillig 2012).

Microplastics have infiltrated numerous environmental settings through a variety of pathways. In terrestrial systems, they accumulate through activities such as the application of plastic protecting in agriculture, the use of soil amendments containing sewage sludge, and littering. These particles can enter the soil from atmospheric deposition, street runoff, and even wind transport (de Souza Machado et al. 2018). Once in the environment, microplastics persist due to their durable nature, with slow rates of natural degradation impairing their accumulation (Rillig 2012). Their presence is not confined to any single environment but spans a diverse range of ecosystems, including agricultural soils, freshwater systems, and marine environments. For example, microplastics have been detected in soils with concentrations as high as 30.7×10^3 particles per kg of dry sludge (Li et al. 2018), and their prevalence is increasingly noted in remote areas, often far from urban centers (Piehl et al. 2018). This widespread distribution underscores the universal nature of plastic pollution and highlights the challenges associated with managing and mitigating its impacts.

The effects of microplastics on terrestrial systems are complex and multifaceted. They can influence soil structure, nutrient dynamics, and microbial communities. For instance, microplastic fibers can affect soil aggregation and water retention, with varying effects depending on the type and concentration of microplastic present (de Souza Machado et al. 2018; Lozano et al. 2021). Similarly, the impact on soil microbial activity can be significant, with some studies showing increased mortality in soil macro-organisms and a decline in bacterial diversity due to microplastic contamination (Huerta et al. 2016). To better understand these effects, recent research has focused on evaluating the impact of different microplastic shapes and types on plant growth and soil health. Controlled experiments have tested how microplastic fibers, films, foams, and fragments of various polymer types and concentrations affect plant biomass, soil aggregation, and microbial activity. These studies aim to provide insights into the complex interactions between microplastics and terrestrial ecosystems, including the potential for microplastic particles to mediate or exacerbate environmental impacts based on their physical and chemical properties (de Souza Machado et al. 2018).

The marine environment is also significantly impacted by microplastics, which enter the ocean through numerous pathways, including runoff, atmospheric deposition, and direct disposal. The scale of marine microplastic pollution is so extensive that these particles are now considered a key marker of the Anthropocene, a term used to describe the current geological age viewed as the period during which human activity has been the dominant influence on climate and the environment (Cózar et al. 2014). Microplastics in the marine environment are particularly concerning due to their potential to be ingested by aquatic animals across various trophic levels, from plankton to larger marine animals. The accumulation of microplastics in marine habitats can cause physical harm to organisms, affect nutrient cycles, and lead to the transfer of toxic pollutants up the food chain (Teuten et al. 2009). The ecological impacts of microplastics in marine environments are profound and necessitate the need for extensive research to assess and mitigate their effects (Zarfl et al. 2011).

Microplastics exhibit a high degree of physical and chemical diversity. They vary in shape, ranging from fibers and films to foams and fragments, and in polymer type, including polyethylene (PE), polypropylene (PP), and polyvinyl chloride (PVC) (Khalid et al. 2021). Each polymer type can contain different additives and monomers that may pose specific hazards to environmental and human health. For example, polyurethane (PU), used extensively in flexible foams, is composed of monomers known to be highly toxic (Lithner et al. 2011). These polymers can affect soil biota and plant health through their chemical leachates and physical presence in the soil matrix. The degradation process of these materials further complicates the environmental impact, as plastics weather into smaller particles that can absorb various pollutants, including heavy metals and organic chemicals. The interaction between microplastics and these contaminants can enhance the toxicity of the particles, posing additional risks to both terrestrial and aquatic ecosystems (Hodson et al. 2017; Rochman et al. 2015).

Classification of Microplastics

Microplastics are categorized into different classes based on type of origin (primary and secondary microplastics), size, shape, and chemical nature.

Classification of Microplastics Based on Types

Microplastics are broadly classified into primary and secondary microplastics based on whether they are originally manufactured of that size (primary) or formed from the breakdown of large plastics (secondary).

Primary microplastics Primary microplastics are small plastic particles having less than 5 mm in size, manufactured with a specific purpose. These are not formed by the degradation of large plastics but are purposefully made small plastic particles. They can be found in various consumer and industrial products and reach the environment/living system through their use or disposal. They include microbeads, nurdles, and microfibers.

Microbeads are tiny plastic particles commonly used in personal care products such as exfoliating scrubs, toothpaste, and facial cleansers. Their primary function is to provide texture and abrasive properties. They are washed down and end up in aquatic environments and oceans. In response to environmental concerns, several countries have implemented bans on microbeads, but their legacy preserves (Liu et al. 2021). *Nurdles* are small plastic pellets used as raw material in the manufacturing of plastic products (pre-production plastics). These are often spilled during transport or production and are regarded as a major source of microplastic pollution. They are often lost during manufacturing, handling, and transportation processes. Spills of nurdles can lead to significant environmental pollution, particularly in marine environments where they are frequently found on beaches and in water. *Microfibers* are synthetic fibers, such as those from polyester, nylon, or acrylic materials, shed from textiles during manufacturing, processing, and washing. These microplastic fibers are released into wastewater systems and are often too small to be filtered out, leading to their accumulation in water bodies (Zhang et al. 2021).

Secondary microplastics Secondary microplastics originate from the degradation of larger plastic items. They result from the breakdown of plastic products over time due to environmental factors. Larger plastic items such as bottles, bags, packaging, etc. are brokendown into smaller particles through physical weathering, UV irradiation, and chemical degradation. Plastic fragments are the larger plastic debris, like bottles, bags, and containers, which break down into smaller pieces through exposure to sunlight, wave action, and physical wear (Zhang et al. 2021). These fragments contribute significantly to the growing microplastic content in oceans and soil. Foams are the expanded polystyrene (EPS) used in packaging and insulation that degrades into smaller foam particles. These are particularly harmful in marine environments, where they can be mistaken for food by marine organisms. Films are the plastic films used in agriculture (e.g., covering films) or packaging that degrade into smaller pieces when exposed to sunlight and environmental weather. These films can persist in soil and aquatic ecosystems for longer periods of time. This fragmentation process generates secondary microplastics that contribute to pollution in both terrestrial and aquatic environments (Lebreton et al. 2017).

Classification of Microplastics Based on Size

Microplastics are generally defined as plastic particles smaller than 5 mm. They are further categorized into mesoplastics, microplastics, and nanoplastics based on their size. Though their classification is controversial, this is generally an accepted classification (Ali et al. 2024).

Mesoplastics These are plastic particles with the size range of 0.5–5 cm. These particles are larger and more visible compared to other microplastics. They typically originate from the breakdown of larger plastic debris, such as plastic bags, bottles, and other consumer products. Mesoplastics are easily ingested by aquatic animals, such as small fish and seabirds, which can mistake them for food. This leads to physical harm and potential chemical contamination within the food chain/web.

Microplastics These are the plastics that have a range from 1 μm to 1 mm. These particles are more challenging to detect and are often generated through the wear and tear of larger plastic products, as well as microbeads and synthetic fibers. The major sources of microplastics are coming from the washing of synthetic textiles, personal care products containing microbeads, and weathering of larger plastics. These smaller particles can be ingested by plankton and other small aquatic organisms, leading to potential bioaccumulation in the food chain/web.

Nanoplastics These are plastic particles smaller than 1000 nm. Due to their tiny size, nanoplastics are not easily detectable with standard equipment, but their presence in the environment is increasingly being recognized. Nanoplastics often originate from the degradation of larger plastics or are manufactured intentionally for use in industrial processes. These nanoplastics can penetrate biological tissues and cells, which is a raising concern about their potential toxicity to aquatic life and human health. They are particularly concerning due to their ability to interact with living cells at the molecular level. Moreover, the classification of microplastics by size is critical because smaller particles tend to have more significant environmental and health impacts (Ali et al. 2024).

Classification of Microplastics Based on Shape

Microplastics are classified into six categories based on their shapes such as fragments, fibers, pellets, foams, films, and beads.

Fragments are the irregular shaped particles resulting from the breakdown of larger plastic items, typically through mechanical, thermal, or photodegradation processes. They are originated from the larger plastic products, such as containers, packaging, consumer goods, etc. Due to their sharp and jagged edges, fragments can physically harm aquatic organisms and are often mistaken for food (Cole et al. 2011).

Fibers/filaments are the thread-like, thin strands derived from synthetic textiles or fishing gear that are predominantly invented from washing synthetic fabrics, such as polyester and nylon, and from the wear and tear of ropes and fishing nets. These fibers are widely distributed in aquatic environments and easily ingested by plankton and small fish, causing physiological and biological impacts (Browne et al. 2010).

Pellets/nurdles are the spherical or cylindrical microplastics used as raw materials in the plastic manufacturing industry, and they are released to the environment during transportation or spillage in industrial processes. These are abundantly found along shorelines and marine environments, where they can be consumed by marine animals and birds, often causing ingestion-related issues (Thompson et al. 2004).

Foams are the lightweight, porous materials, often used for packaging or insulation, that break down into small, spongy particles, which are originated due to the breakdown of expanded polystyrene foam like Styrofoam and from disposable food containers and packaging. They are highly buoyant and float in surface waters, contributing to marine debris (Hidalgo-Ruz et al. 2012).

Films are the thin, flexible sheets or membranes of plastic that tear into smaller fragments while maintaining their flat shape. They are commonly invented from single-use plastic bags, agricultural films, and plastic wraps. These are the reasons for the blocking of light and gas exchange of the surface water (Thompson et al. 2004).

Beads are the small, spherical microplastics commonly found in personal care products and industrial abrasives, and they are the exfoliating agents in cosmetics, toothpaste, and other hygiene products. These are very dangerous and can pass through water filtration systems, entering aquatic ecosystems due to small size (Fendall and Sewell 2009).

Classification of Microplastics Based on Chemical Nature

Understanding the chemical composition of microplastics is critical to assessing their environmental fate and impact. Different types of plastics have unique chemical structures and composition that affect how they degrade, interact with pollutants and accumulate in ecosystems. Some of the major keys which help to identify the microplastics based on their chemical nature. They are thermoplastics, thermosetting plastics, elastomers, and biodegradable plastics (Zhang et al. 2021).

Thermoplastics are the plastics that can be repeatedly melted and reformed upon heating. They are characterized by their linear or branched polymer chains, which allow them to soften when heated. They are one of the keys to examine the chemical nature of microplastics. They make up the majority of plastic waste found in the environment, such as polyethylene (PE), polypropylene (PP), polyethylene terephthalate (PET), polystyrene (PS), and polyvinyl chloride (PVC). Due to their widespread use, thermoplastics are the most commonly found type of microplastic. Their degradation in the environment is slow, and they are resistant to chemical breakdown. Thermoplastics can absorb and transport hydrophobic pollutants, acting as vectors for other harmful chemicals in ecosystems (Andrady 2011; Geyer et al. 2017).

Thermosetting plastics are the plastics once hardened, cannot be melted or reformed, and the high cross-link and inflexible structure make them resistant to heat and chemical degradation. They are typically more rigid and durable than thermoplastics. Epoxy resins, polyurethane (PUR), and phenol formaldehyde (PF) are some examples of thermosetting plastics. Often due to their chemical stability, thermosetting microplastics may remain in the environment for long periods, contributing to long-term pollution and potential ingestion by aquatic animals.

Elastomers (rubber-like polymers) are the flexible, elastic materials that can stretch and return to their original shape. Elastomers like styrene-butadiene rubber (SBR), natural rubber, and silicone are widely used in the production of tires, footwear, and industrial products. They have a cross-linked polymer structure that provides the elastic properties. They are less rigid than thermoplastics and thermosets but are more resistant to mechanical wear and tear. These are commonly ejected to the environment from tires and contribute significantly to microplastic pollution in urban and roadside environments. Tire wear particles are one of the largest sources of microplastics in the environment (Kole et al. 2017).

Biodegradable plastics are designed to degrade more quickly in the environment, typically through microbial activity or under specific conditions. Polylactic acid (PLA), Polyhydroxyalkanoates (PHA), and starch-based plastics are some main examples for the biodegradable plastic. These are often made from renewable resources like cornstarch and are intended to reduce environmental impact. However, their degradation often requires specific industrial composting conditions, and they may persist in natural environments (Tokiwa et al. 2009; Napper and Thompson 2019).

Sources of Microplastics

Microplastics from various sources are released into the environment through the degradation of various commonly used plastic items. They contribute to the primary or secondary microplastic sources (Anik et al. 2021). Some common sources include:

- *Degradation of larger plastics*: The breakdown of larger plastics of any type/source over time due to biotic or abiotic processes poses a significant contribution to the global microplastics release.
- *Plastic packaging*: Commonly used plastic packaging materials, such as films, containers, and wrappers, degrade into microplastics, mainly by sunlight exposure and environmental stressors.
- *Large plastic items*: Bottles, cans, cups, containers, and bags are widely used, and they break down into smaller particles and microplastics due to weathering, sunlight exposure, and mechanical forces.
- *Cosmetics and personal care products*: Many commonly used cleaning, cosmetics, and personal care products, such as toothpaste, carry microbeads made of plastic. After their use and wash-off, they finally end up in water bodies.
- *Paints and coatings*: Paints, polishes, and coatings are widely used in almost all items, and they degrade and deterioration over time, particularly after disposal, forms microplastics. Paints used on ships and boats also contributes to the microplastic formation.
- *Textiles*: Synthetic textiles and fabrics release shreds of fibers like polyester and nylon during manufacturing, processing, sewing and washing.
- *Vehicle tyres*: The synthetic rubber in tyres releases microplastic particles over time, or its disposal significantly contributes to the airborne microplastics.
- *City dust*: City dust comes from a variety of sources exposed to weathering, abrasion, and detergents that create a significant amount of microplastics.
- *Agricultural plastics*: Plastic coverings, films, and textiles used in agricultural fields can degrade into microplastics because of exposure to agricultural activities, weather conditions, and improper disposal.

In addition to the above, natural disasters, medical applications, like dentist tooth polishing, artificial grasses (turf) in the football pitch, unwise dumping of plastic garbage and landfill, etc., also significantly contribute to the production and release of microplastics into the environment (Anik et al. 2021).

Formation of Microplastics

Plastics are resistant to degradation and are known to have a longevity of hundreds or even thousands of years, depending on environmental conditions (Zhang et al. 2021). Still, various biotic and abiotic degradation processes do occur to form plastic fragments and microplastics, which are illustrated in Fig. 1.

Biotic Degradation of Microplastics

This is the deterioration of plastic due to the activities of microorganisms or biological products such as enzymes, resulting in the formation of microplastics. Microorganisms such as fungi and bacteria, and insects substantially contribute to this process. Physical degradation by biting, cutting, chewing, or digestive processes and biological degradation by various biochemical processes. Hydrolysis and enzymatic oxidation lead to chain scission and fragmentation of plastic. The plastic degraded by abiotic processes facilitates and increases the biotic degradation process and the formation of microplastics.

Abiotic Degradation of Microplastics

This is the change in physical or chemical properties of plastic by abiotic factors and processes resulting in the formation of microplastics. They include chemicals, heat, water, photodegradation, UV irradiation, and mechanical stress resulting in the oxidation, cross-linking, dehydrochlorination, chain scission, fragmentation and ablation processes, contributing to plastic degradation and microplastic formation.

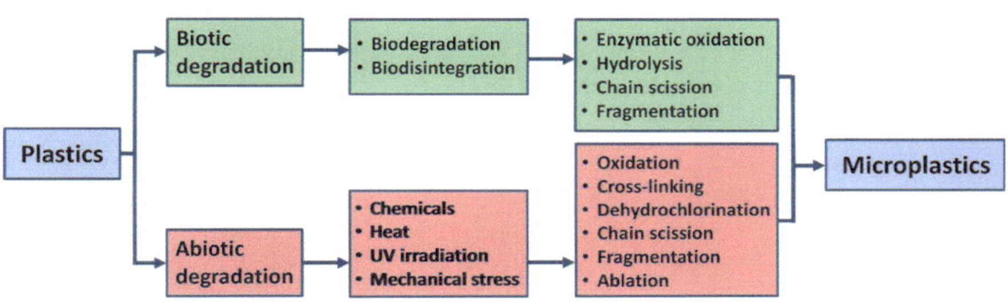

Fig. 1 Schematic diagram showing the processes involved in the biotic and abiotic degradation of plastic

Among these, photo degradation by solar irradiation is regarded as the most important contributing factor. The exposure of plastic to high temperature leads to a thermo-oxidative reaction causing the plastic to break down, which is related to the availability of oxygen and thermal properties of plastic. The photochemical reactions due to repeated exposure cycles (day-night cycle) for longer periods substantially contribute to the degradation process. In addition, the plastic degraded by biotic processes also facilitates the abiotic degradation process.

Entry of Microplastics to the Human Body

Microplastics can enter living organisms through several pathways, primarily through ingestion, inhalation, and dermal contact. These entry routes raise concerns about the bioaccumulation and potential toxic effects of microplastics and the chemicals they carry in the living system (Zhao et al. 2024). The potential routes of entry of microplastics and their distribution in the body have been illustrated in Fig. 2.

Ingestion

This is the entry of microplastics via the oral route by terrestrial and marine organisms, including plankton, fish, larger animals, and humans. These particles

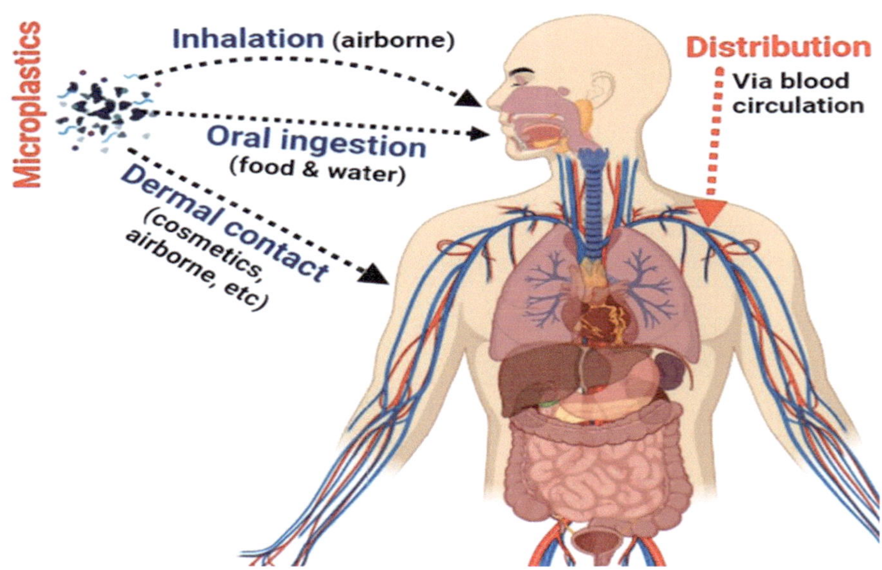

Fig. 2 Schematic diagram showing the pathways of entry of microplastics and their distribution in human body

are mistaken for food due to their small size and resemblance to prey (Cole et al. 2011). Once ingested, microplastics can accumulate in the gastrointestinal tracts of animals, leading to physical blockages, reduced feeding, and impaired growth. More alarmingly, microplastics can bioaccumulate and move up the food chain/web, affecting larger predators, including humans who consume seafood. Studies have documented the presence of microplastics in fish and shellfish, which are part of the human diet, raising concerns about their transfer into human bodies (Galloway et al. 2017). Once absorbed from the gastrointestinal tracts, they are distributed by the blood circulatory system to various parts of the body.

Inhalation

This is the entry of airborne microplastic particles, from synthetic clothing or dust from degraded plastics, via inhalation by humans and animals. Studies have shown that microplastics are present in indoor and outdoor air, especially in urban environments and near industrial activities (Dris et al. 2016). These particles can reach the respiratory system, leading to potential respiratory issues, especially with long-term exposure. In addition, the microplastics are absorbed from the lungs and respiratory tracts; they are distributed to various parts of the body by the blood circulatory system and make it available for exposure to other tissues as well.

Dermal contact

This is the absorption/entry of microplastics via penetrating the skin or absorption through open wounds. This could be active (using products that contain microplastics) or passive entry (via unintentional exposure). The presence of microplastics in air, dust, cosmetics, personal care products, etc., makes its way to the human body via skin contact. Once absorbed/entered, they are distributed by the blood circulatory system to various parts of the body (Zhao et al. 2024). The potential toxicity is, therefore, not just in the skin but to the entire body, the details of which have been given in the preceding section.

Potential Toxicity and Impact on Health

Impact on Animals

The intake of microplastics making a serious impact on the living organism is mainly based on their physical and chemical nature. Many animals, particularly aquatic organisms, mistake microplastics for food and ingest them. Ingested microplastics can cause physical blockages in the digestive system, leading to reduced feeding, impaired digestion, and malnutrition. Studies on marine species like fish, seabirds, and turtles have documented the accumulation of plastic particles in their stomachs, which can reduce the ability to digest the food and absorb nutrients and affect their survival (Lusher et al. 2017). In smaller organisms, such as plankton, microplastics can physically obstruct the feeding apparatus, reducing energy intake and growth (Cole et al. 2013). Moreover, microplastics can adsorb toxic chemicals, such as persistent organic pollutants (POPs), heavy metals, and other hazardous substances from the environment. When animals ingest microplastics, these absorbed chemicals can be released into their bodies, leading to potential toxic effects. Research has shown that microplastics can transfer harmful chemicals through the food chain, potentially impacting predator species and, ultimately, humans who consume contaminated seafood (Teuten et al. 2009). Some certain chemicals associated with plastics, such as bisphenol A (BPA) and phthalates, are known endocrine disruptors. When animals are exposed to these chemicals, they can interfere with hormonal regulation, affecting growth, reproduction, and development (Gallo et al. 2018).

Microplastic exposure has been linked to reduced reproductive success in animals. Studies have shown that invertebrates like oysters and fish species may suffer from reduced egg production, lower hatching success, and impaired larval development after exposure to microplastics (Sussarellu et al. 2016). These effects are likely due to both physical interference from microplastics and chemical toxicity. Also the intake of microplastics can cause physical damage to tissues when ingested or inhaled by animals. Studies have found that microplastics can lead to inflammation and oxidative stress in various organs, including the liver, intestines, and gills. For instance, in fish, microplastic exposure has been linked to liver stress and reduced immune responses (Lu et al. 2016). Inflammation and tissue damage caused by microplastics can impair an animal's overall health, making them more susceptible to diseases. Bioaccumulation and biomagnification are other serious issues for the animals due to the intake of microplastics. Microplastics and the chemicals they carry can accumulate in an animal's body over time, a process known as bioaccumulation. As smaller animals are eaten by larger predators, the concentration of microplastics and toxic chemicals can increase through biomagnification. This process is particularly concerning for top predators in marine and terrestrial ecosystems, including humans, as they can end up with high levels of toxicants due to their consumption of contaminated preys (Carbery et al. 2018). Also microplastics are causing behavioral changes in some animal species. Research on fish has shown that exposure to microplastics can alter normal feeding behaviors, making them less efficient at capturing prey or leading them to consume less nutritious food. These behavioral changes can reduce their overall fitness and survival chances (Savoca et al. 2016).

Impact on Humans

Microplastics are known to have potential toxicity to multiple organ systems in humans, which have been briefly described below and illustrated in Fig. 3. Studies

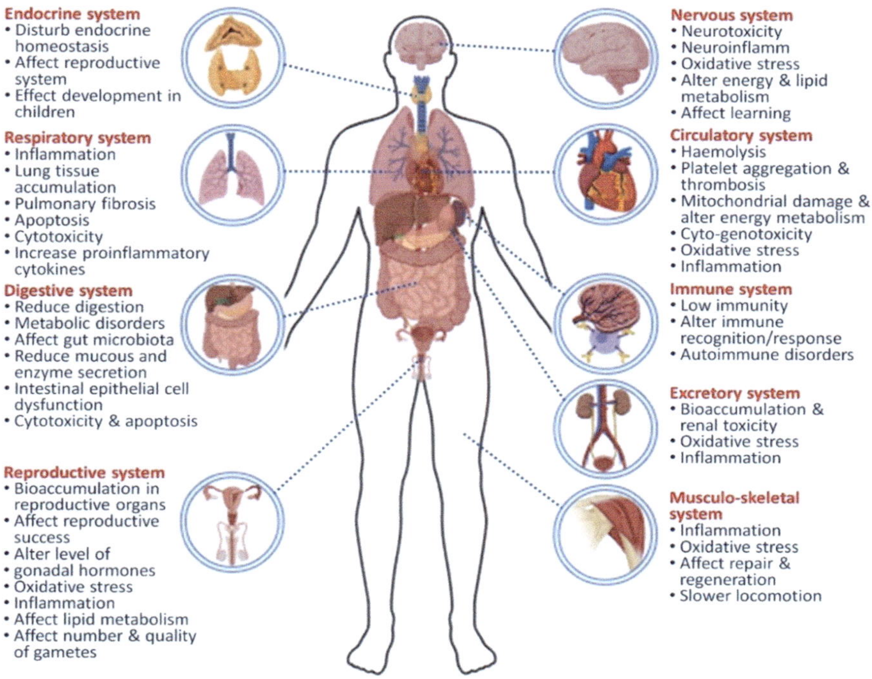

Fig. 3 Potential toxic effects of microplastics and impact on various physiological systems and human health

on naturally exposed animals and experimental conditions reported the toxicity to various physiological systems associated with the exposure to microplastics (Ali et al. 2024; Zhao et al. 2024).

Digestive System

The gastrointestinal tract serves as a predominant site of entry and accumulation of orally ingested microplastics along with the food. As discussed earlier, a wide variety of food materials serve as a potential source of microplastics and their entry to the living system. Microplastics liberated from the digested food can induce toxicity right within the intestine itself and other parts of the body after its absorption and distribution via blood circulation. The effects of microplastics associated with the digestive system include low mucus secretion in the intestine, intestinal barrier/epithelial cell dysfunction, hepatic metabolic disorders, inflammation, altered intestinal mucosal immunity, affected gut microbiome, oxidative stress, DNA damage, genotoxicity, and apoptosis. These effects were widely studied in a number of species, both naturally exposed animals and those under controlled experimental conditions.

Respiratory System

The airborne microplastics constantly enter the human body during respiration and absorb into the blood of alveolar capillaries and distribute to various parts of the body via systemic circulation. The immune system clears much of these entered microplastics but still retains a part of it, particularly the larger particles, which persist for longer periods. They can induce effects in the respiratory system and elsewhere. They can induce or worsen respiratory airway (nasal cavity, trachea, and bronchi) and interstitial lung diseases, such as asthma, bronchitis, emphysema, pneumonia, interstitial fibrosis, granuloma, and even cancer. Experimental studies showed that the inhaled microplastics cause lung injury and affect the integrity of the epithelial cell barrier, inflammation, ventilatory airway dysfunction, fibrosis, enhanced mucus secretion, oxidative stress, genotoxicity, and even death of epithelial cells.

Circulatory System

Microplastics enter to human body via inhalation, oral ingestion, and dermal contact and finally reach the cardiovascular system, and therefore, significant toxicity is obvious. Experiments conducted in different animals following microplastic exposure showed cardiovascular dysfunction, impaired cardiac output, pericardial edema, disturbed angiogenesis, endothelial dysfunction, generation of ROS, oxidative stress, systemic inflammation, interstitial hyperplasia, collagen deposition and apoptosis in cardiomyocytes, mitochondrial dysfunction, promote aortic inflammation, vascular injury, and coagulation dysfunction. These toxic effects of microplastic exposure observed in zebra fish, rat, and chicken could be correlated with human physiology and therefore similar effects are apparent.

Reproductive System

The detection of microplastics in the gonads is a clear indication that microplastics can cross the blood barrier within the gonads (Zhao et al. 2024). The effects on gonads and the whole reproductive system in males include affecting reproductive fitness, declining the number and quality and mortality of sperm, malformation, DNA damage, and abnormal testosterone levels. Toxicity to the female reproductive system includes impaired follicular growth, decreased follicular cell production and oocyte maturation, necrosis, apoptosis, inflammation, and oxidative stress.

Nervous System

Experiments conducted in animals showed microplastics accumulation in the cerebral cortex hippocampus, and cerebellum after oral exposure and consequently showed effects on the normal functioning of the nervous system. These include reduced learning and memory, low production of neurotransmitters and synaptic proteins, neuroinflammation, behavioral changes, neurodegeneration, affected protein and energy metabolism, neuronal damage, variation in cytokine production, and apoptosis.

Immune System

It has widely been reported that the microplastics compromise the immune system. It causes a local immune response in the intestine to the orally ingested microplastics, excessive mucus secretion, affects lymphocyte and neutrophil activation, alters gut microbiome and inflammatory disorders, alters serum cytokine levels, affects the complement system, inhibits T cell activation and differentiation, etc. Moreover, the affected immune system compromises the normal immunity and can cause other infections and diseases as a secondary consequence of the microplastics exposure.

In addition, the exposure of microplastics can affect the endocrine system, both at the level of gland and target tissue, by altering the hormone production and action, affecting hormone receptors, affecting hormone gene expression and overall endocrine homeostasis. Moreover, plastic monomer additives such as phthalates and bisphenol A are known endocrine disruptors. In the excretory system, renal toxicity, oxidative stress, bioaccumulation, and inflammation are reported, which again affect the overall physiology of the body. Effects on the musculoskeletal system include inflammation, oxidative stress, altered repair and regeneration processes and lower locomotion.

Interaction with Organic Pollutants, Heavy Metals, and Pathogens

Microplastics can also interact with organic pollutants, heavy metals, and pathogens and cause secondary effects too. Due to strong hydrophobicity and large surface area, they can adsorb and act as vectors to carry the pollutants and pathogens to humans. Microplastics and organic pollutants often co-exist in the same environment, and therefore, they simultaneously enter the living system together, and consequent bioaccumulation and toxic effects are obviously far higher. Heavy metals such as arsenic, cadmium, and lead can adsorb on microplastics, and they enter the human body together. Consequently, their additive toxic effects are extremely high. Studies suggest that microplastics carry and transport viral (e.g., SARS-Cov-2) and bacterial (e.g., *Helicobacter pylori*, causing peptic ulcers) pathogens, indicating the capacity of microplastics to transmit infectious diseases too (Zhao et al. 2024).

Impact on Ecosystem and Environment

Microplastics have become a universal pollutant in both aquatic and terrestrial ecosystems, posing significant risks to the environment. The environmental health issues caused by microplastics stem from their persistence, ability to transport harmful chemicals, and their detrimental effects on ecosystems and organisms. They are commonly found in soils and terrestrial ecosystems due to activities like plastic waste dumping, agricultural practices, and the use of plastic covering films. When microplastics accumulate in the soil, they can alter its physical and chemical properties. Microplastics can change soil structure by affecting porosity and water retention capacity. They can reduce water infiltration, which impacts soil aeration and water availability for plants. Earthworms, insects, and soil microbes are essential for nutrient cycling and soil health. Studies have shown that microplastics can negatively affect earthworms by reducing their growth and reproductive capacity (Rillig et al. 2017). Previous studies show that polyethylene microplastics reduce the survival and burrowing behavior of earthworms, which can affect soil aeration and nutrient availability (Huerta et al. 2016). Some microplastics are affecting plant growth by altering nutrient availability and by causing physical barriers to root growth. Some studies suggest that plants exposed to microplastics may exhibit stunted growth, reduced seed germination, and lower yields (Boots et al. 2019).

The most significant and well-documented impact of microplastics occurs in marine and freshwater ecosystems. Microplastics are now ubiquitous in oceans, rivers, and lakes, and their persistence in these environments has far-reaching consequences. Many aquatic organisms, from plankton to fish and marine mammals, mistake microplastics for food. Ingestion of microplastics can lead to blockages in the digestive system and reduced feeding and energy intake, which can impair growth and reproduction. The ingestion of microplastics has been observed in species as diverse as zooplankton, shellfish, and seabirds (Lusher et al. 2017). This has cascading effects on the food web, as compromised species can reduce the availability of prey for predators. Also microplastics, particularly floating types, can accumulate on the surface of water bodies, forming floating "plastic soups." These floating particles can block sunlight penetration, disrupting photosynthesis in aquatic plants and phytoplankton. Reduced sunlight can affect the entire aquatic ecosystem, leading to decreased oxygen production and altered food chains (Thompson et al. 2004).

Microplastics not only pose physical threats to ecosystems but also act as vectors for harmful chemicals. These particles can absorb/adsorb pollutants from their surrounding environment, including heavy metals, persistent organic pollutants, and endocrine-disrupting chemicals. When organisms ingest microplastics, these absorbed toxins can enter their bodies and accumulate in tissues, leading to bioaccumulation and biomagnification through the food chain (Teuten et al. 2009). Moreover, microplastics in the marine environments have been shown to carry chemicals such as polychlorinated biphenyls and polycyclic aromatic hydrocarbons. These chemicals are linked to a range of adverse effects, including immune suppression, reproductive issues, and developmental problems in marine life (Koelmans et al. 2015). As microplastics are consumed by smaller organisms, they can accumulate and concentrate in the bodies of these organisms. Predators that feed on these contaminated species may, in turn, ingest larger amounts of microplastics and the toxic chemicals they carry. This process, known as biomagnification, causes the concentration of microplastics and toxins to increase as they move up the food chain, potentially affecting larger predators such as fish, birds, and even humans (Carbery et al. 2018). The presence of microplastics in natural habitats can contribute to a decline in biodiversity. As species ingest microplastics or experience habitat degradation, their populations may suffer, leading to reduced biodiversity. Some species may be more vulnerable than others, and microplastic pollution could push already threatened species toward extinction. Decreased biodiversity weakens ecosystem resilience, making ecosystems less capable of withstanding additional environmental stressors such as climate change (Galloway et al. 2017).

Microplastics are not limited to terrestrial and aquatic environments; they have also been found in the atmosphere. Wind and air currents can carry microplastic fibers and fragments across vast distances, depositing them in remote regions, including mountains and polar areas. Airborne microplastics may contribute to air pollution and be inhaled by both animals and humans, leading to respiratory issues (Dris et al. 2016). Moreover, the abundance of microplastics in the atmosphere does cause climate change. Plastics, when exposed to sunlight, release greenhouse gases such as methane and ethylene. As microplastic pollution increases, the amount of plastic exposed to sunlight in oceans and terrestrial ecosystems may contribute to rising greenhouse gas emissions, exacerbating climate change (Royer et al. 2018).

Control Measures

The widespread presence of microplastics in both terrestrial and aquatic environments poses significant threats to ecosystems and human health. Effective control measures are essential to reduce their impact, focusing on preventing plastic waste generation, improving waste management systems, and implementing policy and technological innovations.

Decrease Plastic Manufacture and Use

A crucial strategy to control microplastic pollution is reducing the overall production and use of plastic materials, particularly single-use plastics. By minimizing the amount of plastic that enters the environment, the generation of microplastics can be significantly reduced. Banning single-use plastics is one of the easy methods to reduce the microplastic accumulation. Many governments have already enacted bans on single-use plastic products like plastic bags, straws, and cutlery. These items are major contributors to plastic waste, which eventually degrades into microplastics. For instance, the European Union implemented the Single-Use Plastics Directive, which aims to reduce the consumption of such plastics across member states. Along with banning, encouraging the development and use of biodegradable or compostable materials as alternatives to conventional plastics can mitigate the formation of microplastics. Biodegradable plastics derived from natural sources, such as starch and cellulose, degrade more quickly in the environment and produce fewer harmful byproducts (Napper and Thompson 2019). Extended producer responsibility (EPR) is the program that holds producers accountable for the end-of-life management of their products, incentivizing companies to design products that are easier to recycle or that generate less waste. This can help reduce the overall amount of plastic waste that becomes microplastic over time (Organization for Economic Cooperation and Development (OECD) 2016).

Enhanced Waste Management Systems

Improving waste management infrastructure is vital for reducing the release of microplastics into the environment. Ineffective waste management, particularly in low- and middle-income countries, leads to the breakdown of plastic waste into microplastics. The advanced recycling technologies, such as chemical recycling, can improve the recycling rates of plastic waste and reduce the amount of plastic that ends up in landfills or the environment. Chemical recycling breaks down plastics into their chemical components, which can be reused to create new plastics without producing microplastics (Geyer et al. 2017). Moreover, implementing better landfill containment measures, such as proper lining and covering, can help reduce the leakage of plastic particles into the environment. Additionally, regulating plastic waste disposal and promoting waste diversion through recycling and composting can significantly decrease the amount of plastic that decomposes into microplastics (Browne et al. 2010).

Upgrading Wastewater Treatment Plants

Wastewater treatment plants are critical in preventing microplastics from entering aquatic ecosystems. However, current filtration technologies are often insufficient to capture the smallest microplastic particles. Technologies such as membrane bioreactors (MBRs), sand filtration, and nanofiltration are more effective at capturing microplastics from wastewater before it is released into rivers and oceans (Carr et al. 2016). Incorporating these technologies into existing wastewater treatment plants can significantly reduce microplastic pollution. Also, governments can implement policies that limit the permissible levels of microplastics in treated wastewater. Some countries, such as Germany, have already set limits on the release of microplastics from wastewater treatment plants (Magnusson and Norén 2014).

Control of Microplastic Pollution from Textiles

The synthetic fibers from textiles, such as polyester, are a significant source of microplastic pollution, especially when released during washing. Installing filters in washing machines to capture microfibers is an effective way to reduce the release of microplastics from synthetic textiles. Some countries are exploring regulations that would require new washing machines to be equipped with microfiber filters (De Falco et al. 2018). Manufacturers can design textiles that shed fewer microfibers during washing by using tighter weaves or blended natural fibers. Research has shown that garments with tighter construction and natural materials produce significantly fewer microplastic fibers compared to synthetic alternatives (Hartline et al. 2016).

Regulation on the Use of Microplastics in Cosmetics

Microbeads, commonly used in personal care products like exfoliants and toothpastes, are a well-known source of microplastic pollution. Countries like the United States, Canada, and the United Kingdom have enacted bans on the use of microbeads in cosmetics and personal care products. These bans are effective in preventing microbeads from entering aquatic systems, where they are difficult to remove (Oberoi and Garg 2021).

Public Awareness and Education

Educating the public about the environmental and health impacts of microplastics can drive behavior change and increase support for plastic reduction initiatives. The governments, NGOs, and environmental organizations can launch campaigns that inform consumers about the sources of microplastics and how to minimize plastic use. Public pressure can encourage companies to reduce plastic packaging and adopt more sustainable practices (Oberoi and Garg 2021). Labeling products that are free from microplastics or that contain sustainable materials can help consumers make informed choices. Eco-labeling schemes, such as the Blue Angel label in Germany, guide consumers toward environmentally friendly products (Gasperi et al. 2018).

Future Perspectives of Research

Future research on microplastic control is likely to focus on several innovative approaches aimed at preventing, mitigating, and remediating plastic pollution (Anik et al. 2021). One promising avenue is the development of advanced biodegradation technologies, such as the use of plastic-degrading enzymes like PETase, which could be engineered to break down microplastics more effectively in natural and engineered environments (Austin et al. 2018). Further research into nanotechnology and catalytic processes may also yield new methods for accelerating the degradation of microplastics at a molecular level, turning them into non-toxic substances. Improving wastewater filtration systems to capture smaller plastic particles, such as nanoplastics, is another priority, with innovations in membrane filtration and adsorbent materials showing potential for more efficient microplastic removal. Additionally, interdisciplinary research exploring the ecotoxicological impacts of microplastics on ecosystems and human health will be essential to inform regulatory frameworks. Ultimately, addressing microplastic pollution will require a combination of technological innovation, policy development, and public awareness, making it a vital area for future scientific inquiry.

Conclusion

Microplastics pose significant environmental and health risks, primarily due to their persistence, small size, and ability to carry toxic chemicals. Classified into primary and secondary microplastics, they originate from a variety of sources, including personal care products, industrial processes, and the breakdown of larger plastics. Microplastics are further categorized by size, shape, and chemical composition, influencing their behavior in ecosystems and living organisms. They enter the food chain through ingestion, inhalation, or skin contact, leading to bioaccumulation and toxic effects in animals and potentially humans. Microplastics also disrupt ecosystems by affecting soil health, marine life, and biodiversity. Effective control measures, such as reducing plastic waste, improving waste management, and implementing policy changes, are crucial to mitigating the impacts of microplastics on the environment and human health.

References

Ali N, Liu W, Zeb A et al (2024) Environmental fate, aging, toxicity and potential remediation strategies of microplastics in soil environment: current progress and future perspectives. Sci Total Environ 906:167785

Andrady AL (2011) Microplastics in the marine environment. Mar Pollut Bull 62(8):1596–1605

Anik AH, Hossain S, Alam M et al (2021) Microplastics pollution: a comprehensive review on the sources, fates, effects, and potential remediation. Environ Nanotechnol Monit Manag 16: 100530

Austin HP, Allen MD, Donohoe BS et al (2018) Characterization and engineering of a plastic-degrading aromatic polyesterase. Proc Natl Acad Sci USA 115(19):E4350–E4357

Boots B, Russell CW, Green DS (2019) Effects of microplastics in soil ecosystems: above and below ground. Environ Sci Technol 53(19):11496–11506

Browne MA, Galloway TS, Thompson RC et al (2010) Spatial patterns of plastic debris along estuarine shorelines. Environ Sci Technol 44(9):3404–3409

Carbery M, O'Connor W, Palanisami T et al (2018) Trophic transfer of microplastics and mixed contaminants in the marine food web and implications for human health. Environ Int 115:400–409

Carr SA, Liu J, Tesoro AG et al (2016) Transport and fate of microplastic particles in wastewater treatment plants. Water Res 91:174–182

Cole M, Lindeque P, Halsband C et al (2011) Microplastics as contaminants in the marine environment: a review. Mar Pollut Bull 62(12):2588–2597

Cole M, Lindeque P, Fileman E et al (2013) Microplastic ingestion by zooplankton. Environ Sci Technol 47(12):6646–6655

Cózar A, Echevarría F, González-Gordillo JI et al (2014) Plastic debris in the open ocean. Proc Natl Acad Sci USA 111(28):10239–10244

De Falco F, Gullo MP, Gentile G et al (2018) Evaluation of microplastic release caused by textile washing processes of synthetic fabrics. Environ Pollut 236:916–925

Dris R, Gasperi J, Saad M et al (2016) Synthetic fibers in atmospheric fallout: a source of microplastics in the environment? Mar Pollut Bull 104(1–2):290–293

Fendall LS, Sewell MA (2009) Contributing to marine pollution by washing your face: microplastics in facial cleansers. Mar Pollut Bull 58(8):1225–1228

Gallo F, Fossi C, Weber R et al (2018) Marine litter plastics and microplastics and their toxic chemicals components: the need for urgent preventive measures. Environ Sci Eur 30(1):13

Galloway TSM, Cole M, Lewis C et al (2017) Interactions of microplastic debris throughout the marine ecosystem. Nat Ecol Evol 1(5):0116
Gasperi J, Wright SLR, Collard F et al (2018) Microplastics in air: are we breathing it in? Curr Opin Environ Sci Health 1:1–5
Geyer R, Jambeck JR, Law KL et al (2017) Production, use, and fate of all plastics ever made. Sci Adv 3(7):e1700782
Hartline NL, Bruce NJ, Karba SN et al (2016) Microfiber masses recovered from conventional machine washing of new or aged garments. Environ Sci Technol 50(21):11532–11538
Hidalgo-Ruz V, Gutow L, Thompson RC et al (2012) Microplastics in the marine environment: a review of the methods used for identification and quantification. Environ Sci Technol 46(6): 3060–3075
Hodson ME, Duffus-Hodson CA, Clark A et al (2017) Plastic bag derived-microplastics as a vector for metal exposure in terrestrial invertebrates. Environ Sci Technol 51(8):4714–4721
Huerta LE, Gertsen H, Gooren H et al (2016) Microplastics in the terrestrial ecosystem: implications for *Lumbricus terrestris* (Oligochaeta, Lumbricidae). Environ Sci Technol 50(5):2685–2691
Khalid N, Aqeel M, Noman A et al (2021) Interactions and effects of microplastics with heavy metals in aquatic and terrestrial environments. Environ Pollut 290:118104
Koelmans AA, Besseling E, Shim WJ et al (2015) Nanoplastics in the aquatic environment. Critical review. In: Marine anthropogenic litter. SpringerOpen, Cham, pp 325–340
Kole PJ, Löhr AJ, Van Belleghem FG et al (2017) Wear and tear of tyres: a stealthy source of microplastics in the environment. Int J Environ Res Public Health 14(10):1265
Li J, Liu H, Chen JP et al (2018) Microplastics in freshwater systems: a review on occurrence, environmental effects, and methods for microplastics detection. Water Res 137:362–374
Lithner D, Larsson Å, Dave G et al (2011) Environmental and health hazard ranking and assessment of plastic polymers based on chemical composition. Sci Total Environ 409(18):3309–3324
Lebreton LC, Van Der Zwet J, et al (2017) River plastic emissions to the world's oceans Abstract Nature Communications 8(1). https://doi.org/10.1038/ncomms15611
Liu S, Shi J, Wang J et al (2021) Interactions between microplastics and heavy metals in aquatic environments: a review. Front Microbiol 12:652520
Lozano YM, Aguilar-Trigueros CA, Onandia G et al (2021) Effects of microplastics and drought on soil ecosystem functions and multifunctionality. J Appl Ecol 58(5):988–996
Lu Y, Zhang Y, Deng Y et al (2016) Uptake and accumulation of polystyrene microplastics in zebrafish (*Danio rerio*) and toxic effects in liver. Environ Sci Technol 50(7):4054–4060
Lusher A, Hollman P, Mendoza-Hill J et al (2017) Microplastics in fisheries and aquaculture: status of knowledge on their occurrence and implications for aquatic organisms and food safety. Available via https://openknowledge.fao.org/server/api/core/bitstreams/4bcd8844-c17b-42ff-97da-d2e540733118/content. Accessed on 12 Dec 2024
Magnusson K, Norén F (2014) Screening of microplastic particles in and down-stream a wastewater treatment plant. Avaialable via https://naturvardsverket.diva-portal.org/smash/get/diva2:773505/FULLTEXT01.pdf. Accessed on 12 Dec 2024
Napper IE, Thompson RC (2019) Environmental deterioration of biodegradable, oxo-biodegradable, compostable, and conventional plastic carrier bags in the sea, soil, and open-air over a 3-year period. Environ Sci Technol 53(9):4775–4783
Oberoi G, Garg A (2021) Single-use plastics: a roadmap for sustainability? Supremo Amicus 24: 585
OECD E P R A (2016) Extended producer responsibility: updated guidance for efficient waste management. OECD, France. Available via https://www.oecd.org/content/dam/oecd/en/publications/reports/2016/09/extended-producer-responsibility_g1g6742c/9789264256385-en.pdf. Accessed on 12 Dec 2024
Piehl S, Leibner A, Löder MG et al (2018) Identification and quantification of macro-and microplastics on an agricultural farmland. Sci Rep 8(1):17950
Pilapitiya PNT, Ratnayake AS (2024) The world of plastic waste: a review. Clean Mater 11:100220

Rillig MC (2012) Microplastic in terrestrial ecosystems and the soil? Environ Sci Technol 46:6453–6454
Rillig MC, Ziersch L, Hempel S (2017) Microplastic transport in soil by earthworms. Sci Rep 7(1):1362
Rochman CM, Tahir A, Williams SL et al (2015) Anthropogenic debris in seafood: plastic debris and fibers from textiles in fish and bivalves sold for human consumption. Sci Rep 5(1):1–10
Royer SJ, Ferrón S, Wilson ST (2018) Production of methane and ethylene from plastic in the environment. PLoS One 13(8):e0200574
Savoca MS, Wohlfeil ME, Ebeler SE et al (2016) Marine plastic debris emits a keystone infochemical for olfactory foraging seabirds. Sci Adv 2(11):e1600395
de Souza Machado AA, Kloas W, Zarfl C et al (2018) Microplastics as an emerging threat to terrestrial ecosystems. Glob Chang Biol 24(4):1405–1416
Sussarellu R, Suquet M, Thomas Y et al (2016) Oyster reproduction is affected by exposure to polystyrene microplastics. Proc Natl Acad Sci 113(9):2430–2435
Teuten EL, Saquing JM, Knappe DR et al (2009) Transport and release of chemicals from plastics to the environment and to wildlife. Philos Trans R Soc Lond B Biol Sci 364(1526):2027–2045
Thompson RC, Olsen Y, Mitchell RP et al (2004) Lost at sea: where is all the plastic? Science 304(5672):838–838
Tokiwa Y, Calabia BP, Ugwu CU, Aiba S et al (2009) Biodegradability of plastics. Int J Mol Sci 10(9):3722–3742
Wang J, Liu X, Li Y et al (2019) Microplastics as contaminants in the soil environment: a mini-review. Sci Total Environ 691:848–857
Zarfl C, Fleet D, Fries E et al (2011) Microplastics in oceans. Mar Pollut Bull 62:1589–1591
Zhang S, Wang J, Yan P et al (2021) Non-biodegradable microplastics in soils: a brief review and challenge. J Hazard Mater 409:124525
Zhao J, Zhang H, Shi L et al (2024) Detection and quantification of microplastics in various types of human tumor tissues. Ecotoxicol Environ Saf 283:116818

Environmental Chemicals and Risk of Polycystic Ovary Syndrome (PCOS)

Mehjbeen Javed, Suramya, Humaira Naaz Bhutto, Shaesta Shahid, Mohd. Anas Saifi, Shahzad Ahmad, and Sheikh Raisuddin

Abstract

Polycystic ovary syndrome (PCOS) is a multifactorial disorder affecting premenopausal women. PCOS is characterized by oligo/anovulation, hyperandrogenism, and polycystic ovaries. According to Rotterdam criteria, the presence of any two features indicates the manifestation of PCOS. Environmental exposure to toxicants, particularly endocrine-disrupting chemicals (EDCs), has become unavoidable and has taken a toll on female fertility. EDCs are a major class of exogenous compounds that encompasses bisphenol A (BPA), BPA analogues, phthalates, di(2-ethyl hexyl) phthalate (DEHP), tributylin, heavy metals, pesticides, diethylstilbestrol (DES), and anti-androgens. These compounds are found in plastic products, plasticizers, cosmetics, personal care products, pharmaceuticals, antifouling agents, beverages, food containers, seafood products, contraceptive pills, etc. Several in vivo, in vitro, and epidemiological studies have established that EDCs mimic natural hormones and affect the folliculogenesis, granulosa, and theca cell function, resulting in decreased fertility in women. This review showcases the PCOS-inducing effects of EDCs. However,

Suramya, Humaira Naaz Bhutto and Shaesta Shahid contributed equally to this work.

M. Javed · S. Ahmad · S. Raisuddin (✉)
Department of Medical Elementology and Toxicology, Jamia Hamdard, School of Chemical and Life Sciences, New Delhi, Delhi, India
e-mail: mehjabeenjaved_sch@jamiahamdard.ac.in; satox@jamiahamdard.ac.in; sraisuddin@jamiahamdard.ac.in

Suramya · H. N. Bhutto · S. Shahid · M. Anas Saifi
Molecular Toxicology Laboratory, Department of Medical Elementology and Toxicology, School of Chemical and Life Sciences, Jamia Hamdard, New Delhi, India
e-mail: suramya_sch@jamiahamdard.ac.in; hnbhutto_sch@jamiahamdard.ac.in; shaestashahid_sch@jamiahamdard.ac.in

S. Raisuddin et al. (eds.), *Environmental Toxicants and Lifestyle Diseases*,
https://doi.org/10.1007/978-981-95-0575-3_2

its exact etiology has not yet been clearly defined. PCOS is a lifestyle disease, and therefore, it can be managed through lifestyle changes like regular exercise, weight management, and a healthy diet.

Keywords

Endocrine disrupting chemicals · Heavy metals · BPA · Phthalates

Introduction

Polycystic ovary syndrome (PCOS) is a complex multifactorial endocrine disorder that affect 4–18% of premenopausal women (Zeng et al. 2022). It has been defined in several ways however, the most accepted one is the Rotterdam criteria according to which any two features indicate PCOS condition: oligo-anovulation, hyperandrogenism and presence of polycystic ovaries. Azziz et al. (2004) reported that about 80–85% women who were diagnosed with clinical hyperandrogenism were found to have PCOS. However, hyperandrogenism was not observed in some other ethnic groups. Further, there have been reports which stated that genetic susceptibility is linked to PCOS and exposure to environmental toxicants, geographical region, and dietary factors play significant role in the expression of related genetic traits. The environmental toxicants which are associated with PCOS manifestation include endocrine disrupting chemical (EDCs) (Suramya et al. 2024). EDCs are exogenous substances which interfere with the metabolism, synthesis, binding, secretion, and transportation of natural hormones (Rashid et al. 2018; Dabeer et al. 2020). The most common EDCs include bisphenol A (BPA), phthalates, di(2-ethyl hexyl) phthalate (DEHP), tributylin, heavy metals, and pesticides. These EDCs are commonly found in our daily used products including dietary sources and water (Raisuddin and Sharma 2017). Other EDCs include per and polyfluoroalkayl substances (PFAS), phytoestrogens, perchlorate, dioxins, etc. Among heavy metals arsenic (As), cadmium (Cd), thallium (Th), copper (Cu), lead (Pb), mercury (Hg) is prominent in interfering natural hormones. All these chemical substances affect female reproductive system and consequently lead to infertility. Several clinical, case control and cohort studies showed the presence of EDCs in the serum, follicular fluid and urine of PCOS women (De Felip et al. 2004; Meeker et al. 2009; Kandaraki et al. 2011; Vagi et al. 2014). Huge amount of lilterature is available on EDCs influence on PCOS. Still the pathophysiology and etiology of PCOS is not completely understood yet. Researchers have shown that EDCs interfere the natural processes by mimicking the hormones and thereby alter the signaling pathways resulting in the PCOS pathogenesis. Rutkowska and Rachoń (2014) revealed that EDCs also have the capability to induce epigenetic changes and cause hormonal imbalance. In this chapter, we have reviewed and highlighted the PCOS inducing effects of different EDCs.

PCOS Inducing Effects of EDCs

EDCs are toxic at environmental concentrations as well as low concentrations on chronic exposure. They contribute to the pathogenesis of PCOS in every aspect, like oligo-anovulation, high levels of androgens (hyperandrogenism), and the presence of polycystic ovaries. In general, the PCOS pathogenesis involves the reduced function of granulosa cells and increased response from the theca cells. Investigations of Cohn et al. (2003) reported that DDT (dichlorodiphenyl trichloroethane), a pesticide, affects the reproductive system of females via the gonadotropin signaling pathway and calcium-dependent pathways and impairs the functions of ovarian granulosa and theca cells. Ragusa et al. (2021) found microplastics in the placenta of humans, which were found to affect folliculogenesis. The impact of some common EDCs is shown in Table 1, which encompasses some of the in vivo studies performed to evaluate the effect of commonly known environmental contaminants on the pathogenesis of PCOS.

A study conducted by Fernandez et al. (2010) demonstrated neonatal exposure to BPA and found a correlation to elevated serum testosterone and estradiol levels and reduced progesterone in adulthood. They also reported a large number of cysts, infertility, and altered hypothalamus–pituitary axis at higher doses, while low doses had reduced fertility without the changes in the number of oocytes. Alternatively, another study performed by Javed et al. (2025) revealed that BPA may not be the only causative agent for PCOS alone, but it could be associated with some metabolic implications. Low-dose BPA disrupted the estrus cycle, altered hormone levels, and also affected the histomorphology of ovary and uterine tissue, elevated the homeostasis model assessment of insulin resistance (HOMA-IR) and HOMA-β, leading to PCOS-like symptoms. In daily life, it is uncommon to get exposed to a single EDC as we encounter a mixture of chemicals present in the environment we are exposed to in our daily life. To investigate this issue, a pilot study was performed by Yang et al. (2019) in which they exposed Sprague-Dawley rats to tributyltin and BPA both and found irregular estrus cycles, disrupted ovarian development, and altered hormone levels at high doses and combined exposure. The mixture also caused PCOS-like symptoms with the presence of more atretic follicles and cysts. They concluded that the combined exposure at the neonatal stage leads to PCOS in adult hood. Another endocrine disrupter involving a similar estrogenic nature as BPA has drawn attention among researchers. DEHP, a popularly used plasticizer added to increase flexibility, has been linked to toxic effects on the reproductive system (Rattan et al. (2018). Pocar et al. (2017) examined environmental dose exposure of DEHP to gestational and lactational exposure across three generations (F0, F1, F2, and F3). From gestation to lactation, F0 dams were administered 0, 0.5, and 5 mg/kg/day. F1 females revealed disrupted ovarian function, reduced oocyte quality, and altered gene expression. Similar effects were found in F2 and F3, indicating transgenerational effects of adult onset of diseases. To further confirm the multigenerational and transgenerational effect of DEHP, another study conducted by Brehm et al. (2018) found that gestational exposure to DEHP at the doses of 20 ug/kg/d–750 mg mg/kg/d disrupted the estrous cycle (at all the doses used in the study)

Table 1 Effects of environmental EDCs exposure indicating the risk of PCOS

S. No.	Chemical name	Species	Dose	Route & exposure duration	Findings	References
1	BPA	Sprague Dawley rat	50 and 500 μg/kg/d	Subcutaneous injection PND1–PND10	Elevated serum testosterone, estradiol and reduced progesterone, cystic follicles, lesser corpora lutea, high number of atretic follicles, and lower antral follicles	Fernandez et al. (2010)
2	BPA	Wistar rat	50 and 500 μg/kg/d	Subcutaneous injection (21 days)	Increased serum hormone levels, androgen levels, cystic in ovarian follicles, theca cells loosely attached, necrosis, elevated HOMA-IR & HOMAβ	Javed et al. (2025)
3	Perfluorohexane sulphonate (PFHxS)	Wistar rat	5 ppm & 25 ppm	Drinking water (90 days)	Irregular estrous cycle, increased percent disorder rate, hormonal disruption, increased metabolic dysfunction, enhanced vacuolation, reduced corpus luteum, thinning of endometrium lining	Suramya et al. (2025)
4	Tributyltin and BPA	Sprague Dawley rat	TBT (10 and 100 ng/kg/d) BPA (50 μg/kg/d)	Subcutaneous injection from PND1-PND16	Irregular estrus cycle, increased serum testosterone and LH hormone, less corpora lutea, antral follicle, more atretic and cystic follicles	Yang et al. (2019)
5	Di(2-ethyl hexyl) phthalate (DEHP)	Mouse (CD-1) (transgenerational study)	Prenatal DEHP exposure 0.05 and 5 mg/kg/d from gestational day 0.5-PND21 pups	Diet (chow)	Early onset of puberty, altered estrus cycle, less primordial follicles (DEHP 0.05, 5 mg/kg/d) in the F1–F3 generations-	Rattan et al. (2018)

					more number of preantral follicles (DEHP 0.05, 5 mg/kg/d) in the F1–F3 generations less antral follicles in the F1 and F2 generations (DEHP 0.05 mg/kg/d), decreased fertility related indices (DEHP; 5 mg/kg/d) in the F1 generation	
6	DEHP	CD-1 mice	20 and 200 μg/kg/d & 500 and 750 mg/kg/d	Oral (from gestational day 11 to birth)	Disrupted estrus cycle, hormone imbalance, prenatal exposure resulted reduction follicle counts in F1 generation while F2 and F3 generations showed increase follicular counts and ovarian cysts	Brehm et al. (2018)
7	Polystyrene microplastic	Wistar rat	0.015 and 0.15 mg/kg/d	Drinking water (90 days)	Altered antioxidant enzymes, apoptosis in granulosa cells, & induction of pyroptosis, thinning of granulose layer of secondary follicles, decreased primary follicles	Hou et al. (2021)

reduced follicle number and increased ovarian cysts in the F1 generation at high dose, whereas the F2 and F3 generations demonstrated altered folliculogenesis and hormone imbalance that collectively directed toward multigenerational and transgenerational effects of female reproduction under the provided experimental condition used in the study.

Some other emerging contaminants, such as microplastics, etc., are also persist in the environment. A study by Hou et al. (2021) performed a mechanistic evaluation of chronic exposure to polystyrene microplastic and found apoptosis in granulosa cells and induction of pyroptosis, which is a type of cell death that causes inflammation, triggered by microbial infection via the NLRP3/Caspase-1 signaling pathway. They also found decreased antioxidant enzymes and an increase in malondialdehyde (MDA), which could cause adverse effects to ovarian tissue, leading to a risk of infertility.

Heavy Metal-Induced PCOS Like Symptoms

Lately, heavy metals are also recognized as EDCs, the prime environmental and occupational pollutants that can be shown to generate symptoms similar to those depicted in the case of PCOS. Recent studies have shown how the exposure to these heavy metals disrupts many physiological processes, contributing to the development and aggravation of PCOS. Additionally, these heavy metals also directly impair the reproductive tissues, exacerbating the clinical manifestations of PCOS.

Cadmium (Cd)

Cd is known to act like estrogen in the body and could therefore cause characteristic hormonal imbalances of PCOS (Abudawood et al. 2021). The evidence suggests that Cd may affect androgen production and interfere with metabolic pathways characteristic of PCOS, thus playing a contributing role in the development and progression of the disease (Kim et al. 2021). Even short-term exposure to Cd can perturb metabolic and hormonal-pituitary-gonadal axis functions, thereby interfering with features similar to PCOS. This includes dyslipidemia, insulin resistance, polycystic ovaries, ovarian depletion, abnormal hypothalamic regulation, increased LH levels, and reduced AMH (antimullerian hormone) levels (da Costa et al. 2021). A recent study has shown that among mixtures of metals (Ar, Cd, Hg, and Ba), Cd is a significant factor in influencing metabolic and hormonal imbalances, particularly in patients with PCOS. The research demonstrated a robust association between Cd exposure and insulin resistance, as evidenced by elevated levels of the HOMA-IR. This indicates that Cd may exacerbate insulin resistance, a key feature of PCOS, by impairing insulin signaling pathways (Wang et al. 2024). Further, exposure to Cd during gestation and lactation was found to affect the reproductive development of female offspring of rats. The findings demonstrated that in utero Cd exposure increased the production of steroid hormones in offspring. The manifestation includes early age at puberty, more rapid development of follicles, and larger litter size (Li et al. 2018). Thus, together, these results showed the possibility of long-lasting effects due to endocrine disruption caused by Cd exposure during pregnancy across generations. This might be additive in PCOS, in which normal hormonal regulation and ovarian function are changed. It leads to the excessive production of steroid hormones, hence early puberty. This stage is also characterized by imbalances in the hormonal axis that control the menstrual cycle and ovarian follicles, which could, in turn, manifest with classic features of PCOS such as irregular periods, high levels of androgens, and polycystic ovaries. Recently, a study by da Costa et al. (2021) demonstrated that subacute exposure to Cd in drinking water disrupted the hypothalamus-pituitary axis (HPG), resulting in PCOS and primary ovarian insufficiency (POI), increased accumulation, metabolic disturbances, altered gene expression, irregular estrous cycles, hormonal imbalances, and cystic follicles were some of the key results of this study. Later on, chronic exposure to Cd was demonstrated by Nasiadek et al. (2019) and revealed altered plasma hormone levels at high doses, which later caused irregular estrus cycles. Cd exposure also caused endometrial edema independent of dose and resulted in damage to the ovary and uterus. Merlo et al. (2019) explored another heavy metal mercury and evaluated its effect on PCOS-like symptoms in Wistar rats. Histology revealed structural alterations in ovarian tissues such as thickened theca albuginea, increased connective tissues, and increased follicular borders.

Lead (Pb)

Steroid hormones are excessively produced due to the fact that Pb is generally toxic to reproductive tissues, thus disrupting ovarian function and hormone production. Pb-induced oxidative stress also contributes to the aggravation of PCOS-related symptoms (Abudawood et al. 2021). The Pb levels were found to be significantly higher in PCOS subjects than in the non-affected females. A positive correlation existed between Pb levels and fasting glucose, hence suggesting that its exposure might be associated with impaired glucose metabolism in PCOS patients. It means that Pb might add to the metabolic disturbances associated with PCOS, particularly insulin resistance. Moreover, it has the ability to change the synthesis and regulation of steroid hormones, worsening the hormonal imbalance typical of PCOS. Further, the accumulation of Pb in the body induces oxidative stress and inflammation, which aggravate cellular damage and metabolic dysfunction. Taken together, these mechanisms suggest that lead exposure may play a role in the development and progression of PCOS through effects on both metabolic and hormonal systems (Kirmizi et al. 2020). A study by Nakade et al. (2015) investigated the lead acetate exposure in experimental rats via drinking water at different doses and found a significant weight loss and decreased uterine relative weight at high dose. Interestingly, blood and bone levels of lead acetate got increased at the median dose but decreased at high-dose exposure. Histopathological evaluation from this study revealed the dose-dependent uterine inflammation, narrow lumen, thickened endometrium, glandular damage, and vacuolar degeneration.

Copper (Cu)

Cu has a dual role in the setting of PCOS, specifically in respect to its influence on oxidative stress and hormonal activity. In high blood glucose conditions, PCOS is associated with an increased rate of formation of ROS, resulting in cell membrane, protein, and DNA damage, eventually leading to the death of the cell and tissue degeneration (González et al. 2006). Cu is an integral factor in the function of various vital enzymes involved in electron transport and other significant biological processes. Chronic exposure to trace elements like Cu has been related to the increase in oxidative stress, inflammation, and cardiovascular disease (Uriu-Adams and Keen 2005). It increased oxidative stress by stimulating the generation of ROS and by reducing glutathione and may further interfere with hormone function due to its property of binding to estrogen receptors (Klotz et al. 2003). According to the literature, high levels of Cu are harmful to reproductive health since they damage ovarian cells and hamper follicular development. Interestingly, while low doses of Cu may have some beneficial effects, high doses and long-term exposure can be toxic and adversely affect sex hormone secretion and reproductive functions (Jiang et al. 2021).

Iron (Fe)

In women with PCOS, changes in a number of proteins linked to iron biology alter blood levels, indicating a likely mild iron overload in the condition. These key proteins in this interrelation are ferritin, haptoglobin, hepcidin, and α2-macroglobulin. Changes in these proteins have to be understood in order to be able to link PCOS with iron metabolism (Mathew et al. 2023). This will bring out one striking phenomenon in the complex landscape of obesity and PCOS: abnormal accumulation of iron within the fatty tissues. The accumulation disrupts normal tissue function and is a major contributor to metabolic problems in these patients with generalized metabolic complications. As ferritin levels elevated it indicates increased iron storage, a clear pattern of insulin resistance begins to take shape. The resistance shows a very tight correlation with ferritin, a major marker of iron in the body, depicting a picture of metabolic imbalance. The problem is also entwined with the function of hepcidin, a hormone that plays a very major role in regulating iron absorption. In women with PCOS, the levels of this hormone are known to be very low. This diminution will allow an increase in the absorption of iron, thus aggravating the overload. This interplay of low hepcidin and high androgen and insulin resistance simply shows a complex network of hormonal and metabolic dysregulation in which each component influences the other in a cycle, which makes the management of PCOS so complicated (Al-Akabi and Hafth 2023). Studies showed that hepcidin, a key iron regulatory hormone, is significantly bound to α2-macroglobulin in the circulation. This complexed form of hepcidin with α2-macroglobulin has more potent effects on reducing the cell surface density of the iron exporter protein ferroportin. In PCOS subjects with low α2-macroglobulin levels, this may further reduce hepcidin's ability to regulate iron export and contribute to an increased iron level in the circulation (Huang et al. 2013). Haptoglobin is a protein in plasma that complexes with hemoglobin, and its low levels indicate excessive iron in the body. For example, in hemolytic anemia, increased amounts of cell-free hemoglobin bind to haptoglobin, which, upon clearance by the liver, releases iron into the circulation. This relationship between the haptoglobin level and iron regulation is of importance in diseases like PCOS, in which decreased levels of haptoglobin can increase intestinal iron absorption and induce oxidative stress. A common genetic variation in the gene encoding the α-chain of haptoglobin is associated with PCOS and results in decreased expression of haptoglobin (Mathew et al. 2023).

Mercury (Hg)

Mercury can induce PCOS-like symptoms by its multifaceted approach that influences reproductive and metabolic health. The metal disrupts the regulation of hormonal balance, with an impact on the hypothalamic-pituitary-ovarian axis, leading to an imbalance in FSH vs. LH and producing irregular menstrual cycles and anomalous ovarian development. Hg damages ovarian follicles, reducing their number of primordial follicles and increasing atretic ones. Mercuric exposure causes insulin resistance and hyperglycemia metabolically and thereby aggravates glucose intolerance seen in PCOS. Direct mercurial-induced inflammation in reproductive tissues, by promoting edema, necrosis, and damaging the uterine lining, further exaggerates the disturbances to menstrual regularity and fertility. Chronic exposure further retards puberty, with effects on the maturity of oocytes, reflecting broader reproductive health problems (Massányi et al. 2020). Exposures to concentrations of Hg relevant to humans have been reported to cause reproductive and metabolic disturbances similar to those reported in PCOS. This would then cause a wide array of disorders similar in symptoms to PCOS. These include anovulatory dysfunction and abnormal development of the ovaries. It also causes atrophy of the uterus with associated inflammation. Moreover, Hg exposure is capable of impairing glucose metabolism, causing high blood sugar levels and thus giving rise to insulin resistance (Merlo et al. 2019).

Thallium (Tl)

Thallium is an extremely toxic heavy metal that exerts its toxicity toward cellular function through two major mechanisms. Firstly, Tl could modulate cellular events by acting as a potassium ion mimic (K^+). Potassium is an essential element for the proper activity of an enzyme involved in the maintenance of the ionic equilibrium across the cellular membrane called the Na^+/K^+ -ATPase. By replacing potassium in this enzyme, TI disturbs the enzyme's function and, as a result, the normal flux of sodium and potassium ions. Such disruption can give rise to various cellular dysfunctions because the Na^+/K^+-ATPase is essential for the maintenance of cellular homeostasis, volume regulation, and electrical excitability. Second, TI can combine with thiol groups in proteins and other cellular molecules. Thiol groups containing the sulfur atom play a vital role in biochemical functions, including maintenance of the redox status and enzyme activity balance. Binding of TI to thiol groups may inhibit some of the vital enzymatic reactions associated with metabolic processes. This interference results in oxidative stress due to the decrease in the levels of major antioxidants like glutathione, important for scavenging ROS and protecting the cells from damage. Under such oxidative stress, ovarian function is affected, further exacerbated by inflammation. Thus, Tl becomes an arguably key player in the pathogenesis of PCOS as a result of breaking the balance between oxidative stress and redox status (Abudawood et al. 2023).

Arsenic (As)

Animal studies involving female rats have shown that the physiological and hormonal changes induced by arsenic are quite significant, ranging from reduced ovarian and uterine weights to disrupted estrous cycling, and finally, to key reproductive hormone alterations, including decreased estradiol, progesterone, and testosterone and increased levels of LH and FSH. Experiments demonstrated that, at a molecular level, As exposure can down-regulate crucial proteins involved in steroidogenesis, such as FSH receptor, steroidogenic acute regulatory protein, CYP17A1, HSD3B1, and CYP19A1. Moreover, this disruption further extends to some vital signaling pathways, particularly to the protein kinase A and extracellular signal-regulated kinases, c-Jun N-terminal kinase, and c-JUN. Meanwhile, other pathways, including AKT and cAMP response element-binding protein, remain unaffected. These biochemical and molecular disruptions then translate to compromised reproductive outcomes (Chen et al. 2022). Thus, As exposure highlights potential mechanisms that can also be relevant to understanding PCOS. Although these parallels may suggest that some common mechanisms between As-induced reproductive impairments and features of PCOS exist, one should point out that PCOS is a heterogeneous condition influenced by genetic predispositions as well as by environmental factors like diet or life style. Further studies are required in order to reveal the direct link between chronic environmental exposures to As as a potential trigger for conditions like PCOS in humans.

Heavy Metal Action Mechanism as EDCs

A detailed understanding of the mechanisms through which heavy metals are involved in PCOS is required to assess the wider implications of environmental toxins on reproductive health and to devise effective preventive and therapeutic strategies.

Thus, heavy metals can induce polycystic ovary syndrome (PCOS)-like symptoms through several key mechanisms, each affecting hormonal and metabolic functions in distinct ways (Fig. 1).

1. *Hormonal disruption:* Heavy metals disrupt the hormonal balance of PCOS patients by several mechanisms. They are capable of mimicking actions like estrogens, binding to hormone receptors, and altering the hormonal activities, thereby interfering with androgen and other hormone regulation. Impaired physiology of the HPO axis and disruption of the secretion of FSH and LH could lead to disordered menstrual cycles and ovarian insufficiency. They also interfere with other mechanisms integrating centers of gonadal steroidogenesis and endocrine signaling pathways, which leads to disturbances in the synthesis and secretion of these reproductive hormones. These metals can also affect the sensitivity, as well as the function of, hormone receptors, leading to further alteration in the levels of such hormones and enhanced imbalances (Kirmizi et al. 2020).
2. *Oxidative stress and inflammation:* Metals generate reactive oxygen species that cause oxidative damage to cellular components, inferring cellular dysfunction,

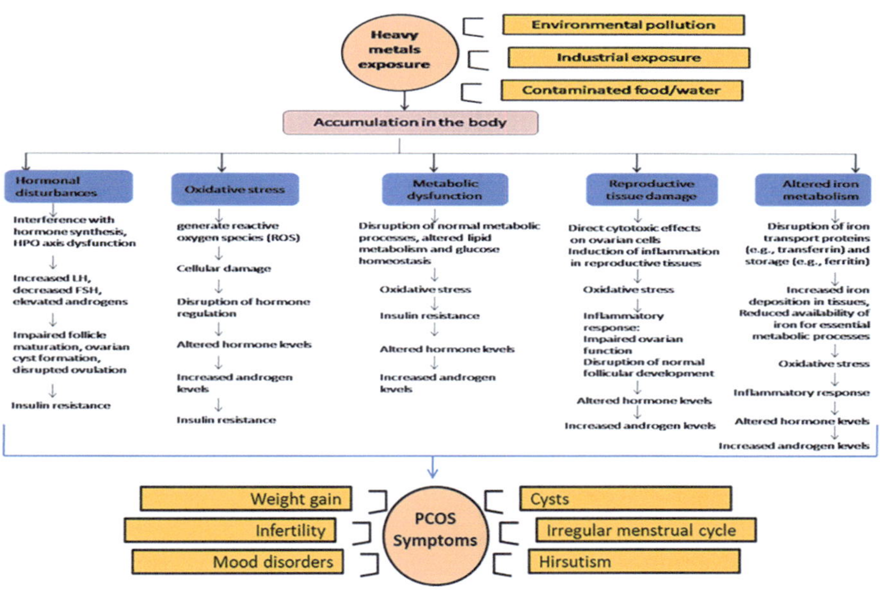

Fig. 1 Mechanism of heavy metals-induced PCOS-like symptoms

apoptosis, and impairment of normal ovarian function. Consequently, impaired follicular development ensues, with damage in ovarian and uterine tissues. Furthermore, the accumulation of heavy metals activates the signaling pathways of inflammatory responses, which promotes increased release of pro-inflammatory cytokines and infiltration of inflammatory cells. This sets off a chain reaction in which chronic inflammation further exacerbates oxidative damage, worsens impairment of the function of reproductive tissue, and promotes PCOS progression. Subsequently, the interplay between oxidative stress and inflammation may possibly enter into a vicious cycle that further aggravates these typical hormonal and metabolic imbalances characteristic of PCOS (Srnovršnik et al. 2023).

3. *Metabolic dysfunction:* Heavy metals induce metabolic dysfunction in polycystic ovary syndrome by inhibiting the major metabolic pathways and insulin signaling. The metals act to interfere with glucose metabolism by oxidation of insulin, leading to insulin resistance, a common feature in PCOS. A good example is the fact that the metals are capable of reducing the effectiveness of insulin to lower blood glucose levels by the alteration function of insulin receptors and related signaling molecules. This disturbance favors hyperglycemia and worsens the usual metabolic syndrome that accompanies PCOS. Also, toxic heavy metals can interfere with lipid metabolism through disturbances in its synthesis and catabolism, which gives way to dyslipidemias characterized by altered cholesterol and triglyceride metabolism. The combined effect of insulin resistance and dyslipidemia contributes to the development of obesity and other metabolic complications observed in PCOS. These mechanisms represent how heavy metals worsen metabolic dysfunction and insulin resistance, both of which are at the center of the pathophysiology of PCOS (Herman et al. 2023).
4. *Reproductive tissue damage:* Heavy metals have induced damage to reproductive tissue in PCOS through several detrimental mechanisms. The metals directly act on ovarian and uterine tissues, leading to cellular toxicity and structural damage. Heavy metals, for example, may induce apoptosis and necrosis of follicles directly in the ovaries, thus reducing follicular count and impairing follicular development, which is characteristic of PCOS. It can also act to the detriment of the integrity of the uterine lining by producing inflammation and edema, contributing to menstrual irregularity and diminished fertility. Furthermore, heavy metals have the ability to induce both oxidative stress and inflammation in an organism, resulting in increased tissue damage by the release of ROS and proinflammatory cytokines. These chronic inflammation and oxidative damages compromise the normal functioning of reproductive tissues and lead to pathologies such as polycystic ovaries and atrophy of the uterus. The cumulative result of these mechanisms has been shown to explain a significant degree of reproductive dysfunctions observed in PCOS (Mukherjee et al. 2022).
5. *Altered iron metabolism:* Heavy metals disturb the regulation of iron homeostasis and further worsen associated metabolic disturbances in polycystic ovary syndrome. Heavy metals can impede the action of hepcidin, an integral hormone that regulates absorption and distribution of iron and consequently results in

increasing intestinally absorbed iron and high serum ferritin. Increased iron levels elevate the state of oxidative stress, which is attained by the generation of reactive oxygen species, extremely promoting damage in dry ovarian tissue and other tissues. What is more, heavy metals may affect the level of proteins responsible for iron metabolism, such as ferritin and α2-macroglobulin, thus leading to the loss of balance of iron homeostasis. Thus, the caused dysregulation in the body conveys the further resultant metabolic imbalance and inflammatory processes that enhance the state of insulin resistance and problems with other symptoms of PCOS. This makes the interaction of heavy metals with exposure, disturbed iron metabolism, and oxidative stress intensify the sorrow of hormonal and metabolic imbalances that are specific to PCOS (Mathew et al. 2023).

Pharmaceutical Agents as EDCs

The pharmaceutical agents, specifically those having estrogenic or anti-androgenic characteristics, can also function as EDCs. For example, diethylstilbestrol (DES), antiandrogens, and other estrogen like compounds. Therefore, such products have the potential to contribute PCOS pathogenesis. Besides, valproate, may elevate androgen synthesis in ovaries, leading to a form of PCOS.

- *Diethylstilbestrol (DES):* It is the artificial estrogen and recognized as EDC. Research studies have associated exposure to DES with various reproductive health-related issues, including increased susceptibility to PCOS, through its enduring impacts on estrogen signaling pathways.
- *Anti-androgens:* They work by blocking androgen receptors and are used to treat conditions like male pattern baldness and hirsutism (excessive hair growth). However, their off-target effects possibly lead to interference with the endocrine system, potentially impacting ovarian function and contributing to PCOS by disrupting steroidogenesis and normal follicular development.
- *Other estrogen-like compounds:* Certain drugs with estrogen-like activity, for instance, contraceptive pills, can imitate estrogen action and disrupt the delicate hormonal balance that contributes to PCOS.

Epidemiological Data

There is a paucity of epidemiological data related to environmental contaminants and PCOS. Li et al. (2011) analyzed serum of PCOS and non-PCOS women, where no significant differences were observed in phthalic acid esters and BPA concentrations, whereas octylphenol was linked with insulin resistance. Kandaraki et al. (2011) reported significantly higher BPA (1.2 ng/mL) levels in serum of PCOS patients as compared to non-hyperandrogenic patients. Likewise, in adolescent girls, serum BPA levels were 1.1 ng/mL who were showing PCOS symptoms (Akın et al. 2015). Rashidi et al. (2017) assessed higher BPA (3.34 ng/mL) in the urine of PCOS women than healthy women. Another study on BPA concentrations in serum correlated it with testosterone and free androgen index (Konieczna et al. 2018). Perfluorooctanoate, perfluorooctane sulfonate, and monobenzyl phthalate are also found to be linked with PCOS pathophysiology (Vagi et al. 2014). Triclosan was also found in the urine of patients showing PCOS symptoms and also positively correlated with LH and LH/FSH ratio (Ye et al. 2018). The presence of DEHP was also reported in the follicular fluid (1.68 ng/mL) of PCOS patients (Jin et al. 2019). Gu et al. (2019) positively correlated urine organic UV filter-octocrylene concentrations with that of PCOS.

Mitigation of PCOS Symptoms

PCOS is a multifactorial disorder: likewise, responses among patients are also variable. Several medications like metformin, clomiphene, letrozole, etc., are present in the market, which shows improvement in the symptoms of PCOS. However, they also have some side effects. Therefore, the PCOS symptoms can be managed through lifestyle changes, which include regular exercise, a balanced diet, and stress management. The lifestyle changes can help in hormone regulation, weight management, and thereby reduction in the risk of PCOS complications.

Conclusion

The effects of EDCs and other xenobiotics on PCOS have been extensively studied, but the pathophysiology and etiology of PCOS are not yet well understood. The epidemiological and preclinical studies have revealed the involvement of multiple factors in the pathogenesis or exacerbation of PCOS. Therefore, further studies are warranted on the development of a suitable animal model of PCOS to understand the molecular and cellular mechanism behind the disease and also to develop a specific molecular marker that could be considered while making a therapeutic strategies.

References

Abudawood M, Tabassum H, Alanazi A et al (2021) Antioxidant status in relation to heavy metals induced oxidative stress in patients with polycystic ovarian syndrome (PCOS). Sci Rep 11(1): 22935

Abudawood M, Alnuaim L, Tabassum H et al (2023) An insight into the impact of serum tellurium, thallium, osmium and antimony on the antioxidant/redox status of PCOS patients: a comprehensive study. Int J Mol Sci 24(3):2596

Akın L, Kendirci M, Narin F et al (2015) The endocrine disruptor bisphenol a may play a role in the aetiopathogenesis of polycystic ovary syndrome in adolescent girls. Acta Paediatr 104:e171–e177

Al-Akabi D, Hafth H (2023) Physiological effect of iron status on patients with polycystic ovary syndrome in Basrah city. J Med Biochem 42(3):530–535

Azziz R, Sanchez LA, Knochenhauer ES et al (2004) Androgen excess in women: experience with over 1000 consecutive patients. J Clin Endocrinol Metab 89(2):453–462

Brehm E, Rattan S, Gao L et al (2018) Prenatal exposure to di(2-ethylhexyl) phthalate causes long-term transgenerational effects on female reproduction in mice. Endocrinology 159(2):795–809

Chen P, Luo Q, Lin Y et al (2022) Arsenic exposure during juvenile and puberty significantly affected reproductive system development of female SD rats. Ecotoxicol Environ Saf 242: 113857

Cohn BA, Cirillo PM, Wolff MS (2003) DDT and DDE exposure in mothers and time to pregnancy in daughters. Lancet 361:2205–2206

da Costa CS, Oliveira TF, Freitas-Lima LC et al (2021) Subacute cadmium exposure disrupts the hypothalamic-pituitary-gonadal axis, leading to polycystic ovarian syndrome and premature ovarian failure features in female rats. Environ Pollut 269:116154

Dabeer S, Afjal MA, Ahmad S et al (2020) Transgenerational effect of parental obesity and chronic parental bisphenol A exposure on hormonal profile and reproductive organs of preadolescent Wistar rats of F1 generation: a one-generation study. Hum Exp Toxicol 39:59–76

De Felip E, Di Domenico A, Miniero R et al (2004) Polychlorobiphenyls and other organochlorine compounds in human follicular fluid. Chemosphere 54:1445–1449

Fernandez M, Bourguignon N, Lux-Lantos V et al (2010) Neonatal exposure to bisphenol a and reproductive and endocrine alterations resembling the polycystic ovarian syndrome in adult rats. Environ Health Perspect 118(9):1217–1222

GonzáLez F, Rote NS, Minium J et al (2006) Reactive oxygen species-induced oxidative stress in the development of insulin resistance and hyperandrogenism in polycystic ovary syndrome. J Clin Endocrinol Metab 91(1):336–340

Gu J, Yuan T, Ni N et al (2019) Urinary concentration of personal care products and polycystic ovary syndrome: a case-control study. Environ Res 168:48–53

Herman R, Sikonja J, Jensterle M et al (2023) Insulin metabolism in polycystic ovary syndrome: secretion, signaling, and clearance. Int J Mol Sci 24(4):3140

Hou J, Lei Z, Cui L et al (2021) Polystyrene microplastics lead to pyroptosis and apoptosis of ovarian granulosa cells via NLRP3/Caspase-1 signaling pathway in rats. Ecotoxicol Environ Saf 212:112012

Huang MLH, Austin CJ, Sari MA et al (2013) Hepcidin bound to A2-macroglobulin reduces ferroportin-1 expression and enhances its activity at reducing serum iron levels. J Biol Chem 288(35):25450–25465

Javed M, Suramya MA et al (2025) Bisphenol A-induced polycystic ovary syndrome (PCOS) with hormonal and metabolic implications in rats. Reprod Toxicol 131:108750

Jiang Q, Zhang F, Han L et al (2021) Serum copper level and polycystic ovarian syndrome: a meta-analysis. Gynecol Obstet Investig 86(3):239–246

Jin Y, Zhang Q, Pan JX (2019) The effects of di(2-ethylhexyl) phthalate exposure in women with polycystic ovary syndrome undergoing in vitro fertilization. J Int Med Res 47:6278–6293

Kandaraki E, Chatzigeorgiou A, Livadas S et al (2011) Endocrine disruptors and polycystic ovary syndrome (PCOS): elevated serum levels of bisphenol A in women with PCOS. J Clin Endocrinol Metab 96:E480–E484

Kim K, Pollack AZ, Nobles CJ et al (2021) Associations between blood cadmium and endocrine features related to PCOS-phenotypes in healthy women of reproductive age: a prospective cohort study. Environ Health 20(1):64

Kirmizi DA, Baser E, Turksoy VA et al (2020) Are heavy metal exposure and trace element levels related to metabolic and endocrine problems in polycystic ovary syndrome? Biol Trace Elem Res 198(1):77–86

Klotz L, Kröncke K, Buchczyk DP et al (2003) Role of copper, zinc, selenium and tellurium in the cellular defense against oxidative and nitrosative stress. J Nutr 133(5):1448S–1451S

Konieczna A, Racho'n D, Owczarek K et al (2018) Serum bisphenol A concentrations correlate with serum testosterone levels in women with polycystic ovary syndrome. Reprod Toxicol 82: 32–37

Li T, Xu L, Chen Y et al (2011) Effects of eight environmental endocrine disruptors on insulin resistance in patients with polycystic ovary syndrome: a preliminary investigation. Nan Fang Yi Ke Da Xue Xue Bao 31(10):1753–1756

Li Z, Li T, Leng Y et al (2018) Hormonal changes and folliculogenesis in female offspring of rats exposed to cadmium during gestation and lactation. Environ Pollut 238:336–347

Massányi P, Massányi M, Madeddu R et al (2020) Effects of cadmium, lead, and mercury on the structure and function of reproductive organs. Toxics 8(4):94

Mathew M, Sivaprakasam S, Phy JL et al (2023) Polycystic ovary syndrome and iron overload: biochemical link and underlying mechanisms with potential novel therapeutic avenues. Biosci Rep 43(1):BSR20212234

Meeker JD, Missmer SA, Altshul L et al (2009) Serum and follicular fluid organochlorine concentrations among women undergoing assisted reproduction technologies. Environ Health 8:32

Merlo E, Schereider IRG, Simões MR et al (2019) Mercury leads to features of polycystic ovary syndrome in rats. Toxicol Lett 312:45–54

Mukherjee AG, Wanjari UR, Renu K et al (2022) Heavy metal and metalloid – induced reproductive toxicity. Environ Toxicol Pharmacol 92:103859

Nakade UP, Garg SK, Sharma A et al (2015) Lead-induced adverse effects on the reproductive system of rats with particular reference to histopathological changes in uterus. Indian J Pharmacol 47(1):22–26

Nasiadek M, Danilewicz M, Klimczak M et al (2019) Subchronic exposure to cadmium causes persistent changes in the reproductive system in female Wistar rats. Oxidative Med Cell Longev 2019:6490820

Pocar P, Fiandanese N, Berrini A et al (2017) Maternal exposure to di(2-ethylhexyl) phthalate (DEHP) promotes the transgenerational inheritance of adult-onset reproductive dysfunctions through the female germline in mice. Toxicol Appl Pharmacol 322:113–121

Ragusa A, Svelato A, Santacroce C et al (2021) Plasticenta: first evidence of microplastics in human placenta. Environ Int 146:106274

Raisuddin S, Sharma S (2017) Endocrine-disrupting chemicals in food and their toxicological implications. In: Sachan A, Hendrich S (eds) Food toxicology: current advances and future challenges. Apple Academic Press Inc., Oakville, pp 199–250

Rashid H, Sharma S, Beigh S et al (2018) Bisphenol A-induced endocrine toxicity and male reprotoxicopathy are modulated by the dietary iron deficiency. Endocr Metab Immune Disord Drug Targets 68:626–636

Rashidi HB, Amanlou M, Lak TB et al (2017) The association between bisphenol A and polycystic ovarian syndrome: a case-control study. Acta Med Iran 55:759–764

Rattan S, Brehm E, Gao L et al (2018) Di(2-ethylhexyl) phthalate exposure during prenatal development causes adverse transgenerational effects on female fertility in mice. Toxicol Sci 163(2):420–429

Rutkowska A, Rachoń D (2014) BisphenolA (BPA) and its potential role in the pathogenesis of the polycystic ovary syndrome (PCOS). Gynecol Endocrinol 30:260–265

Srnovršnik T, Virant-Klun I, Pinter B (2023) Heavy metals and essential elements in association with oxidative stress in women with polycystic ovary syndrome: a systematic review. Antioxidants 12(7):1398

Suramya JM, Pratap S et al (2024) Global research trends in endocrine-disrupting chemicals- a two-decade (2003–2023) bibliometric perspective. Biol Bull Rev 14(Suppl 3):S233–S253

Suramya JM, Mangla A et al (2025) Dietary protein deficiency exacerbates perfluorohexane sulfonate (PFHxS)-induced reproductive abnormalities and metabolic disruptions in female rats. Reprod Toxicol 135:108921

Uriu-Adams JY, Keen CL (2005) Copper, oxidative stress, and human health. Mol Asp Med 26(4–5):268–298

Vagi SJ, Azziz-Baumgartner E, Sjödin A et al (2014) Exploring the potential association between brominated diphenyl ethers, polychlorinated biphenyls, organochlorine pesticides, perfluorinated compounds, phthalates, and bisphenol a in polycystic ovary syndrome: a case-control study. BMC Endocr Disord 14:86

Wang X, Zhang Y, Peng J et al (2024) Association between exposure to multiple toxic metals in follicular fluid and the risk of PCOS among infertile women: the mediating effect of metabolic markers. Biol Trace Elem Res 203(2):775–789

Yang Z, Shi J, Guo Z et al (2019) A pilot study on polycystic ovarian syndrome caused by neonatal exposure to tributyltin and bisphenol A in rats. Chemosphere 231:151–160

Ye J, Zhu W, Liu H et al (2018) Environmental exposure to triclosan and polycystic ovary syndrome: a cross-sectional study in China. BMJ Open 8:e019707

Zeng LH, Rana S, Hussain L et al (2022) Polycystic ovary syndrome: a disorder of reproductive age, its pathogenesis, and a discussion on the emerging role of herbal remedies. Front Pharmacol 13:874914

Microplastics and Lung Diseases: An Overview

Hina Javed and Mohd. Anas Saifi

Abstract

Microplastics (MPs) are microscopic fragments that reach air, water, and soil. These particles formed from the breakdown of larger plastic objects, such as bottles, bags, and fishing nets, as well as from the discharge of synthetic materials like fibers from clothing and microbeads found in personal care products. From here, MPs migrate to the human body via inhalation and ingestion. However, research showed that the inhalation rate of MPs exceeds that of ingestion. Therefore, breathing in the microplastic-contaminated air may result in inflammation, oxidative stress, cellular damage, and apoptosis, which may lead to infections in the lungs such as chronic obstructive pulmonary disease (COPD), asthma, pneumoconiosis, and fibrosis. Further, the MPs can absorb and carry other harmful chemicals and pollutants, which could be released into the lungs. Thus, MP exposure worsens the existing respiratory condition. This overview presents in brief the sources of MP emission, negative effects on the respiratory system, the mechanism of underlying lung damage, and techniques for biomonitoring and assessment of MPs in the biological and environmental samples. Despite growing concerns about MP pollution, more research is necessary to better understand the processes of microplastic deposition, their distribution in lung tissue, and the long-term health risks.

Keywords

Environmental contaminant · COPD · Asthma · Oxidative stress

H. Javed (✉)
Department of Chemistry, Aligarh Muslim University, Aligarh, Uttar Pradesh, India

M. Anas Saifi
Molecular Toxicology Laboratory, Department of Medical Elementology and Toxicology, School of Chemical and Life Sciences, Jamia Hamdard, New Delhi, India

S. Raisuddin et al. (eds.), *Environmental Toxicants and Lifestyle Diseases*,
https://doi.org/10.1007/978-981-95-0575-3_3

Introduction

Plastic pollution has been a matter of great concern worldwide. The global increase in plastic production has significantly contributed to the rise in plastic pollution (Aragaw 2020). Each year, more than 311 million tons of plastic waste are produced worldwide, with about 8 million tons reach to the oceans, causing extensive harm to marine ecosystems (Khamis et al. 2025). In the year 2018, global plastic production exceeded 360 million tons, and it is projected to triple by 2050. However, in 2023, global plastic production reached 400.3 million metric tons (Rhodes 2018; Winiarska et al. 2024). A published report highlights that Asia is the largest producer and consumer of plastic products, with China alone contributing 32% of the "white pollution" (Plastics Europe 2020). Further, India is one of the world's largest plastic producers, with an annual production of 9.4 million tons of plastic waste (TOI 2023). Plastic pollution poses a significant global threat, as it is closely linked to numerous United Nations Sustainable Development Goals (SDGs), including SDG 3 (Good Health and Well-being), SDG 6 (Clean Water and Sanitation), SDG 11 (Sustainable Cities and Communities), SDG 12 (Responsible Consumption and Production), SDG 13 (Climate Action), SDG 14 (Life Below Water), and SDG 15 (Life on Land).

Larger plastic products can be degraded to smaller fragments with a diameter size of <5 mm, which are classified as microplastics (MPs) (Browne et al. 2015). These particles result from the breakdown of larger plastic objects, such as bottles, bags, and fishing nets, and the discharge of synthetic materials like fibers from clothing and microbeads found in personal care products. Microplastics vary greatly in type, structure, shape, and form. The most commonly encountered types in water and the environment are polyvinyl chloride (PVC), polypropylene (PP), polyester (PES), polyethylene (PE), polystyrene (PS), polyethylene terephthalate (PET), polyurethane (PUR), and polyamide (PA). The specific composition of microplastics affects their physical and chemical properties, surface characteristics, and additives. One key polymeric feature, crystallinity, plays a role in determining the density of microplastics (Ahmed et al. 2022; Siwach et al. 2022). Despite the establishment of stringent guidelines by the World Health Organization and other agencies for the proper disposal of solid waste, issues such as improper disposal in unauthorized dumps, lack of waste segregation, and insufficient infrastructure have made waste management more challenging. Due to this, MPs accumulate in marine habitats, where they are consumed by marine species such as fish and shellfish, eventually entering the food chain. This presents potential health risks due to the transfer of toxic pollutants, as well as causing inflammation and oxidative stress. Therefore, immediate action is needed to minimize their influence on marine life (Khamis et al. 2025). However, the adverse effects of MPs on human health are not fully inferred yet. Humans are exposed to MPs through food and water as well as inhalation. Moreover, the inhalation of MPs exceeds the ingestion rate (Cox et al. 2019), suggesting that inhalation is the main exposure pathway (Fournier et al. 2020).

In an investigation by D.J. Christopher, Professor of Pulmonary Medicine at Christian Medical College (CMC), Vellore, reported that there is a high prevalence of respiratory problems among Indians, with shortness of breath found in 49.3% and cough in 27.1% of the cohort. It is explicit from the study that lung function is affected more in the Indian population compared to data from other countries across every category of disease severity," (The Hindu, Feb 27, 2024). Further, Tetsuya Yamada, Managing Director of Omron Health Care, India, has said that unfortunately, Delhi has become the "asthma capital" of the nation with 1 million dying of asthma every year (CNBC, November 22, 2023). In another news report published in The Guardian (April 6, 2022), of 13 lung patients who underwent surgery, 11 were found to have MPs in their lungs (The Guardian 2022). Further, in highly populated cities (Paris), the average airborne MP particle amount would reach as high as 110 ± 96 particles/m^2/day (Dris et al. 2016). Another study in northern China found that airborne MP concentrations were 358 ± 132 items/m^3 (Zhu et al. 2021). Therefore, amid high MP pollution, extensive research is warranted on PS-MPs-induced respiratory disorders so that some regulation guidelines and prevention strategies can be made. This chapter examines how microplastics affect respiratory health, focusing on the pathogenic pathways and their involvement in lung illnesses.

Classification of MPs Based on Origin

There are two main types of microplastics: primary and secondary microplastics (Javed 2022).

Primary microplastics Primary microplastics are particles made for commercial purposes, such as those found in personal care products, cosmetics, and apparel (toothpaste, face cleansers, scrubbers, and creams). They are also utilized in biological research and dental products.

Secondary microplastics They are produced when big plastic materials break down and fragment due to various physical and chemical processes. There can be shock and abrasion, oxidative and thermal destruction, water cavitation, microbial/mechanical degradation, or bio/photodegradation.

Sources and Features of Inhaled MPs

Human and environmental health are increasingly at risk from industrial emissions, household dust, and vehicle emissions of MPs. They can be present in a variety of items and can enter the air, soil, and water.

Production MPs are discharged as plastic dust and other industrial manufacturing byproducts (Laskar and Kumar 2019).

Sewage sludge As sewage sludge is utilized as fertilizer, microplastics may pollute agricultural land.

Household dust Households and other interior spaces may contain MP dust, a kind of indoor air pollution. MPs are minute plastic particles composed of a range of substances, such as PE, PP, and PS (Zhang et al. 2020a, b).

Vehicle tires and brakes Tire wear is a necessary byproduct of use, and the composition, design, speed, acceleration, brake use, and texture of the road all affect how quickly tires abrasively wear down (Lehner et al. 2019). About 2–3 g of MPs is produced by cars, and 5–7 g is produced by truck tires for every 100 km (Prenner and Allesch 2022). There is currently no quantitative evaluation of the pollution caused by brake pads, which are also contributing MPs (Kazimirova et al. 2016).

Effects of Industrial Emissions of MPs

Ecosystem Humans, animals, and ecosystems can all be harmed by microplastic exposure.

Human health DDT and hexachlorobenzene are two harmful substances that microplastics can carry and harm cells and tissues (Laskar and Kumar 2019).

Food Food packaging and environmental contamination can both introduce microplastics into food.

MPs and Lungs

MP pollution is ubiquitous across the planet, making human exposure unavoidable, so there is an increasing concern regarding the hazards to human health (Yang et al. 2022). Recently, a report published in the newspaper stated that there is a surge in respiratory cases in India with high rates of lung function damage and shortness of breath, a worrying finding, and experts attribute it to various reasons, including co-morbidities and pollution (The Hindu 2024). MPs can directly harm lung epithelial cells by apoptosis, cytotoxicity, and cellular integrity disruption (Winiarska et al. 2024). Human exposure to airborne MPs is mainly influenced by their size. These particles are mainly divided into four groups: those larger than 10 μm, ultrafine particles under 0.1 μm, and those smaller than 10 μm. If the particles are inhaled, those exceeding 10 μm are typically trapped in the upper airways, while particles smaller than 10 μm can reach the bronchioles. Additionally, particles under 2.5 μm and ultrafine particles can infiltrate deep into the alveoli (Saha and Saha 2024).

The respiratory epithelium is a key target for inhaled MPs, whose biological reactivity relies on their physical and chemical characteristics. In lungs the area of alveoli covers about 150 m^2 and has a tissue barrier thinner than 1 μm, allowing nanoparticles and MPs to penetrate capillaries and enter the bloodstream, leading to systemic distribution (Lehner et al. 2019). Autopsy studies of human lung tissue have confirmed the appearance of MPs, with polymer particles and fibers found in

13 of 20 samples (Amato-Lourenço et al. 2021). To study the absorption of airborne PS, a method was developed that assesses the concentration of styrene (StyU) and its metabolites—mandelic acid (MA) and phenylglyoxylic acid (PGA)—in urine. MA and PGA, which are key urinary metabolites in humans, act as biomarkers for styrene exposure. Elevated levels of StyU and these metabolites have been found in workers exposed to high amounts of inhaled polystyrene (Persoons et al. 2018).

A study utilizing human biomonitoring has provided evidence of plastic pieces in lung tissue, suggesting that MPs from the air can deposit in the lungs (Pauly et al. 1998). Prolonged exposure to these plastic pieces has been linked to respiratory conditions, including asthma and pneumoconiosis (Atis et al. 2005; Turcotte et al. 2013). Additionally, fibrous particles from synthetic fabrics are pervasive in the environment, and individuals can inhale them. Once inhaled, most of these particles are likely to be trapped by the fluid lining the lungs. However, some particles may bypass the lung's natural defense mechanisms (Wright and Kelly 2017). Zhang et al. (2020a, b) note that inhalation is considered the primary route of exposure to MPs, with humans potentially inhaling airborne microplastics due to their small size.

Islam et al. (2023) observed the deposition of MPs in the lung airways. They found that cylindrical microplastics with diameters of 2.56 and 5.56 μm exhibited greater deposition, whereas 1.6 μm showed lower deposition rates at a flow rate of 7.5 L/min. In contrast, spherical and tetrahedral microplastics showed similar deposition rates at the same flow rate. However, the overall deposition efficiency at 7.5 L/min was higher than at 30 L/min for all microplastic sizes. This difference is attributed to the longer residence time at lower flow rates, which affects the movement of microplastics through the upper airway. Factors such as gravitational sedimentation and Brownian diffusion are significant, with Brownian diffusion being more noticeable at lower flow rates and decreasing as the flow rate increases.

Chronic Obstructive Pulmonary Disease (COPD)

The largest new threat to world health is chronic illness. Non-communicable diseases are responsible for two of every three fatalities globally (Lozano et al. 2013). Chronic obstructive pulmonary disease (COPD) is currently the third most common cause of mortality worldwide. COPD mostly affects the lungs and may be due to long-term exposure to irritants like cigarette smoke or hereditary risk factors (Christenson et al. 2022). According to current findings, smoking is the main risk factor for the development of COPD, and cigarette smoke contains MPs that range from 20 to 500 μm (Lu et al. 2022). Further, in this same study, smokers' bronchoalveolar lavage fluid (BALF) included more distinct MPs than non-smokers. Huang et al. (2022) reported that people with COPD and smokers had higher amounts of MPs in their sputum than people without the disease and smokers, respectively. Exposure to PS-MPs was observed to reduce the protein expression of α1-antitrypsin in a bronchial epithelial cell line, and lack of this protein is known to increase the likelihood of developing COPD (Dong et al. 2020).

Asthma

A long-term lung condition called asthma irritates the airways. It can appear early in infancy, making breathing challenging. Coughing, wheezing, and pressure in the chest are among the symptoms of asthma. Even though asthma is a highly diverse illness, two categories can be identified: allergic and non-allergic asthma. People who were exposed to polyvinyl chloride experienced abnormal radiography as early as the 1970s; there were reports of dyspnea, malignancy, and/or pulmonary function impairment (Arnaud et al. 1978). These symptoms were first referred to by Sokol et al. as "meat-wrapper's asthma" (Sokol et al. 1973). In animal models, eosinophils and lymphocytes were more prevalent in the bronchoalveolar lavage fluid of mice exposed to 1–5 μm amino-resin spheres, as well as more macrophages in lung tissue than in the control group. Another study shows that in mice given ovalbumin, another type of allergic asthma, the intake of spherical polystyrene nanoparticles exacerbated the symptoms of allergic asthma (Han et al. 2023).

Pneumoconiosis

This lung disorder is associated with occupational MP dust inhalation. In pneumoconiosis, inflammation and severe damage to the lungs occur, which sometimes leads to cancer. Workers who are involved in the processing, manufacturing, and recycling of plastic goods are more vulnerable to this condition. Different sizes and types of plastics are linked to specific pneumoconiotic conditions.

Mechanisms of MP-Induced Lung Toxicity

Oxidative Stress and Inflammation

Oxidative stress is one of the ways through which MPs affect the respiratory system. The pathophysiology of lung illnesses, like asthma, COPD, and acute respiratory distress syndrome (ARDS), is mainly influenced by oxidative stress (Hecker 2018). Antioxidants in simulated lung fluid were depleted after exposure to MPs isolated from the environment (Abbasi et al. 2019). In mouse lung tissue, exposure to MPs generated ROS, which induced inflammation, the first step that causes tissue damage (Woo et al. 2023). Exposure to MPs changes bronchial epithelial cells' oxidative state and antioxidant capacity (Lu et al. 2022), further dramatically lowering the dose-dependent GSH-Px, CAT, and SOD activities, linked to oxidative stress (Li et al. 2022).

Inflammation is the main reaction in any respiratory problem, and inflammasomes play an important role in the innate immune system and the body's response to cellular damage (Yao et al. 2024). Their activation triggers a type of inflammatory cell death known as pyroptosis (He et al. 2016). Among inflammasomes, the nucleotide-binding oligomerization domain (NOD) receptor containing pyrin domain-3 (NLRP3) inflammasome stands out as the most extensively researched. This can be triggered by various stimuli and has been involved in the development of numerous inflammatory conditions, such as cryopyrin-associated periodic syndromes, Alzheimer's disease, gout, autoinflammatory diseases, and atherosclerosis (Kelley et al. 2019). Prior research has demonstrated the involvement of the NLRP3 inflammasome in idiopathic pulmonary fibrosis (IPF) (Lasithiotaki et al. 2016). NLRP3 interacts with an adaptor protein called apoptosis-associated speck-like protein containing a caspase-recruitment domain (ASC), facilitating the recruitment of pro-caspase-1 (CASP1) to form the inflammasome complex. Active CASP1 then catalyzes the cleavage of pro-interleukin (pro-IL)-1β and pro-IL-18 into their mature, biologically active forms (Fernandes-Alnemri et al. 2007). Therefore, the potential of the NLRP3 inflammasome to ameliorate microplastic-induced damage to healthy lung tissue demands further investigation. Respiratory epithelial cells serve as the body's initial defense against harmful invaders and actively participate in initiating and perpetuating inflammation. They express several pattern recognition receptors capable of promptly detecting pathogen-associated molecular patterns and damage-associated molecular patterns released by injured tissue, eliciting a cascade of cytokine and chemokine release responses (Lu et al. 2022). Chronic inflammation can raise the chance of developing long-term respiratory diseases like asthma and COPD and damage lung tissue (Saha and Saha 2024). Hence, investigating the function of inflammasomes in respiratory structure during PS-MP exposure would offer novel insights into preventing epithelial dysfunction. PS particles were found to damage alveolar epithelium and impact lung function (Yang et al. 2021). MPs also decreased cell proliferation and major morphological changes (Goodman et al. 2021). Moreover, pulmonary inflammation, aggregation of macrophages in BALF, elevated plasma IgG1, and high mucus production were reported in mice (Lu et al. 2021). Therefore, the adverse effects and potential threats of PS-MPs cannot be ignored. However, there is scarce information related to the toxic mechanisms of MPs on the human respiratory system. Gautam et al. (2022) demonstrated that polyethylene microplastics caused oxidative stress and decreased cell viability at the maximum concentration of 1000 μg/mL; they had little effect on lung and intestinal cell viability.

Cellular Damage and Apoptosis

The MPs induced cellular damage, and programmed cell death involves several signaling pathways and molecular mechanisms, which include apoptosis, necroptosis, pyroptosis, ferroptosis, and autophagy. The programmed cell death is influenced by the characteristics of MP particles, such as functional group, surface potential, size, and aging. The MPs, once entered the human body, alter the cellular fate and homeostasis. Chen et al. (2022) showed that PS-MPs induced both autophagy and apoptosis at the same time by evidence of mitochondrial membrane potential depolarization and autophagosome formation. MP treatment resulted in a decrease in sphingosine levels, proving that MPs can considerably impair the metabolism of sphingolipids, which changes cellular apoptosis (Zhang et al. 2021). There is evidence that links the toxicity of MPs to inflammation, apoptosis, and oxidative stress (Cao et al. 2023). MPs derived Fe and lipid damages (ferroptosis) reported in several lung illnesses, including COPD, pulmonary injury, lung fibrosis, and lung cancer (Shen et al. 2023; Yang et al. 2022; Zhang et al. 2021). Recently, histological alterations like alveolar epithelium, infiltration of inflammatory cells into the perivascular/parenchymal layer, and macrophage accumulation in the mouse lungs were observed following inhalation of PP-MPs (Woo et al. 2023). Another research finding showed that when PE-MPs were taken orally for longer than 4 weeks, the alveoli of ICR mice developed granulomatous inflammation (Lee et al. 2022).

Identification and Monitoring of MPs

Several methods and techniques have been developed so far for the detection, identification, and characterization of MPs that affect lung health.

Fourier Transform Infra Red (FTIR) Spectroscopy By this spectroscopic method, the infrared absorption spectra of MPs can be examined to determine their chemical composition (Jenner et al. 2022).

Raman Spectroscopy This is one of the important techniques that uses how light enters matter and then determines the chemical makeup of MPs (Jianpeng et al. 2023).

Microscopy A very simple, practical, and famous technique that can be used for the identification of MPs is stereomicroscopy. It can examine particle sizes between hundreds of microns. Further, scanning electron microscopy (SEM) and transmission electron microscopy (TEM), more advanced techniques, can be utilized for morphological analysis and MP imaging within lung tissues.

Conclusion

MPs are an environmental and public health concern due to their universal existence in the air and their function to absorb unfavorable substances. Their inhalation is associated with severe lung illnesses, including but not limited to COPD, asthma, pneumoconiosis, fibrosis, etc. They pose a significant threat to the respiratory system via several mechanisms, such as oxidative stress, inflammation, cellular damage, and apoptosis. However, limited experimental and epidemiological data is available on the deleterious consequences of MPs. Therefore, more research is warranted on the long-term impact of MP manifestation on the respiratory system and human health.

References

Abbasi S, Keshavarzi B, Moore F et al (2019) Distribution and potential health impacts of microplastics and microrubbers in air and street dusts from Asaluyeh County, Iran. Environ Pollut 244:153–164

Ahmed R, Hamid AK, Krebsbach SA et al (2022) Critical review of microplastics removal from the environment. Chemosphere 293:133557

Amato-Lourenço LF, Carvalho-Oliveira R, Júnior GR (2021) Presence of airborne microplastics in human lung tissue. J Hazard Mater 416:126124

Aragaw TA (2020) Surgical face masks as a potential source for microplastic pollution in the COVID-19 scenario. Mar Pollut Bull 159:111517

Arnaud A, De SP, Garbe L et al (1978) Polyvinyl chloride pneumoconiosis. Thorax 33:19–25

Atis S, Tutluoglu B, Levent E et al (2005) The respiratory effects of occupational polypropylene flock exposure. Eur Respir J 25(1):110–117

Browne MA, Underwood AJ, Chapman MG et al (2015) Linking effects of anthropogenic debris to ecological impacts. Proc Biol Sci 282:20142929

Cao J, Xu R, Geng Y et al (2023) Exposure to polystyrene microplastics triggers lung injury via toll like receptor 2 and activation of the NF-κB signal in mice. Environ Pollut 320:121068

Chen YC, Chen KF, Lin KYA (2022) The nephrotoxic potential of polystyrene microplastics at realistic environmental concentrations. J Hazard Mater 427:127871

Christenson SA, Smith BM, Bafadhel M et al (2022) Chronic obstructive pulmonary disease. Lancet 399:2227–2242

CNBC (2023, November 22). Available via https://www.cnbctv18.com/healthcare/respiratory-disease-environmental-pollution-clean-air-managing-indias-huge-disease-burden-18379991.htm. Accessed on 12 Feb 2025

Cox KD, Covernton GA, Davies HL et al (2019) Human consumption of microplastics. Environ Sci Technol 53:7068–7074

Dong CD, Chen CW, Chen YC et al (2020) Polystyrene microplastic particles: in vitro pulmonary toxicity assessment. J Hazard Mater 385:121575

Dris R, Gasperi J, Saad M et al (2016) Synthetic fibers in atmospheric fallout: a source of microplastics in the environment? Mar Pollut Bull 104:290–293

Fernandes-Alnemri T, Wu J, Yu JW et al (2007) The pyroptosome: a supramolecular assembly of ASC dimers mediating inflammatory cell death via caspase-1 activation. Cell Death Differ 14: 1590–1604

Fournier SB, D'Errico JN, Adler DS et al (2020) Nanopolystyrene translocation and fetal deposition after acute lung exposure during late-stage pregnancy. Part Fibre Toxicol 17:55

Gautam R, Jo J, Acharya M et al (2022) Evaluation of potential toxicity of polyethylene microplastics on human derived cell lines. Sci Total Environ 838:156089

Goodman KE, Hare JT, Khamis ZI (2021) Exposure of human lung cells to polystyrene microplastics significantly retards cell proliferation and triggers morphological changes. Chem Res Toxicol 34:1069–1081

Han Q, Gao X, Wang S et al (2023) Co-exposure to polystyrene microplastics and di-(2-ethylhexyl) phthalate aggravates allergic asthma through the TRPA1-p38 MAPK pathway. Toxicol Lett 384:73–85

He Y, Hara H, Nunez G (2016) Mechanism and regulation of NLRP3 inflammasome activation. Trends Biochem Sci 41:1012–1021

Hecker L (2018) Mechanisms and consequences of oxidative stress in lung disease: therapeutic implications for an aging populace. Am J Physiol Lung Cell Mol Physiol 314:L642–L653

Huang S, Huang X, Bi R et al (2022) Detection and analysis of microplastics in human sputum. Environ Sci Technol 56:2476–2486

Islam MS, Rahman MM, Larpruenrudee P et al (2023) How microplastics are transported and deposited in realistic upperways? Phys Fluids 35:063319

Javed H (2022) Microplastics: an overview. In: Ahmad I, Mahamood M, Javed M, Alhewairini SS (eds) Toxicology and human health. Springer, Singapore, pp 75–82

Jenner LC, Rotchell JM, Bennett RT et al (2022) Detection of microplastics in human lung tissue using μFTIR spectroscopy. Sci Total Environ 831:154907

Jianpeng AO, Xu G, Wu H (2023) Fast detection and 3D imaging of nanoplastics and microplastics by stimulated Raman scattering microscopy. Cell Rep Phys Sci 4(10):101623

Kazimirova A, Peikertova P, Barancokova M et al (2016) Automotive airborne brake wear debris nanoparticles and cytokinesis block micronucleus assay in peripheral blood lymphocytes: a pilot study. Environ Res 148:443–449

Kelley N, Jeltema D, Duan Y et al (2019) The NLRP3 inflammasome: an overview of mechanisms of activation and regulation. Int J Mol Sci 20:3328

Khamis F, Hegab HM, Hasan SW (2025) Green and sustainable pH-responsive adsorptive membrane using aminated-MXene mangrove-filled polylactic acid ultrafiltration membrane for enhanced water permeability and microplastic removal. Desalination 594:118268

Lasithiotaki I, Giannarakis I, Tsitoura E et al (2016) NLRP3 inflammasome expression in idiopathic pulmonary fibrosis and rheumatoid lung. Eur Respir J 47(3):910–918

Laskar N, Kumar U (2019) Plastics and microplastics: a threat to environment. Environ Technol Innov 14:100352

Lee S, Kang KK, Sung SE et al (2022) Toxicity study and quantitative evaluation of polyethylene microplastics in ICR mice. Polymers (Basel) 14(3):402

Lehner R, Weder C, Petri-Fink A et al (2019) Emergence of nanoplastic in the environment and possible impact on human health. Environ Sci Technol 53:1748–1765

Li Z, Chang X, Hu M et al (2022) Is microplastic an oxidative stressor? Evidence from a meta-analysis on bivalves. J Hazard Mater 423:127211

Lozano R, Naghavi M, Foreman K et al (2013) Global and regional mortality from 235 causes of death for 20 age groups in 1990 and 2010: a systematic analysis for the Global Burden of Disease Study 2010. Lancet 380:2095–2128

Lu K, Lai KP, Stoeger T et al (2021) Detrimental effects of microplastic exposure on normal and asthmatic pulmonary physiology. J Hazard Mater 416:126069

Lu K, Zhan D, Fang Y et al (2022) Microplastics, potential threat to patients with lung diseases. Front Toxicol 4:958414

Pauly JL, Stegmeier SJ, Allaart HA et al (1998) Inhaled cellulosic and plastic fibers found in human lung tissue. Cancer Epidemiol Biomarkers Prev 7(5):419–428

Persoons R, Richard J, Herve C et al (2018) Biomonitoring of styrene occupational exposures: biomarkers and determinants. Toxicol Lett 298:99–105

Plastics Europe (2020) Plastics–the facts Avenue E. van Nieuwenhuyse 4/3, 1160 Brussels. Plastics Europe, Belgium. Available via https://www.plast icseurope.org/de/resources/publications/4312-plastics-facts-2020. Accessed on 12 Feb 2025

Prenner S, Allesch A (2022) How tyre wear of trucks and passeneger cars gets into the environment and the human body. EUropainfo. Das Magazin des EU-Umweltburos. Available via https://forschung.boku.ac.at/en/publications/144655. Accessed on 12 Feb 2025

Rhodes CJ (2018) Plastic pollution and potential solutions. Sci Prog 101(3):207–260

Saha SC, Saha G (2024) Effect of microplastics deposition on human lung airways: a review with computational benefits and challenges. Heliyon 10(2):e24355

Shen K, Wang X, Wang Y et al (2023) miR-125b-5p in adipose derived stem cells exosome alleviates pulmonary microvascular endothelial cells ferroptosis via Keap1/Nrf2/GPX4 in sepsis lung injury. Redox Biol 62:102655

Siwach S, Bharti M, Yadav S (2022) Unveiling the ecotoxicological impact of microplastics on organisms – the persistent organic pollutant (POP): a comprehensive review. J Contam Hydrol 266:104397

Sokol WN, Aelony Y, Beall GN (1973) Meat-wrapper's asthma. A new syndrome? JAMA 226: 639–641

The Guardian (2022, April 6). Available via https://www.theguardian.com/environment/2022/apr/06/microplastics-found-deep-in-lungs-of-living-people-for-first-time. Accessed on 12 Feb 2025
The Hindu (2024, February 27). Available via https://www.thehindu.com/sci-tech/health/study-finds-high-rate-of-lung-damage-in-indians-after-acute-covid-19-experts-decode-why/article67890978.ece. Accessed on 10 Feb 2025
Times of India (TOI) (2023, April 16). Available via https://timesofindia.indiatimes.com/blogs/scientifically-trended/india-and-microplastics-redefining-indias-one-of-the-most-crucial-environmental-problems/. Accessed on 5 Feb 2025
Turcotte SE, Chee A, Walsh R et al (2013) Flock worker's lung disease: natural history of cases and exposed workers in Kingston, Ontario. Chest 143(6):1642–1648
Winiarska E, Jutel M, Zemelka-Wiacek M (2024) The potential impact of nano and microplastics on human health: understanding human health risks. Environ Res 251(Pt 2):118535
Woo JH, Seo HJ, Lee JY et al (2023) Polypropylene nanoplastic exposure leads to lung inflammation through p38-mediated NF-κB pathway due to mitochondrial damage. Part Fibre Toxicol 20:2
Wright SL, Kelly FJ (2017) Plastic and human health: a micro issue? Environ Sci Technol 51(12):6634–6647
Yang S, Cheng Y, Chen Z et al (2021) In vitro evaluation of nanoplastics using human lung epithelial cells, microarray analysis and co-culture model. Ecotoxicol Environ Saf 226:112837
Yang X, Man YB, Wong MH et al (2022) Environmental health impacts of microplastics exposure on structural organization levels in the human body. Sci Total Environ 825:154025
Yao J, Sterling K, Wang Z et al (2024) The role of inflammasomes in human diseases and their potential as therapeutic targets. Signal Transduct Target Ther 9:10
Zhang J, Wang L, Kannan K (2020a) Microplastics in house dust from 12 countries and associated human exposure. Environ Int 134:105314
Zhang Q, Xu EG, Li J et al (2020b) A review of microplastics in table salt, drinking water, and air: direct human exposure. Environ Sci Technol 54(7):3740–3751
Zhang H, Jiao W, Cui H (2021) Combined exposure of alumina nanoparticles and chronic stress exacerbates hippocampal neuronal ferroptosis via activating IFN-γ/ASK1/JNK signaling pathway in rats. J Hazard Mater 411:125179
Zhu X, Huang W, Fang M et al (2021) Airborne microplastic concentrations in five megacities of Northern and Southeast China. Environ Sci Technol 55:12871–12881

Occurrence, Fate, Effects, Risk Assessment, Biodegradation, and Bio-transformation of Polycyclic Aromatic Hydrocarbons (PAHs) and Health Hazards of Their Metabolites

Annu Khatri, Krishan Kumar, and Indu Shekhar Thakur

Abstract

Polycyclic aromatic hydrocarbons (PAHs) are a group of organic compounds with multiple aromatic rings, known for their persistence in the environment and potential health hazards. They originate from both natural sources, such as forest fires, and anthropogenic activities, including industrial processes and vehicle emissions. Environmental contaminants known as PAHs are typically produced as a consequence of insufficient burning of natural sources (e.g., petrol, wood, coal, and oil). According to research, PAH pollution is very hazardous, mutagenic, carcinogenic, teratogenic, and immune-toxicogenic to a variety of living forms. This review addresses the occurrence, fate, effects, risk assessment, and biodegradation of PAHs, alongside the health hazards posed by their metabolites. Furthermore, it explores microbial degradation that includes bacteria, algae, plants, fungi, and rhizosphere microorganisms to mitigate PAH contamination. Despite the effectiveness of these biological processes in reducing PAH levels, the metabolites produced can sometimes be more toxic than the parent compounds, posing additional health hazards. The review integrates current knowledge on the occurrence, fate, and effects of PAHs, the risk assessment framework, and the complex interplay between biodegradation processes and metabolite toxicity. It underscores the importance of advancing research in omics techniques and conducting thorough risk assessments to safeguard environmental and public health.

A. Khatri · K. Kumar · I. S. Thakur (✉)
Amity School of Earth and Environmental Sciences, Manesar, Gurugram, Haryana, India
e-mail: isthakur@ggn.amity.edu

S. Raisuddin et al. (eds.), *Environmental Toxicants and Lifestyle Diseases*,
https://doi.org/10.1007/978-981-95-0575-3_4

Keywords

Persistent organic pollutants · Environmental contaminants · Metabolites · Biodegradation · Microbial remediation

Introduction

Environmental pollution's health impacts on the population have become a major topic of concern in recent decades. According to the World Health Organization, pollution causes one-third of all diseases that affect humans. Scientists have identified numerous chemical contaminants that are toxic and persist in the environment, bioaccumulative, and prone to long-distance atmospheric transboundary movement and deposit since World War II (Swati and Thakur 2019). These pollutants are expected to have major adverse effects on people, animals, and marine life both near and far from their source of emission. POPs are generally man-made chemical substances designed for use in diverse fields, such as agriculture and industry and garbage incineration. POPs are divided into different types, including organo-chlorinated pesticides OCP, polychlorinated biphenyls (PCBs), perfluorinated compounds (PFCs), polyaromatic hydrocarbons (PAHs), dioxins, brominated compounds (BFR), and furans. E-waste, like old electronic gadgets, can contain substances such as PAHs that are harmful to the environment. When e-waste is not properly disposed of or recycled, PAHs can be released into the air, soil, and water, causing pollution and potential health risks. The presence of PAHs in e-waste underscores the importance of managing electronic waste responsibly to prevent the spread of these harmful compounds into the environment. Most of these chemicals are caused by humans in nature. PAHs are common pollutants found throughout the natural environment. Biodegradation and biotransformation processes are natural mechanisms by which microorganisms metabolize PAHs into less harmful compounds. During biodegradation, microorganisms use PAHs as sources of energy and carbon, decomposing them into simpler substances such as carbon dioxide, water, and biomass. Biotransformation involves structural changes in PAHs catalyzed by microorganisms, leading to the formation of less toxic compounds. These processes are crucial for mitigating the environmental impact of PAHs (Mishra et al. 2022). Figure 1 shows the classification of persistent organic compounds based on their origin.

Occurrence and Types of Persistent Organic Pollutants

The commercialization and usage of persistent organic pollutants are used for the control of diseases and pests dates back to the early 1900s, when the industrial application of persistent organic pollutants began. Researchers have spent a lot of time studying the toxicity, bioaccumulation, and persistence factor of common POPs, such as organochlorine pesticides and polychlorinated biphenyls. Despite

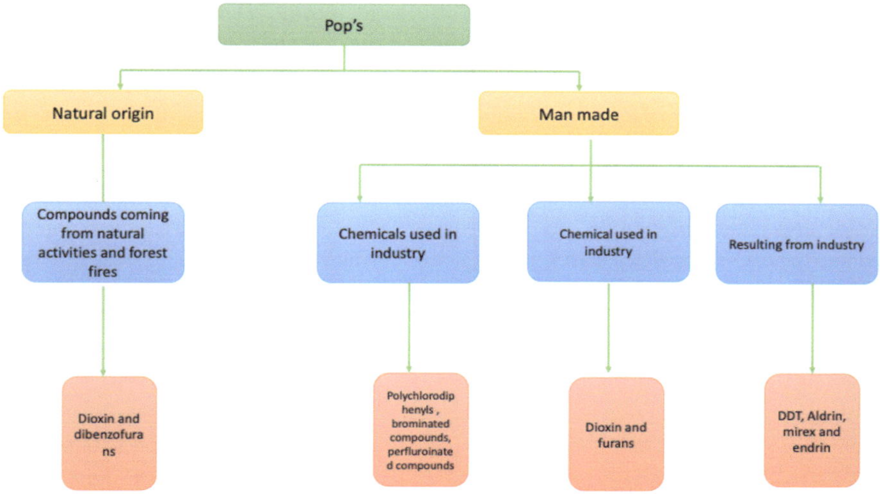

Fig. 1 Persistent organic compounds classification based on origin

many of them being banned due to their harmful effects, persistent organic pollutants are still found in significant amounts in the environment worldwide. Persistent organic pollutants concentrations have been found to be notably higher in several matrices, such as biota and sediment, soil, surface water, and drinking water, according to several studies conducted in the study area. It is no doubt that the rapid increase in human population, urbanization and industrialization has had a great impact on the rapid increase of the persistent organic pollutants in the environment. Moreover, farming practices such as the discharge of pesticides and fertilizers into the environment also lead to a significant increase of persistent organic pollutants in the environment. In 2001, 92 nations ratified the Stockholm Convention on persistent organic pollutants, seeking to minimize or eradicate the emission of the "dirty dozen" pollutants or the original 12 persistent organic pollutanting substances. More contaminants have been found, but the initial 12 still pose the biggest threat. These contaminants include the two accidentally manufacture chemicals polychlorinated dibenzo-p-dioxins, polychlorinated dibenzofurans and as well as the ten chemicals intentionally produced: Polychlorinated biphenyls, dicholoro diphenyl trichloroethane, Aldrin, mirex, chlordane, dieldrin, endrin, heptachlor, toxaphene, and aromatic hydrocarbons that are polycyclic (Patel et al. 2020).

Intentionally POPs

These chemicals will produce from different chemical reactions involving chlorine as desired byproducts. These organic molecules having connected Organo-chlorine chemicals have chlorine atoms, are very lipophilic, and frequently have severe neurotoxicity chlorinated pesticides such as dicholoro diphenyl trichloroethane and polychlorinated biphenyls are examples of organochlorines. They possess a number of substances that can be dividing into two groups: organochlorine pesticides and industrial chemicals (Gbeddy et al. 2020).

Industrial Chemicals

Polychlorinated biphenyls, which were first identified as environmental contaminants, were discovered in the world's water, avian tissue, sediments, and fish tissue in 1966. A sizable specific garbage category is comprised of these materials. Among the chemical compounds known as polychlorinated biphenyls, the biphenyl molecule contains two to ten chlorine atoms linked. These substances make up a significant class of special wastes. Polychlorinated biphenyls are a class of chemical substances in which the biphenyl molecule has 2–10 chlorine atoms bonded. Polychlorinated biphenyls are frequently described as monochlorinated biphenyls, which contain one chlorine atom linked to the biphenyl molecule (Maheshwari et al. 2022).

Organochlorine Pesticide

Biological or chemical agents employed to eradicate or manage pests are known as pesticides. They can be classified into three categories: fungicides, herbicides, and insecticides. Additionally, given that acaricides, molluscicides, rodenticides, and nematicides. Most of these chemicals are synthetic organic compounds created by humans, although some exceptions come from plants or naturally occurring inorganic elements. In the seventeenth century, nicotine, derived from tobacco leaf extracts, was the first natural insecticide used to manage lace beetles and plum curculios. Insecticides containing chlorinated hydrocarbons were first developed in large quantities in the 1940s, but it wasn't until the 1950s that they were widely used. Examples of common endrins are aldrin, dieldrin, heptachlor, and aldrin. Unfortunately, despite their initial promise, these organochlorine insecticides are now used relatively less because of their negative environmental impacts categories of persistent organic pollutants are mentioned in Fig. 2.

Unintentionally POPs

They were unintended consequences of combustion or chemical reactions that took place when chlorine compounds were present. They fall into three categories: dioxins, furan compounds, and PAHs.

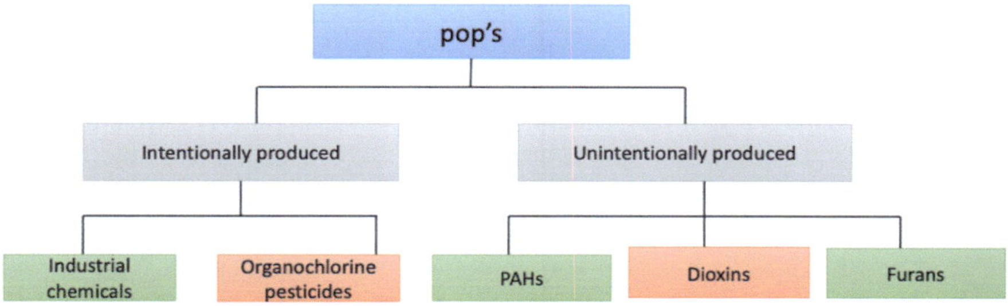

Fig. 2 Categories of persistent organic pollutants

Dioxins and Dibenzofurans

They can grow naturally in the environment, primarily following prairie and forest fires or as an effect of plant synthesis. However, human activities such as the combustion of fossil fuels and the production of industrial waste are what release the most PAHs into the environment. Cooking and agricultural fires both have the potential to release PAHs. Combustion processes, chemical processes, and secondary sources are the three primary categories of manmade polychlorinated dibenzo-p-dioxins and polychlorinated dibenzo-p-difuran sources. In Hong Kong, crematoria, incinerators for animal carcasses, landfill gas incinerators, chemical waste incinerators, incinerators for clinical waste or municipal waste, and coal combustion are the primary producers of polychlorinated dibenzofurans and polychlorinated dibenzo-p-dioxins.

Polycyclic Aromatic Hydrocarbons

PAHs are composed of two or more benzene rings that are fused together in a linear arrangement, angularly, or in clusters and only contain carbon and hydrogen. Stable carbon-carbon bonds serve as the glue that holds the main molecular structure together (Gupta et al. 2020). The United States Environmental Protection Agency identifies 16 PAHs as major pollutants because they are thought to be either probable or possible human carcinogens in the soil, air, and sediments. They can grow in the environment naturally, usually resulting from plant synthesis or following grassland and forest fires. However, human activities such as the combustion of fossil fuels and producing industrial waste are what release the most PAHs into the environment. Cooking and agricultural fires both have the potential to release PAHs. The majority of PAHs are white, colorless, or pale yellow-green solids with a light, pleasant smell when they are present in their purest chemical state. They are essentially non-polar organic compounds and are commonly made up of hydrogen and carbon atoms. Insufficient combustion of organic materials such as coal, oil, tar, and gas, as well as car exhaust, smoked food, tobacco, or during either industrial and other human activities or even during geothermal processes linked to the processing of fossil fuels and minerals, is the main reason of PAHs creation. They develop in nature as a result of bacterial and plant responses, volcanic eruptions, or forest fires. Different forms of combustion produce various distributions of PAHs, both in terms of the relative amounts of the different PAHs and their isomers. When organic materials undergo heating to high temperatures (pyrolysis), PAHs are produced; the composition of these products is heavily influenced by both the nature of the starting material and the transformation temperature. Moreover, fossil fuels include massive amounts of PAHs that are inadvertently released into the atmosphere through the transportation, use, or disposal of petroleum products, or as a result of uncontrolled emissions. PAHs are often present in the air, soil, aquatic ecosystems, sediments, surface water, and ground water.

Natural and human sources, as well as global transport phenomena, all are responsible for the worldwide dispersion of PAHs. As a result, PAHs are spread from the atmosphere to plants, where they eventually bioaccumulate in living things. The ring structure of PAHs is another factor used to categories them: (a) Non-Alternant (b) Alternant. Phenanthrene, Phenanthrene, and Chrysene are examples of alternative PAHs that are created from benzene by the fusing of extra 6-membered benzoid rings and have fewer rings than benzoid rings. Fluoranthene is a non-alternant PAHs that also has a 6-membered ring and rings with less than 6 carbon atoms. Bioavailability of PAHs a good bioremediation procedure requires physical contact between the bacteria or their enzymes and the organic pollutant. Bioavailability and bio accessibility are determined by the properties of the soil as well as the kind of pollutant (Thakur and Srivastava 2011). The percentage of a substance that is absorbed by a cell and results in harmful consequences is known as bioavailability. While "bio accessibility" refers to the portion of soil that may be available to biota, also known as "environmental availability," the bioremediation of PAHs in soil takes place over the course of two processes. They are (a) microorganisms absorbing the PAHs and (b) the PAHs being broken down (metabolized) the rate at which PAHs are absorbed in the environment affects their breakdown, transport, and sequestration. Chrysene is made up of four fused, six-carbon benzene rings. Fluoranthene, on the other hand, is a result of lower temperature and less efficient combustion because it consists of a benzene unit connected by a five-membered ring (in the center of the structure). PAHs enter the environment through a variety of methods and are often discovered as a mixture of two or more of these compounds. Some PAHs are produced in the industry. The mechanism of toxicity is believed to involve interaction with cellular membrane function and enzyme systems connected to the membrane. PAHs have been shown to be mutagenic, carcinogenic, and effective immune suppressants. The effects have been verified. Rapid industrialization and urbanization have resulted in numerous human activities, including the release of PAHs and other pollutants into the environment (Rathour et al. 2018).

Sources of Polycyclic Aromatic Hydrocarbons

Polycyclic aromatic hydrocarbons are organic pollutants that are formed of two or more bonded benzene rings of hydrogen and carbon. Aromatic rings can be arranged in space using molecular configurations that are straight, angular, or crowded (Abdel-Shafy and Mansour 2016).

PAHs are divided into two sections on the basis of aromatic rings number they contain: PAHs with a low molecular mass PAHs; containing two or three rings and PAHs with a high molecular mass PAHs; having four or more aromatic rings). They produce mainly gaseous low molecular mass PAHs or particulate high molecular mass PAHs depending on their molecular weight. The two primary categories of sources of PAHs contamination are sources of manmade emissions and sources of organic emissions. Natural emission sources that are negligible or less valuable include forest fires, volcanic eruptions that occur naturally, and lightning-caused fires on moorland. Different type of PAH emission sources are in Fig. 3. Human-caused emissions from four types of sources—industrial, mobile, home, and agricultural emission sources—are the primary contributors to PAHs contamination (Maliszewska-Kordybach 1999). Sources of PAHs are Industrial sources, Mobile sources, Domestic sources, Agricultural sources. Incomplete combustion is the major source of PAHs emissions from various industrial activities, such as waste incineration, iron and steel production, aluminum production, cement, coal-tar pitch production, dyes, asphalt industries, tyre rubber manufacturing, fungicide and insecticide production, refinery exhaust, and power generation. Another source of industrial emission includes diesel engines, gasoline-powered engines in heavy machinery, electric arc furnaces, oxygen furnaces, and gasification of coal (Lee 2010).

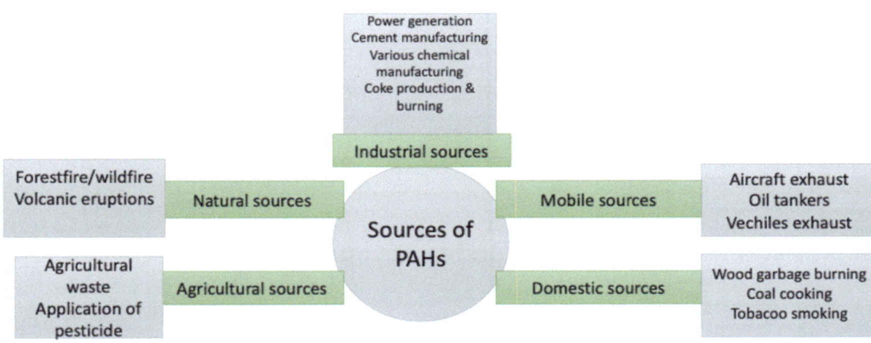

Fig. 3 Different types of polycyclic aromatic hydrocarbon emission sources

Transport and Fate of PAHs in the Environment

Several processes cause PAHs move between land and aquatic environments while dispersing in the air. PAHs exposure can occur orally, topically, or by direct inhalation.

PAH Emission and Deposition in the Atmosphere

The majority of the PAHs environmental problem is received by the atmosphere, making it the most significant method of PAHs dispersal. As a result, PAHs are omnipresent in the environment. The insufficient combustion of organic compound is the major cause of PAHs emissions into the environment. Combustion can be triggered by either natural or human-induced causes. Among the natural causes are forest fires and volcanoes (Ravindra et al. 2008).

PAHs in Surface Soils

These PAHs are released from a variety of sources, including vehicle exhaust from nearby highways. Different distances have been covered by other PAHs as they originate from farther-off sources. If fill materials include PAHs, these may be introduced to soils in the meantime. When PAHs are dumped on the surface of the earth, they become mobile. The size of the activated particle and the pore throat of the soils are the crucial variables affecting the mobility of PAHs of particulates in the subsurface since most of the soil's PAHs will be associated with soil particles. The tiniest aperture between two dirt grains is what is meant by a pore throat. Because PAHs tend to adhere to particles, their mobility is limited if the soil particles they attach to cannot penetrate through the soil (Låg et al. 2020).

PAHs in Sediments

Similar mechanisms that regulate how dirt is accumulating on the surface are used to deposit PAHs into the sedimentary environment. The PAHs that adsorb to rural regions may experience either dry or wet air particle deposition on the surface of streams, lakes, and seas. They gradually blend in with the silt after being scattered by the currents. Apart from this, the atmospheric deposition of PAHs has an impact on the sediments close to metropolitan areas. Roadway runoff, sanitary sewage effluents, and storm water runoff are additional sources of PAHs for them. Certain PAHs will eventually bond to particles will settle and get absorbed into the geologic record. Because of their non-polar properties, PAHs are remaining stationary once they have been integrated into sediments because they cannot dissolve in water. However, PAHs, especially the smaller molecular weight PAHs are not entirely soluble. Small quantities of PAHs do, therefore, they disintegrate and end up in ground water, where they are soluble. Even though PAHs will be sorbeted onto these organic colloids, the presence of ground water organic colloids may raise PAHs concentrations above their aqueous solubility. These are then conveniently carried via the sediment's pore spaces. To collide and obtain sorption can thereby improve the bioavailability and mobility of PAHs in sediments. Ecological harm caused by PAHs through both photo-oxidation and metabolism has an influence on aquatic creatures and the toxicity of PAHs (Lyytikäinen et al. 2007). They are generally more hazardous when UV radiation is present. For aquatic and avian life, acute toxicity of PAHs ranges from mild to severe. With the exception of extremely contaminated soil, PAHs in soil have the potential to harm terrestrial invertebrates. The three main ways that mammals can absorb PAHs are by ingestion, cutaneous contact, and inhalation. However, plants have the ability to take up PAHs from the soil through their roots and transfer them to anther plant components. In general, concentration, water solubility, physicochemical condition, and soil type all affect absorption rates. PAHs effects on plants are uncommon. The database on this still lacks complete information. Plants with certain compounds can fend against the effects of PAHs. Other plants are able to produce PAHs, among which are growth hormones. PAHs are bio accumulative and only in the surroundings with a medium level of persistence. Fish and shellfish should contain substantially greater levels of PAHs than the environment from which they were harvested. Invertebrates from the land have also demonstrated bioaccumulation. The metabolism of PAHs, however, is sufficient to stop bio magnifications (Choi et al. 2010).

PAH in Food Occurrence

PAH levels shouldn't typically be high in raw meals. In places far from populated areas or industries, the presence of PAHs in unprocessed foods is a reflection of background contamination. In addition to the naturally occurring emissions from volcanoes and forest fires, these PAHs are frequently transported across large distances by aircraft carrying polluted particles. Vegetation may be 10 times more polluted than in rural regions in the vicinity of industrial sites or along roadways. Food processing, including drying and smoking, as well as heating food at more temperatures (grilling, frying and frying), are key sources of PAHs. Smoked fish and pork contain PAHs concentrations as high as 200 lg/kg. 130 lg/kg of PAHs has been discovered grilling beef (Sun et al. 2019). Because heavier PAHs end to combine with particle matter, air fallout is a major pathway for contamination. Nearly, all PAHs with two or three rings are in the vapor phase is at a middle position with four rings. Therefore, plants with big leaves, grazing creatures, or poultry that could consume soil contaminated particles are all at risk of being contaminated by PAHs that have attached themselves to particles. Vegetable and fruit surfaces that are waxy have the ability to concentrate lower molecular weight PAHs mostly via adsorption on surface. As a whole, plant surfaces (peel, outside leaves) have higher PAHs contents than inside tissue (Zelinkova and Wenzl 2015).

Degradation of Polycyclic Aromatic Hydrocarbons

The degradation of PAHs in the environment involves a combination of abiotic and biotic processes that work together to transform these potentially hazardous compounds into less toxic forms. Degradation of PAHs involves various methods, such as physical, chemical, and biological methods.

Physical Degradation Method

Physical degradation technique for polycyclic aromatic hydrocarbons is essential for the remediation of contaminated environments, utilizing various physical principles to reduce the concentration and toxicity of PAHs. Photo degradation leverages sunlight, particularly ultraviolet (UV) radiation, to break down PAHs molecules, making it effective for surface water and soils exposed to direct sunlight. Thermal desorption involves heating contaminated soil or sediment to volatilize and remove PAHs, which are then captured and treated, making it suitable for soils with high levels of contamination. Different types of physical degradation techniques for poly aromatic hydrocarbon shown in Table 1. Soil washing uses water or solvents to extract PAHs from soils, enhancing the removal of these pollutants when combined with other remediation techniques. Sorption methods, such as activated carbon adsorption, biochar adsorption, and the use of clay minerals and synthetic polymers, play a significant role in PAHs degradation. Activated carbon, known for its high surface area and adsorptive properties, is commonly used in water and air purification systems to trap PAHs. Biochar, derived from the pyrolysis of biomass, offers a sustainable option for soil and water remediation by adsorbing PAHs and improving soil quality. Clay minerals, with their large surface area and ion exchange capacities, are effective in soil remediation efforts. Synthetic polymers, designed with specific functional groups, can be tailored to adsorb particular PAHs contaminants, providing versatility in water and soil remediation applications. These physical degradation methods, often used in conjunction, form a comprehensive approach to managing and mitigating PAHs pollution in the environment (Kronenberg et al. 2017).

Chemical Degradation Method

Chemical degradation methods for polyaromatic hydrocarbons involve the use of various chemical processes to break down these persistent organic pollutants into less harmful compounds. Chemical oxidation is one of the most widely used techniques, which employs oxidizing agents such as ozone, hydrogen peroxide, and permanganate to convert PAHs into more water-soluble and less toxic molecules. Fenton's reagent, a mixture of hydrogen peroxide and an iron catalyst,

Table 1 Different types of physical degradation techniques for poly aromatic hydrocarbon

Method	Description	Applications
Photodegradation	Use of sunlight (UV radiation) to break down PAH molecules into smaller, less complex compounds.	Effective in surface water and soil exposed to sunlight; can be enhanced with artificial UV light sources
Thermal desorption	Heating contaminated soil or sediment to volatilize and remove PAHs, capturing and treating the vapoors	Used for treating soils and sediments contaminated with volatile and semi-volatile PAHs
Soil washing	Using water or solvents to extract PAHs from contaminated soil, separating PAHs from soil particles	Suitable for soils with high PAH concentrations; often combined with other remediation techniques
Solid-phase extraction (SPE) Biochar adsorption	Concentrating and purifying PAHs from liquid samples by passing the liquid through a solid adsorbent material. Utilizing biochar, a carbon-rich material produced from the pyrolysis of biomass, to adsorb PAHs from soil and water	Primarily used for sample preparation and analysis in environmental monitoring and research Effective for soil and water remediation; enhances soil quality while adsorbing contaminants
Clay minerals adsorption	Using clay minerals like bentonite or kaolinite to adsorb PAHs due to their large surface area and ion exchange capacities	Effective for soil remediation and can be combined with other methods for enhanced effectiveness
Synthetic polymers adsorption	Employing synthetic polymers with high surface areas and tailored functional groups to adsorb PAHs from contaminated environments	Used in water and soil remediation; can be engineered for specific PAH contaminants

generates highly reactive hydroxyl radicals that can effectively degrade PAHs. Ozonation, which uses ozone gas to oxidize PAHs, is particularly effective in treating contaminated water and air. Photo catalysis involves the use of light and a catalyst, such as titanium dioxide to produce reactive species that break down PAHs. Different types of chemical degradation techniques for poly aromatic hydrocarbon shown in Table 2. Advanced oxidation processes, which combine multiple oxidation methods like ozone, hydrogen peroxide, and UV light, provides a robust approach for treating highly contaminated environments. Catalytic degradation employs specific catalysts to enhance the breakdown of PAHs through chemical reactions. These chemical degradation methods are essential for the remediation of PAHs-contaminated sites, offering effective solutions to reduce the environmental and health risks associated with these pollutants (Gupta et al. 2020). This table provides a concise overview of various chemical degradation methods for PAHs, including a brief description of each method and its applications (Bleeker et al. 2003).

Table 2 Different types of chemical degradation techniques for poly aromatic hydrocarbon

Method	Description	Applications
Chemical oxidation	Use of oxidizing agents like ozone, hydrogen peroxide, or permanganate to transform PAHs into less toxic compounds	Effective for soil, sediment, and groundwater remediation
Fenton's reagent	Solution of hydrogen peroxide and an iron catalyst generating hydroxyl radicals to degrade PAHs	Widely used for soil and groundwater remediation
Ozonation	Use of ozone gas to oxidize PAHs, breaking down aromatic rings	Suitable for treating contaminated water and air.
Photocatalysis Advanced oxidation processes (AOPs)	Employing light (usually UV) and a catalyst (such as TiO2) to produce reactive species that degrade PAHs Combination of oxidation methods (e.g., ozone, hydrogen peroxide, UV light) to enhance PAH degradation	Effective for treating contaminated water and air Used for highly contaminated water and industrial effluents
Catalytic degradation	Use of specific catalysts to enhance the chemical breakdown of PAHs	Suitable for treating both soil and water contaminated with PAHs

Biological Degradation Method

PAHs Degradation by Bacteria

Organisms from the *Pseudomonas* genus, *Microbulbifer, Sphingomonas Micrococcus, Cellulomonas, Gordonia, Micrococcus Marinobacter, Mycobacterium, Haemophilus, Rhodococcus, Arthrobacter, Paenibacillus, Bacillus, Aeromonas, Burkholderia, Alcanivorax Xanthomonas, Acinetobacter, Dietzia, Corynebacterium* and effective PAH chemical degradation by *Enterobacter* has been demonstrated. Several bacterial species have been found to use naphthalene as their only source of carbon and energy, according to numerous studies. They include *Bacillus firmus-APIS272, Bacillus subtilis-SBS526, Pseudomonas alcaligenesDAFS311* and are members of the genera *Alcaligenes, Burkholderia, Mycobacterium, Polaromonas, Pseudomonas, Ralstonia, Rhodococcus, Sphingomonas*, and *Streptomyces*. A number of bacteria, including *Bacillus firmus-APIS272, Bacillus subtilis-SBS526, Bacillus licheniformis, Beijerinckia* sp., *Sphingomonas* sp., *Burkholderia cepacia-DAFS11, Pseudomonas alcaligenesDAFS311, Nocardia* sp., *Rhodococcus* sp., *Bacillus subtilis-SBS526* and *Mycobacterium* sp. *Etc* (Maheshwari et al. 2022).

PAHs Degradation by Fungi

Species of fungi are excellent PAHs detoriate due to they can thrive in harsh environments like high temperatures and low pH levels and can tolerate high concentrations of recalcitrant compounds. Fungi can also expand and propagate by means of a dense matrix by secreting extracellular enzymes, which degrade PAHs in polluted areas (in situ). This is made possible by the large, branching mycelia that they form. They have enzyme groups or systems that effectively oxidize PAHs, such as versatile peroxidases, lignin peroxidase, manganese-dependent peroxidases, phenoloxidases (tyrosinases, lacases), enzymes that produce cytochrome P450 monooxygenase, hydrogen peroxide and enzymes that hydrolyze epoxides, furthering their competence. The three organisms *Trichocladium canadense, Aspergillus* sp., and *Fusarium oxysporum* were found to degrade low molecular weight, PAHs most thoroughly. *Aspergillus* sp., *Trichocladium canadense*, *Achremonium* sp., and *Verticillium* sp., showed the greatest degradation of high molecular weight PAHs. The analysis also demonstrated that fungi were highly capable of reducing a wide range of PAHs in low-oxygen environments. Fungal species are able to adsorb PAHs onto their hydrophobic cell walls in addition to biodegrading and mineralizing PAHs (Cerniglia 1993).

PAHs Degradation by Algae

Prokaryotic and eukaryotic algae strains have a considerable potential to break down PAHs chemicals since aquatic settings are frequent repositories of PAHs pollutants. Under photoautotrophic circumstances, prokaryotic and eukaryotic marine algae, such as cyanobacteria (blue-green), green algae, and diatoms (red and brown), oxidize PAHs to produce hydroxylated intermediates. According to reports, cyanobacteria can oxidize naphthalene and phenanthrene into metabolites that resemble those made by mammals and fungi. It has been demonstrated that the marine *cyanobacterium Oscillatoria* sp., strain Japan collection of microorganisms may oxidise naphthalene by using an intermediate arene oxide that isomerizes and shifts no enzymatically at the same time. Another study showed that the marine *cyanobacterrium Agmenellum quadruplicatum PR6* converted phenanthrene into trans9,10-dihydroxy, 1- methocyphenanthrene 9 and 10-dihydrophenanthrene. The potential for using algal-bacterial microcosms to biodegrade PAHs pollutants has been investigated. When some algae co-occur in a consortium, particularly with bacteria, it has been observed that the elimination of fluoranthene and pyrene is enhanced. Phenanthrene is influenced by the synergistic relationship between bacteria and algae. It was discovered that *Chlorella sorokiniana*, a kind of green alga improved the biodegradation of the contaminant when combined with the bacterial species *Ralstonia basilensis* and *Acinetobacter haemolyticus* (Lee 2010).

PAHs Degradation by Plant

An emerging technique called phytoremediation makes use of different plants to break down, extract, contain, or immobilize contaminants. There are different mechanisms used to remove pollutants from air, soil, and water, including volatilization, rhizoremediation, phytotransformation, phytostabilization, and hydraulic control. Photo degradation, also known as phytotransformation, refers to the breakdown of organic pollutants into simpler molecules or their incorporation into plant tissues. Research indicates that, different grasses and leguminous plants can phytodegrade organic compounds. According to reports, switch grass (*Panicum virgatum*) and tall fescue grass (*Festuca arundinacea*) can break down 38% of pyrene in 190 days (Folwell et al. 2016).

Rhizoremediation

It is a type of specific phytoremediation, in which cleanup of contaminated soils is done by plant-associated rhizosphere microorganisms. Rhizoremediation shows great demand for PAHs degradation, and success is based on a proper collaboration of plant and microorganisms with degradation capabilities. For microbial development and repair of contaminants in the soil up to 10–15 m deep, plants provide a large root surface area. Plant roots provide nutrients like carbohydrates, amino acids, flavonoids, and organic acids for the growth and all the activities of PAHs-degrading microbes in rhizoremediation, whereas microbes help in overcoming stress caused by pollutants and reduce the risk of contamination by assisting the plants (Bisht et al. 2015).

Vermiremediation

For PAHs removal from fine soil (pores size 0.1 m), Vermiremediation is employed alone or in combination with microorganisms or plants. PAHs in fine soil pores are not bioavailable or bio accessible to plant root hairs (size 15–17 m) or microorganisms (size 1–10 m). During Vermiremediation, burrowing activity of earthworms widens the pore size of the soil leading the degrading bacteria and plant root enter inside the soil, able to proliferate and eventually able to destroy concealed PAHs. Earthworms by cutaneous absorption or intestinal digestion, remove PAHs from the which bio transforms or biodegrades them into innocuous chemicals. Vermiremediation benefits includes the improvement of physical and biological soil quality, the excretion of nutritional elements as vermicasts, and the growth of beneficial organisms. Earthworms reproduce quickly and require little or no energy, which may increase PAHs removal in a very small period of time, showing Vermiremediation to be very cost-effective, eco-friendly, and sustainable. After 5 weeks, *Eisenia andrei* is introduced into a sewage sludge bioreactor after pre-composting leads to improved PAHs reduction. The only disadvantage of Vermiremediation is that it is only suitable in low and medium contaminated sites where earthworms can survive and grow (Rorat et al. 2017).

Omics Techniques for Investigating Sequence-Based and Function-Based PAH Bioremediation

Bioremediation has found new directions due to innovations like Next-generation sequencing. We can now see essential bio degradative pathways of bacteria that are essential to the environment. We have gained a better insight of the structure of microbial communities, their interactions with one another and their opposites, and their participation in the complex network of interactions that exists between microbes, plants, and the environment because of techniques like metagenomics, meta-transcriptomics, meta-proteomics, and other omics methods. Omics techniques and their components are shown in Fig. 4. Metagenomics method shows two peculiar approaches—the sequence-based metagenomics technique that can be utilized to examine microbial diversity, its evolution, the microbial entity interactions, its entire degradation pathway, the degrading genes present, and so on. The function-based metagenomics technique may be useful in the discovery of new metabolising genes. Microbial communities and its functional capacities can be studied at several levels, including meta-transcriptomics, meta-proteomics, and also metabolomics. Decoding of the messenger ribonucleic acid profile for up-down regulation of degradative microorganisms and the associated genes using met transcriptomics is the major used technique nowadays. Met proteomics is the study of pollution caused by proteins by degrading microorganisms. Metabolomics involves the study of metabolites produced by degradative bacteria in order to cope up with contaminants. Meta transcriptomics techniques are beneficial not only for predicting the functions of abundant members of the organism but also for predicting the functions of low-abundance members that constitutes as little as 0.05% of the total population. A comprehensive list of microorganisms that might help in the

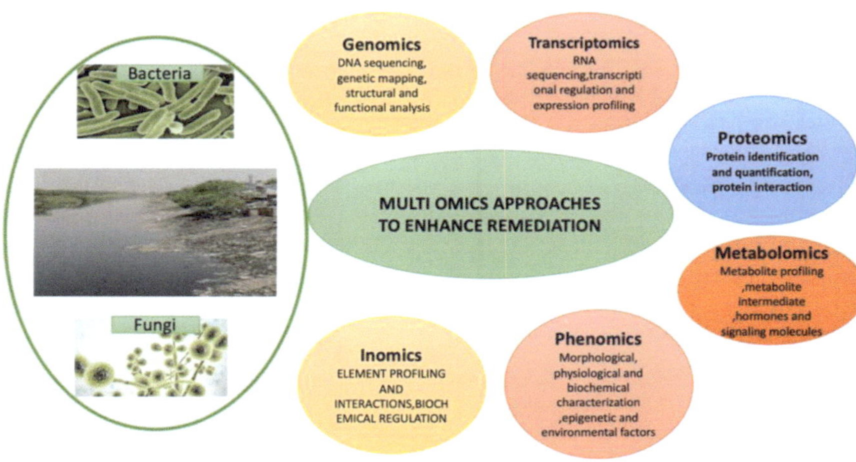

Fig. 4 Omics techniques and their components

Table 3 Various microorganisms used in bioremediation of poly aromatic hydrocarbon

Microorganisms	Compound/ heavy metal	Properties/application
Rhodococcus sp. IN306	PCBs	Encodes the enzyme responsible for breaking down hydrocarbons
Pseudomonas sp. (strain CB-3) and *Comamonas* sp. (strain CD-2)	PCBs	The biphenyl metabolism gene cluster is found in the CB-3 genome. A metabolic gene cluster for dechlorination and protocatechuate (PCA) is present in CD-2
Mycolicobacterium frederiksber- gense IN53	Total petroleum hydrocarbons (TPH)	Encodes the enzyme responsible for breaking down hydrocarbons
Rhodococcus erythropolis strain CD 167	TPH	Higher levels of genes related to the breakdown of TPH
Alphaproteobacterium Rhizo- bium sp. NT-26	Arsenic	Have different genes that allow it to break down arsenite
Marine bacterium *Achromobac- ter* sp. HZ01	Petroleum hydrocarbon	Contains genes involved in the production of biosurfactants and breaks down aromatic compounds and n-alkanes through the catechol and terminal oxidation pathways, respectively
Trichoderma atroviride	Dichlorvos pesticide	The T23 strain was transformed using the plasmid pV2's hph (hygromycin B phosphotransferase) gene, which improved its degradation capacity
Cupriavidus necator strain JMS34	PCBs	Strain that has been genetically altered to include the Burkholderia xenovorans LB400 bph locus, which codes for the PCB breakdown pathway

future development of practical strategies for the bioremediation of polluted areas is provided in Table 3. (Kumar and Thakur 2022).

Amplicon-based metagenomics helps with the creation of a sustainable system for determining the accuracy of bioremediation in soil or in the sediment by identifying appropriate microbial species as targets. Microbial taxa with pollutant degradation ability and the ability to survive at a polluted site can act as a biofertilizers, whereas microbial taxa shows the characteristic features such as present in abundant amount at site, Growing quickly and being easily cultivated, non-pathogenicity and the ability to survive and metabolize in high concentrations of pollutants that can be also be used for bio augmentation. Polyromantic hydrocarbons are long-lasting in nature and frequently immiscible in water and molecularly stable. Their limited water solubility reduces the amount of them that microorganisms can utilise. Thus, microorganisms that are capable of degrading polycyclic aromatic hydrocarbons typically display a strong enzymatic apparatus for cleaving and dissolving these substrates. Recent research on the metabolism of lipophilic PAHs such as naphthalene, pyrene, phenanthrene, and benzo[a]pyrene is included in this section. Among PAHs, the molecular weight with the lowest value is naphthalene and the most straightforward structure. The gentisic and salicylic acid pathways are the most frequent ones via which naphthalene breaks down. *Burkholderia* sp. *BC1* is used to identify the gentisic and salicylic acid-mediated degradation of naphthalene. *Burkholderia* sp. *BC1* not only decomposed naphthalene but also 2-hydroxy-1-naphthoic acid, an intermediate seen during the degradation of phenanthrene, via 2-naphthol and gentisic acid. This catabolic pathway for the breakdown of hydroxynaphthoic acid is different from the traditional gentisic and salicylic acid pathways due to the involvement of 2H1NA decarboxylase in the conversion of 2H1NA to 2-naphthol. PAHs and heavy metals have frequently been discovered together. Phenanthrene, a molecule with three fused rings, was broken down by strain *B. fungorum FM-2* throughout a broad pH range (pH 3–9) and in the presence of heavy metal complexes. Pyrene is a molecule with four fused rings that can be broken down by the strain *B. fungorum T3A13001* over a temperature range of 25 to 37 °C and in a pH range of 5 to 9. Lipophilic PAHs degradation involves genes including naphthalene dioxygenase subunit A, protocatechuate-degrading bacterial, naphthalene dioxygenase subunit A3 and phenanthrene dioxygenase subunit A2. Additionally, these genes were found in *B. fungorum*, which showed 100% fluoranthene degradation in 24 days and 99% pyrene degradation in 16 days (Morya et al. 2020).

In this specific example, focus is done on studying the enzymes and proteins produced by *Burkholderia* bacteria when PAHs are breaking down. Proteins from *Burkholderia* cultures that are actively breaking down PAHs must be extracted as the initial step. Next, these proteins are examined utilizing sophisticated proteomic methods such as mass spectrometry. The specific enzymes, such as dioxygenases and monooxygenases, play crucial roles in the initial oxidation of PAHs molecules. These enzymes introduce oxygen atoms into the PAHs structure, making them more reactive and easier to break down. As the degradation process progresses, other enzymes come into play, further modifying the PAHs molecules until they undergo a conversion into simpler, less toxic compounds. Another example of proteomic analysis in the breakdown of PAHs involves studying the enzymes released by *Pseudomonas putida*, a common soil bacterium known for its ability to degrade pollutants like PAHs. In this example, a protein is extracted from *Pseudomonas putida* cultures that were degrading PAHs and subjected them to mass spectrometry analysis. Using this method, they were able to pinpoint specific enzymes, such as cytochrome P450 monooxygenases and catechol dioxygenases, which are crucial for the initial oxidation and subsequent breakdown of PAHs molecules. By analyzing the proteome of *Pseudomonas putida* during PAHs degradation, the metabolic pathways involved in the process were better understood by researchers (Liu et al. 2021).

Mycobacterium vanbaalenii degrades polycyclic aromatic hydrocarbons through a process known as biodegradation. This bacterium has developed specific enzymatic pathways to break down complex PAHs compounds into simpler and more easily metabolizable forms. The degradation of PAHs by *Mycobacterium vanbaalenii* typically involves the following steps:

Mycobacterium vanbaalenii first attaches itself to the PAH compound present in the environment. Once attached, *Mycobacterium vanbaalenii* produces enzymes such as dioxygenases and monooxygenases that oxidize the PAHs compounds. These enzymes produce oxygen into the PAHs structure, breaking down the complex aromatic rings into smaller, more manageable molecules. The oxidized PAHs molecules undergo ring cleavage, where the aromatic rings are opened up into simpler structures. This step is essential for further degradation of PAHs into metabolites that can be utilized by the bacterium as a carbon source. The ring cleavage produce metabolites are further metabolized by *Mycobacterium vanbaalenii* through various enzymatic reactions. These reactions involve the conversion of the PAHs breakdown products into central metabolites that can enter the bacterium's metabolic pathways for energy production and growth. Through a series of enzymatic reactions and metabolic processes, *Mycobacterium vanbaalenii* continues to degrade the PAHs compounds until they are completely converted into simple, non-toxic end products. This complete breakdown of PAHs helps in reducing the environmental impact of these pollutants. By employing a combination of enzymatic activities and metabolic pathways, *Mycobacterium vanbaalenii* efficiently degrades PAHs compounds, contributing to the bioremediation of environments contaminated with these hazardous pollutants (Kim et al. 2008).

Metabolites of Polycyclic Aromatic Hydrocarbon

Organic substances having many fused aromatic rings are referred to as polycyclic aromatic hydrocarbons, or PAHs. They are commonly found in the environment, arising from both natural processes (like wildfires) and human activities (such as burning fossil fuels, tobacco smoke, and industrial processes). PAHs can be absorbed by the human body through ingestion, inhalation, and skin contact. Once inside the body, PAHs undergo metabolic processes to become metabolites, which can then be excreted in urine and feces. These metabolites are often employed as biomarkers to assess exposure to PAHs.

Polycyclic aromatic hydrocarbons undergo a metabolic pathway that transforms them from hydrophobic and potentially toxic compounds into more water-soluble and excretable forms, primarily in the liver. This process begins with Phase I metabolism, where enzymes from the cytochrome P450 family (cytochrome P450 family 1 subfamily A member 1, cytochrome P450 family 1 subfamily A member 2, and cytochrome P450 family 1 subfamily B member 1) oxidize PAHs, introducing reactive functional groups. For instance, benzo[a]pyrene is converted into benzo[a]pyrene-7,8-epoxide, a highly reactive intermediate. These epoxides can further react to form dihydrodiols, such as benzo[a]pyrene-7,8-dihydrodiol, which can further oxidize to produce dihydrodiol epoxides, substances recognized for their carcinogenic properties. In Phase II metabolism, the reactive intermediates from Phase I are conjugated with endogenous molecules to increase their solubility. Enzymes such as Uridine Diphosphate -glucuronosyltransferases and sulfotransferases add glucuronic acid and sulfate groups, respectively, to the hydroxylated PAHs, producing sulfate

Table 4 Some important and signature molecule of PAH metabolites and their parent compounds

Parent compound	PAH metabolite	Relevance as biomarker
Pyrene	1-Hydroxypyrene (1-OHP)	General PAH exposure
Benzo[a]pyrene	3-Hydroxybenzo[a]pyrene (3-OHBaP)	Carcinogenic potential assessment
2-Hydroxyfluorene (2-OHF) Fluorene	2-Hydroxyfluorene (2-OHF)	PAH exposure indicator
Phenanthrene	2-Hydroxyphenanthrene (2-OHPhe)	General PAH exposure
Phenanthrene	9-Hydroxyphenanthrene (9-OHPhe)	General PAH exposure
Phenanthrene	Phenanthrene 3-Hydroxyphenanthrene (3-OHPhe)	General PAH exposure
Chrysene	1-Hydroxychrysene (1-OHChr)	PAH exposure indicator
Chrysene	Chrysene 6-Hydroxychrysene (6-OHChr)	PAH exposure indicator
Benzo[a]anthracene	3-Hydroxybenzo[a]anthracene (3-OHBaA)	Carcinogenic potential assessment
Naphthalene	Naphthalene 1-Hydroxynaphthalene (1-OHNap)	General PAH exposure
Naphthalene	2-Hydroxynaphthalene (2-OHNap)	General PAH exposure

and glucuronide conjugates. Glutathione S-transferases also play a crucial role by conjugating glutathione to reactive intermediates, aiding in their detoxification. These types of metabolites are then excreted from the body through urine and feces, effectively reducing the toxic burden. Some important and signature molecules of PAH metabolites, their parent compounds, and their typical use as biomarkers are shown in Table 4 (Yebra-Pimentel et al. 2015). A few of the most important metabolites that are used to measure exposure to PAHs are included in Table 4. These metabolites' presence in biological samples, including urine, can reveal information about an individual's level and source of PAHs exposure (Chen et al. 2021).

Health Hazards and Toxicity of PAH Metabolites

Once PAHs enter the body through ingestion, inhalation, or skin contact, they undergo metabolic activation primarily by cytochrome P450 enzymes, transforming into more reactive intermediates such as epoxides and dihydrodiol epoxides. Deoxyribonucleic acid (DNA) adducts can be formed by these metabolites, which are complexes where the metabolites bind to DNA, leading to mutations that can initiate carcinogenesis. Such DNA adducts are highly linked to an elevated risk of lung and bladder cancers, among other cancers, and skin cancers. Furthermore, the metabolic metabolism of PAHs compounds produces reactive oxygen species, which can lead to oxidative stress. This oxidative stress can harm lipids, proteins, and DNA within the cell, which can lead to inflammatory reactions and the emergence of chronic illnesses such as cardiovascular disease and respiratory disorders. Moreover, the ability of PAHs metabolites to disrupt endocrine function and weaken the immune system further exacerbates their health risks, potentially leading to reproductive issues and increased susceptibility to infections. Populations at high risk include workers in industries with significant PAHs exposure, smokers, urban residents, and weaker demographics like pregnant women and children. Therefore, the health hazards associated with PAHs metabolites underscore the importance of minimizing exposure through regulatory measures, personal protective equipment, and public health initiatives (Lodovici et al. 1998).

Short-Term Health Effects (Acute)

The impact of PAHs on human health are mostly influenced by the length and method of exposure, as well as the quantity of PAHs exposed, as well as the overall toxicity of the PAHs. Additional factors, such as previous health condition and aging, can also have an effect on medical results. Eye irritation, nausea, vomiting, and diarrhea were caused by occupational exposure to high concentration of PAHs containing pollutant mixtures. Even so, it is questionable which combination components caused these outcomes, and other chemicals frequently associated with PAHs might also be to responsible. PAHs mixtures are also linked inflammation and skin irritation. Direct skin irritants include benzo(a)pyrene, anthracene, and naphthalene. However, benzo(a)pyrene and anthracene have been recognized to irritate skin shown to be skin sensitizers, causing allergic reactions in both people and animals. Acute and chronic health effects are shown in Fig. 5.

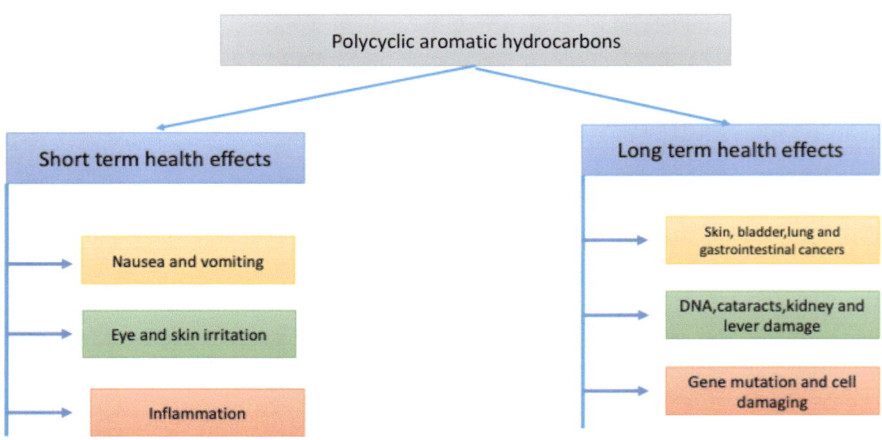

Fig. 5 Acute and chronic health effects

Long-Term Health Effects (Chronic)

Long-term health consequences or chronic exposure to PAHs may result in impaired immunological function, cataracts, breathing difficulties liver and kidney damage (e.g., jaundice), Asthmatic symptoms, and disorders in respiratory symptoms. Furthermore, continuous skin contact may cause redness and irritation. Analysis has shown that certain PAHs are carcinogenic, meaning they have the ability to cause cancer in humans. Long-term exposure to high levels of PAHs has been connected to a higher risk of getting certain kinds of cancer, including lung, skin, and bladder cancer. A specific PAHs, naphthalene, if ingested in high quantity it breakdown red blood cells. The toxic effects that may happen if human is exposed to PAHs are primarily dependent on the method of exposure. As a group, PAHs are considered since they have been defined as the most dangerous in terms of exposure potential and negative impacts on human health. Because of the widespread diffusion of these substances and their importance in terms of toxicology, there is special interest in biological monitoring of PAHs exposure. Because of their toxicological effect and on-going presence in environmental media configurations, even in remote geographic locations, Persistent organic pollutants pose a serious risk. Persistent organic pollutants Eco toxicological effects on the environment have recently raised significant concerns, and as a result, many nations have restricted or outright banned the use of these chemicals. Polychlorinated biphenyls and other organic pollutants that currently, contamination of the aquatic environment Organochlorine pesticides is a great issue (Mohammadi et al. 2022). Due to their great persistence and resistance to biological degradation, the marine environment still harbours a sizable number of these hazardous organic contaminants. Persistent organic pollutants can be found in humans along with other living things such as plants and animals. Public concern over Persistent organic pollutant pollution has recently increased with the discovery of numerous Persistent organic pollutants as hormone disruptors, which can affect the normal functioning of the endocrine and reproductive systems in both people and wildlife. Environmental presence of certain chemicals with a number of risks and effects, and none of them are beneficial for the environment. When these contaminants are released into the environment, as They can stay in the body for years, leading to problems including cancer, birth deformities, learning difficulties, immunological, behavioral, and neurological variations in people and other animal species, as well as concerns with reproduction. Persistent organic pollutants accumulate in the food chain, polluting both food and water, and affect a variety of species, including eagles, killer whales, polar bears and humans. Various researches indicate that many people nowadays may have a sufficient quantity of Persistent organic pollutants in their body fat to seriously harm their health, including disease and death. Persistent organic pollutants have been linked to cancer, immune system dysfunction, neurobehavioral disorders, and endocrine disruption in laboratory studies and ecological impact research in the field (Guo et al. 2019).

Extraction and Analysis of PAH

Different extraction methods are employed to isolate PAHs from different samples. Some common extraction methods include: Soxhlet extraction, ultrasonic agitation/ sonication, mechanical agitation, accelerated solvent extraction /fluid extraction, microwave-assisted extraction, solid-phase extraction, liquid–liquid extraction, and accelerated solvent extraction. The Soxhlet extraction method involves continuous extraction using a Soxhlet apparatus. The sample is placed in a thimble, and a solvent is continuously circulated through the sample, extracting the PAHs. In SPE, a solid-phase material is used to separate and concentrate PAHs from the sample matrix. The sample is passed through a solid-phase cartridge, and the PAHs are retained while unwanted compounds are washed away. Liquid–liquid extraction involves partitioning PAHs between two immiscible solvents. The sample is mixed with a solvent that can extract PAHs, and after separation, the PAHs-containing solvent is collected. Accelerated Solvent Extraction: it is an automated method that uses high pressure and temperature to extract PAHs efficiently. The sample is placed in a cell with a solvent, and the mixture is heated and pressurized to enhance extraction. These methods are commonly used in environmental and analytical chemistry to extract and analyse PAHs from various samples. Each method has its benefits and limitations, and the choice of technique depends on the sample type, PAHs properties, and analytical requirements (Lau et al. 2010). Fluorescence spectroscopy and gas chromatography-mass spectrometry are two distinct methodologies employed for the examination of PAHs. Let's delve into the details of each method:

Fluorescence Spectroscopy

Fluorescence spectroscopy is a technique that measures the fluorescent properties of substances. When PAHs are exposed to light of a specific wavelength, they absorb this light energy and then re-emit light at a longer wavelength, which is detected and measured. As in the instance of PAHs analysis, certain PAHs compounds exhibit fluorescence properties that are suitable for identification and quantification. Advantages of fluorescence spectroscopy for PAHs analysis include high sensitivity, rapid analysis, and non-destructive.

GC-MS (Gas Chromatography-Mass Spectrometry)

Gas chromatography-mass spectrometry is a potent analytical method that combines the ability to separate materials with the detection and identification capabilities of mass spectrometry. In the context of PAHs analysis, gas chromatography separates the PAHs compounds based on their chemical properties, and then the separated compounds are identified and quantified by mass spectrometry. In summary, fluorescence spectroscopy is useful for its sensitivity and speed in PAHs analysis, while gas chromatography offers high specificity and quantitative capabilities. The choice of method depends on the specific requirements of the analysis, such as sensitivity.

Risk Assessment of PAH

The risk assessment process for PAHs involves several key steps: identifying the presence and types of PAHs, evaluating the dose-response relationship to understand the health impact of various exposure levels, assessing human exposure through many channels, and characterizing the overall risk. Standards for regulations have been set to limit PAHs concentrations in the environment, including air and water quality standards, and soil contamination thresholds. Effective risk management requires a multifaceted approach, including stricter regulatory measures to control emissions, technological innovations to reduce PAHs production, public education to raise awareness about exposure risks, and continuous environmental monitoring to ensure compliance and protect public health.

Future Perspective

Pure habitats and human health have been adversely impacted by the several above-mentioned factors that might cause bio-toxic PAHs pollution. This study showed how several technologies and methods for removing PAHs pollution from the environment have evolved dramatically. However, there are still numerous problems that need to be fixed and PAH site rehabilitation is still a difficult undertaking. Numerous examples, thorough explanations and diverse mechanistic insights into the degradation of low molecular weight and low molecular weight, PAHs have been presented by laboratory and the pilot-reactor scale investigations. Low molecular weight PAHs has been demonstrated to be rather easy to degrade, despite the fact that most carcinogenic PAHs are high molecular weight and PAHs with more compact rings and are difficult to degrade. Additionally, the translation of laboratory data into applications for the field appears to be insufficient, according to the cumulative evidence from years of research. It was evident from the scholarly literature this study looked at that the bioremediation methods offered a tenable answer to the problem of PAH cleanup. A fresh approach to PAH treatment involves revitalizing established systems including phytoremediation, land farming, and composting. More significantly, many aerobic PAHs-degrading bacteria (many of which are found in the *Pseudomonas* and *Bacillus genera*) have the ability to express bio surfactants, which makes them perfect for use in bio augmentation. Despite the paucity of anaerobic data, there is evidence for the effectiveness of this strategy in anaerobic PAHs biodegradation. For instance, the use of the surfactant Linear Alkylbenzene Sulfonate increased fluorescence lifetime biodegradation in the Streptomyces venezuelae study. Surfactant use may therefore be a tactic to enhance the anaerobic biodegradation of PAHs.

In addition, biodegradation is now being applied to new, cutting-edge technologies that use may therefore be a tactic to enhance the anaerobic biodegradation of PAHs. In addition, biodegradation is now being applied to new, cutting-edge technologies that use nanoparticles or nanomaterial. Nanomaterial's' enhanced ability to adsorb substrates due to their higher surface-to-volume ratio also enhances biodegradation. This method has been tested in methanogenic, sulphate-reducing and aerobic biodegradation environments. Future applications of these techniques in the anaerobic biodegradation of HMW-PAHs are awaited. In situ bioremediation of PAHs has made good use of the synergistic interactions between microorganisms and plants. New integrated methods, such as bio nanoparticle-based photochemical degradation followed by microbial degradation, bioaugmentation-biostimulation, and eco-engineered bioremediation of PAHs. PAHs pollution in the environment may now be remedied much more effectively thanks to the immobilization of microbial consortia in microcapsules and other forms. Future study must focus on identifying and understanding the processes of PAHs degradation in soil and marine environments. The bioavailability of the emitted PAHs is greatly decreased since they are frequently trapped in coal tar and black-clayish carbon particles under the soil environment. As a result, it is a significant obstacle to the effective PAHs bioremediation. The success of PAHs remediation strategies depends on how well dissolved oxygen content, accessible nutrients, co-contaminants (such as heavy metals, phenolic, etc.), and abiotic factors interact. Understanding the synergistic hydrocarbon degradation mechanisms is essential for the proper implementation and scale-up of in situ biostimulation and bio augmentation approaches because the natural environment is an interactive and interdependent system for nutrient cycling by microbial communities. Notably, recent developments in the creation of value-added by-products during PAHs remediation (such as biofuels, biosurfactants, and bioenergy) indicate a feasible technique that must be examined in future research.

Conclusion

PAHs have been identified as potential human carcinogens and represent a significant class of environmental chemical carcinogens. To exert their mutagenic or carcinogenic effects, PAHs must undergo activation to form electrophilic metabolites. PAHs can be degraded through various mechanisms, including adsorption, volatilization, photolysis, and chemical degradation. However, microbial degradation is the most critical process for PAHs degradation. Microorganisms such as algae, bacteria, and fungi play a pivotal role in this process, which involves the biotransformation of organic compounds into simpler metabolites and their mineralization into inorganic compounds like water and carbon dioxide (under aerobic conditions), or methane (under anaerobic conditions). Bioremediation harnesses the ability of microorganisms to transform hazardous substances into less toxic or non-toxic forms with minimal chemical, energy, and time inputs. The degradation of PAHs is influenced by several factors, including environmental conditions, the abundance and diversity of microorganisms, and the chemical nature and composition of the PAHs. Given their mutagenic and carcinogenic properties, the environmental fate of PAHs is of significant concern. Microorganisms capable of degrading

PAHs convert or mineralize these compounds into carbon dioxide and water. Recent studies indicate that effective depollution of PAHs-contaminated areas using physical, chemical, mixed bioremediation, and phytoremediation techniques are both economically viable and environmentally friendly. Therefore, it can be concluded that the use of microorganisms represents a green and efficient method for the remediation of PAHs-contaminated sites.

Recommendations for Health and the Environmental Protection

Given the extensive prevalence of PAHs in the environment and their significant toxicological implications, the assessment of PAHs exposure is critical for both occupational and general populations. Individuals employed in industries such as metallurgical coke production, aluminum and carbon electrode manufacturing, and asphalt work, as well as the general populace, should be systematically evaluated for PAHs exposure. The biological impact of total PAHs intake via respiratory, dermal, and gastrointestinal pathways must be rigorously assessed. Due to the immunotoxic and carcinogenic effects of PAHs on coke-oven workers, occupational exposure to these compounds should be minimized or eliminated. This can be achieved by reducing emissions as much as possible. In instances where emission reduction is impractical, the provision of effective personal protective equipment is imperative. Enhancing public awareness and education regarding the sources and health effects of PAHs exposure is essential.

Promoting the use of vented indoor spaces for fireworks should be encouraged in many countries. Outdated combustion methods should be replaced with more efficient, versatile, and well-ventilated combustion units. Global emphasis on protecting non-smokers from the risks of passive PAHs exposure through second hand smoke is crucial. Industrial emissions must be rigorously filtered, and adherence to industrial wastewater treatment protocols is particularly critical in developing nations. The enforcement of catalytic converters and particle traps on motor vehicles should be stringent to mitigate health deterioration associated with PAHs exposure. In summary, proactive measures such as emission reduction, effective use of personal protective equipment, public education, technological advancements in combustion and industrial processes, and stringent regulatory enforcement are essential to mitigating PAHs exposure and its associated health risks.

Credit Authorship Contribution Statement

Annu Khatri wrote original draft and carried out conceptualization and investigation. Krishan kumar performed investigation, writing, data curation and corrected the MS properly. Indu Shekhar Thakur carried out initial planning, conceptualization, supervision, visualization, writing, editing, and reviewing.

Acknowledgments Authors and co-authors are grateful to Prof. P.B. Sharma, Vice Chancellor, Amity University Haryana, Gurugram, for the necessary help and support related to the completion of research work.

References

Abdel-Shafy HI, Mansour MS (2016) A review on polycyclic aromatic hydrocarbons: source, environmental impact, effect on human health and remediation. Egypt J Pet 25(1):107–123

Bisht S, Pandey P, Bhargava B et al (2015) Bioremediation of polyaromatic hydrocarbons (PAHs) using rhizosphere technology. Braz J Microbiol 46:7–21

Bleeker EA, Wiegman S, Droge ST et al (2003) Towards an improvement of the risk assessment of polycyclic (hetero) aromatic hydrocarbons. In: Report 2003–01 of the Department of Aquatic Ecology and Ecotoxicology. University of Amsterdam, Amsterdam

Cerniglia CE (1993) Biodegradation of polycyclic aromatic hydrocarbons. Curr Opin Biotechnol 4(3):331–338

Chen YY, Kao TW, Wang CC et al (2021) Polycyclic aromatic hydrocarbon metabolites and mortality risk in an adult population. Environ Geochem Health 43:37–45

Choi H, Harrison R, Komulainen H et al (2010) Polycyclic aromatic hydrocarbons. In WHO guidelines for indoor air quality: selected pollutants. https://www.ncbi.nlm.nih.gov/books/NBK132015/ Accessed on 10 Feb 2024

Folwell BD, McGenity TJ, Whitby C (2016) Biofilm and planktonic bacterial and fungal communities transforming high-molecular-weight polycyclic aromatic hydrocarbons. App Environ Microbiol 82(8):2288–2299

Gbeddy G, Goonetilleke A, Ayoko GA et al (2020) Transformation and degradation of polycyclic aromatic hydrocarbons (PAHs) in urban road surfaces: influential factors, implications and recommendations. Environ Pollut 257:113510

Guo W, Pan B, Sakkiah S et al (2019) Persistent organic pollutants in food: contamination sources, health effects and detection methods. Int J Environ Res Public Health 16(22):4361

Gupta B, Puri S, Thakur IS et al (2020) Enhanced pyrene degradation by a biosurfactant producing *Acinetobacter baumannii* BJ5: growth kinetics, toxicity and substrate inhibition studies. Environ Technol Innov 19:100804

Kim SJ, Kweon O, Jones RC et al (2008) Genomic analysis of polycyclic aromatic hydrocarbon degradation in *Mycobacterium vanbaalenii* PYR-1. Biodegradation 19(6):859–881

Kronenberg M, Trably E, Bernet N et al (2017) Biodegradation of polycyclic aromatic hydrocarbons: using microbial bioelectrochemical systems to overcome an impasse. Environ Pollut 231:509–523

Kumar V, Thakur IS (2022) Omics insights in environmental bioremediation. Springer, Singapore

Låg M, Øvrevik J, Refsnes M et al (2020) Potential role of polycyclic aromatic hydrocarbons in air pollution-induced non-malignant respiratory diseases. Respir Res 21:1–22

Lau EV, Gan S, Ng HK (2010) Extraction techniques for polycyclic aromatic hydrocarbons in soils. Int J Anal Chem 2010(1):398381

Lee BK (2010) Sources, distribution and toxicity of polyaromatic hydrocarbons (PAHs) in particulate matter. In: Villanyi V (ed) Air pollution. IntechOpen, London

Liu Y, Hu H, Zanaroli G et al (2021) A *Pseudomonas* sp. strain uniquely degrades PAHs and heterocyclic derivatives via lateral dioxygenation pathways. J Hazard Mater 403:123956

Lodovici M, Akpan V, Giovannini L et al (1998) Benzo [a] pyrene diol-epoxide DNA adducts and levels of polycyclic aromatic hydrocarbons in autoptic samples from human lungs. Chem Biol Interact 116(3):199–212

Lyytikäinen M, Pehkonen S, Akkanen J et al (2007) Bioaccumulation and biotransformation of polycyclic aromatic hydrocarbons during sediment tests with oligochaetes (*Lumbriculus variegatus*). Environ Toxicol Chem 26(12):2660–2666

Maheshwari N, Thakur IS, Srivastava S (2022) Enhanced microbial degradation of pyrene using biosurfactant isolated from *Bacillus* sp. SS105. Syst Microbiol Biomanufact 2(4):723–731

Maliszewska-Kordybach B (1999) Sources, concentrations, fate and effects of polycyclic aromatic hydrocarbons (PAHs) in the environment. Part A: PAHs in air. Pol J Environ Stud 8:131–136

Mishra A, Kumari M, Swati et al (2022) Persistent organic pollutants in the environment: risk assessment, hazards, and mitigation strategies. Bioresour Technol Rep 19:101143

Mohammadi MJ, Goudarzi G, Dehaghi BF et al (2022) Polycyclic aromatic hydrocarbons and their effects on the occurrence of chronic obstructive pulmonary disease (COPD): a review. Jundishapur J Chronic Dis Care 11(3):e122852

Morya R, Salvachúa D, Thakur IS (2020) Burkholderia: an untapped but promising bacterial genus for the conversion of aromatic compounds. Trends Biotechnol 38(9):963–975

Patel AB, Shaikh S, Jain KR et al (2020) Polycyclic aromatic hydrocarbons: sources, toxicity, and remediation approaches. Front Microbiol 11:562813

Rathour R, Gupta J, Tyagi B et al (2018) Biodegradation of pyrene in soil microcosm by *Shewanella* sp. ISTPL2, a psychrophilic, alkalophilic and halophilic bacterium. Bioresour Technol Rep 4:129–136

Ravindra K, Sokhi R, Van Grieken R (2008) Atmospheric polycyclic aromatic hydrocarbons: source attribution, emission factors and regulation. Atmos Environ 42(13):2895–2921

Rorat A, Wloka D, Grobelak A et al (2017) Vermiremediation of polycyclic aromatic hydrocarbons and heavy metals in sewage sludge composting process. J Environ Manag 187:347–353

Sun Y, Wu S, Gong G (2019) Trends of research on polycyclic aromatic hydrocarbons in food: a 20-year perspective from 1997 to 2017. Trends Food Sci Technol 83:86–98

Swati GP, Thakur IS (2019) Biodegradation of pyrene by *Pseudomonas* sp. ISTPY2 isolated from landfill soil: process optimisation using box-Behnken design model. Bioresour Technol Rep 8: 100329

Thakur IS, Srivastava S (2011) Bioremediation and bioconversion of chromium and pentachlorophenol in tannery effluent by microorganisms. Int J Technol 2(3):224–233

Yebra-Pimentel I, Fernández-González R, Martínez-Carballo E et al (2015) A critical review about the health risk assessment of PAHs and their metabolites in foods. Crit Rev Food Sci Nutr 55(10):1383–1405

Zelinkova Z, Wenzl T (2015) The occurrence of 16 EPA PAHs in food–a review. Polycycl Aromat Compd 35(2–4):248

Navigating Nanochemicals: Understanding the Landscape of Recent Emerging Contaminants

Amrit Krishna Mitra and Santanu Chakrabarti

Abstract

Nanotechnology has recently made significant inroads into diverse areas such as consumer products, healthcare, and environmental management. Nanochemicals exhibit distinctive properties that depend on their size, shape, composition, and degree of aggregation. These materials can be classified into organic-based, metal-based, and composite nanoparticles. As nanochemical production and usage increase, so does the potential for human and environmental exposure, raising concerns about their toxicity. This chapter explores the characteristics, uses, and toxicity of both metal- and non-metal-based nanochemicals. Such systems can accumulate in vital organs such as the heart, brain, kidneys, spleen, and liver following skin exposure, inhalation, and ingestion. Research, both in vitro and in vivo, indicates that nanochemicals can produce reactive oxygen species (ROS), which play a significant role in their toxic effects. Elevated ROS levels can cause inflammation, oxidative stress, and damage to membranes, cellular proteins, and DNA. Factors influencing ROS production include the surface properties, size, aggregation, shape, composition, solubility, and cellular uptake. Additionally, the toxicity of nanochemicals can vary based on their oxidation state, surface coatings, solubility, and physical form, and various environmental and health conditions.

A. K. Mitra (✉)
Department of Chemistry, Government General Degree College, Singur, Hooghly, West Bengal, India

S. Chakrabarti (✉)
Department of Zoology, Government General Degree College, Singur, Hooghly, West Bengal, India

S. Raisuddin et al. (eds.), *Environmental Toxicants and Lifestyle Diseases*,
https://doi.org/10.1007/978-981-95-0575-3_5

Keywords

Nanochemicals · Nanoparticles · In vitro and in vivo toxicity

Introduction

Nanotechnology deals with particles or materials between 1 and 100 nanometers in size. To put it into perspective, the diameter of a human hair is roughly 80,000 nanometers, DNA measures between 2 and 12 nanometers wide, a red blood cell is about 7000 nanometers across, and a water molecule spans nearly 0.3 nanometers (Mekuye and Abera 2023). This field is advancing rapidly, with diverse applications in medicine, agriculture, and defense. Particles within this scale, known as nanoparticles (NPs), nanochemicals (NCs), or nanomaterials (NMs), are synthesized and manipulated for a wide range of uses. With increasing applications, NC production is expanding daily, and global investment in nanotechnology is surging (De et al. 2008). Global development and urbanization are exacerbating environmental issues, raising concerns over toxic pollutants. Advanced materials, particularly NCs derived from carbon, metals, and metal-organic frameworks, offer promising solutions to these challenges (Buzea et al. 2007). NCs possess unique properties that make them suitable for a wide range of applications and are already widely used across various industries (Iqbal et al. 2012). However, their high reactivity, due to a greater surface area-to-volume ratio, can also lead to increased toxicity. Factors such as solubility, charge, and form further influence their potential hazards to ecosystems. As new NCs and NPs emerge, thorough toxicity assessments are crucial to ensure environmental safety (Poirier et al. 2014). When NCs enter the bloodstream, they may be recognized as foreign substances, triggering an immune response. Understanding the extent of in vivo toxicity of NCs is crucial for developing NC-based drug delivery and imaging technologies with minimal side effects (Thakur et al. 2022). Numerous toxicological studies on NCs have been conducted using various cells, tissues, and animal models (Kumar et al. 2012; Roco 2005; Helland et al. 2008; Shang et al. 2019; Iavicoli et al. 2017). Once in the bloodstream, NCs may be recognized as antigens, triggering immune responses. The binding of complement proteins to NCs can activate macrophages and induce toxicity. NMs made from NCs may release these particles over time, raising safety concerns due to uncertainties in their production, use, and disposal (Kumar et al. 2012). The rapid growth of nanotechnology and commercialization of its products often outpaces the assessment of ethical concerns and toxicity risks. A lack of specific regulations for nanoscale materials allows products to be marketed without thorough life cycle testing or disclosing the presence of nanochemicals on labels (Roco 2005). This highlights the need for ethical considerations and proper regulations to ensure the safe and responsible use of nanotechnology in consumer products (Helland et al. 2008). This chapter aims to summarize current knowledge on the toxicity and behavior of several nanochemicals in animals.

Applications of Nanochemicals: Current Trends and Emerging Uses

Agriculture has long been vital to both self-sustaining and developing economies, supplying essential raw materials for food industries that support nearly 7.5 billion people. It plays a key role in economic growth, environmental restoration, technological progress, biodiversity, globalization, and food security. However, the sector faces major challenges such as nutrient deficiencies, climate change, soil degradation, water scarcity, reduced crop yields, and labor shortages. These issues, driven by industrialization and population growth, emphasize the need for safer and more sustainable practices. Nano-agrochemicals, including nanofertilizers, nanopesticides, and nanosensing materials, offer potential solutions to enhance environmental sustainability in agriculture (Shang et al. 2019).

NCs play a key role in developing nanofertilizers by improving their synthesis, stability, and performance. Nanofertilizers, considered "smart" carriers for macro- and micronutrients, enhance nutrient delivery to plants, boosting crop yields and vegetation growth. Macronutrients like phosphorus, nitrogen, and magnesium, along with micronutrients such as zinc and iron, can be encapsulated in nanoparticles for efficient delivery (Iavicoli et al. 2017). Nanofertilizers promote sustainability by reducing costs, minimizing nutrient loss, and providing slow nutrient release through materials like chitosan, carbon nanotubes, and metal oxides. For example, urea-hydroxyapatite NCs significantly improve nitrogen use efficiency in plants (Okeke et al. 2022).

Recent studies emphasize the use of calcium phosphate NCs doped with urea and NPK nutrients for efficient nutrient delivery. Micronutrient nanofertilizers commonly use nanoparticles like iron, manganese, copper oxide, and zinc oxide (Avila-Quezada et al. 2022). For example, ZnO NCs outperform bulk zinc compounds, improving yields, germination, root and shoot growth, and biofortification in crops such as maize and chickpeas. Iron nanoparticles consistently enhance plant growth and yield, while CuNPs and MnNPs surpass their bulk forms in boosting crop productivity. Copper nanowires improve alfalfa traits, and manganese zinc ferrite nanoparticles enhance squash yield and mineral content. Nano-enhanced fertilizers like silicon nanoparticles, zeolites, GO, TiO_2, and MWCNTs improve nutrient absorption without directly supplying them. Zeolites, for example, aid in absorbing macronutrients like potassium and ammonium due to their large surface area and high cation exchange capacity.

A range of nanopesticides exhibits bactericidal, insecticidal, and fungicidal properties, including polymeric nanoparticles, solid lipid nanoparticles, and inorganic nanoparticles like silver, alumina, silica, copper, and TiO_2. Carbon materials such as MWCNTs and graphene, along with nanoemulsions and nanogels, enhance their effectiveness. Nanoemulsions, with droplet sizes of 20 to 200 nm, facilitate targeted pesticide delivery. Effective pest control is essential for increasing crop yields, and nano-insecticides, especially inorganic nanoparticles and nanosilicate-alumina, offer viable solutions. For instance, nanostructured alumina (NSA) effectively manages pests from genera like Coleoptera and Diptera. Titanium and silver nanoparticles are known for their antibacterial properties, significantly inhibiting bacterial growth in crops. Inorganic formulations of aluminum and copper exhibit insecticidal and antibacterial effects, with copper nanoparticles acting as fungicides against *Fusarium solani* and *Fusarium oxysporum*. Additionally, cobalt and nickel ferrite nanoparticles have shown efficacy against these fungi (dos Santos et al. 2024). Zinc oxide and silica nanoparticles have been utilized as foliar sprays against tobacco mosaic virus, showcasing the potential of nanomaterials for innovative antiviral strategies. Researchers have also investigated engineered nanomaterials like MWCNTs, TiO_2, magnetite, and chitosan to enhance plant resistance to viral infections (Mosa et al. 2024) (Table 1).

Fungal diseases pose a growing threat to crops globally, impacting yields and national economies. Nanofungicides, with broad-spectrum antifungal properties, enhance pesticidal efficacy by improving solubility and controlled release at targeted sites. Nanoparticles of copper, silver, and zinc overcome resistance to chemical fungicides, offering potent antimicrobial action while remaining non-toxic. Chitosan-based nanochemicals (Ch-NCs) effectively suppress pathogens like *Macrophomina phaseolina*, *Alternaria alternata*, *Rhizoctonia solani*, and *Fusarium solani* (Table 2) (Mosa et al. 2024). Copper-chitosan NCs, chitosan-boehmite alumina nanocomposite films, and thyme oil have shown efficacy against *Monilinia laxa*, while silver nanoparticles (AgNPs) demonstrate strong fungicidal properties against *Sclerotinia minor*, *S. sclerotiorum*, and *R. solani*. AgNPs significantly inhibit fungal growth and spore germination. Additionally, AgNPs combined with zineb (Zb) combat *Neoscytalidium dimidiatum* and *Colletotrichum gloeosporioides*, while AgNPs and selenium NPs (SeNPs) boost crop stress tolerance by enhancing chlorophyll, biomass, soluble sugars, seed germination, and antioxidant enzyme activity under heat stress (Table 3).

Assessing the In Vitro Toxicity of Metal Oxide Nanoparticles: Insights and Implications

Assessing nanoparticle (NP) risks often relies on in vitro cell-based assays and in vivo animal studies. However, the expanding range of NPs makes in vivo tests less feasible due to ethical and financial constraints. This highlights the urgent need for in vitro assays that accurately predict in vivo toxicity for safe nanotechnology applications (Attarilar et al. 2020).

Toxicity of Co_3O_4-Based NPs

Co-based NPs are used in catalysis, biomedicine, energy storage, sensing, and drug delivery. However, cobalt oxide (Co_3O_4) NPs can cause ROS-dependent toxicity in HepG2 and ECV-304 cell lines, with effects increasing by concentration and exposure time. Rapid cellular uptake may contribute to this toxicity. Occupational

Table 1 [a]Effective nanoformulations for plant growth promotion

Nanoformulation	Mode of application	Target crop	Effect on plant physiological process		
			Growth phases	Enzyme activities	Nutrient uptake and release
SiO_2NPs	Seed treatment	*L. esculentummill*	Enhanced seed germination	No visible effect on the activity of enzyme	–
Chitosan Cu- NPs	Seed treatment	*Z. mays*	Promote seeding growth	Enhance activity of α-amylase and protease	–
Nano-chitosan	Seed treatment	*C. areitinum*	Promotes total biomass, germination and vigor index up to 57%	Increase activities of catalyst and superoxide dismutase	–
Au NPs	Seed imbibition	*Z. mays*	Promote germination and seeding growth	Promote activity of superoxide dismutase peroxidase, and catalase	Increase nutrient uptake of maize excluding iron
Chitosan poly (methacrylic) acid NPK-NPs	Foliar Spray	*P. vulgaris*	Enhance plant growth and total biomass	Enhance the antioxidant enzyme activities	Promote nutrient uptake and accumulation

[a]Data taken from Okeke et al. (2022); Avila-Quezada et al. (2022); dos Santos et al. (2024); Mosa et al. (2024)

Table 2 Successful application of nanocomposites for biotic stress tolerance (Tortella et al. 2023)

Pathogen type	Nanoparticle used	Mechanism of action
Bipolaris sorokiniana	AgNPs biosynthesized with *Serratia* sp.	Enhances lignifcation of vascular bundles
Fusarium graminearum	Multiwalled carbon nanotubes, graphene oxide, reduced graphene oxide	Inhibits spore germination of *Fusarium graminearum*
Gloeophyllum abietinum	Green-synthesized AgNPs extracted with turnip leaf	Inhibits the conidia development
Xanthomonas perforans	Photochemically active TiO_2NPs	Due to high photocatalytic activity, reduction in bacterial spot is observed
Bipolaris sorokiniana	AgNPs biosynthesized with *Serratia sp*	Inhibits conidial germination
Xanthomonas alfalfae	Synthesized $Mg(OH)_2$NPs	Significantly decreases the activity of *X. alfalfae*
Aspergillus spp.	SiNPs	Inhibits fungus proliferation

Table 3 Successful application of nanocomposites for abiotic stress tolerance (Javed et al. 2022)

Stress type	Nano particles	Plant species	Mode of application	Results
Drought	TiO_2	*Linum usitatissimum*	Foliar spray	Enhances chlorophyll and carotenoid content as well as lowers the activity of MDA and H_2O_2
	TiO_2	*Triticum aestivum*	Amended soil	Increases seedling growth, antioxidant enzymes, total chlorophyll, and carotenoid content
	Fe_2O_3	*Mentha piperita*	Hoagland solution	Increases activity of antioxidant enzymes
	CNTs, graphene	*Gossypium hirsutum*	Seed priming	Increase seedling growth and biomass
Salinity	ZnO and Si	*Mangifera indica*	Foliar spray	Enhance nutrient uptake, carbon assimilation in plants
	SiO_2	*Fragaria ananassa*	Soil application	Enhances growth, proline, and chlorophyll content, and relative water content
	Fe_2O_3	*Helianthus annuus*	Foliar spray	Enhances dry weight, leaf area, chlorophyll content
	CNTs, graphene	*Catharanthus roseus.*	Murashige and Skoog medium	Enhance the number of leaves and flowers
Heavy metal	Fe	*Triticum aestivum*	Soil application	Increases the rate of photosynthesis, chlorophyll content, and plant growth
UV-B	Si	*Triticum aestivum*	Nutrient solution	Improves the antioxidant defense system
Cold	TiO_2	*Cicer arietinum*	Amended soil	Decreases the MDA levels and electrolyte leakage index

exposure to Co-based NPs has been linked to rhinitis, asthma, dermatitis, and cardiomyopathy. Additionally, it has been found that Co_3O_4 NPs cause DNA damage, apoptosis, and oxidative stress, especially in A549 cells, posing respiratory risks while showing minimal toxicity to other systems. Controlled exposure may allow safe use in industry and medicine, but further studies are needed (Magaye et al. 2012).

Toxicity of TiO_2-Based NPs

Titanium dioxide (TiO_2) NPs are commonly used in cosmetics, food, and pharmaceuticals, but their health effects remain uncertain (Li et al. 2008). TiO_2 NPs have been linked to red blood cell abnormalities, including hemagglutination and hemolysis. Research by Huang et al. demonstrated that TiO_2 NPs disrupt chromosome segregation, cytokinesis, and cell cycle regulation, causing chromosomal instability and cell transformation at higher doses (Huang et al. 2009). TiO_2 NPs can induce apoptosis via the mitochondrial pathway and necrosis in human lymphocytes and U937 cells. Titanate nanofilaments like nanotubes and nanowires with $Na_xTiO_2{}^+{}_\delta$ composition show strong, dose-dependent cytotoxicity toward H596 lung tumor cells, with toxicity increasing due to structural defects from Na^+ to H^+ substitution. In contrast, CNTs with similar morphology are less toxic. TiO_2 NPs affect aquatic animals, causing oxidative stress in *Mytilus* mussels and behavioral changes at low concentrations. Though TiO_2 NPs do not affect *Daphnia magna*, fullerenes (C_{60}) and $C_{60}H_xC_{70}H_x$ NPs increased hopping and appendage movement, potentially raising predation risk and lowering reproduction.

Toxicity of CuO-Based NPs

Copper oxide nanoparticles (CuONPs) have diverse applications in nanomedicine, with their properties shaped by the synthesis method, which is crucial for ensuring biocompatibility. Concerns about CuONP toxicity have gained attention, particularly in vitro studies showing dose-dependent toxicity in Hep-2 cells at concentrations of 1–40 μg/ml. This toxicity is linked to oxidative stress, increased ROS and lipid peroxidation, and reduced glutathione, leading to apoptosis. Similar findings have also been reported in Hep-2 cells. It has been found that CuONPs induce oxidative stress and DNA damage in H4 neuroglioma and A549 lung cells, suggesting factors beyond Cu ion release may drive toxicity (Karlsson et al. 2008).

Toxicity of Iron Oxide-Based NPs

Iron oxide (Fe_2O_3/Fe_3O_4) nanoparticles coated with SiO_2 and carbon nanotubes (CNTs) are widely used in biomedical applications for their magnetic properties. Fe_3O_4 nanoparticles (20–30 nm) show dose-dependent toxicity in normal lung epithelial cells (L-132) and tumor cell lines, including HeLa and A549 (Choi et al. 2009). Fe_3O_4 NPs coated with oleic acid are highly toxic to mouse hepatoma and human fibrosarcoma cells, with lower toxicity in normal 3 T3 cells, likely due to tumor cell traits like colony formation (Zablotskaya et al. 2009). Polyvinyl alcohol-coated superparamagnetic iron oxide NPs display shape- and size-dependent toxicity in primary mouse connective tissue cells, with nanobeads being more toxic than nanoworms and nanospheres. Smaller particles may be more easily internalized, increasing cytotoxicity.

Toxicity of MnO-Based NPs

Manganese oxide (MnO_2) nanoparticles are used in MRI contrast agents, drug delivery, mass spectrometry, wastewater treatment, and batteries. With increased production, there are rising concerns about potential risks to workers and the environment. Most toxicity studies focus on respiratory effects. Hussain et al. (2006) reported that MnO_2 nanoparticles cause dose-dependent toxicity in PC-12 cells, increasing ROS levels and depleting dopamine, a specific effect of MnO_2 NPs, not related to nanosize. While 15-nm AgNPs were more toxic, they only depleted dopamine at higher doses (Hussain et al. 2006).

Toxicity of ZnO-Based NPs

ZnO nanoparticles are valued for their versatile properties and applications. Green methods help stabilize and control their size and shape. In agriculture, ZnO NPs boost crop growth, while in biomedicine, they are used in bioimaging, drug delivery, biosensors, and gene delivery. They also offer alternatives to antibiotics, combating drug-resistant microorganisms. ZnO nanomaterials' semiconducting, piezoelectric, and optical properties suit them for energy storage, nanosensors, cosmetics, and nanoelectronics. Biodegradable and low in toxicity, ZnO NPs release Zn^{2+} ions under acidic and basic conditions, stressing cells. The solubility of ZnO NPs is key to their toxicity, though the exact mechanisms remain unclear.

In 2014, Vicario-Parés et al. found ZnO NPs to be less toxic than ionic zinc in zebra fish embryos, and it was also reported that ZnO NPs toxicity is dose-dependent (Vicario-Parés et al. 2014). Green-synthesized ZnO NPs had antibacterial activity and did not harm zebra fish embryos at 10 mg/ml. ZnO NPs can interact with DNA, causing damage in human epidermal cells via oxidative stress and lipid peroxidation. Huang et al. reported that ZnO NPs increase oxidative stress, levels of Ca^{2+}, and gene expression changes in BEAS-2B cells, leading to cytotoxicity (Huang et al. 2010).

A Comparative Analysis of Nanoparticle Toxicity

Comparative toxicity studies indicate that ZnO, TiO_2, and MgO nanoparticles induce apoptosis and necrosis in human astrocytoma U87 cells and fibroblasts, with toxicity ranked as ZnO > TiO_2 > MgO (Lai et al. 2008). Al_2O_3 and TiO_2 NPs are less toxic than multiwalled carbon nanotubes (MWCNTs) in A549 human lung cells. Higher solubility in metal oxide NPs correlates with increased toxicity in human mesothelioma (MSTO-211H) and rodent 3 T3 fibroblast cell lines. In human laryngeal epithelial cells (HEp-2), only CuO NPs caused dose-dependent cytotoxicity, generating reactive oxygen species (ROS) and inhibiting antioxidant defenses, while high doses of SiO_2 and Fe_2O_3 also produced ROS (Fahmy and Cormier 2009). Fe_2O_3, Fe_3O_4, and TiO_2 NPs are generally nontoxic to primary human bone marrow hematopoietic progenitor cells, but antimony oxide (Sb_2O_3) NPs are toxic to erythroid cells during colony formation. Although NP accumulation in cells has not been linked to cytotoxicity, they may affect cell membranes (Bregoli et al. 2009).

The toxicity of NPs depends on their chemical properties. Hussain et al. reported the toxicity of low doses of molybdenum (MoO_3) NPs (5–50 μg/ml) to rat liver-derived BRL cells, while TiO_2 and Fe_3O_4 NPs (30, 47 nm) showed minimal toxicity. Higher doses (100–250 μg/ml) caused significant cytotoxic effects, including irregular cell shape and shrinkage. Auffan et al. (2009) reported that stable metallic oxide NPs like Fe_2O_3 are nontoxic under physiological conditions. Conversely, NPs with strong oxidative (e.g., CeO_2, Mn_3O_4, Co_3O_4) or reductive (e.g., Fe^0, Fe_3O_4, Ag^0, Cu^0) potentials tend to induce cytotoxicity and genotoxicity due to redox reactions. Blinova et al. observed the toxicities of CuO and ZnO NPs to crustaceans *Daphnia magna* and *Thamnocephalus platyurus* in both artificial and natural freshwater. In contrast, Wiench et al. noted that the toxicities of CuO and ZnO NPs arising primarily from soluble ion discharge, with TiO_2 NPs being less toxic than ZnO NPs, where Zn^{2+} ions contribute more to toxicity than the NPs themselves (Wiench et al. 2009). Environmental contaminants can enhance NP toxicity. For instance, arsenic (As) adsorbs strongly onto NP surfaces, increasing its accumulation in carp (*Cyprinus carpio*) by 44% after 25 days of exposure, with accumulation order: viscera > gills > skin and scales > muscle. This underscores the importance of assessing NP release and toxicity before environmental use or discharge.

Assessing the In Vivo Toxicity of Metal Oxide Nanoparticles: Insights into Organ-Specific Damage and Systemic Effects

NPs behave differently in living organisms than outside, making it essential to evaluate their acute toxicity before release. While in vitro studies are easier, in vivo studies provide more accurate toxicity assessments. Kobayashi et al. found that TiO_2 nanoparticles induced varying toxic responses in rat lungs, with smaller particles causing more inflammation short-term, but these responses were reversible over time (Kobayashi et al. 2009). In contrast, Warheit et al. reported low inflammation and lung toxicity in their studies (Warheit et al. 2007). Wang et al. observed that anatase TiO_2 NPs were more toxic than rutile, leading to greater CNS damage and inflammatory responses in female mice (Wang et al. 2008). Park et al. noted that TiO_2 instillation in mice triggered chronic inflammation and elevated immune-related gene expression (Wang et al. 2008). Additionally, tumor-targeted superparamagnetic iron oxide nanoparticles (SPION) showed promise as antitumor agents in mice, though their effectiveness waned after 5 days. In rats, ZnO NP exposure caused inflammation similar to metal fume fever, with significant but reversible effects. Zhu et al. found that ZnO NP aggregates reduced hatching rates and caused pericardial edema in zebrafish embryos, linked to Zn^{2+}-dependent ROS production (Zhu et al. 2009).

Overall, while metal oxide NPs like SiO_2, MgO, and TiO_2 generally show low in vivo toxicity, NiO and specific alumina types can cause significant lung inflammation due to their surface reactivity. Additionally, certain NPs impact reproductive capabilities and mortality in nematodes.

Assessing the In Vitro Toxicological Impact of Metallic Nanoparticles

In Vitro Toxicity Based on Silver Nanoparticles

AgNPs have gained significant interest for engineering applications and are often incorporated into fabrics for their antimicrobial properties. Given their direct contact with skin, assessing their cytotoxicity is crucial. A safety evaluation of Ag NP-coated cotton fabric found it safe for guinea pigs, with no skin irritation reported. Ag NPs were also shown to be non-toxic to human skin carcinoma and fibrosarcoma cell lines, although they induced cytotoxicity in human hepatoma cells via oxidative stress. These nanoparticles tend to aggregate in the cytoplasm and nuclei, increasing oxidative stress and regulating mRNA levels of related genes differently from Ag^+ ions. Limited information exists on the genotoxicity of Ag NPs. It was found that polysaccharide-functionalized Ag NPs induced more apoptosis in mouse embryonic fibroblasts than uncoated NPs. Both 15-nm and 100-nm Ag NPs reduced glutathione (GSH) levels in rat liver BRL 3A cells, leading to increased reactive oxygen species (ROS) accumulation, although the mechanism of GSH depletion remains unclear (Ahamed et al. 2008). AgNPs (45 nm) demonstrated concentration-dependent cytotoxicity in coronary endothelial cells, where lower doses inhibited proliferation by altering endothelial nitric oxide synthase, while higher doses enhanced it. Additionally, Ag NPs caused vasoconstriction at low concentrations and vasodilation at high concentrations in rat aortic rings. Liu et al. reported that Ag NPs (32–380 nm) led to neurotoxic effects in rat hippocampal CA1 neurons, resulting in persistent sodium current changes and neuronal dysfunction (Liu et al. 2009).

In Vitro Toxicity Based on Gold Nanoparticles

Gold nanoparticles (AuNPs) are used in chemical sensing and imaging, exhibiting unique toxicity profiles compared to bulk gold. Distinguishing between cytotoxicity and cellular damage is essential, as non-toxic nanoparticles can still cause harm (Murphy et al. 2008). Au NPs do not show genotoxicity in the HeLa human cell line and do not alter gene expression. The cytotoxicity and cellular uptake of NPs depend on their material and nanoscale surface interactions. Chemically synthesized metallic NPs can be cytotoxic, especially when surfactants from the synthesis process remain attached. While green synthesis may enhance NP production, it often lacks control over size and shape (Alkilany et al. 2009). For example, gold nanorods coated with CTAB may exhibit cytotoxicity toward human colon cancer cells due to free CTAB release, although the nanorods themselves are less toxic, particularly with polymer coatings that reduce CTAB desorption. Surface charge significantly impacts Au NP toxicity. Cationic Au clusters have shown toxicity to mammalian Cos-1 cells and red blood cells, while those coated with cationic and anionic particles demonstrate variable toxicity based on concentration, likely due to electrostatic binding leading to cell lysis (Goodman et al. 2004). Wang et al. (2008) found that Au nanorods coated with CTAB were highly toxic to human skin cells, whereas spherical Au NPs of various sizes were reported as non-toxic.

In Vitro Toxicity Based on Copper Nanoparticles

The use of copper oxide nanoparticles (CuONPs) has rapidly increased in fields such as industrial catalysts, gas sensors, electronic materials, biomedicines, and environmental remediation due to their versatile properties and high surface area-to-volume ratio. This widespread application raises concerns about short- and long-term toxicity, making their environmental release a key research focus. Factors influencing CuONPs' toxicity include particle shape, size, surface functionalization, dose, exposure duration, and the choice of animal or cell models. CuONPs can generate ROS, causing oxidative stress, inflammation, cytotoxicity, genotoxicity, and immunotoxicity (Ingle et al. 2014). Due to their small size, CuONPs can be easily ingested, inhaled, or absorbed through the skin, traveling to somatosensory neurons in the dorsal root ganglia (DRG). Prabhu et al. reported that Cu NPs induced neurotoxic effects in rat DRG neuron cultures, leading to vacuole formation, detachment from the substratum, and disruption of the neurite network. Toxicity varied with nanoparticle size and concentration, with smaller sizes and higher concentrations resulting in greater effects (Prabhu et al. 2010).

In Vitro Toxicity Based on Platinum Nanoparticles

Platinum nanoparticles (PtNPs) are widely used in industrial applications and medical diagnostics. They can be absorbed through the respiratory and digestive systems, and while they may penetrate the skin, significant absorption has not been proven. PtNPs primarily accumulate in the liver and spleen but can also affect the heart, kidneys, and lungs (Pedone et al. 2017). Their toxicity is size-dependent; PtNPs are generally non-toxic to normal and cancerous human cell lines and can protect against ROS-induced cell death. Larger nanoparticles are more effective at scavenging superoxide anions, suggesting their potential as innovative antioxidants. Intratracheal exposure to PtNPs has been shown to increase proinflammatory cytokines in the lungs, while oral exposure can induce oxidative stress and inflammation. Intravenous delivery may harm the kidneys and liver. Although the cancer-causing potential of PtNPs has not been studied, they can induce DNA damage and cell death without significant cytotoxic effects. No fetal or maternal toxicity has been reported, but there are concerns about increased mortality and stunted growth in offspring. Despite possible skin penetration, evidence of significant absorption is lacking. Due to limited research, a precise occupational exposure limit for PtNPs has not been established, so caution is advised for those handling them. PtNP exposure reduced cellular GSH levels and damaged DNA integrity in a human colon carcinoma cell line. Interestingly, larger PtNPs exhibited decreased toxicity under the same conditions, while longer incubation periods heightened the harmful effects.

In Vitro Toxicity Based on Aluminum Nanoparticles

Aluminum (Al) poses toxicity risks to both humans and animals through various exposure routes, including inhalation, skin contact, ingestion, and medical procedures. Effects of Al exposure include oxidative stress, inflammation, genetic damage, enzymatic dysfunction, and cell death, contributing to health issues such as pneumonia, granulomas, stroke, Alzheimer's disease, infertility, and various cancers. Ultrafine particulate air pollutants (under 100 nm) exacerbate cardiovascular risks by causing endothelial dysfunction. Oesterling et al. observed that aluminum nanoparticles (AlNPs) induced inflammation and atherosclerosis, increasing inflammatory markers (VCAM-1, ICAM-1, ELAM-1) and enhancing monocyte adhesion in endothelial cells. Al-NPs also exhibited dose-dependent toxicity in rat liver-derived BRL cell lines, with higher doses (100–250 μg/ml) leading to significant cell abnormalities, while lower doses (5–50 μg/ml) caused minimal effects (Oesterling et al. 2008).

In Vitro Toxicity Based on Palladium Nanoparticles

The increasing use of palladium nanoparticles (PdNPs) in various technologies has raised concerns about potential health risks, especially for workers with frequent exposure. Their toxicokinetics and dynamic properties remain largely unclear. Studies indicate that PdNPs may disrupt ecosystem processes, cause cytotoxic and pro-inflammatory effects in vitro, and trigger early changes in target organs in vivo. While Pd-NPs show minor effects on human colon adenocarcinoma Caco-2 cells and HaCaT keratinocytes, they induce concentration-dependent cytotoxicity in primary bronchial epithelial cells (PBECs) and lung carcinoma cells (A549), with PBECs being more adversely affected. Apoptosis was observed in PBECs in a dose-dependent manner, while A549 cells did not exhibit this response (Leso and Iavicoli 2018).

Assessing the In Vivo Toxicological Impact of Metallic Nanoparticles

In Vivo Toxicity of Silver-Based Nanoparticle

In vivo studies of silver nanoparticles (AgNPs) show dose-dependent cytotoxicity. Sung et al. observed that inhaling Ag-NPs at concentrations up to 100 $\mu g/m^3$ for 90 days was generally safe for Sprague-Dawley rats; however, higher doses led to inflammatory responses primarily affecting the lungs and liver. Rahman et al. reported that intraperitoneal injection of Ag-NPs in mice induced oxidative stress, altered gene expression, and caused apoptosis in brain tissue, indicating neurotoxicity. Additionally, Roh et al. found that Ag-NPs smaller than 100 nm caused toxicity in *C. elegans* by inducing oxidative stress and resulting in reproductive failure (Roh et al. 2009).

In Vivo Toxicity of Gold-Based Nanoparticle

Exposure to AuNPs can lead to acute inflammatory responses, tissue distribution changes, and histopathological alterations. While AuNPs are generally considered safe and widely used in biomedical applications, their cellular effects are not well understood. Some studies indicate potential cytotoxicity; for example, it was found that a single intravenous injection of 4-nm and 100-nm AuNPs altered apoptosis rates and disrupted the cell cycle, triggering inflammatory and metabolic responses in mice's livers. Additionally, Tedesco et al. demonstrated that exposure to Au-citrate nanoparticles caused oxidative stress in the digestive gland and gills of the aquatic filter-feeder *Mytilus edulis* after 24 h (Tedesco et al. 2008).

In Vivo Toxicity of Copper-Based Nanoparticle

Chen et al. (2006) found that both nano-sized (23.5 nm) and ionic copper ($CuCl_2{\bullet}2H_2O$) particles exhibited moderate toxicity, causing significant damage to the kidneys, liver, and spleen in mice, with male mice showing more severe effects than females. In contrast, micro-sized copper particles (17 μm) were deemed non-toxic. The toxicity of nanoparticles is also influenced by environmental conditions. Meng et al. discovered that copper nanoparticles react more actively in the stomach than their micron-sized counterparts, as nanoscale particles interact more rapidly with hydrogen ions, leading to increased oxidation and the formation of highly toxic cupric ions, which raise mortality rates in experimental mice (Meng et al. 2007).

In Vivo Toxicity of Nickel-Based Nanoparticle

Nickel (Ni) is utilized in various fields such as catalysis, sensing, and electronics, highlighting the need for thorough toxicity assessments of Ni nanoparticles (NPs) from both environmental and biological perspectives. The effects of different sizes of NiNPs-30, 60, and 100 nm in diameter- on zebra fish embryos. Among these, the dendritic clusters formed by 60-nm particles were the most toxic. Notably, all Ni based NPs are less harmful compared to soluble Ni forms. While Ni based NPs led to thinning of the intestinal epithelium and separation of skeletal muscle fibers, soluble Ni only causes muscle fiber separation without affecting the intestines.

Toxicity of Non-metallic Nanoparticles

Toxicity of Silica-Based Nanoparticles

Silica-based nanoparticles are gaining attention due to their tunable pore sizes, large surface area-to-volume ratios, biocompatibility, and ease of modification. These properties make them valuable in biomedical and biotechnological applications, such as photodynamic therapy, diagnostics, drug delivery, and bioimaging. Silica nanoparticles induce ROS production and autophagy in human hepatocellular carcinoma cells, leading to autophagic cell death. In addition to that, silica nanoparticles of different sizes (10, 25, 50, 100 nm) can trigger apoptosis in human umbilical vein endothelial cells. Zhou et al. assessed the genotoxicity of these nanoparticles, revealing significant increases in DNA damage and micronucleus frequencies, along with elevated ROS levels and decreased glutathione in endothelial cells (Zhou et al. 2019).

Toxicity of Quantum Dot-Based Nanoparticles

Quantum dots (QDs) are semiconductor nanoparticles, measuring 2–10 nm in diameter, with their size influencing optical properties. QDs can be linked to bioactive molecules for targeted applications, such as labeling cancer cells and cellular structures. However, their toxicity depends on factors like size, concentration, charge, and surface coating. For instance, neutral polyethylene glycol-coated QDs did not affect cell morphology at concentrations up to 300 μg/mL, while negatively charged carbon dots enhanced oxidative stress and triggered cell cycle arrest at the G2/M phase without entering the nucleus (Bali et al. 2020). In contrast, positively charged dots were highly cytotoxic and penetrated the nucleus, significantly affecting the G0/G1 phase. Cadmium sulfide QDs produced minimal nuclear DNA damage but increased ROS and activated apoptotic pathways. 3-Mercaptopropionic acid-coated QDs were biocompatible and reduced lysosomal activation, whereas polyethylene glycol-coated QDs were toxic due to ROS production and lysosomal damage.

Fullerene, especially C_{60}, is used in various fields but tends to accumulate in tissues like bones and liver. C_{70} fullerene increased ROS levels in human cells at 25.2 μg/mL. Hepatic toxicity at doses of 10 mg/kg, while no significant cytotoxicity was observed at lower concentrations. Fullerene aggregates were non-toxic below 1.65 mg/L, but higher concentrations induced cytotoxicity through mitochondrial dysfunction and ROS production. Waissi-Leinonen et al. (2012) exposed *Chironomus riparius* to C_{60} fullerene, finding significant gut damage and inhibited larval growth due to morphological changes.

Toxicity of Polymer-Based Nanoparticles

Polymeric nanoparticles are essential in cancer therapy due to their advantageous pharmacokinetic properties, such as effective drug loading, controlled release, structural stability, and degradation. They can be conjugated or encapsulated to enhance nanomedicine with sustained release and good biocompatibility. Most biodegradable polymeric nanoparticles are non-toxic, non-inflammatory, and do not activate neutrophils. Voigt et al. used in vivo confocal neuroimaging to track rhodamine-labeled polybutylcyanoacrylate nanoparticles in rats for 5 weeks, observing no neuronal death at effective concentrations, even with potentially harmful surfactant coatings. They concluded these nanoparticles were safe for central nervous system drug delivery, whereas positively charged nanoparticles showed toxicity to various cell types (Voigt et al. 2014). The toxicity of poly(lactide-co-glycolic acid) nanoparticles was evaluated on THP-1 macrophages. At concentrations below 0.1 mg/mL, the nanoparticles exhibited minimal toxicity, but above 1 mg/mL, those with stabilizers showed increased cytotoxicity, highlighting the significant impact of surface coatings on nanoparticle toxicity and pharmacokinetics.

Molecular Mechanisms of Nanoparticle-Induced Toxicity

As NP production rises, human and environmental exposure increases through inhalation, ingestion, skin contact, and injection. Once inside the body, nanoparticles can traverse cellular barriers and accumulate in critical organs like the lungs, liver, and kidneys, potentially causing mitochondrial damage, DNA mutations, and cell apoptosis (Sengul and Asmatulu 2020). Nanoparticle toxicity primarily arises from ROS production, which leads to oxidative stress, inflammation, and cellular damage. Factors such as nanoparticle uptake, size, shape, composition, and surface properties affect ROS generation. For example, 10-nm Ag-NPs exhibited cytotoxicity in human lung cells starting at 20 μg/mL, with no significant difference between citrate and polyvinylpyrrolidone coatings (Gliga et al. 2014). TiO_2-NPs (21 nm) significantly impacted liver function and oxidative stress markers at high doses (500 mg/kg) and prolonged exposure. The shape of AgNPs also influences their toxicity, with gold nanostars being more cytotoxic than nanospheres. Excess ROS can lead to oxidative DNA damage and mitochondrial membrane damage, resulting in necrosis and cell death. Additionally, incomplete phagocytosis by neutrophils and macrophages generates ROS, triggering inflammatory responses by activating cell signaling pathways and promoting the expression of chemokines and cytokines (Manke et al. 2013).

Inhalation poses significant risks to respiratory and cardiovascular health, while dermal exposure is less concerning unless the skin is compromised. Ingestion is generally viewed as the least hazardous route. However, exposure can lead to various health issues, including respiratory conditions, neurodegenerative diseases, gastrointestinal disorders, and organ-specific diseases, as well as potential autoimmune disorders (Buzea et al. 2007).

Evaluating the Safety of Nanochemicals: Risks and Mitigation Strategies

Advancements in nanotechnology have introduced a wide range of NCs into environmental and soil systems. In soil, NCs may undergo transformations based on their properties and soil composition, affecting their stability, reactivity, and toxicity. Comprehensive studies are needed to understand the behavior of NCs in soil, as they can enter human food chains through plants or processed foods. Their small size and large surface area allow NCs to cross cellular barriers, posing potential toxicity risks to humans and livestock. Before applying NCs in agriculture and food processing, it is essential to investigate NP–plant interactions, bio-distribution, entry into food chains, and toxicological effects (Bergamaschi et al. 2015). Despite the promising opportunities for sustainable agriculture, challenges such as limited understanding of plant-NM interactions, inadequate delivery methods, and potential health risks from NCs in the food chain hinder their commercialization.

To optimize nano-agrochemicals for improving plant function, understanding how NCs interact with leaf cuticles and chloroplasts is crucial. Evaluating NC properties such as size, dose, surface chemistry, and accumulation is vital. More in situ studies are needed to identify beneficial mechanisms and develop smarter NCs to enhance photosynthesis and manage plant stress (Isigonis et al. 2019). Modifying the charge and size of NCs or coating them with biomolecules can improve their uptake and targeted delivery to plant cell compartments. NC applications should occur under optimal environmental conditions to mitigate stress, while careful monitoring is necessary to manage risks in nano-agriculture. Concerns about human health, safety, and ecological impacts remain largely unaddressed, highlighting the need for collaboration among experts in plant biology, nanochemistry, and plant pathology to fully leverage NC potential in agriculture (Thomas et al. 2009).

Conclusion and Future Perspectives

Advancements in nanotechnology are crucial for progress across various industries, but the potential risks to human health and the environment require careful assessment. This chapter reviews existing research on the toxicity of different nanomaterials, including metal oxides, metallic nanoparticles, quantum dots, silica nanoparticles, and carbon nanotubes, which are already widely used. In vitro and in vivo studies indicate that most nanomaterials exhibit toxicity in animals, influenced by factors like charge, size, surface coating, shape, and reactivity. The interactions between nanoparticles and their environments remain unclear. Future research should focus on assessing the safety of complex nanomaterials and developing validated models to track their release, transport, and accumulation in the environment. These models should connect the properties of nanomaterials to their behavior, facilitating a comprehensive impact assessment on exposed populations. Discussions about the risks and benefits of nanotechnology must extend beyond the scientific community. Toxicity also varies based on dose, administration route, and exposure level, underscoring the need for rigorous assessments before deeming any nanoparticle safe for general use. A standardized testing procedure for evaluating nanomaterial toxicity is essential, along with effective communication strategies to convey crucial information to policymakers and consumers.

In conclusion, reaching a global consensus on the risks of nanotechnology is vital for ensuring equitable business practices in developing safe technologies. Despite the progress made, numerous research challenges remain. Prioritizing education in nanotechnology and sharing knowledge with a broader audience should be key future initiatives.

Acknowledgement Amrit Krishna Mitra would like to express gratitude to the Department of Science & Technology and Biotechnology, Government of West Bengal, India (Memo No. 860 (Sanc.)/STBT-11012 (25)/5/2019-ST SEC dated 03/11/2023) for the financial support.

Conflict of Interests The authors declare that there is no conflict of interest.

References

Ahamed M, Karns M, Goodson M et al (2008) DNA damage response to different surface chemistry of silver nanoparticles in mammalian cells. Toxicol Appl Pharmacol 233(3):404–410

Alkilany AM, Nagaria PK, Hexel CR et al (2009) Cellular uptake and cytotoxicity of gold nanorods: molecular origin of cytotoxicity and surface effects. Small 5(6):701–708

Attarilar S, Yang J, Ebrahimi M et al (2020) The toxicity phenomenon and the related occurrence in metal and metal oxide nanoparticles: a brief review from the biomedical perspective. Front Bioeng Biotechnol 8:822

Auffan M, Rose J, Wiesner MR et al (2009) Chemical stability of metallic nanoparticles: a parameter controlling their potential cellular toxicity in vitro. Environ Pollut 157(4):1127–1133

Avila-Quezada GD, Ingle AP, Golińska P et al (2022) Strategic applications of nano-fertilizers for sustainable agriculture: benefits and bottlenecks. Nanotechnol Rev 11(1):2123–2140

Bali AS, Sidhu GPS, Kumar V (2020) Root exudates ameliorate cadmium tolerance in plants: a review. Environ Chem Lett 18(4):1243–1275

Bergamaschi E, Murphy F, Poland CA et al (2015) Impact and effectiveness of risk mitigation strategies on the insurability of nanomaterial production: evidences from industrial case studies. Wiley Interdiscip Rev Nanomed Nanobiotechnol 7(6):839–855

Bregoli L, Chiarini F, Gambarelli A et al (2009) Toxicity of antimony trioxide nanoparticles on human hematopoietic progenitor cells and comparison to cell lines. Toxicology 262(2):121–129

Buzea C, Pacheco II, Robbie K (2007) NMs and NPs: sources and toxicity. Biointerphases 2(4): MR17

Chen Z, Meng H, Xing G et al (2006) Acute toxicological effects of copper nanoparticles in vivo. Toxicol Lett 163(2):109–120

Choi SJ, Oh JM, Choy JH (2009) Toxicological effects of inorganic nanoparticles on human lung cancer A549 cells. J Inorg Biochem 103(3):463–471

De M, Ghosh PS, Rotello VM (2008) Applications of NPs in biology. Adv Mater 20(22): 4225–4241

dos Santos OAL, dos Santos MS, Antunes Filho S et al (2024) Nanotechnology for the control of plant pathogens and pests. Plant Nano Biol 8:100080

Fahmy B, Cormier SA (2009) Copper oxide nanoparticles induce oxidative stress and cytotoxicity in airway epithelial cells. Toxicol In Vitro 23(7):1365–1371

Gliga AR, Skoglund S, Odnevall Wallinder I et al (2014) Size-dependent cytotoxicity of silver nanoparticles in human lung cells: the role of cellular uptake, agglomeration and Ag release. Part Fibre Toxicol 11:1–17

Goodman CM, McCusker CD, Yilmaz T et al (2004) Toxicity of gold nanoparticles functionalized with cationic and anionic side chains. Bioconjug Chem 15(4):897–900

Helland A, Scheringer M, Siegrist M et al (2008) Risk assessment of engineered nanomaterials: a survey of industrial approaches. Environ Sci Technol 42(2):640–646

Huang BR, Chen PY, Huang CY et al (2009) Bioavailability of magnetic nanoparticles to the brain. J Magn Magn Mater 321(10):1604–1609

Huang CC, Aronstam RS, Chen DR et al (2010) Oxidative stress, calcium homeostasis, and altered gene expression in human lung epithelial cells exposed to ZnO nanoparticles. Toxicol In Vitro 24(1):45–55

Hussain SM, Javorina AK, Schrand AM et al (2006) The interaction of manganese nanoparticles with PC-12 cells induces dopamine depletion. Toxicol Sci 92(2):456–463

Iavicoli I, Leso V, Beezhold DH et al (2017) Nanotechnology in agriculture: opportunities, toxicological implications, and occupational risks. Toxicol Appl Pharmacol 329:96–111

Ingle AP, Duran N, Rai M (2014) Bioactivity, mechanism of action, and cytotoxicity of copper-based nanoparticles: a review. Appl Microbiol Biotechnol 98:1001–1009

Iqbal MA, Md S, Sahni JK et al (2012) Nanostructured lipid carriers system: recent advances in drug delivery. J Drug Target 20(10):813–830

Isigonis P, Hristozov D, Benighaus C et al (2019) Risk governance of nanomaterials: review of criteria and tools for risk communication, evaluation, and mitigation. Nano 9(5):696

Javed T, Shabbir R, Hussain S et al (2022) Nanotechnology for endorsing abiotic stresses: a review on the role of nanoparticles and nanocompositions. Funct Plant Biol 50(11):831–849

Karlsson HL, Cronholm P, Gustafsson J et al (2008) Copper oxide nanoparticles are highly toxic: a comparison between metal oxide nanoparticles and carbon nanotubes. Chem Res Toxicol 21(9): 1726–1732

Kobayashi N, Naya M, Endoh S et al (2009) Comparative pulmonary toxicity study of nano-TiO2 particles of different sizes and agglomerations in rats: different short-and long-term post-instillation results. Toxicology 264(1–2):110–118

Kumar V, Kumari A, Guleria P et al (2012) Evaluating the toxicity of selected types of nanochemicals. Rev Environ Contam Toxicol 215:39–121

Lai JC, Lai MB, Jandhyam S et al (2008) Exposure to titanium dioxide and other metallic oxide nanoparticles induces cytotoxicity on human neural cells and fibroblasts. Int J Nanomedicine 3(4):533–545

Leso V, Iavicoli I (2018) Palladium nanoparticles: toxicological effects and potential implications for occupational risk assessment. Int J Mol Sci 19(2):503

Li SQ, Zhu RR, Zhu H et al (2008) Nanotoxicity of TiO2 nanoparticles to erythrocyte in vitro. Food Chem Toxicol 46(12):3626–3631

Liu Z, Ren G, Zhang T et al (2009) Action potential changes associated with the inhibitory effects on voltage-gated sodium current of hippocampal CA1 neurons by silver nanoparticles. Toxicology 264(3):179–184

Magaye R, Zhao J, Bowman L et al (2012) Genotoxicity and carcinogenicity of cobalt, nickel and copper-based nanoparticles. Exp Ther Med 4(4):551–561

Manke A, Wang L, Rojanasakul Y (2013) Mechanisms of nanoparticle-induced oxidative stress and toxicity. Biomed Res Int 2013(1):942916

Mekuye B, Abera B (2023) NMs: an overview of synthesis, classification, characterization, and applications. Nano Select 4(8):486–501

Meng H, Chen Z, Xing G et al (2007) Ultrahigh reactivity and grave nanotoxicity of copper nanoparticles. J Radioanal Nucl Chem 272(3):595–598

Mosa MA, Ibrahim DS, El-Tabakh MA et al (2024) Recent advances in using nanotechnology in the management of soilborne plant pathogens. In: Rai M, Avila-Quezada G (eds) Nanotechnology in plant health. CRC Press, Boca Raton, pp 311–334

Murphy CJ, Gole AM, Stone JW et al (2008) Gold nanoparticles in biology: beyond toxicity to cellular imaging. Acc Chem Res 41(12):1721–1730

Oesterling E, Chopra N, Gavalas V et al (2008) Alumina nanoparticles induce expression of endothelial cell adhesion molecules. Toxicol Lett 178(3):160–166

Okeke ES, Ezeorba TPC, Mao G et al (2022) Nano-enabled agrochemicals/materials: potential human health impact, risk assessment, management strategies and future prospects. Environ Pollut 295:118722

Pedone D, Moglianetti M, De Luca E et al (2017) Platinum nanoparticles in nanobiomedicine. Chem Soc Rev 46(16):4951–4975

Poirier M, Simard JC, Antoine F et al (2014) Interaction between silver nanoparticles of 20 nm (AgNP20) and human neutrophils: induction of apoptosis and inhibition of de novo protein synthesis by AgNP20 aggregates. J Appl Toxicol 34(4):404–412

Prabhu BM, Ali SF, Murdock RC et al (2010) Copper nanoparticles exert size and concentration dependent toxicity on somatosensory neurons of rat. Nanotoxicology 4(2):150–160

Roco MC (2005) International perspective on government nanotechnology funding in 2005. J Nanopart Res 7:707–712

Roh JY, Sim SJ, Yi J et al (2009) Ecotoxicity of silver nanoparticles on the soil nematode Caenorhabditis elegans using functional ecotoxicogenomics. Environ Sci Technol 43(10): 3933–3940

Sengul AB, Asmatulu E (2020) Toxicity of metal and metal oxide nanoparticles: a review. Environ Chem Lett 18(5):1659–1683

Shang Y, Hasan MK, Ahammed GJ et al (2019) Applications of nanotechnology in plant growth and crop protection: a review. Molecules 24(14):2558

Tedesco S, Doyle H, Redmond G et al (2008) Gold nanoparticles and oxidative stress in Mytilus edulis. Mar Environ Res 66(1):131–133

Thakur P, Verma Y, Thakur A (2022) Toxicity of nanomaterials: an overview. In: Thakur A, Thakur P, Khurana SMP (eds) Synthesis and applications of nanoparticles. Springer, Singapore, pp 535–544

Thomas T, Bahadori T, Savage N et al (2009) Moving toward exposure and risk evaluation of nanomaterials: challenges and future directions. Wiley Interdiscip Rev Nanomed Nanobiotechnol 1(4):426–433

Tortella G, Rubilar O, Pieretti JC et al (2023) Nanoparticles as a promising strategy to mitigate biotic stress in agriculture. Antibiotics 12(2):338

Vicario-Parés U, Castañaga L, Lacave JM et al (2014) Comparative toxicity of metal oxide nanoparticles (CuO, ZnO and TiO2) to developing zebrafish embryos. J Nanopart Res 16:1–16

Voigt N, Henrich-Noack P, Kockentiedt S et al (2014) Toxicity of polymeric nanoparticles in vivo and in vitro. J Nanopart Res 16:1–13

Waissi-Leinonen GC, Petersen EJ, Pakarinen K et al (2012) Toxicity of fullerene (C60) to sediment-dwelling invertebrate Chironomus riparius larvae. Environ Toxicol Chem 31(9): 2108–2116

Wang J, Liu Y, Jiao F et al (2008) Time-dependent translocation and potential impairment on central nervous system by intranasally instilled TiO2 nanoparticles. Toxicology 254(1–2): 82–90

Warheit DB, Hoke RA, Finlay C et al (2007) Development of a base set of toxicity tests using ultrafine TiO2 particles as a component of nanoparticle risk management. Toxicol Lett 171(3): 99–110

Wiench K, Wohlleben W, Hisgen V et al (2009) Acute and chronic effects of nano-and non-nano-scale TiO2 and ZnO particles on mobility and reproduction of the freshwater invertebrate *Daphnia magna*. Chemosphere 76(10):1356–1365

Zablotskaya A, Segal I, Lukevics E et al (2009) Water-soluble magnetic nanoparticles with biologically active stabilizers. J Magn Magn Mater 321(10):1428–1432

Zhou F, Liao F, Chen L et al (2019) The size-dependent genotoxicity and oxidative stress of silica nanoparticles on endothelial cells. Environ Sci Pollut Res 26:1911–1920

Zhu X, Wang J, Zhang X et al (2009) The impact of ZnO nanoparticle aggregates on the embryonic development of zebrafish (*Danio rerio*). Nanotechnology 20(19):195103

Human Health Dimension of Exposure to Copper Oxide Nanoparticles

Naveed Raza, Shabana Naz, Sajida Batool, and Rifat Ullah Khan

Abstract

The toxicity of copper oxide nanoparticles (CuO NPs) has become a significant concern. There is a growing need for educational investigations on exposure to nanomaterials, including CuO NPs, to assess their potential effects. Environmental and energy concerns have hindered societal progress, while nanoscience and nanotechnology have advanced rapidly, leading to the extensive use of nanosized metal and metal oxide particles. CuO NPs, in particular, have gained attention due to their unique properties and widespread applications, which pose risks to organisms and the environment. This review examines the toxicity of CuO NPs, focusing on their synthesis methods, modes of toxicity, and human exposure. We conducted a literature review using PubMed, Google Scholar, and Science Direct to access recent and comprehensive materials. CuO NPs exposure causes cytotoxicity, genotoxicity, immunotoxicity, ROS generation, inflammation, and oxidative stress. The toxicity of CuO NPs is influenced by their physicochemical properties, concentration, animal models, modes of exposure, and assessment methods. This review summarizes the toxicity of CuO NPs by discussing preparation methods, underlying toxicity mechanisms, modes of exposure, and factors affecting toxicity, highlighting the need for more attention to their effects even at low concentrations.

N. Raza · S. Naz
Department of Zoology, Government College University, Faisalabad, Pakistan
e-mail: drshabananaz@gcuf.edu.pk

S. Batool · R. U. Khan (✉)
College of Veterinary Sciences, Faculty of Animal Husbandry & Veterinary Sciences, The University of Agriculture, Peshawar, Pakistan
e-mail: rukhan@aup.edu.pk

S. Raisuddin et al. (eds.), *Environmental Toxicants and Lifestyle Diseases*,
https://doi.org/10.1007/978-981-95-0575-3_6

Keywords

Nanotechnology · Heavy metals pollution · Reactive oxygen species · Cytotoxicity · Genotoxicity

Nanotechnology

Human dreams and creativity often give birth to new advancements in science and technology. These aspirations have paved the way for the frontier of nanotechnology in the twenty-first century. These materials have dimensions smaller than 100 nanometers (Ahamed et al. 2010). Nanoparticles (NPs) have attracted much notice due to their unique characteristics and wide-ranging applications in various fields. These tiny particles, typically measuring less than 100 nanometers in size, display unique characteristics in contrast to their counterparts in bulk. The exceptional features of nanoparticles make them valuable in diverse areas such as medicine, energy, environmental science, and electronics. One of the primary reasons for the growing interest in nanotechnology is that the surface area to volume ratio of nanoparticles experiences a substantial boost, resulting in a significant increase in their ratio, at the nanoscale. This phenomenon exposes unique behaviors associated with quantum mechanics, resulting in the creation of cutting-edge materials with enhanced mechanical, electrical, and optical properties. The distinctive capabilities of nanomaterials make them highly attractive for an extensive range of applications such as medicine, electronics, materials science and engineering, and energy, to name a few. In today's technological world, CuO-NPs are among the most commonly produced nanoparticles used in various industrial applications. Due to their distinctive qualities and adaptability, these nanoparticles have become very popular. The nanotechnology market has seen significant growth since the early 2000s, and the worldwide market has been worth 39.2 billion USD in 2016. It is projected that the global nanotechnology market will continue to expand, with a forecast of reaching 90.5 billion USD in 2021 (McWilliams 2016).

Nanoparticles

Particles having dimensions that are fewer than 100 nanometers in one dimension and having characteristics that are different from those of bulk materials are referred to as nanoparticles (Letchumanan et al. 2021). Three sorts of sources for nanomaterials can be determined based on their origin: (I) Accidental nanomaterials are formed as a byproduct of industrial operations. For example, car exhaust, welding fumes, and combustion processes can all produce nanoparticles. Even certain natural processes, like forest fires, can generate nanomaterials unintentionally. (II) Designed nanomaterials are intentionally created by humans. They are built with specific features to serve particular purposes. Scientists and engineers carefully design these nanomaterials to have the desired properties for their intended uses. (III)

Both natural processes and the bodies of living things contain naturally created nanomaterials. They may be present in various biological systems, such as insemination. Additionally, certain natural events like forest fires can also generate nanomaterials. The differences between naturally occurring, accidental, and artificial NPs, on the other hand, are often obscured. Incidental NMs, for example, might be regarded as a subclass of natural NMs in certain instances (Jeevanandam et al. 2018). Since NPs are used in so many ways, scientists are paying more attention to them, and they have become a prominent topic of study recently. Once released into the soil, NPs have the potential to change soil chemical and physical characteristics, interact with other contaminants, produce new hazardous chemicals, and disrupt soil microbial functioning, performance, and plant development (Rajput et al. 2018).

Although nanoparticles have provided tremendous benefits in many sectors, they also pose major concerns about human and environmental exposure and detrimental effects. In recent years, several in vivo and in vitro readings have been started to enhance our understanding of the toxicological consequences and potential risks associated with nanoparticle exposures on both humans and the environment. It's true that there's still a lot we don't know about the toxicity of nanoparticle exposure. It's an important area that needs more research. Hopefully, with continued studies, we'll be able to fill in those knowledge gaps and better understand the potential risks.

Nanoparticle Applications

Nanotechnology has been grabbing a ton of attention lately because of its super cool and innovative uses in a variety of disciplines, from consumer goods to medical to environmental improvement. Nanoparticles are essential in nanotechnology not only because of their specific physicochemical features, but also because they are regarded as the basic building blocks of many applications. Carbon nanotubes, fullerenes, graphite nanoparticles, and other carbon-based particles have definitely made a big impact in the field of nanotechnology. They have unique properties that make them incredibly versatile for various applications. It's fascinating to see how these particles are being utilized in different industries (Patel et al. 2019). Because of their variable forms and different features like as electrical and thermal conductivity, these carbons based particles have numerous of tenders, encompassing fields such as biomedicine, nanoelectronics, and mechanical engineering. Their versatility allows them to be utilized in various industries for different purposes (Ema et al. 2016). Metallic nanoparticles are widely utilized in numerous applications, making them highly prevalent in various industries. Metal-based nanoparticles have found their way into a wide range of applications, Using UV blockers in sunscreen, solar cells, paints, coatings, cosmetics, and environmental cleaning initiatives. Their unique properties make them valuable in these areas (EPA 2017). Metallic nanoparticles are a desirable candidate for a therapeutic medication due to their physiochemical properties, high reactivity, high stability, photothermal, and plasmonic capabilities (Ema et al. 2016). Another type of nanoparticle is an organic one. These branching components, which make up these nanoscale polymers, can be tailored to carry out specific chemical functions (Roy and Bhattacharya 2015).

Polymers, liposomes, dendrimers, and micelle production are examples of organic-based nanoparticles that are crucial for drug administration, therapeutic agents, polymer materials, chemical sensors, biosensors, and DNA transfecting agents for prion diseases. It is incredible how these organic nanoparticles have opened up new possibilities in different fields. Composite-based nanoparticles, which can be a combination of carbon-carbon, inorganic-inorganic, or carbon-inorganic nanoparticles, are indeed used to improve the mechanical properties of automobile components and packaging materials. Nanoparticles have indeed found incredible applications in drug delivery and cancer detection. Their unique properties and ability to target specific cells make them valuable tools in delivering medications directly to the desired site and improving the effectiveness of treatments. Additionally, nanoparticles can be utilized in cancer detection methods, such as imaging and sensing techniques, to help diagnose and monitor the progression of the disease (Jeevanandam et al. 2018).

Metal Oxide Nanoparticles Toxicity

The development of metal and metal oxide nanoparticles has indeed led to remarkable advancements in the field of biomedicine. These nanoparticles have proven to be highly valuable for biosensing, imaging, diagnosis, and treatment purposes. They offer enhanced sensitivity and specificity, enabling more accurate and targeted approaches in healthcare (Jeffree et al. 2019). However, there are worries regarding potential negative impacts on human health when created nanoparticles are released into the environment. Compared to iron (Fe), titanium (Ti), or silver (Ag) oxides, CuO-NPs have been observed to cause more discomfort. It has been demonstrated that CuO-NPs affect the lungs in certain ways. CuO-NPs may have an effect on lung inflammation and injury, as evidenced by the increased recruitment of neutrophils and total cells, as well as elevated levels of total protein and lactate dehydrogenase activity in bronchoalveolar lavage (BAL) fluid. The aforementioned results underscore the possible inflammatory impacts of CuO-NPs in contrast to alternative metal oxides (Salata 2004; Jeffree et al. 2019). In fact, studies have shown that oxidative stress contributes significantly to CuO NPs' cytotoxicity against human airway epithelial cells. Cellular damage may result from an imbalance between antioxidant defenses and reactive oxygen species. This oxidative stress can contribute to the detrimental effects observed in cells exposed to CuO NPs. The fast expansion of NP production and its many applications pose additional hazards to the environment and also to humans (Salata 2004).

Copper Oxide NPs

An essential micronutrient for the correct operation of living things is copper. Important biological processes like the synthesis of neurotransmitters and enzyme function are influenced by it (Linder and Hazegh-Azam 1996). Cu proenzymes are a class of enzymes that require copper as a cofactor. These cu proenzymes are essential for the creation of connective tissue, energy, neuropeptide activation, iron metabolism, and neurotransmitters, among other biological functions. Cu's involvement in these processes highlights its importance in maintaining proper cellular function and overall health. Cu is required in the body in a trace quantity, just alongside the enzymes that are absolutely necessary. Cu output worldwide in 2015 was at 18.7 million metric tonnes, with an estimated 39 percent going to electrical and electronics, 30 percent to pipes, and the rest to equipment, cars, and consumer goods. Cu-based nanoparticles were produced from a mere few hundred tons of the overall output. This trace element may be found in enzymes with names like superoxide dismutase, cytochrome oxidase, and tyrosinase. Superoxide dismutase helps protect cells from damage caused by reactive oxygen species, cytochrome oxidase is involved in cellular respiration, and tyrosinase plays a role in melanin production. These enzymes rely on copper as a crucial component for their proper functioning. Cu-free ions have the ability to cause injury to the human body at the cellular, organ, and systemic levels. As a result, there is a need for controls on the amount of Cu ions found in living organisms (Mendhulkar and Yadav 2017). Copper nanoparticles are very susceptible to oxidation, which results in the formation of inorganic NPs known as copper oxides (CuOs). Natural CuO NPs are widely distributed and inexpensive materials. These NPs, which are found in Cu0, CuI, CuII, and CuII oxidation states, exhibit intriguing physical and chemical characteristics (Bhanushali et al. 2015). The use of copper and CuO NPs as anticancer, antibacterial, and antioxidant agents is widespread (Naz et al. 2023). Because CuO NPs are frequently used as catalysts, photovoltaic cells, gas sensors devices, heat transfer nano fluids, and antimicrobial agents in commercial applications so there is much need to focuses on their toxicity. CuO NPs have been shown to be efficient against a wide spectrum of microbes (Mohammed et al. 2017).

In addition to this, they may also be employed as an anti-fungal agent when they are included into plastics, textiles, coatings, and other similar products. Because of their electrochemical characteristics, CuO NPs are suitable for usage in graphite surface coating, where they may help enhance the material's capacitance qualities (Pendashteh et al. 2013). Recent research has indicated that CuO NPs may be more toxic than carbon nanotubes, highlighting the need for a deeper understanding of their fate and involvement in the environment. Studies on the toxic effects of CuO NPs have been conducted over the past 15 years. These studies have revealed that the toxicity of nanoparticles is influenced not only by their size but also by the type of material. For example, CuO NPs have been found to be highly potent in terms of cytotoxicity and DNA damage, and this cannot be solely attributed to the dissolved ions released by the nanoparticles (Ahamed et al. 2010). Furthermore, research has shown that CuO NPs release copper ions into solutions, which contributes to their toxic effects (Rasheed et al. 2021). Due to a growing interest in the impact of NPs, metal and metal oxide NPs, in particular CuO are receiving more focus.

Synthesis Methods of CuO NPs

Numerous researchers have reported different methods that can be used to synthesize CuO NPs (Fig. 1). Some of these methods include:

1. Ultrasound irradiation hydrothermal: This method involves the use of ultrasound waves in a hydrothermal reaction to synthesize CuO nanoparticles.
2. Biosynthesis approaches: These methods utilize biological agents, such as plants or microorganisms, to facilitate the synthesis of CuO nanoparticles.
3. Electron beam lithography: In this technique, a focused electron beam is used to selectively remove or deposit material on a substrate, consequentially in the construction of CuO nanoparticles.
4. Solid-state reaction: This technique includes the direct response among solid copper compounds and oxygen at high temperatures to produce CuO nanoparticles.
5. Template methods using surfactants: Surfactants are used as models to control the size and shape of the CuO nanoparticles during synthesis.
6. Sol-gel technique: This procedure involves the formation of a colloidal solution (sol) followed by gelation to produce CuO nanoparticles.
7. Microwave-assisted protocol: Microwaves are used to heat the reaction mixture, accelerating the synthesis of CuO nanoparticles.
8. Decomposition of copper acetate: Copper acetate is thermally decomposed to produce CuO nanoparticles.
9. Sonochemical synthesis: This method utilizes high-intensity ultrasound waves to induce chemical reactions and synthesize CuO nanoparticles.

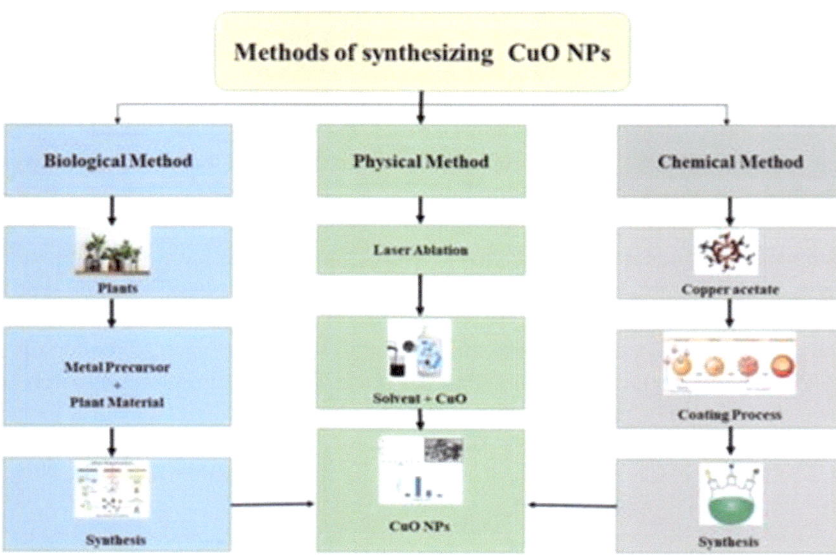

Fig. 1 Methods of synthesizing CuO NPs

Table 1 Different methods of preparation of CuO NPs along with their advantages and disadvantages

Methods	Advantages	Disadvantages	References
Biological	This procedure is easy, simple, safe, eco-friendly, cost-effective and does not require harmful components	Single microorganisms are not attractive because aseptic culture is required and because industrial-scale manufacturing costs are rising	Anchan et al. (2019)
Chemical	It increases the efficiency of large-scale manufacturing	Non-ecofriendly goods are produced and hazardous solvents are used as reducing and stabilizing agents	Dipankar and Murugan (2012)
Physical	CuO NPs with uniform, regulated sized, and high purity may be produced with controlled crystallinity, shape, and manufacturing	Demand substantial financial investments and large amounts of energy	Ahamed et al. (2010)

These different methods offer researchers flexibility in monitoring the properties, size and form of the CuO nanoparticles for various applications (Table 1; Azam et al. 2012). In addition, it has been revealed that the manner of CuO NPs production influences their morphological features and toxicological behavior (Azam et al. 2012).

Characteristics of CuO NPs

CuO, being the most basic member of the Cu compound family, displays some captivating physical features. It has been found to exhibit high-temperature superconductivity, which refers to the ability to conduct electricity without resistance at relatively high temperatures. This property has attracted significant attention in the field of materials science and has potential applications in various technological advancements. Additionally, CuO shows interesting spin dynamics, which involve the behavior of the intrinsic angular momentum of electrons. This aspect contributes to the magnetic properties and potential applications of CuO. Furthermore, electron correlation effects, which involve the interactions between electrons in a material, play a crucial role in shaping the electronic and magnetic behavior of CuO. These intriguing physical features make CuO a subject of extensive research and exploration (Namburu et al. 2007). CuO, with its semiconducting nature and monoclinic crystal structure, has attracted significant interest in the field. The low band gap of CuO crystal formations makes it particularly suitable for applications in photovoltaics and photoconductivity. These properties enable CuO to efficiently absorb and convert light energy into electrical energy, making it a promising material for solar cells and other optoelectronic devices (Xu et al. 1999). In addition,

CuO NPs have the ability to alter the viscosity of fluids and to increase their thermal conductivity (Xu et al. 1999). CuO NPs have been used in multiple applications, like gas sensors field emission emitters, temperature superconductors (Dar et al. 2008) industrial catalysis (Jammi et al. 2009), high batteries (Zhang et al. 2005) and solar energy conversion (Yin et al. 2005). CuO NPs may be able to take the role of noble metal catalysts in industrial catalysis for the oxidation of carbon monoxide (Zhou et al. 2006). This would save manufacturing costs while also increasing catalytic efficiency as utilized in machine tools as a heat transfer fluid. CuO is far less expensive than silver oxide, and its easier to combine with polymers to create hybrids with unique physical and chemical characteristics (Battez et al. 2008). This would save manufacturing costs while also enhancing catalytic effectiveness. The suspension's high thermal conductivity makes it perfect for machine tools as a heat transfer fluid (Chang et al. 2005). Copper oxide is far less expensive than silver oxide, and it is simpler to combine copper oxide with polymers in order to produce composites with novel chemical and physical characteristics. Lubricants, polymers, metallic coatings plastics, and all employ CuO NPs as an addition because of their ability to lessen the amount of friction that occurs and to repair surfaces that have been worn (Battez et al. 2010). Furthermore, the radically different crystal morphologies and relatively large surface areas of CuO NPs contribute to antibacterial activity more efficiently. The strains of *Escherichia coli* in a dose-dependent manner are inhibited, but they do not inhibit *Salmonella typhimurium* strains. This discovery lays the possibility for the creation of a new antibacterial agent that targets a particular pathogen (Ren et al. 2009).

Toxicity of Copper Oxide NPs

It is anticipated that as nanoparticle production and application grow, so will human and environmental exposure. The most frequent ways that humans are exposed to nanoparticles are by eating, skin contact, inhalation through the respiratory tract, and injection (Wu and Tang 2018). When CuO NPs enters the human body, nanoparticles have the potential to traverse multiple cellular barriers and target organs that are particularly susceptible, such as the lungs, liver, and kidneys. This can result in detrimental effects, including mitochondrial damage, DNA alterations, and ultimately, cell apoptosis or death (Bahadar et al. 2016). Figure 2 highlights different factors involved in CuO NPs toxicity.

Cu is one of the essential components for sustaining homeostasis in organisms (Galhardi et al. 2005). Toxic effects may result from Cu ion concentrations that are higher than our bodies can withstand physiologically. When Cu ions are above the physiological tolerance range in vivo, poisoning may result (Galhardi et al. 2004). As a result, both the general public and scientific experts are concerned about the potential health impacts and toxicity of CuO NPs. Studies that have been conducted to evaluate toxicity have, for the most part, concentrated their attention on examining the consequences of various exposure pathways, such as the gastrointestinal tracts or

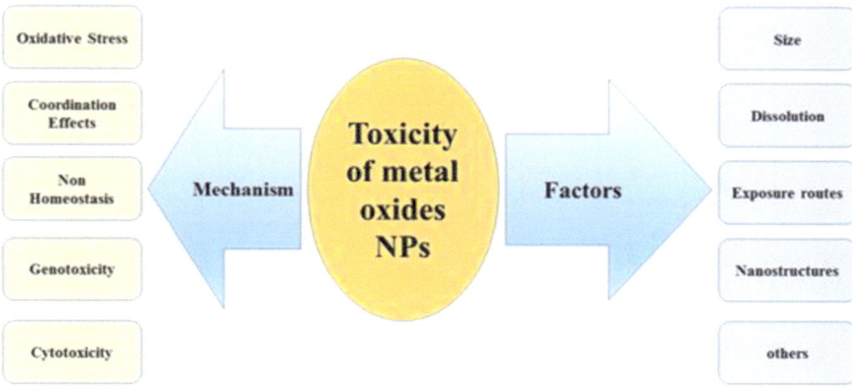

Fig. 2 Different factors and mechanism involved in toxicity of CuO NPs

respiratory. Lung carcinogenic bioassays were performed after intratracheal instillation of CuO NPs by (Yokohira et al. 2009). BY DNA oxidative lesions and damage were found using the comet assay method. To establish the existence of oxidative lesions inside the cells, the generation of ROS as assessed by an oxidation-sensitive fluoroprobe probe was utilized (Karlsson et al. 2008). Reactive oxygen species were tracked as they formed inside the cells (DCFH-DA). Oxidative stress was the principal harmful consequence by comparing the link between ROS formation and DNA damage (Wang et al. 2017). When compared to cells that had been cultivated in normal media, cells that had been exposed to CuO NPs showed decreased activity of the enzymes glutathione reductase and catalase, while showing enhanced action of glutathione peroxidase (GPx1). The observation that there was an increase in the ratio of oxidation to total glutathione led researchers to hypothesize that CuO NPs not only produced ROS but also inhibited the body's natural antioxidant defense mechanisms (Fahmy and Cormier 2009). Research conducted on mice exposed to copper and iron nanoparticles through inhalation demonstrated that copper nanoparticles induced increased inflammatory responses in the mice 3 weeks post-exposure (Pettibone et al. 2008). In vitro toxicity of CuO NPs was much higher than that of a large number of other types of metal oxide NPs and nanotubes. When Cu in the form of water was presented to branchial chloride cells, the proportion of necrotic and apoptotic chloride cells rose, gaps within cells widened, and a high number of white blood cells invaded the cells (Li et al. 1998).

Depending on their size, CuO NPs can be extremely harmful to lung epithelial cells (Wongrakpanich et al. 2016). CuO NPs with sizes of 4 and 24 nanometers were used in the study, and the outcomes demonstrated that the 24 nm oxide nanoparticles were significantly more detrimental to cells than the 4 nm ones. According to their hypothesis, larger nanoparticles after internalization undergo more detrimental effects than smaller ones due to processes including ionic form disintegration and reactive oxygen release (Wongrakpanich et al. 2016). Furthermore, CuO NPs cause dose dependent genotoxicity in human lung epithelial cells via inducing ROS

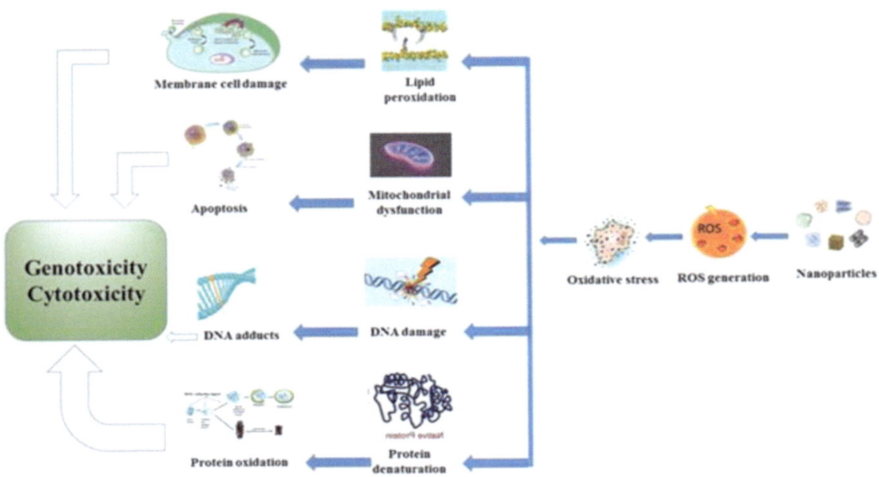

Fig. 3 Genotoxicity and cytotoxicity mechanism of CuO NPs

production (Akhtar et al. 2016; Fig. 3). Similarly, CuO NPs were cytotoxic to Chinook salmon cells, changing their shape and generating oxidative stress (Srikanth et al. 2016). CuO NPs can cause neurotoxicity and hepatotoxicity, in addition to genotoxicity and cytotoxicity (Sun et al. 2016). Studies have indeed shown that CuO NPs can have toxic effects on astrocytes in the brain. The nanoparticles have been found to accumulate in astrocytes through endocytosis, leading to increased cellular copper content. This accumulation can trigger the production of ROS, which can be harmful to the cells. As a result, cell viability may be reduced, and disturbances in brain copper balance can potentially contribute to the development of diseases (Bulcke et al. 2017). All of these investigations support the increased cytotoxicity of CuO NPs and underline the importance of size, shape, dosage, and concentration in causing cytotoxicity. Furthermore, cytotoxicity is caused by the breakdown of metal ions into their ionic forms, metal ions concentrate in the cells, causing oxidative stress and lipid peroxidation. Copper ions may behave as heavy metals in certain situations and have a trojan horse like mechanism where they connect to cell organelles, including genetic material, and hinder cell growth (Bulcke et al. 2017). When it comes to cytotoxicity and DNA damage, CuO nanoparticles were the most effective NPs.

The possessions of CuO NPs on species, particularly those in watery settings, have been the focus of most environmental studies on their toxicity. Algae and zebrafish are the commonly used experiment organisms, with their toxicity and growth used as environmental significance markers. CuO NPs toxicity in algal *Pseduokirchneriella subcapitata* was investigated using a bulk metal oxide formulation as a control. The investigation revealed that soluble copper ions had a significant impact on both the bulk and nanoscale toxicities of copper oxide (Wang et al. 2017). These findings were comparable to Grosell (Grosell et al.

2007) and Griffitt's (Griffitt et al. 2007) conclusions. The soluble Cu forms were shown in both investigations to be extremely toxic to fish. CuO NP suspensions have also been demonstrated to alter plasma metal levels, which injure gill lamellae and obstruct epithelial cell growth (Griffitt et al. 2007) and also the number and width of chloride cells (Pelgrom et al. 1995). As a consequence, it is hypothesized that CuO NPs may accumulate in the mussel digestive gland and cause poisoning (Gomes et al. 2012). The toxicity of CuO NPs is significantly higher compared to ionic Cu and that they decrease the duckweed's chlorophyll content (Shi et al. 2011). This is because of the enhanced toxicity of CuO NPs caused by the increased absorption of Cu produced by NPs. The investigation on fish exposed to soluble and nano-Cu solution demonstrated that the changes in gill shape and transcription were not only brought about by the dissolution of Cu NPs. It suggests that there might be other factors at play in the toxicity of nano Cu (Griffitt et al. 2007). As a consequence, Gomes reasoned that the digestive gland of the mussel may potentially accumulate CuO NPs, which would lead to toxicity (Gomes et al. 2012).

Oxidative Stress

The primary mechanism responsible for causing toxicity and inducing oxidative stress, inflammation, disruption of cell membranes, and consequent damage to proteins and DNA is the production of ROS, which are simply defined as reactive species that include oxygen (Fard et al. 2015). The ROS are essential for many physiological functions, including signaling, cell death, differentiation, and the synthesis of inflammatory mediators. They are outcomes of the oxidative metabolism of cells. Maintaining appropriate cellular function requires an understanding of the balance and management of ROS (Touyz 2005). It has been demonstrated that oxidative stress and ROS generation are frequent and significant ways that NPs harm cells (Yang et al. 2009). Massive levels of ROS might be produced in cells by adding even little amounts of CuO NPs (Dudev and Lim 2008). A quantity of variables, including form, size, surface characteristics, aggregation, composition, particle absorption, solubility, and the presence of transition metals and mutagens, might affect how ROS are produced by nanoparticles. This emphasizes how crucial it is to take into account the properties of nanoparticles when determining their potential toxicity (Jeevanandam et al. 2018; He et al. 2015). The physical and chemical characteristics of nanoparticles do in fact affect all of the NP-mediated processes, including fibrosis, genotoxicity, inflammation, and carcinogenesis. Size, charge, surface area, and chemical structure are a few examples of factors that are important in defining the type and extent of ROS-related effects (Shvedova et al. 2012). Different nanoparticle kinds may induce the generation of ROS via different ways. Factors including the makeup of nanoparticles, their surfaces, and their interactions with biological systems can all have an impact on these pathways. The manner in which these differences add to the various consequences of NP-induced ROS generation is intriguing (Huang et al. 2010).

Dissolution

It is believed that the release of Cu^{2+} from the surface of CuO NPs played a significant role in their enhanced toxicity to organisms. This assumption stemmed from the understanding that Cu^{2+} ions can induce oxidative stress and other harmful effects (Gunawan et al. 2011). Large amounts of Cu^{2+} were able to leak out of CuO NPs in cell medium and suspension, creating massive amounts of OH via Fenton reactions and inflicting lipid, protein, and nucleic acid damage (Sharma and Hartwig 2013). Moreover, extracellular CuO NPs may release more Cu^{2+}, leading to oxidative damage and DNA damage, when they enter acidic organelles like lysosomes or come into contact with acidic substances like nucleic acid (Cuillel et al. 2014). Particle dissolving can undoubtedly be influenced by the characteristics of the particles, including their size, surface area, and chemical makeup, as well as by the surrounding conditions, including pH, temperature, and organic matter. It's fascinating how these factors interact to affect the behavior of particles in different environments. Cu^{2+} plays a vital role in maintaining cellular homeostasis. It helps regulate enzymatic reactions, cellular respiration, and antioxidant defense mechanisms. Keeping Cu^{2+} levels in check is crucial for proper cellular function (Galhardi et al. 2005). Once the Cu^{2+} produced by CuO NPs exceeds the physiological tolerance limit, the metal cation homeostatic systems will be compromised. Cu^{2+} may increase intracellular Ca^{2+} concentrations (Xia et al. 2008) and cause the endoplasmic reticulum plasma membrane to experience an increase in Ca^{2+} (Hoyal et al. 1998). Numerous cellular functions, including the synthesis of nitric oxide, depend critically on intracellular Ca^{2+} (Raddassi et al. 1994). Cell injury and disturbance of the mitochondria may result from these mechanisms (Xia et al. 2008).

Human Exposure

As the production and usage of nanoparticles grows, individuals are increasingly probable to be visible occupationally or through consumer items and the environment. Understanding the toxicity of CuO NPs is crucial for assessing their possible impact on human health and the environment (Fig. 4). Although human exposure to nanoparticles has existed throughout history, it grew considerably during the age of industrialization. NPs may readily pass through cell walls and films due to their tiny mass (Anreddy 2018). Occupational workers are the demographic cohort that has the greatest likelihood of being exposed to Cu NP or CuO NP by inhalation (Midander et al. 2009). For instance, industrial facilities that produce asphalt and rubber are known to release nanoparticles of copper throughout the manufacturing process. Nanomaterials can enter the human body by the mouth, nose, or throat, and then be absorbed through the skin or gastrointestinal tract. These are the three main routes of exposure that can potentially introduce nanomaterials into the body.

CuO NPs have a strong antibacterial effect, and their usage in a variety of biomedical applications has raised concerns regarding the possible harmful effects of CuO NPs on human skin. However, just a few researchers have looked at the

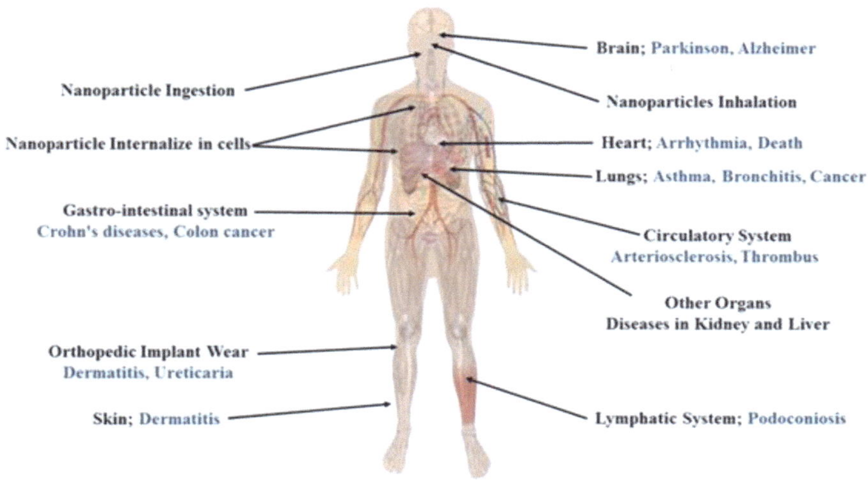

Fig. 4 Human exposure effects of nanoparticles on different organ and organ systems

toxicity of CuO NPs in humans, with a particular emphasis on the skin diffusion route. The skin's ability to absorb topically applied CuO NPs was examined using a Franz static diffusion cell model. Both intact and injured skins to determine static dissolution and colloidal behavior of commercial bare CuO NPs. As the receiving phase, used physiological solutions and for the donor phase, CuO NPs were mixed into artificial sweat and put on the skin's surface (0.11 mg cm2) for 24 h. CuO NPs absorption over undamaged skin was shown to be insignificant in the experiments. However, much increase in copper penetration in injured skin. Capacity of CuO NPs to enter the in vitro diffusion cell setup, pass through the outermost skin layers, release Cu^{2+} ions into the stratum corneum, and eventually reach the receptor fluid (Zanoni et al. 2019).

Nanoparticles have the ability to disperse more rapidly in the air compared to larger particles, and their small size and unique structure allow them to penetrate deeper into the respiratory system. This can potentially pose health risks, especially if there is prolonged exposure to high concentrations of nanoparticles. Because of the principal effects of particle nanomaterials, inhalation may be quite dangerous. The major focus is on the effects on the respiratory and cardio-vascular systems. When it comes to dermal entrance, healthy skin outperforms the respiratory system in terms of blockage. The capacity to hinder may be restricted by wounds, mechanical strain, or tiny nanoparticles (5–10 nm). The least harmful mode of introduction is by ingestion (Bonner 2010).

CuO NPs-Induced Toxicity: What's Next?

Nanotechnology has indeed revolutionized various industries, including food and cosmetics. The use of nanomaterials in these sectors has expanded rapidly, offering benefits such as improved manufacturing processes, enhanced packaging, extended shelf life, and increased bioavailability. Additionally, zinc oxide quantum dot nanoparticles have shown promising antimicrobial efficacy against food-borne microorganisms (Jin et al. 2009), and nanoparticles are currently utilized as food sensors to determine the quality and safety of food (Weiss et al. 2006). In essence, as compared to their bulk counterparts, nano-sized Cu particles boost copper release and inflict toxicity. The existing research shows that a variety of variables impact CuO NPs-induced toxicity According to the studies by Chang et al. (2005) and Liu et al. (2016), it has been suggested that exposure to CuO nanoparticles at certain dosages may lead to epigenetic changes in human small airway epithelial cells. These changes are associated with the generation of ROS and may pose potential health risks during long-term exposure. It emphasizes the importance of implementing proper safeguards in the use of CuO nanoparticles and products.

Conclusion

The toxicity profiling of NMs has gained popularity as a research issue in recent years all around the world. Natural nanomaterials have been present in the environment for a long period, and living organisms have developed mechanisms to mitigate their potential harm. Medical nanomaterials hold great promise in various areas of healthcare. They can help improve medication delivery, allowing for more targeted and efficient treatments. Antimicrobial coatings on medical equipment can reduce the risk of infections. Nanomaterials can also aid in decreasing inflammation and promoting faster tissue repair after surgeries. Additionally, they have the potential to enhance the diagnosis of circulating cancer cells, leading to earlier detection and more effective treatment. However, because of the absence of trustworthy toxicity data, the potential of NPs to harm human health remains a big worry. Current studies have indeed exposed that nanoparticles can have acute harmful effects on biological systems. It's crucial to recognize that both naturally occurring nanomaterials and those produced through human activities can potentially induce hazardous consequences in living beings. This review paper highlights the importance of extensive cytotoxicity testing for emerging nanoparticles, such as viral nanoparticles and nanozymes, to ensure safe application procedures and appropriate dose levels. To effectively detect and prevent the harmful effects of nanoparticles, extensive research in the field of nanotoxicology is necessary, along with robust regulatory regulations. Regarding the usage of nanomaterials, it is imperative to put human and environmental safety first. In a variety of applications, CuO NPs have become a significant class of nanomaterials with possible hazards to living things and the environment. The ecological destiny of CuO NPs must be thoroughly studied, and requirements for long-term uses in many industries must be developed. We recommend additional investigation into the genotoxicity exhibited by CuO NPs because it is an understudied topic. In order to comprehensively assess the risks associated with the utilization, dispersion, and environmental release of CuO NPs, it is imperative to conduct thorough studies within the existing risk assessment framework. These studies should focus on understanding the various uses of CuO NPs, as well as their dispersion patterns and potential impact on the environment. By conducting such research, we can gain valuable insights into the potential risks associated with CuO NPs and make informed decisions accordingly.

The toxicity of CuO NPs in relation to human exposure is indeed a topic of significant concern. It's crucial to thoroughly understand the potential risks associated with these NPs and take appropriate measures to ensure the safety of individuals who may come into contact with them. By conducting comprehensive research and implementing proper safeguards, we can better assess and mitigate any potential harmful effects. It's important to prioritize the well-being of individuals and address any concerns related to the toxicity of CuO NPs. Understanding the generation of ROS, the dilution process, and the potential dangers according to previous literature helps shed light on the risks associated with copper oxide nanoparticle exposure. Further research and future perspectives will undoubtedly contribute to ensuring the safety and well-being of individuals exposed to these nanoparticles.

Recommendations

We looked at the possible toxic effects of CuO NPs, the pros and cons of the method of preparation, mechanisms that cause toxicity, such as oxidative stress, generation of ROS and dilutions, and possible human exposure. However, there are a number of information gaps that need to be addressed in order to have a thorough understanding of the toxicity of CuO NPs and the processes that are associated with it. In the first place, the characterization of the circumstances under which the CuO NPs were exposed is insufficient. The toxic effects of CuO NPs are influenced by various factors such as particle dissolution, agglomeration, and precipitation. These processes depend on factors like particle size, surface properties, agglomerate size, medium chemistry, and exposure pathways. Therefore, future toxicity studies should describe exposure settings in detail, which is critical for confirming literature comparisons and interpreting toxicity results. The third factor that contributes to the induction of toxicity is the dissolution of Cu^{2+} from CuO NPs. Techniques need to be developed in order to discriminate between the toxicities caused by dissolved Cu^{2+} and those caused by CuO NPs. Fourthly, efforts to reduce the toxicity of CuO NPs and understand the underlying mechanisms to ensure their safe usage. Regulating the particle diameter, adjusting surface characteristics, choosing the appropriate exposure route, and controlling the release of Cu^{2+} from CuO NPs are important strategies. It's important to prioritize the evaluation of long-term impacts to ensure the responsible and sustainable use of CuO NPs.

References

Ahamed M, Siddiqui MA, Akhtar MJ et al (2010) Genotoxic potential of CuO NPs in human lung epithelial cells. Biochem Biophys Res Commun 396:578–583

Akhtar MJ, Kumar S, Alhadlaq HA et al (2016) Dose-dependent genotoxicity of CuO NPs stimulated by reactive oxygen species in human lung epithelial cells. Toxicol Ind Health 32(5):809–821

Anchan S, Pai S, Sridevi H et al (2019) Biogenic synthesis of ferric oxide nanoparticles using the leaf extract of *Peltophorum pterocarpum* and their catalytic dye degradation potential. Biocatal Agric Biotechnol 20:101251

Anreddy RNR (2018) CuO NPs induces oxidative stress and liver toxicity in rats following oral exposure. Toxicol Rep 5:903–904

Azam A, Ahmed AS, Oves M et al (2012) Size-dependent antimicrobial properties of CuO nanoparticles against gram-positive and-negative bacterial strains. Int J Nanomedicine 7: 3527–3535

Bahadar H, Maqbool F, Niaz K et al (2016) Toxicity of nanoparticles and an overview of current experimental models. Iran Biomed J 20(1):1–11

Battez AH, González R, Viesca JL et al (2008) CuO, ZrO2 and ZnO nanoparticles as antiwear additive in oil lubricants. Wear 265(3–4):422–428

Battez AH, Viesca JL, González R et al (2010) Friction reduction properties of a CuO nanolubricant used as lubricant for a NiCrBSi coating. Wear 268(1–2):325–328

Bhanushali S, Ghosh P, Ganesh A et al (2015) 1D copper nanostructures: progress, challenges and opportunities. Small 11(11):1232–1252

Bonner JC (2010) Nanoparticles as a potential cause of pleural and interstitial lung disease. Proc Am Thorac Soc 7(2):138–141

Bulcke F, Dringen R, Scheiber IF et al (2017) Neurotoxicity of copper. Adv Neurobiol 18:313–343

Chang H, Jwo CS, Lo CH et al (2005) Rheology of CuO nanoparticle suspension prepared by ASNSS. Rev Adv Mater Sci 10(2):128–132

Cuillel M, Chevallet M, Charbonnier P et al (2014) Interference of CuO nanoparticles with metal homeostasis in hepatocytes under sub-toxic conditions. Nanoscale 6(3):1707–1715

Dar MA, Kim YS, Kim WB et al (2008) Structural and magnetic properties of CuO nanoneedles synthesized by hydrothermal method. App Surf Sci 254(22):7477–7481

Dipankar C, Murugan S (2012) The green synthesis, characterization and evaluation of the biological activities of silver nanoparticles synthesized from Iresine herbstii leaf aqueous extracts. Colloids Surf B Biointerfaces 98:112–119

Dudev T, Lim C (2008) Metal binding affinity and selectivity in metalloproteins: insights from computational studies. Annu Rev Biophys 37:97–116

Ema M, Hougaard KS, Kishimoto A et al (2016) Reproductive and developmental toxicity of carbon-based nanomaterials: a literature review. Nanotoxicology 10(4):391–412

EPA U (2017) Technical fact sheet–nanomaterials. https://nepis.epa.gov/ Accessed on 10 Jan 2024

Fahmy B, Cormier SA (2009) CuO NPs induce oxidative stress and cytotoxicity in airway epithelial cells. Toxicol In Vitro 23(7):1365–1371

Fard JK, Jafari S, Eghbal MA et al (2015) A review of molecular mechanisms involved in toxicity of nanoparticles. Adv Pharm Bull 5(4):447–454

Galhardi CM, Diniz YS, Faine LA et al (2004) Toxicity of copper intake: lipid profile, oxidative stress and susceptibility to renal dysfunction. Food Chem Toxicol 42(12):2053–2060

Galhardi CM, Diniz YS, Rodrigues HG et al (2005) Beneficial effects of dietary copper supplementation on serum lipids and antioxidant defenses in rats. Ann Nutr Metab 49(5):283–288

Gomes T, Pereira CG, Cardoso C et al (2012) Accumulation and toxicity of CuO NPs in the digestive gland of *Mytilus galloprovincialis*. Aquat Toxicol 118-119:72–79

Griffitt RJ, Weil R, Hyndman KA et al (2007) Exposure to copper nanoparticles causes gill injury and acute lethality in zebrafish (*Danio rerio*). Environ Sci Technol 41(23):8178–8186

Grosell M, Blanchard J, Brix KV et al (2007) Physiology is pivotal for interactions between salinity and acute copper toxicity to fish and invertebrates. Aquat Toxicol 84(2):162–172

Gunawan C, Teoh WY, Marquis CP et al (2011) Cytotoxic origin of copper (II) oxide nanoparticles: comparative studies with micron-sized particles, leachate, and metal salts. ACS Nano 5(9): 7214–7225

He X, Aker WG, Fu PP et al (2015) Toxicity of engineered metal oxide nanomaterials mediated by nano–bio–eco–interactions: a review and perspective. Environ Sci Nano 2(6):564–582

Hoyal CR, Giron-Calle J, Forman HJ et al (1998) The alveolar macrophage as a model of calcium signaling in oxidative stress. J Toxicol Environ Health B Crit Rev 1(2):117–134

Huang YW, Wu CH, Aronstam RS et al (2010) Toxicity of transition metal oxide nanoparticles: recent insights from in vitro studies. Materials 3(10):4842–4859

Jammi S, Sakthivel S, Rout L et al (2009) CuO nanoparticles catalyzed C− N, C− O, and C− S cross-coupling reactions: scope and mechanism. J Org Chem 74(5):1971–1976

Jeevanandam J, Barhoum A, Chan YS et al (2018) Review on nanoparticles and nanostructured materials: history, sources, toxicity and regulations. Beilstein J Nanotechnol 9(1):1050–1104

Jeffree AI, Karman S, Ibrahim S et al (2019) Biosensors approach for lung cancer diagnosis-a review. Lect Notes Mech Engg 2019:425–435

Jin T, Sun D, Su JY et al (2009) Antimicrobial efficacy of zinc oxide quantum dots against Listeria monocytogenes, *Salmonella enteritidis*, and *Escherichia coli* O157: H7. J Food Sci 74(1):M46–M52

Karlsson HL, Cronholm P, Gustafsson J et al (2008) CuO NPs are highly toxic: a comparison between metal oxide nanoparticles and carbon nanotubes. Chem Res Toxicol 21(9):1726–1732

Letchumanan D, Sok SP, Ibrahim S et al (2021) Plant-based biosynthesis of copper/CuO NPs: an update on their applications in biomedicine, mechanisms, and toxicity. Biomolecules 11(4):564

Li J, Quabius ES, Bonga SW et al (1998) Effects of water-borne copper on branchial chloride cells and Na+/K+-ATPase activities in Mozambique tilapia (*Oreochromis mossambicus*). Aquat Toxicol 43(1):1–11

Linder MC, Hazegh-Azam M (1996) Copper biochemistry and molecular biology. Am J Clin Nutr 63(5):797S–811S

Liu Y, Liang J, Wang Q et al (2016) Copper nanoclusters trigger muscle cell apoptosis and atrophy in vitro and in vivo. J Appl Toxicol 36(3):454–463

McWilliams A (2016) The maturing nanotechnology market: products and applications. PR Newswire. https://www.prnewswire.com/news-releases/the-maturing-nanotechnology-market-products-and-applications-300369337.html Accessed on 10 Feb 2024

Mendhulkar VD, Yadav A (2017) Anticancer activity of camellia Sinensis mediated copper nanoparticles against HT-29, MCF-7, and MOLT-4 human cancer cell lines. Asian J Pharm Clin Res 10:82–88

Midander K, Cronholm P, Karlsson HL et al (2009) Surface characteristics, copper release, and toxicity of nano-and micrometer-sized copper and copper (II) oxide particles: a cross-disciplinary study. Small 5(3):389–399

Mohammed WM, Mubarak TH, Al-Haddad RM et al (2017) Effect of CuO nanoparticles on antimicrobial activity prepared by sol-gel method. Int J Appl Engg Res 13:10559–10562

Namburu PK, Kulkarni DP, Misra D et al (2007) Viscosity of CuO NPs dispersed in ethylene glycol and water mixture. Exp Thermal Fluid Sci 32(2):397–402

Naz S, Gul A, Zia M, Javed R et al (2023) Synthesis, biomedical applications, and toxicity of CuO nanoparticles. Appl Microbiol Biotechnol 107(4):1039–1061

Patel KD, Singh RK, Kim HW et al (2019) Carbon-based nanomaterials as an emerging platform for theranostics. Mater Horiz 6(3):434–469

Pelgrom SMGJ, Lock RAC, Balm PHM et al (1995) Integrated physiological response of tilapia, Oreochromis mossambicus, to sublethal copper exposure. Aquat Toxicol 32(4):303–320

Pendashteh A, Mousavi MF, Rahmanifar MS et al (2013) Fabrication of anchored CuO NPs on graphene oxide nanosheets via an electrostatic coprecipitation and its application as supercapacitor. Electrochim Acta 88:347–357

Pettibone JM, Adamcakova-Dodd A, Thorne PS et al (2008) Inflammatory response of mice following inhalation exposure to iron and copper nanoparticles. Nanotoxicology 2(4):189–204

Raddassi K, Berthon B, Petit JF et al (1994) Role of calcium in the activation of mouse peritoneal macrophages: induction of NO synthase by calcium ionophores and thapsigargin. Cell Immunol 153(2):443–455

Rajput VD, Minkina T, Suskova S et al (2018) Effects of copper nanoparticles (CuO NPs) on crop plants: a mini review. BioNano Sci 8(1):36–42

Rasheed T, Hussain T, Anwar MT et al (2021) Hybrid nanofluids as renewable and sustainable colloidal suspensions for potential photovoltaic/thermal and solar energy applications. Front Chem 9:737033

Ren G, Hu D, Cheng EW et al (2009) Characterisation of CuO NPs for antimicrobial applications. Int J Antimicrob Agents 33(6):587–590

Roy A, Bhattacharya J (2015) Nanotechnology in industrial wastewater treatment. IWA Publishing, London

Salata OV (2004) Applications of nanoparticles in biology and medicine. J Nanobiotechnol 2(1): 1–6

Sharma A, Hartwig JF (2013) Enantioselective functionalization of allylic C–H bonds following a strategy of functionalization and diversification. J Am Chem Soc 135(47):17983–17989

Shi J, Abid AD, Kennedy IM et al (2011) To duckweeds (Landoltia punctata), nanoparticulate copper oxide is more inhibitory than the soluble copper in the bulk solution. Environ Pollut 159(5):1277–1282

Shvedova AA, Pietroiusti A, Fadeel B et al (2012) Mechanisms of carbon nanotube-induced toxicity: focus on oxidative stress. Toxicol Appl Pharmacol 261(2):121–133

Srikanth K, Pereira E, Duarte AC et al (2016) Evaluation of cytotoxicity, morphological alterations and oxidative stress in Chinook salmon cells exposed to CuO NPs. Protoplasma 253(3): 873–884

Sun Y, Zhang G, He Z et al (2016) Effects of CuO NPs on developing zebrafish embryos and larvae. Int J Nanomedicine 11:905–918

Touyz RM (2005) Molecular and cellular mechanisms in vascular injury in hypertension: role of angiotensin II–editorial review. Curr Opin Nephrol Hypertens 14(2):125–131

Wang Y, Andrukhov O, Rausch-Fan X (2017) Oxidative stress and antioxidant system in periodontitis. Front Physiol 8:910

Weiss J, Takhistov P, McClements DJ et al (2006) Functional materials in food nanotechnology. J Food Sci 71(9):R107–R116

Wongrakpanich A, Mudunkotuwa IA, Geary SM et al (2016) Size-dependent cytotoxicity of CuO NPs in lung epithelial cells. Environ Sci Nano 3(2):365–374

Wu T, Tang M (2018) Review of the effects of manufactured nanoparticles on mammalian target organs. J Appl Toxicol 38(1):25–40

Xia T, Kovochich M, Liong M et al (2008) Comparison of the mechanism of toxicity of zinc oxide and cerium oxide nanoparticles based on dissolution and oxidative stress properties. ACS Nano 2(10):2121–2134

Xu JF, Ji W, Shen ZX et al (1999) Preparation of CuO and characterization nanocrystals. J Solid State Chem 147(2):516–519

Yang H, Liu C, Yang D et al (2009) Comparative study of cytotoxicity, oxidative stress and genotoxicity induced by four typical nanomaterials: the role of particle size, shape and composition. J Appl Toxicol 29(1):69–78

Yin M, Wu CK, Lou Y et al (2005) Copper oxide nanocrystals. J Am Chem Soc 127(26): 9506–9511

Yokohira M, Hashimoto N, Yamakawa K et al (2009) Lung carcinogenic bioassay of CuO and TiO2 nanoparticles with intratracheal instillation using F344 male rats. J Toxicol Pathol 22(1): 71–78

Zanoni I, Crosera M, Ortelli S et al (2019) CuO nanoparticle penetration through intact and damaged human skin. New J Chem 43(43):17033–17039

Zhang DW, Yi TH, Chen CH et al (2005) Cu nanoparticles derived from CuO electrodes in lithium cells. Nanotechnology 16(10):2338

Zhou K, Wang R, Xu B et al (2006) Synthesis, characterization and catalytic properties of CuO nanocrystals with various shapes. Nanotechnology 17(15):3939

Glyphosate Herbicide: An Omnipresent Environmental Toxicant and Its Potential Human Health Impacts

Reenu Singh, Sabreen Bashir, and Anand Mohan

Abstract

The increasing application of agrochemicals for the purpose of weed control has emerged as a defining characteristic of contemporary industrial agriculture over the course of last few decades. The widespread application of chemical herbicides has been a source of concern in the agricultural sector all over the world, which has led to the over-expression of worry for the environment and human health. N-phosphomethyl glycine (glyphosate) is the herbicide that is most successful, celebrated, and widely utilized to date. Nevertheless, its value is currently in jeopardy due to the advent of weed species that are resistant to glyphosate and the harmful effects that it exhibits on the environment and human health. This herbicide has also developed resistance in a number of weed species, which has resulted in a decrease in crop production. The excessive application of glyphosate in agricultural regions and the runoff of glyphosate from the surface into groundwater sources are the primary causes of the existence of glyphosate intermediates in many systems, including soil ecosystems and water bodies. Given its widespread presence, glyphosate poses a threat to various non-target species, including humans and animals, within the environment.

Keywords

Agrochemicals · Glyphosate · Toxicity · Herbicides · Environment · Human health

R. Singh · S. Bashir · A. Mohan (✉)
School of Bioengineering and Biosciences, Lovely Professional University, Phagwara, Punjab, India

S. Raisuddin et al. (eds.), *Environmental Toxicants and Lifestyle Diseases*,
https://doi.org/10.1007/978-981-95-0575-3_7

Introduction

Agrochemicals have been extensively employed to augment crop yield in agricultural regions. Agrochemicals are simultaneously used for the purpose of insect and rodent control, as well as for the management of microbiological infections. A diverse variety of insecticides, herbicides, and pesticides are utilized in contemporary land cultivation methods. The application of pesticides in agricultural practices first began as a means to enhance food production for the benefit of humanity and society. However, the relentless pursuit of higher crop yields driven by increasing human demands and greed has resulted in the excessive and indiscriminate use of these pesticides (Taylor et al. 2007). The identification of *N-(phosphonomethyl) glycine* was a key breakthrough in the world of agriculture (Gill et al. 2018). Glyphosate is widely used as an herbicide worldwide (Valavanidis 2018), with an estimated global demand of around 500,000 tonnes per year (Sharma et al. 2019). Glyphosate is a versatile synthetic herbicide that is enormously used in both agricultural and non-agricultural sectors to eliminate unwanted plants (Benbrook 2016). In 1950, Henri Martin, a researcher from a Swiss pharmaceutical business, was one of the pioneers in synthesizing *N-(phosphonomethyl) glycine* (Finley and Duke 2020). Monsanto conducted herbicidal activity tests of glyphosate and assessed its effectiveness against perennial weeds (Nandula 2010). In 1970, Monsanto first synthesized and tested *N-phosphonomethyl glycine,* which was later popularly referred to as glyphosate (krimsky 2022). Monsanto claimed and received the patents that resulted from the re-synthesis and successfully introduced glyphosate to the market as an herbicide, marketing it under the brand name Roundup (Mendelson 1998). Upon its introduction to the market, glyphosate quickly established itself as a leading pesticide due to several factors, including its unique mechanism of action, ability to resist contaminants from other herbicides, and relatively swift translocation within plants (Shaner 2006). The adoption of glyphosate in agricultural practices was limited during the late 1980s and early 1990s due to various other herbicides in the market that provided viable alternatives and posed competition. However, by 2015, it had emerged as the most actively consumed herbicide globally, with a 15-fold surge in worldwide consumption from 1990 to 2014 (Benbrook 2016). Glyphosate is the only ideal herbicide that hinders aromatic amino acid synthesis by blocking 5-enolpyruvylshikimate-3-phosphate synthase (EPSPS) which is a popular and essential enzyme associated with the shikimate pathway, which involves the synthesis of shikimic acid (Haslam 2014). Plants, fungi, and bacteria possess EPSPS, whereas vertebrates lack this enzyme (Kishore and Shah 1988). Glyphosate has emerged as a widely recognized efficient agricultural tool and an environmentally friendly herbicide applied to farming businesses for more than 10 years. Initially, glyphosate was proven to validate minimal toxicity to humans, with few evidence of causing impairment of genes or being cancer-inducing in mammals (De Roos et al. 2005). In 1974, the United States after its introduction, the usage of the herbicides, glyphosate, and its chemical derivatives significantly increased, driven by the idea that it does not have any adverse effects on mammals. However, the abrupt usage and inappropriate dosage of this compound have caused concern due to its potential negative effects on both the environment and human health. Significant evidence indicates that glyphosate-based herbicides may harm health, requiring additional scrutiny (Benachour and Séralini 2009). In a 2015 evaluation, the World Health Organisation (WHO) classified glyphosate as a potential toxin carcinogen. Thus, it is urgent to thoroughly examine the potential hazards linked to glyphosate, acknowledge the significance of community concerns, and effectively weigh apprehension against the advantages in agriculture (Gandhi et al. 2021). The chapter aims to analyze and assess the potential repercussions associated with the herbicide emphasizing its impacts on the natural ecosystem and human well-being, leading us towards gaining a deeper understanding of the challenges concerning the consumption of glyphosate and guide us towards making better conscious decisions regarding its further usage in near future.

Glyphosate: Chemical Properties

Glyphosate acts as a zwitterion (Soares et al. 2021) that possesses phosphonate, carboxylate, and amine functionalities, as depicted in Fig. 1. This characteristic pertains to its chemical structure, which favors its capacity to form complexes with both trivalent and quadrivalent metallic elements (Singh and Kumar 2017).

Glyphosate possesses a number of physical and chemical properties as listed in Table 1, including higher adsorption, substantial water solubility, and integration with other types of chemicals. These properties are a result of the covalent link that exists between the carbon atoms and the phosphorus atoms, which is a feature associated with such synthetic chemical substances (Singh and Kumar 2017).

Glyphosate: X-Ray Crystallography Structure

Glyphosate was discovered to exhibit a monoclinic structure arrangement using a P21/c space group, as elucidated by a technique popularly known as X-Ray crystallography, illustrated in Table 2. The subsequent measurements represent the sizes of the cells in the crystal: The values are as follows: $a = 8.682(5)$ Å, $b = 7.973(8)$ Å, $c = 9.875(5)$ Å, $\beta = 105.74(4)°$, $V = 675.9(8)$ Å3 and $Z = 4$. In glyphosate structure, bond lengths (Å) are documented as given: C-P bond has a length of 1.823 Å, and C-H and N-H bonds have lengths ranging from 0.914 Å to 1.118 Å. The O-H bonds in phosphono groups (−P (=O) OH2) and carboxyl groups (COOH) have lengths of 0.711 Å and 0.743 Å, respectively. In glyphosate structure, bond lengths of the two oxygen (O) atoms in the phosphonate group specifically P-O1 = 1.500 Å and

Fig. 1 Chemical structure of glyphosate

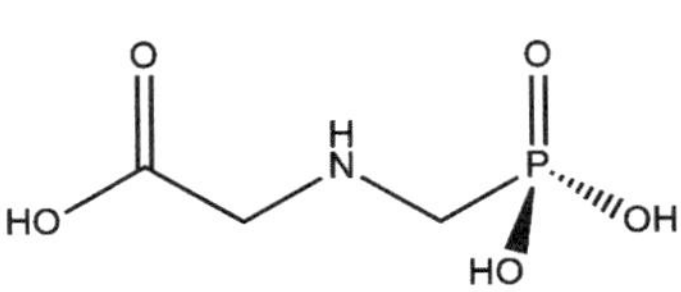

Table 1 Glyphosate chemical and physical properties

S. no.	Properties	Description
1.	Chemical formula	$C_3H_8NO_5P$
2.	CAS registry number	1071-83-6
3.	Chemical name	*N*-(phosphonomethyl) glycine
4.	Appearance	White crystalline solid
5.	Molecular mass	169 g/mol
6.	Density	1.74 g/mL
7.	Stability	32 days at 25 °C; pH 5,7,9
8.	pH (1% solution)	4.4 (acidic)
9.	Melting point	189.5 °C
10.	Amphoteric nature	Complex, polyprotic molecule
11.	Solubility	Soluble in water
12.	Solubility in other solvents	Soluble in acetone, ethanol, methanol
13.	Oxidizing properties	Non-oxidizing
14.	Corrosivity	Non-corrosive
15.	Vapor pressure	Negligible
16.	Dissociation constant	Four main ionisable functional groups: pKa1 = 0.78 (first phosphonic group), pKa2 = 2.29 (carboxylate group); pKa3 = 5.96 (second phosphonic group), pKa4 = 10.13 (amine group)
17.	Odor	Odorless

P-O2 = 1.501 Å, indicate these are in a state of compatibility with their counterparts. Conversely, the third P-O bonds with a measurement of bond length 1.576 Å possess a singular bond. The molecules of glyphosate are interconnected through hydrogen bonds (Knuuttila et al. 1979).

Glyphosate: Mode of Action

Glyphosate destroys plants and microorganisms via a mode of action that primarily involves its inhibition by binding and blocking the crucial enzyme EPSPS (Muñoz et al. 2021) as illustrated in Fig. 2. Basically, EPSPS promotes the synthesis of chorismate, an intermediate for aromatic amino acids biosynthesis via engaging with the shikimate (shikimic acid) cascade (Agostini et al. 2020). EPSPS enzyme is usually crucial for the biosynthesis of aromatic amino acids such as tryptophan, tyrosine, phenylalanine and in addition to enormous vital plant metabolites in higher plants, algae, bacteria, and fungi (Maeda and Dudareva 2012). Consequently, the inhibitory activity of glyphosate hampers the growth of several plants (Williams et al. 2000), and as a result, insufficient levels of the enzyme lead to a deficiency of

Table 2 Glyphosate X-Ray crystallography properties

Category	Property	Value
Structure	Crystal system	Monoclinic
	Space group	P21/c
Measurement method	Technique	X-ray crystallography
Cell dimensions	A	8.682(5) Å
	B	7.973(8) Å
	C	9.875(5) Å
	B	105.74(4) Å
	Volume (V)	675.9(8) $Å^3$
	Z	4
Bond lengths (Å)	C-P	1.823 Å
	C-H	0.914–1.118 Å
	N-H	0.914–1.118 Å
	O-H (phosphono group)	0.711 Å
	O-H (carboxyl group)	0.743 Å
	P-O1	1.500 Å
	P-O2	1.501 Å
	P-O (third bond)	1.576 Å
Intermolecular	Molecular Interconnection	Hydrogen bonds

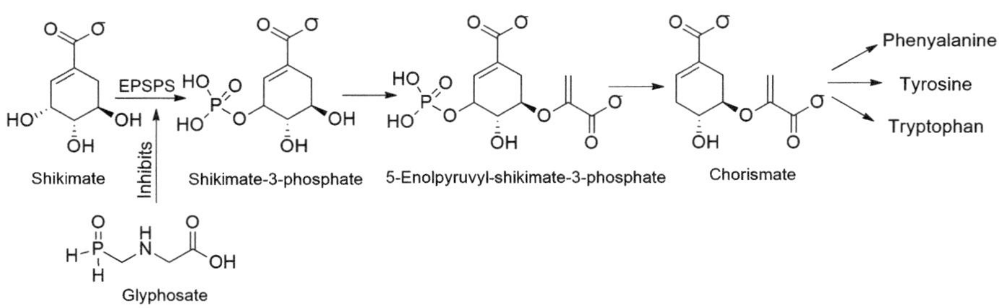

Fig. 2 Mode of action of glyphosate

aromatic amino acids, which consequently affects various metabolic processes in plants and gradually leads to plant's death (Tu et al. 2001).

Glyphosate: Global Consumption

Glyphosate controls several weeds, including broadleaf weeds and crop-side grasses. After the weeds emerge, it is effective and low toxic to mammals (Garthwaite et al. 2008). The United States of America is among the biggest and fastest-growing end-user of glyphosate (Myers et al. 2016) with farmers using 6–8 million pounds in 1987 (Bonny 2008). In 2007, the EPA found that glyphosate use had climbed to 81.6–83.9 million kilograms over 20 years (Heap 2014). Glyphosate is used in nearly 35–39% of Denmark and Germany's agriculture depends on its application (Steinmann et al. 2012). Argentina uses approximately 180–200 million liters of glyphosate annually (Nedelkoska and Low 2004). The United Kingdom utilizes glyphosate on crops for consumption and commercial fruit farms. Glyphosate is applied to 50–60% of sunflower crops in Hungary, Romania, and France each year (Duke 2018). Already known from data that agriculture plays a vital role in India's economy, glyphosate is extensively used in the country because agriculture serves as the primary income source for the population comprising 70%. A survey found that India used 35 kilo tonnes of glyphosate in 2015, and it will increase upto 75 kilo-tonnes by 2024. Glyphosate-resistant crops have forced farmers worldwide to use more of this pesticide. (Woodburn 2000).

Based on an analysis of worldwide glyphosate consumption statistics from various nations, it has been determined that the glyphosate market by 2022 year within the United States is estimated to achieve a value of worth United states dollar (USD) 2.4 billion. China by 2030 year, the second-largest economy globally is projected to achieve a market size of USD 2.2 billion, with a compound annual growth rate (CAGR) of 5.7% from 2022 to 2030. Japan and Canada are anticipated for experiencing growth rates of 3.2% and 3.9%, respectively, from 2022 to 2030, making them significant worldwide markets. The United Kingdom is estimated to acknowledge a CAGR of 2.8% within Europe. By 2030, the Asia-Pacific market region is forecasted to have an amount of USD 1.5 billion, with prominent contributions from countries including India, Australia, and South Korea (Report Linker 2023).

Glyphosate Resistance: Weeds and Plants

Plants belonging to higher taxonomic groups have susceptibility to glyphosate, although they do not possess inherent resistance to the compound. Certain plants exhibit varying degrees of susceptibility in comparison with others based on their specific morphological and biochemical traits. Glyphosate has been known to demonstrate increased resistance against unwanted plants such as bermudagrass (*Cynodon dactylon* (L.) and field bindweed (*Convolvulus arvensis* L) (Westwood and Weller 1997). The widespread usage of glyphosate globally is leading to the development of resistance in invasive species and plants. The initial instance of resistance was observed in Australian fields of annual ryegrass (*Lolium rigidum*) (Pratley et al. 1999). Cultivation of glyphosate-resistant crops is prevalent in major sections of the United States, Argentina, and Brazil. Additionally, this development has contributed to the discovery of a distinct category of glyphosate-resistant weeds. Ragweed (*Ambrosia artemisiifolia*), horseweed (*Conyza*) resistant to glyphosate, and lamb's quarters (*Chenopodium album*) are abundantly and frequently observed across the United States (Powles and Preston 2006).

Glyphosate Resistance: Cropping Systems

In 1996, the introduction of glyphosate-resistant crops to the market was an immediate consequence of the extensive glyphosate application in agriculture (Johnson et al. 2009). Glyphosate-resistant crops, such as soybean, cotton, and corn, are cultivated worldwide (Duke et al. 2012) and CP4 genes from a species of *A. tumefaciens* are usually employed to produce glyphosate-resistant EPSPS by transcription. The CP4 genes and a transcription factor are introduced within the genetic makeup of crops, which tend to display increased levels of glyphosate resistance. Maize genes were subjected to alteration to create a variant of EPSPS that is capable to withstanding the effects of glyphosate (Padgette et al. 2018). Maize genomes were subjected to modification in order to create glyphosate-resistant EPSPS in maize (Vande Berg et al. 2008). Therefore, the adoption of genetically engineered synthetic crops in agriculture has resulted in a surplus boost in consumption of glyphosate, consequently leading to the degradation of cropping systems. Controlling the growth of weeds has been challenging due to their increased resistance to the herbicide (Martins-Gomes et al. 2022). Over time, the usage of glyphosate has led to its widespread presence in waterways and agricultural systems, making it a prominent component. A high use of glyphosate leads to presence of a derivative called aminomethylphosphonic acid (AMPA) in numerous agricultural crops at the period of harvesting, as well as in food that is processed. The abuse of glyphosate has led to the proliferation of resilient crops, which in turn has had a significant detrimental effect on a wide spectrum of different species in the ecosystem (Dill 2005).

Glyphosate Impact on Animals

Glyphosate has been demonstrated to negatively affect an extensive range of organisms across various groups and classes in the animal kingdom, impacting both single-celled and multi-celled animals in aquatic and soil habitats as depicted below in Table 3.

Impact of Glyphosate on Humans

Over the past 40 years, researchers have widely used human cell lines to study the toxic effects of herbicides. Both healthy and cancerous cell lines have proven to be valuable tools in this field.

Cells Several studies have revealed that exposure to glyphosate (GLY) or glyphosate-based herbicides (GBH) can disrupt various cellular processes in these human cell models as illustrated in Fig. 3.

Blood cells Studies on the genotoxic effects of glyphosate and glyphosate-based herbicides (GBHs) on human cell lines have yielded mixed results over the past

Table 3 Various adverse effects of glyphosate across different animal groups

Animal groups	Effects	References
Unicellular	Impairs photosynthesis in *Euglena gracilis* Alters soil microbiomes Inhibits mycorrhizal fungi Affects periphyton photosynthesis	Richardson et al. (1979) Newman et al. (2016) Estok et al. (1989)
Multicellular: Algae	Significant structural changes Reduced photosynthetic activity	Oliveira et al. (2016)
Invertebrates: Nemathelminthes	Reduced infective ability in nematodes	Achiorno et al. (2008)
Annelids	Adverse effects on earthworms	Correia and Moreira (2010)
Crustaceans	Decreased growth, fertility, and increased abortion rates in *Daphnia magna*	Cuhra et al. (2013)
Insects	Reduced acetylcholinesterase activity in honeybees	Boily et al. (2013)
Molluscs	Accumulation in snails; no growth/survival effects	Tate et al. (1997)
Echinoderms	Disrupted cell cycles and development in sea urchins	Marc et al. (2004)
Fish	Varying sensitivity; higher toxicity at high pH and temperatures	Folmar et al. (1979)
Amphibians	Decreased glycogen and lipid levels in tadpoles	Dornelles and Oliveira (2016)
Reptiles	DNA damage in tegu lizards	Schaumburg et al. (2016)
Birds	Disrupted hormone synthesis and reproductive organs in male drakes	Oliveira et al. (2007)
Mammals	Reduced sperm quality and testosterone levels in rats; oxidative stress	Dallegrave et al. (2007)

several decades. The first study examining Roundup's® genotoxicity in human lymphocytes was done by Vigfusson and Vyse (1980) and found it to be weakly mutagenic at high concentrations. Subsequent research on lymphocytes has produced conflicting data. Some studies (Mladinic et al. 2009; Mañas et al. 2009) found no significant genotoxic effects at concentrations relevant to human exposure. However, others (Santovito et al. 2018) reported dose-dependent DNA damage and increased chromosomal aberrations, suggesting potential genotoxicity.

Research performed with peripheral blood mononuclear cells (PBMCs) has shown more consistent toxicological effects. Martínez et al. (2007) observed toxic effects from glyphosate with commercial formulations appearing more toxic due to adjuvants. Kwiatkowska et al. (2016) reported declined cell viability, decreased ATP levels, and improved cell granularity, though only at very high glyphosate concentrations. A follow-up study (Kwiatkowska et al. 2017) discovered that

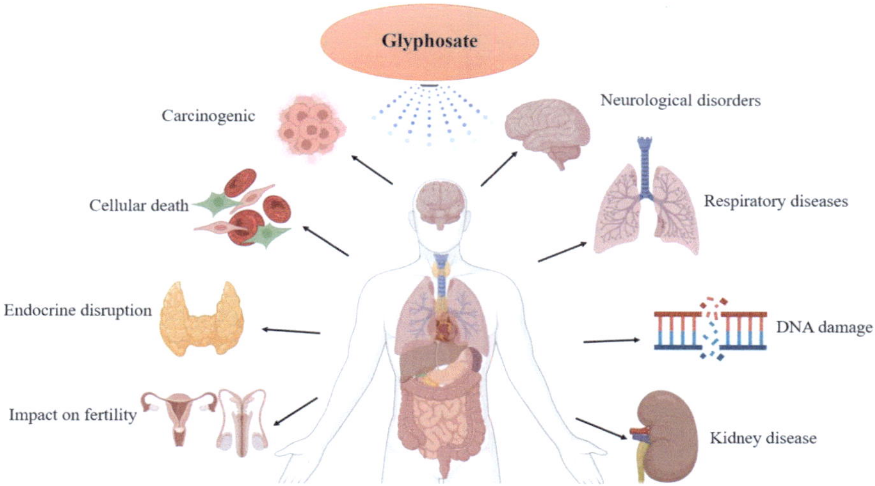

Fig. 3 Impact of glyphosate on humans

glyphosate exposure changed DNA methylation patterns and caused DNA damage in PBMCs.

Regarding human erythrocytes, Pieniężek et al. (2004) reported increased methemoglobin levels and lipid peroxidation post glyphosate exposure. Kwiatkowska et al. (2014) supported that glyphosate shows only minor hemolytic effects, and primarily at high concentrations.

Epithelial cells Studies on human skin keratinocytes (HaCaT) and epithelial cells have provided noteworthy insights into the cellular effects of glyphosate and glyphosate-based herbicides. These studies constantly exhibit that GLY and GBHs can induce oxidative stress and disrupt cellular processes, often with more pronounced impacts observed in commercial formulations compared to GLY alone.

Gehin et al. (2006) reported that GLY caused abnormalities in cellular antioxidant status, including dose-dependent glutathione (GSH) depletion, enzymatic disorders, and increased lipid peroxidation. Heu et al. (2012) observed that glyphosate exposure led to a stiffening of cell membrane and cytoskeleton alterations at low cytotoxic concentrations. At higher concentrations (IC50, ~30 mM), cells exhibited membrane damage and entered late apoptotic or necrotic states. The research also showed that glyphosate-induced disruption of mitochondrial membrane potential was linked to apoptosis in keratinocytes.

A study revealed that exposing HaCaT cells to glyphosate (0.1 mM for 72 h) led to enhanced cell proliferation, decreased calcium levels, and escalated the production of reactive oxygen species (ROS). The researchers proposed that glyphosate stimulates proliferation by altering the equilibrium between calcium levels and oxidative stress, possibly suppressing mitochondrial apoptotic signaling pathways

(George and Shukla 2013). Another study reported on the impacts of GLY formulations on human epithelial type 2 (HEp-2) cells (Coalova et al. 2014). The researchers noted a reduction in cell viability that related with both concentration and duration of exposure. Notably, combinations of glyphosate formulations exhibited greater toxicity compared to their individual components. Both the individual formulations and the mixtures led to a significant elevation in ROS levels.

These studies reveal that GLY and GBHs can induce oxidative stress, alter cell membrane properties, disrupt cellular antioxidant systems and affect cell proliferation and death pathways in human skin and epithelial cells. Importantly, commercial GBH formulations often demonstrate higher toxicity than GLY alone, suggesting that additional components in these formulations may contribute to their overall toxicity (Coalova et al. 2014).

Other cells Mesnage et al. (2013) investigated the impacts of GLY and GBHs on human embryonic kidney cells (HEK 293), observing dose-dependent cytotoxicity for both substances. Their study evaluated membrane degradation, mitochondrial activity, and caspase 3/7 activity, providing a comprehensive view of cellular damage. Benachour and Séralini (2009) expanded on this work, examining four GBHs on both HEK 293 cells and human umbilical vein endothelial cells (HUVECs). The study revealed that all tested formulations induced complete cell death within 24 h, even at concentrations lower than those recommended for agricultural use, and comparable to residual levels found in food. The observed effects included inhibition of the mitochondrial succinate dehydrogenase activity, necrosis, and apoptosis, with the severity of effects correlating with GLY exposure levels and adjuvant composition.

Martinez and Al-Ahmad (2019) provided valuable insights into glyphosate's potential neurological impacts. Using an in vitro blood–brain barrier (BBB) model based on induced pluripotent stem cells (iPSCs), they observed increased BBB permeability at glyphosate concentrations of even 1 μM. Elevated glyphosate concentrations (100 μM) were found to disrupt neuronal cell metabolism and impair glucose uptake in brain microvascular endothelial cells. These findings indicate that high-level glyphosate exposure may potentially result in neurological harm.

Other studies have reported effects on different human cell types. A study observed impaired sperm motility after exposure to 0.36 mg/L of glyphosate, although sperm DNA integrity remained unchanged (Anifandis et al. 2018). These studies collectively demonstrate that GLY and GBHs can induce a range of detrimental effects on human cells, including cytotoxicity, mitochondrial dysfunction, increased apoptosis, and potential neurological impacts.

Carcinogenicity of Glyphosate

To investigate the potential carcinogenic effects of glyphosate exposure, numerous studies have employed various tumor cell lines as models. This approach allows researchers to examine how glyphosate might influence cancer development, progression, or behavior at the cellular level. Using established cancer cell lines, scientists can gain insights into the molecular mechanisms through which glyphosate could potentially contribute to carcinogenesis or affect existing tumors. These studies form a key part of the broader effort to assess the long-term health risks associated with glyphosate exposure and its possible role in cancer etiology.

Breast and hepatic tumor cells Glyphosate and glyphosate-based herbicides have been investigated for their potential as endocrine-disrupting chemicals (EDCs), particularly concerning estrogenic effects and cancer cell proliferation (Mnif et al. 2011). Several studies have provided evidence supporting this hypothesis. Thongprakaisang et al. (2013) reported that glyphosate exhibits estrogenic and/or antiestrogenic effects in T47D breast cancer cells, inducing proliferation and increasing estrogen receptor expression. Similarly, Mesnage et al. (2017) found that glyphosate, but not other glyphosate-based herbicide components, can activate ERα in MCF-7 breast cancer cells at high concentrations. Sritana et al. (2018) demonstrated GLY-induced proliferation in cholangiocarcinoma cells via non-genomic estrogen receptor/ERK1/2 signaling. Beyond estrogenic effects, deregulation of cell cycle and DNA damage repair pathways in both ER-positive and ER-negative breast cancer cells exposed to low glyphosate concentrations was observed (Stur et al. 2019). In hepatotoxicity studies, Gasnier et al. (2009) reported decreased cell viability and DNA damage in HepG2 hepatoma cells exposed to GBHs, with effects varying by formulation. Collectively, these studies suggest that GLY and GBHs may have endocrine-disrupting potential, particularly in relation to estrogen signaling, with observed impacts on cell proliferation, cell cycle regulation, and DNA integrity.

Placental and epithelial tumor cells Studies on glyphosate and glyphosate-based herbicides have extended beyond cancer cell lines to examine effects on other human cell types. A study investigated the impact of glyphosate on human choriocarcinoma cell line (JEG3) cells, a human placental cell line derived from choriocarcinoma (Richard et al. 2005). They found that GBHs reduced cell viability at least twice as effectively as GLY alone, with toxicity increasing over time. Notably, these effects were observed at concentrations below those commonly used in agriculture. In a separate study, the cytotoxic and genotoxic properties of glyphosate in the human buccal epithelial cell line (TR146) cells, derived from buccal epithelial origin, were examined (Koller et al. 2012). It was observed that glyphosate induced DNA instability at concentrations of 10–20 mg/L, with effects including chromosomal aberrations, micronuclei formation, and increased nuclear buds.

Other tumor cells Research on the effects of glyphosate, its metabolite AMPA, and glyphosate-based herbicides has revealed diverse impacts on various cancer cell lines. Research conducted by Li et al. (2013) showed that glyphosate and AMPA, at concentrations reaching 50 mM, suppressed cell growth in various cancer cell lines. The affected lines included those derived from prostate, ovarian, cervical, and lung cancers. They observed decreased cell viability ranging from 17% to 73% across these cell lines when exposed to 50 mM glyphosate. Hao et al. (2019a) further investigated the cytotoxicity of GBHs in HepG2, A549, and SH-SY5Y cell lines, finding significant dose-dependent toxicity in all three cell types. Interestingly, they noted that GLY alone did not exhibit significant toxicity, highlighting potential differences between the active ingredient and formulated products. In a separate study, Hao et al. (2019b) focused on A549 lung cancer cells, reporting that Roundup® induced cytotoxic effects at concentrations ranging from 50 to 125 μg/mL, which are lower than occupational exposure levels. This study observed concentration-dependent increases in DNA damage, apoptosis, and mitochondrial membrane collapse. Therefore, the findings suggest that GLY, AMPA, and particularly GBHs can exert cytotoxic and genotoxic effects on various cancer cell lines, with some formulations showing effects at concentrations relevant to human exposure. However, the differential effects observed between GLY alone and formulated products underscore the complexity of assessing the overall impact of these compounds on human health.

Epidemiological Studies

The global usage of glyphosate and glyphosate-based herbicides has increased exponentially in recent decades, leading to a concurrent rise in human exposure to these compounds. GLY and its primary metabolite, aminomethylphosphonic acid (AMPA), have demonstrated persistence in various environmental compartments, potentially impacting multiple trophic levels within ecosystems (Bai and Ogbourne 2016). Human exposure pathways are diverse, encompassing food consumption, soil contact, water ingestion, and inhalation. These exposures can occur through both direct and indirect routes, including occupational settings in agriculture and forestry, as well as domestic applications in households and gardens (International Agency for Research on Cancer 2015). The widespread use and environmental persistence of these chemicals underscore the importance of comprehensive risk assessments to evaluate potential long-term effects on human health and ecological systems.

Cancer Numerous epidemiological studies have investigated potential associations between glyphosate exposure and various cancers, with mixed results. The Agricultural Health Study (AHS), initiated by Alavanja et al. (1996), has been a significant source of data for many subsequent analyses. While some studies reported moderate associations between GLY exposure and cancer-related complications (Eriksson et al. 2008; Leon et al. 2019), these findings were often limited to univariate analyses and not confirmed in multivariate models. Several studies using the AHS cohort found no significant associations between GLY use and various cancers, including prostate cancer (Alavanja et al. 2003), multiple myeloma (De Roos et al. 2005) colorectal cancer (Lee et al. 2007), and pancreatic cancer (Andreotti et al. 2018). However, some of these studies suggested potential, albeit non-significant, correlations.

Non-Hodgkin's lymphoma (NHL) has been a focus of multiple studies, with several suggesting possible associations but failing to reach statistical significance in multivariate analyses (Eriksson et al. 2008). Another study reported a moderately elevated positive meta-hazard Ratio for diffuse large B-cell lymphoma and GLY exposure (Leon et al. 2019). Other studies found no significant associations between GLY exposure and hairy cell leukemia (Nordström et al. 1998) stomach cancer, or esophageal cancer (Lee et al. 2004). The collective body of research indicates that while some studies indicate potential associations between GLY exposure and certain cancers, particularly NHL, the evidence remains inconclusive because of the lack of consistent, statistically significant results in multivariate analyses across multiple studies. Further investigation is required to clarify these potential associations and their implications for public health.

Endocrine disruption Research studies on glyphosate present a complex picture of potential health impacts. In vitro studies have shown potential endocrine disruption, cytotoxicity, and genotoxicity in various cell types, with formulated products often exhibiting more pronounced effects compared to GLY alone (Thongprakaisang et al. 2013; Hao et al. 2019a, b; Mesnage et al. 2017; Gasnier et al. 2009).

Respiratory diseases Research on the respiratory effects of GLY exposure has produced mixed results. Some studies suggest a link to conditions such as asthma and rhinitis, while others have found no significant associations (Henneberger et al. 2014). The increasing global use of GLY and GBHs, coupled with their environmental persistence (Bai and Ogbourne 2016) underscores the need for ongoing research to clarify long-term effects on human health and ecosystems. Current evidence suggests exposure levels may be below established safety thresholds (Williams et al. 2016) the potential for chronic, low-level exposure warrants continued vigilance. Regulatory bodies should continue to review and update guidelines based on evolving scientific understanding, balancing agricultural benefits with potential risks to human health and the environment (International Agency for Research on Cancer 2015).

Neurological impacts Studies examining the link between glyphosate exposure and neurological disorders have produced varied results. Von Ehrenstein et al. (2019) identified a positive association between prenatal GLY exposure and autism spectrum disorder (ASD) and ASD with intellectual disability in a California-based study. In the context of Parkinson's disease (PD), Caballero et al. (2018) observed a 33% increased risk of premature mortality from PD among individuals exposed to GLY-associated land use in Washington State. Conversely, utilizing the Agricultural Health Study cohort, no significant association between glyphosate exposure and PD prevalence or incidence was found (Kamel et al. 2007). These inconsistent findings highlight the need for additional research to elucidate the potential neurological effects of GLY exposure.

Impact on fertility Several studies have investigated the effects of glyphosate exposure during pregnancy. Parvez et al. (2018) reported a correlation between urinary GLY levels and shortened gestational length in a cohort of 71 women from central Indiana but found no direct correlations with fetal growth indicators. In a region with long-term GLY use, it was found that women took longer to conceive (Sanin et al. 2009).

Arbuckle et al. (2001) found an association between preconception GLY exposure and late spontaneous abortions. However, Aris and Leblanc (2011) reported no detectable GLY levels in maternal or umbilical cord blood, suggesting limited direct fetal exposure. The potential link between maternal proximity to GLY application sites and neural tube defects (NTDs) in California was investigated, adding to the body of research on potential developmental impacts (Rull et al. 2006). Various aspects of GLY exposure during pregnancy and fetal development have also been examined by other studies (Singh et al. 2024).

Conclusively, while some studies indicate potential associations between GLY exposure and adverse pregnancy outcomes, the evidence remains inconsistent. The variability in study designs, exposure assessments, and outcomes measured makes it challenging to draw definitive conclusions. Further research, particularly large-scale prospective studies with robust exposure assessments, is needed to clarify the potential risks of GLY exposure during pregnancy and its impacts on fetal development.

Other health problems Researchers have conducted investigations to examine the potential link between glyphosate (GLY) exposure and a range of non-cancerous health issues. Research findings indicated an increased incidence of chronic kidney disease among Sri Lankan farmers who utilized glyphosate as a pesticide, suggesting a possible connection between glyphosate exposure and kidney health issues (Jayasumana et al. 2015). Using data from the AHS, an increased risk of hypothyroidism was found associated with the ever-use of GLY in a sample of 663 cases (Shrestha et al. 2018).

Regarding metabolic health, Montgomery et al. (2018) conducted a case-control study with 865 cases of diabetes mellitus (DM) and 23,072 controls, reporting an inverse association between GLY exposure and DM. This surprising discovery calls for additional research to uncover the underlying mechanisms at play.

Conclusion

Since its introduction, the global usage of glyphosate has been on a steady rise. Even though its active metabolites are linked with relatively moderate chronic toxicity, numerous other toxicological concerns warrant thorough investigation. The increased application of glyphosate in recent years has amplified worries about its potential health risks, especially regarding its possible carcinogenic effects. The scientific community remains divided on the potential carcinogenicity of glyphosate, with ongoing debate over its long-term health effects. This uncertainty underscores the urgent need for research into the toxicity of glyphosate, with a particular focus on GBHs. These herbicidal formulations may contain additional toxic substances that pose risks to human health.

In certain conditions, some soil microorganisms can employ glyphosate as a nutrient source; however, it can also lead to a raise in the population of pathogenic microbes. Glyphosate can modify soil properties and inhibit the growth of beneficial soil microorganisms. Research has revealed that both glyphosate and its metabolites have the ability to accumulate in soil and contaminate aquatic environments, leading to concerns about their potential environmental impact. Although glyphosate-based herbicides are utilized to eliminate weeds and undesirable plants, their exposure to non-target species can negatively impact ecosystems. This non-target exposure can disrupt the balance of various species and considerably harm biodiversity. Therefore, it is crucial to understand and validate the broader ecological implications of widespread glyphosate application, particularly its long-term effects on soil health, water quality, and the overall functioning of ecosystems as well.

References

Achiorno CL, de Villalobos C, Ferrari L (2008) Toxicity of the herbicide glyphosate to *Chordodes nobilii* (Gordiida, Nematomorpha). Chemosphere 71(10):1816–1822

Agostini LP, Dettogni RS, Dos Reis RS et al (2020) Effects of glyphosate exposure on human health: insights from epidemiological and in vitro studies. Sci Total Environ 705:135808

Alavanja CRM, Sandler DP, McMaster SB et al (1996) The agricultural health study. Environ Health Perspect 104(4):362–369

Alavanja CRM, Samanic C, Dosemeci M et al (2003) Use of agricultural pesticides and prostate cancer risk in the Agricultural Health Study cohort. Am J Epidemiol 157(9):800–814

Andreotti G, Koutros S, Hofmann JN et al (2018) Glyphosate use and cancer incidence in the Agricultural Health Study. J Natl Cancer Inst 110(5):509–516

Anifandis G, Katsanaki K, Lagodonti G et al (2018) The effect of glyphosate on human sperm motility and sperm DNA fragmentation. Int J Environ Res Public Health 15(6):1117

Arbuckle TE, Lin Z, Mery LS (2001) An exploratory analysis of the effect of pesticide exposure on the risk of spontaneous abortion in an Ontario farm population. Environ Health Perspect 109(8): 851–857

Aris A, Leblanc S (2011) Maternal and fetal exposure to pesticides associated to genetically modified foods in Eastern townships of Quebec, Canada. Reprod Toxicol 31(4):528–533

Bai SH, Ogbourne SM (2016) Glyphosate: environmental contamination, toxicity and potential risks to human health via food contamination. Environ Sci Pollut Res 23(19):18988–19001

Benachour N, Séralini GE (2009) Glyphosate formulations induce apoptosis and necrosis in human umbilical, embryonic, and placental cells. Chem Res Toxicol 22(1):97–105

Benbrook CM (2016) Trends in glyphosate herbicide use in the United States and globally. Environ Sci Eur 8:1–5

Boily M, Sarrasin B, DeBlois C et al (2013) Acetylcholinesterase in honey bees (*Apis mellifera*) exposed to neonicotinoids, atrazine and glyphosate: laboratory and field experiments. Environ Sci Pollut Res Int 20:5603–5614

Bonny S (2008) Genetically modified glyphosate-tolerant soybean in the USA: adoption factors, impacts and prospects: a review. Agron Sustain Dev 28:21–32

Caballero M, Amiri S, Denney JT et al (2018) Estimated residential exposure to agricultural chemicals and premature mortality by Parkinson's disease in Washington State. Int J Environ Res Public Health 15(12):1–11
Coalova I, Ríos de Molina MDC, Chaufan G (2014) Influence of the spray adjuvant on the toxicity effects of a glyphosate formulation. Toxicol In Vitro 28(7):1306–1311
Correia FV, Moreira JC (2010) Effects of glyphosate and 2, 4-D on earthworms (Eisenia foetida) in laboratory tests. Bull Environ Contam Toxicol 85(3):264–268
Cuhra M, Traavik T, Bøhn T (2013) Clone-and age-dependent toxicity of a glyphosate commercial formulation and its active ingredient in *Daphnia magna*. Ecotoxicology 22:251–262
Dallegrave E, Mantese FD, Oliveira RT et al (2007) Pre-and postnatal toxicity of the commercial glyphosate formulation in Wistar rats. Arch Toxicol 81:665–673
De Roos AJ, Blair A, Rusiecki JA et al (2005) Cancer incidence among glyphosate-exposed pesticide applicators in the agricultural health study. Environ Health Perspect 113(1):49–54
Dill GM (2005) Glyphosate-resistant crops: history, status and future. Pest Manag Sci 61(3): 219–224
Dornelles MF, Oliveira GT (2016) Toxicity of atrazine, glyphosate, and quinclorac in bullfrog tadpoles exposed to concentrations below legal limits. Environ Sci Pollut Res 23:1610–1620
Duke SO (2018) The history and current status of glyphosate. Pest Manag Sci 74(5):1027–1034
Duke SO, Lydon J, Koskinen WC et al (2012) Glyphosate effects on plant mineral nutrition, crop rhizosphere microbiota, and plant disease in glyphosate-resistant crops. J Agric Food Chem 60(42):10375–10397
Eriksson M, Hardell L, Carlberg M et al (2008) Pesticide exposure as risk factor for non-Hodgkin lymphoma including histopathological subgroup analysis. Int J Cancer 123:1657–1663
Estok D, Freedman B, Boyle D (1989) Effects of the herbicides 2, 4-D, glyphosate, hexazinone and triclopyr on the growth of three species of ectomycorrhizal fungi. Bull Environ Contam Toxicol 42(6):835–839
Finley JW, Duke SO (2020) Agnes Rimando, a pioneer in the fate of glyphosate and its primary metabolite in plants. J Agric Food Chem 68(20):5623–5630
Folmar LC, Sanders HO, Julin AM (1979) Toxicity of the herbicide glyphosate and several of its formulations to fish and aquatic invertebrates. Arch Environ Contam Toxicol 8:269–278
Gandhi K, Khan S, Patrikar M et al (2021) Exposure risk and environmental impacts of glyphosate: highlights on the toxicity of herbicide co-formulants. Environ Chall 4:100149
Garthwaite DG, Thomas MR, Parrish G et al (2008) Pesticide usage survey report 224. In: Arable Crops in Great Britain 2008. Central Science Laboratory, UK
Gasnier C, Dumont C, Benachour N et al (2009) Glyphosate-based herbicides are toxic and endocrine disruptors in human cell lines. Toxicology 262:184–191
Gehin A, Guyon C, Nicod L (2006) Glyphosate-induced antioxidant imbalance in HaCaT: the protective effect of vitamins C and E. Environ Toxicol Pharmacol 22:27–34
George J, Shukla Y (2013) Emptying of intracellular calcium pool and oxidative stress imbalance are associated with the glyphosate-induced proliferation in human skin keratinocytes HaCaT cells. ISRN Dermatol 2013:825180
Gill JP, Sethi N, Mohan A et al (2018) Glyphosate toxicity for animals. Environ Chem Lett 16:401–426
Hao Y, Zhang Y, Ni H et al (2019a) Evaluation of the cytotoxic effects of glyphosate herbicides in human liver, lung, and nerve. J Environ Sci Health B 54:737–744
Hao Y, Chen H, Xu W et al (2019b) Roundup® confers cytotoxicity through DNA damage and mitochondria-associated apoptosis induction. Environ Pollut 252:917–923
Haslam E (2014) The shikimate pathway: biosynthesis of natural products series. Elsevier, Amsterdam
Heap I (2014) Global perspective of herbicide-resistant weeds. Pest Manag Sci 70(9):1306–1315
Henneberger PK, Liang X, London SJ et al (2014) Exacerbation of symptoms in agricultural pesticide applicators with asthma. Int Arch Occup Environ Health 87:423–432

Heu C, Berquand A, Elie-Caille C et al (2012) Glyphosate-induced stiffening of HaCaT keratinocytes, a peak force tapping study on living cells. J Struct Biol 178:1–7
International Agency for Research on Cancer (2015) Glyphosate. In: Some organophosphate insecticides and herbicides: diazinon, glyphosate, malathion, parathion, tetrachlorvinphos. Available: http://monographs.iarc.fr/ENG/Monographs/vol112/index.php. Accessed 12 Apr 2024
Jayasumana C, Gunatilake S, Siribaddana S (2015) Simultaneous exposure to multiple heavy metals and glyphosate may contribute to Sri Lankan agricultural nephropathy. BMC Nephrol 16:1–7
Johnson WG, Davis VM, Kruger GR et al (2009) Influence of glyphosate-resistant cropping systems on weed species shifts and glyphosate-resistant weed populations. Eur J Agron 31(3): 162–172
Kamel F, Tanner CM, Umbach DM et al (2007) Pesticide exposure and self-reported Parkinson's disease in the Agricultural Health Study. Am J Epidemiol 165(4):364–374
Kishore GM, Shah DM (1988) Amino acid biosynthesis inhibitors as herbicides. Annu Rev Biochem 57:627–663
Knuuttila P, Knuuttila H, Nishida T et al (1979) The crystal and molecular structure of N-(phosphonomethyl) glycine (glyphosate). Acta Chem Scand 33:623–626
Koller VJ, Fürhacker M, Nersesyan A et al (2012) Cytotoxic and DNA-damaging properties of glyphosate and Roundup in human-derived buccal epithelial cells. Arch Toxicol 86:805–813
Krimsky S (2022) Glyphosate-based herbicides and public health: making sense of the science. J Agric Environ Ethics 35:3
Kwiatkowska M, Huras B, Bukowska B (2014) The effect of metabolites and impurities of glyphosate on human erythrocytes (in vitro). Pestic Biochem Physiol 109:34–43
Kwiatkowska M, Jarosiewicz P, Michałowicz J (2016) The impact of glyphosate, its metabolites and impurities on viability, ATP level and morphological changes in human peripheral blood mononuclear cells. PLoS One 11:1–13
Kwiatkowska M, Reszka E, Woźniak K (2017) DNA damage and methylation induced by glyphosate in human peripheral blood mononuclear cells (in vitro study). Food Chem Toxicol 105:93–98
Lee WJ, Lijinsky W, Heineman EF et al (2004) Agricultural pesticide use and adenocarcinomas of the stomach and oesophagus. Occup Environ Med 61:743–749
Lee WJ, Sandler DP, Blair A et al (2007) Pesticide use and colorectal cancer risk in the Agricultural Health Study. Int J Cancer 121:339–346
Leon ME, Schinasi LH, Lebailly P et al (2019) Pesticide use and risk of non-Hodgkin lymphoid malignancies in agricultural cohorts from France, Norway and the USA: a pooled analysis from the AGRICOH consortium. Int J Epidemiol 48:1519–1535
Li Q, Lambrechts MJ, Zhang Q et al (2013) Glyphosate and AMPA inhibit cancer cell growth through inhibiting intracellular glycine synthesis. Drug Des Devel Ther 7:635–643
Maeda H, Dudareva N (2012) The shikimate pathway and aromatic amino acid biosynthesis in plants. Annu Rev Plant Biol 63:73–105
Mañas F, Peralta L, Raviolo J et al (2009) Genotoxicity of glyphosate assessed by the comet assay and cytogenetic tests. Environ Toxicol Pharmacol 28:37–41
Marc J, Mulner-Lorillon O, Bellé R (2004) Glyphosate-based pesticides affect cell cycle regulation. Biol Cell 96(3):245–249
Martinez A, Al-Ahmad AJ (2019) Effects of glyphosate and aminomethylphosphonic acid on an isogeneic model of the human blood-brain barrier. Toxicol Lett 304:39–49
Martínez A, Reyes I, Reyes N (2007) Cytotoxicity of the herbicide glyphosate in human peripheral blood mononuclear cells. Biomedica 27:594–604
Martins-Gomes C, Silva TL, Andreani T et al (2022) Glyphosate vs. glyphosate-based herbicides exposure: a review on their toxicity. J Xenobiot 12(1):21–40
Mendelson J (1998) Roundup: the world's biggest-selling herbicide. The Ecologist. Available via https://www.orpheusweb.co.uk/john.rose/round.html. Accessed 15 Apr 2024

Mesnage R, Bernay B, Séralini GE (2013) Ethoxylated adjuvants of glyphosate-based herbicides are active principles of human cell toxicity. Toxicology 313:122–128

Mesnage R, Phedonos A, Biserni M et al (2017) Evaluation of estrogen receptor alpha activation by glyphosate-based herbicide constituents. Food Chem Toxicol 108:30–42

Mladinic M, Berend S, Vrdoljak AL et al (2009) Evaluation of genome damage and its relation to oxidative stress induced by glyphosate in human lymphocytes in vitro. Environ Mol Mutagen 50:800–807

Mnif W, Hassine AIH, Bouaziz A et al (2011) Effect of endocrine disruptor pesticides: a review. Int J Environ Res Public Health 8:2265–2303

Montgomery MP, Kame F, Saldana TM et al (2018) Incident diabetes and pesticide exposure among licensed pesticide applicators: agricultural health study 1993–2003. Am J Epidemiol 167:1235–1246

Muñoz JP, Bleak TC, Calaf GM (2021) Glyphosate and the key characteristics of an endocrine disruptor: a review. Chemosphere 270:128619

Myers JP, Antoniou MN, Blumberg B et al (2016) Concerns over use of glyphosate-based herbicides and risks associated with exposures: a consensus statement. Environ Health 15(1):19

Nandula VK (ed) (2010) Glyphosate resistance in crops and weeds: history, development, and management. Wiley, New Jersey

Nedelkoska TV, Low GC (2004) High-performance liquid chromatographic determination of glyphosate in water and plant material after pre-column derivatisation with 9-fluorenylmethyl chloroformate. Anal Chim Acta 511(1):145–153

Newman MM, Hoilett N, Lorenz N et al (2016) Glyphosate effects on soil rhizosphere-associated bacterial communities. Sci Total Environ 543:155–160

Nordström M, Hardell L, Magnuson A et al (1998) Occupational exposures, animal exposure and smoking as risk factors for hairy cell leukaemia evaluated in a case-control study. Br J Cancer 77:2048–2052

Oliveira AG, Telles LF, Hess RA et al (2007) Effects of the herbicide roundup on the epididymal region of drakes *Anas platyrhynchos*. Reprod Toxicol 23(2):182–191

Oliveira RDC, Boas LK, Branco CC (2016) Assessment of the potential toxicity of glyphosate-based herbicides on the photosynthesis of *Nitella microcarpa* var. *wrightii* (Charophyceae). Phycologia 55(5):577–584

Padgette SR, Re DB, Barry GF et al (2018) New weed control opportunities: development of soybeans with a roundup ready™ gene. In: Duke SO (ed) Herbicide-resistant crops, 1st edn. CRC Press, US, pp 53–84

Parvez S, Gerona RR, Proctor C et al (2018) Glyphosate exposure in pregnancy and shortened gestational length: a prospective Indiana birth cohort study. Environ Health 17:1–12

Pieniężek D, Bukowska B, Duda W (2004) Comparison of the effect of Roundup Ultra 360 SL pesticide and its active compound glyphosate on human erythrocytes. Pestic Biochem Physiol 79:58–63

Powles SB, Preston C (2006) Evolved glyphosate resistance in plants: biochemical and genetic basis of resistance. Weed Technol 20(2):282–289

Pratley J, Urwin N, Stanton R et al (1999) Resistance to glyphosate in *Lolium rigidum*. I Bioevaluation. Weed Sci 47(4):405–411

Report Linker (2023) Global glyphosate market to reach $10.6 billion by 2030. https://finance.yahoo.com/news/global-glyphosate-market-reach-10-142800106.html. Accessed 6 July 2023

Richard S, Moslemi S, Sipahutar H et al (2005) Differential effects of glyphosate and Roundup on human placental cells and aromatase. Environ Health Perspect 113:716–720

Richardson JT, Frans RE, Talbert RE (1979) Reactions of *Euglena gracilis* to fluometuron, MSMA, metribuzin, and glyphosate. Weed Sci 27(6):619–624

Rull RP, Ritz B, Shaw GM (2006) Neural tube defects and maternal residential proximity to agricultural pesticide applications. Am J Epidemiol 163:743–753

Sanin LH, Carrasquilla G, Solomon KR et al (2009) Regional differences in time to pregnancy among fertile women from five Colombian regions with different use of glyphosate. J Toxicol Environ Health A 72:949–960
Santovito A, Ruberto S, Gendusa C et al (2018) In vitro evaluation of genomic damage induced by glyphosate on human lymphocytes. Environ Sci Pollut Res 25:34693–34700
Schaumburg LG, Siroski PA, Poletta GL et al (2016) Genotoxicity induced by Roundup®(Glyphosate) in tegu lizard (*Salvator merianae*) embryos. Pestic Biochem Physiol 30:71–78
Shaner D (2006) An overview of glyphosate mode of action: why is it such a great herbicide. North Central Weed Sci Soc Proc 61:94
Sharma A, Kumar V, Shahzad B et al (2019) Worldwide pesticide usage and its impacts on ecosystem. SN Appl Sci 1:1–6
Shrestha S, Parks CG, Goldner WS et al (2018) Pesticide use and incident hypothyroidism in pesticide applicators in the agricultural health study. Environ Health Perspect 126:1–11
Singh R, Kumar S (eds) (2017) Green technologies and environmental sustainability. Springer, Singapore
Singh R, Shukla A, Kaur G et al (2024) Systemic analysis of glyphosate impact on environment and human health. ACS Omega 9(6):6165–6183
Soares D, Silva L, Duarte S et al (2021) Glyphosate use, toxicity and occurrence in food. Foods 10(11):2785
Sritana N, Suriyo T, Kanitwithayanun J et al (2018) Glyphosate induces growth of estrogen receptor alpha positive cholangiocarcinoma cells via non-genomic estrogen receptor/ERK1/2 signaling pathway. Food Chem Toxicol 118:595–607
Steinmann HH, Dickeduisberg M, Theuvsen L (2012) Uses and benefits of glyphosate in German arable farming. Crop Prot 42:164–169
Stur E, Aristizabal-Pachon AF, Peronni KC et al (2019) Glyphosate-based herbicides at low doses affect canonical pathways in estrogen positive and negative breast cancer cell lines. PLoS One 14(7):e0219610
Tate TM, Spurlock JO, Christian FA (1997) Effect of glyphosate on the development of *Pseudosuccinea columella* snails. Arch Environ Contam Toxicol 33:286–289
Taylor EL, Holley AG, Kirk M (2007) Pesticide development: a brief look at the history. Southern Reg Extension. Forestry FM-010:1–7
Thongprakaisang S, Thiantanawat A, Rangkadilok N et al (2013) Glyphosate induces human breast cancer cells growth via estrogen receptors. Food Chem Toxicol 59:129–136
Tu M, Hurd C, Randall JM (2001) Weed control methods handbook. The nature conservancy. Available via http://tncweeds.ucdavis.edu. Accessed 21 Apr 2024
Valavanidis A (2018) Glyphosate, the most widely used herbicide: health and safety issues. Why scientists differ in their evaluation of its adverse health effects. Scientific Reviews 1:1–41
Vande Berg BJ, Hammer PE, Chun BL et al (2008) Characterization and plant expression of a glyphosate-tolerant enolpyruvyl shikimate phosphate synthase. Pest Manag Sci 64(4):340–345
Vigfusson NV, Vyse ER (1980) The effect of the pesticides, Dexon, Captan and Roundup, on sister-chromatid exchanges in human lymphocytes in vitro. Mutat Res 79:53–57
Von Ehrenstein SO, Ling C, Cui X et al (2019) Prenatal and infant exposure to ambient pesticides and autism spectrum disorder in children: population-based case-control study. BMJ 364:1–21
Westwood JH, Weller SC (1997) Cellular mechanisms influence differential glyphosate sensitivity in field bindweed (*Convolvulus arvensis*) biotypes. Weed Sci 45(1):2–11
Williams GM, Kroes R, Munro IC (2000) Safety evaluation and risk assessment of the herbicide Roundup and its active ingredient, glyphosate, for humans. Regul Toxicol Pharmacol 31(2):117–165
Williams GM, Aardema M, Acquavella J et al (2016) A review of the carcinogenic potential of glyphosate by four independent expert panels and comparison to the IARC assessment. Crit Rev Toxicol 46:3–20
Woodburn AT (2000) Glyphosate: production, pricing and use worldwide. Pest Manag Sci 56(4):309–312

Electronic Waste: An Emerging Pollutant of Human Health Concern and Biological Techniques for Their Treatment

Manglam Bajpai, Mohit Nigam, Lalit Kumar Singh, and Garima Awasthi

Abstract

In the current era, the technologies are rapidly increasing the electronic waste (E-waste) over the world. E-waste contributes a significant hazard of chemical pollution to the environment. The improper management of E-waste poses severe risks to both individual and global health. Addressing this issue necessitates effective recycling and removal systems for discarded electronic products. E-waste contains various metals, including common ones like copper and aluminum, valuable metallic elements like silver, gold, palladium, and platinum and harmful metallic elements like lead, mercury, and arsenic. Biological techniques, particularly those involving microorganisms, offer promising solutions for E-waste recovery and treatment. The various methods, such as biosorption, bioleaching, bioaccumulation, biotransformation, biomineralization, and microbially enhanced chemisorption, can transform harmful waste material into non-hazardous or less harmful forms. Additionally, genetically modified organisms and genetically engineered microorganisms hold the potential for

M. Bajpai
Amity Institute of Biotechnology, Amity University Uttar Pradesh, Lucknow Campus, Lucknow, Uttar Pradesh, India

Department of Biochemical Engineering, School of Chemical Technology, Harcourt Butler Technical University, Kanpur, Uttar Pradesh, India

M. Nigam · L. K. Singh
Department of Biochemical Engineering, School of Chemical Technology, Harcourt Butler Technical University, Kanpur, Uttar Pradesh, India

G. Awasthi (✉)
Amity Institute of Biotechnology, Amity University Uttar Pradesh, Lucknow Campus, Lucknow, Uttar Pradesh, India
e-mail: gawasthi@lko.amity.edu

S. Raisuddin et al. (eds.), *Environmental Toxicants and Lifestyle Diseases*,
https://doi.org/10.1007/978-981-95-0575-3_8

future E-waste degradation. These biological methods are effective as well as very cost-efficient. The current chapter aims to encompass the application of various biological tools and techniques to enhance the bioremediation of E-waste.

Keywords

E-waste · Pollutants · Biodegradation · Bioleaching · Biotransformation

Introduction

The word "waste" is used for the discarded material when it loses the purpose of what it was manufactured for or becomes useless to its owner. A large amount of waste is generated every day, which needs responsiveness for proper treatment. In India, waste is generated, especially electronic waste (E-waste), in huge amounts, but its management is very problematic. Unused electronic devices like office/industrial electronic equipment, household electronic devices, cell phones, TV sets, LEDs, cooling devices, and many more are described as electronic waste (Singh et al. 2018). The old electronics that are considered for sale, recycle, retrieval, or thrown away are also considered E-waste. Consumer-oriented growth has always led to the increasing manufacturing of newer gadgets with better technological advances, hence the previous ones become obsolete and old-fashioned, and finally, their fate is E-waste. The proper handling of E-waste is becoming a major issue for the protection of the environment. Right now, we can see our home as the largest E-waste-producing source. In each house/industry, we can see lots of electronics around us, which will become E-waste when it loses its functionality.

In India, electronic waste is seen so easily in commercial places, residential areas, and industrial areas. As India is a developing country, we all can see Indians moving towards a technology-based lifestyle (Ikhlayel 2018). The disposal of electronic waste is an important topic to think about. Many diseases are occurring just because of the improper treatment of E-waste. Electronic waste includes items containing electronic parts like monitors; television (TV), cell phones, and Wi-Fi routers; cooling devices; and devices for washing purposes, air cooling devices, vacuum cleaners, etc. (Narayanasamy et al. 2017) the measurement of E-waste is a risky arrangement of metals like gold, silver (costly metals), copper, zinc, aluminum (base metals), mercury, chromium (dangerous metals), etc., along with halogens, plastics (unstable substances), and flame retardants (Islam et al. 2020). Few chemical and mechanical methods are quite rampant while the biological strategy is one of the highly encouraging methodologies (Awasthi et al. 2019). Nonetheless, these techniques have to be performed methodically, like the mechanical cycle needs at any rate capital cost venture, while the hydrometallurgy strategy expends synthetic compounds, these substance cycle may cause contamination to nature, while the pyrometallurgy technique require high vitality inputs which makes them relatively overpriced (Ikhlayel 2018). Biological approaches may be used for the recovery and treatment of E-waste. Microorganisms are widely used as valuable and prevalent biological agents for the purpose of treating E-waste.

Distribution of E-waste

E-wastes have various sources which are still unknown. The E-waste is generated by the electronic gadgets that are disowned by their respective owners. When a gadget is used seldom, malfunctions, or becomes unusable, it is classified as electronic garbage (E-waste). These wastes are generated through the various sources in the world. The E-waste distribution is done on the bases of electronics functionality and metal extracted through them. The sources of the E-wastes are described in Fig. 1.

Every class of waste is considered useless when there is not justifiable use or working of these elements (Ahirwar and Tripathi 2021). These types of devices have extraordinary input in economic development; however, they harm the environmental conditions as these wastes cannot be reused or recycled. Various metals could be extracted through bioleaching from E-waste. These metals in environment can be lethal to humans as it may cause various diseases in humans. These metals generated from the E-wastes are listed in Table 1.

Every year there is an increase in pollution via E-waste generation. Every person on the earth is using electronic equipment, which is becoming E-waste (Kumar 2019). The global E-waste production reached around 50.1 million tonnes (Mt) in 2019. The overall development of E-waste is expanding quickly. As the world is slanting towards better innovations, it is moving towards the accumulation of

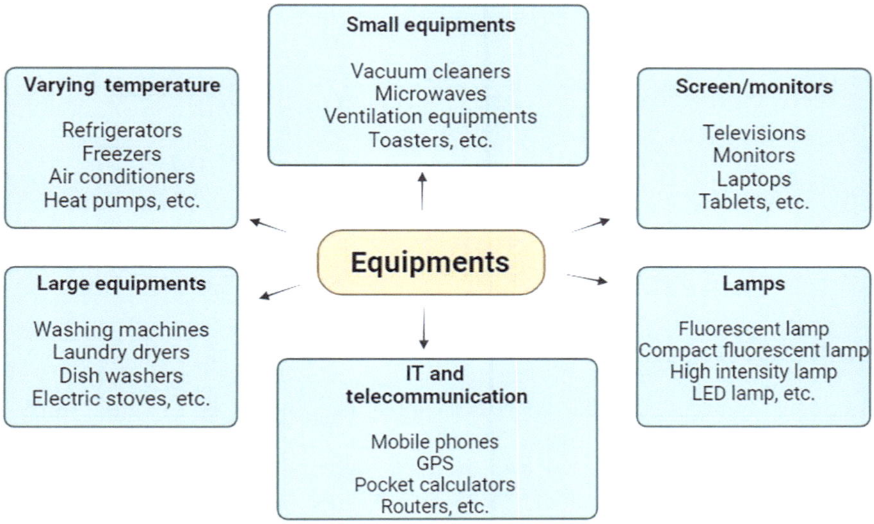

Fig. 1 Sources of the E-wastes

Table 1 List of equipment E-waste and the associated metals (Sharma et al. 2012; Bastiaan et al. 2010)

S. no.	Types of E-waste	Metal present
1.	Printed wiring board	Aluminum, copper, iron, gold, lead, nickel
2.	Printed Circuit boards	Cadmium, beryllium, and lead
3.	Motherboards	Cadmium, barium, and lead oxide
4.	Cathode ray tubes (CRTs)	Mercury
5.	Screen monitors	Cadmium
6.	Capacitors and transformers	Brominated cable
7.	Cable insulation/coating	Bromine
8.	Portable audio scrap	Aluminum, copper, gold, iron, lead, nickel, palladium, silver
9.	Digital versatile disc (DVD) player scrap	Aluminum, copper, gold, iron, lead, nickel, palladium, silver
10	Mobile phone scrap	Aluminum, copper, gold, iron, lead, nickel, palladium, silver
11.	TV board scrap	Aluminum, copper, gold, iron, lead, nickel, palladium, silver
12.	Calculator scrap	Aluminum, Copper, gold, iron, lead, nickel, palladium, silver
13.	TV scrap	Aluminum, Copper, gold, iron, lead, nickel, palladium, silver
14.	Computer scrap	Aluminum, Copper, gold, iron, lead, nickel, palladium, silver

Table 2 List of E-waste quantity generated yearly (Balde et al. 2015; Awasthi et al. 2019)

S. no.	Year wise	Electronic waste (million tonnes)	Population (billion)
1.	2010	33.9	6.8
2.	2011	35.3	6.9
3.	2012	37.9	6.9
4.	2013	39.9	7.1
5.	2014	41.9	7.2
6.	2015	43.9	7.3
7.	2016	45.8	7.4
8.	2017	47.7	7.5
9.	2018	49.9	7.5
10.	2019	50.1	7.6
11.	2020	53.6	7.8

electronic squander and the world is creating huge heaps of E-waste. Table 2 displays the global rise in E-waste growth.

The exponential increase of e-waste is observed relative to the global human population between 2010 and 2020. The magnification of the squander of electronic goods is understandable from the graph; as the population steadily rises, there is a noticeable surge in the creation of electronic waste (Withanage and Habib 2021).

The population is increasing day by day, and the land is limited, whereas the enormous rise of E-waste generation is making it a serious issue, as this will be hazardous to humans as well as the environment.

Electronic Waste Impact Related to Human Health

E-waste has become a critical global concern due to its adverse impacts on human health. E-waste comprises discarded electrical or electronic devices, including computers, mobile phones, televisions, and other household and industrial electronics. The improper disposal and management of e-waste, especially in informal recycling operations, expose humans to a variety of toxic chemicals and hazardous materials. These include heavy metals like lead, mercury, cadmium, and arsenic, as well as persistent organic pollutants (POPs) such as polybrominated diphenyl ethers (PBDEs) used as flame retardants. Additionally, polychlorinated biphenyls and dioxins are often released during the incineration or thermal treatment of e-waste, adding further to environmental contamination (Ghulam and Abushammala 2023).

The human health impacts of e-waste exposure are vast and varied, depending on the specific compounds involved. Lead, commonly found in cathode-ray tubes and circuit boards, is a potent neurotoxin, particularly dangerous to children, as it can impair cognitive development, cause learning disabilities, and result in behavioral issues. Chronic exposure to cadmium, which is used in rechargeable batteries and solar panels, can lead to kidney damage, bone degradation, and increased cancer risk. Mercury, present in fluorescent lamps and older batteries, is another neurotoxin that can cause severe damage to the nervous system, particularly in fetuses and young children. POPs like PBDEs disrupt endocrine systems, interfering with hormone function, reproductive health, and immune system integrity (Rautela et al. 2021).

E-wastes contain numerous destructive metallic contaminants, for example, lead, cadmium, nickel, etc. These material partitions include copper, aluminum, gold, iron, and different metals in E-waste, which is almost greater than 60%, while plastics covers almost 30%, and the perilous contaminants are just about 2.70% in electronic equipment (Widmer et al. 2005). Lead is a deadly element that is easily found in electronic devices for different purposes (Jang and Townsend 2003; Rautela et al. 2021). It gets exposed to natural systems, that is, water, air, and soil, and might cause deadly diseases. Lead causes various health problems and environmental pollution too (Jang and Townsend 2003). Children are more susceptible to these risks compared to adults due to their higher intake of lead from the air and the weakening of their immune system as a result of early exposure (Bathurst et al. 1992). There are various metals apart from lead that causes harm to the biological system as stated in Table 3. Electronic garbage, or e-waste, may inflict harm on the human body and result in a range of illnesses, perhaps leading to fatalities in both people and animals (Gunjal et al. 2020).

In regions where e-waste is dismantled and processed by hand, without protective measures, workers are at a heightened risk of exposure. Informal e-waste recycling

Table 3 List of various metals and their health impacts (Puckett et al. 2002)

S. no.	Metals	Health effects
1	Lead	Nervous system, blood system, kidney damage
2	Cadmium	Kidney and liver, causes neural damage
3	Mercury	Brain, respiratory, skin disorders
4	Chromium	Bronchitis
5	Plastics and PVC	Reproductive and developmental problems
6	Lithium	Pass to breast milk and may harm a nursing baby, lung edema
7	Beryllium	Carcinogenic (lung cancer), chronic beryllium disease or berylliosis

centers, particularly in developing countries, often rely on open burning or acid leaching to recover valuable metals such as gold, silver, and copper from electronic components. These crude methods not only expose workers to harmful chemicals but also pollute the air, soil, and water in surrounding communities. Inhalation of toxic fumes and particulate matter can lead to respiratory problems, chronic bronchitis, asthma, and even lung cancer. Long-term exposure to toxicants from e-waste, such as PCBs and dioxins, is linked to immune system suppression, birth defects, developmental disorders, and carcinogenic effects (Grace et al. 2020).

The threat posed by e-waste extends beyond direct exposure in recycling hubs. Through environmental contamination, pollutants can enter the food chain, affecting broader populations. For instance, heavy metals and toxic chemicals leach into soil and groundwater, contaminating drinking water supplies and agricultural products. These substances bioaccumulate in the tissues of living organisms, traveling up the food chain to humans, where they can cause long-term health effects even at low exposure levels. Pregnant women, infants, and young children are particularly vulnerable, with increased risks of developmental and neurobehavioral disorders (Ghulam and Abushammala 2023).

Given the scale and complexity of e-waste pollution, mitigating its impacts on human health requires an integrated approach. Policymakers, industries, and communities must collaborate to strengthen regulations on e-waste disposal, promote the recycling and recovery of electronic components through safer, more sustainable methods, and raise public awareness of the health risks associated with e-waste (Ghulam and Abushammala 2023). Biological treatment technologies, such as bioremediation, have emerged as promising solutions for managing e-waste. Utilizing microorganisms to degrade, detoxify, and immobilize hazardous compounds can help reduce environmental contamination and prevent human exposure to harmful pollutants. Such biological methods, though still in the early stages of development, offer a potentially eco-friendly alternative to current e-waste management practices, aligning with global efforts toward sustainable environmental and public health protection (Saha et al. 2021).

Electronic Waste Management

Electronic waste management involves the development of measures to protect both individuals and the environment from the dangers posed by electronic waste. This was by and large done to adjust the well-being of mankind and that of nature. Furthermore, it is a step where the earth promises to gain extremely valuable results from the management of these wastes (Mohammed et al. 2013). Microorganisms can consume the important parts in processing pollutants, including's metals, and it is a key contaminant in E-waste soil contamination (Gollakota et al. 2020). The decent variety of most microbial taxa can shift with the substantial metal focus to various levels of a different degree (Jiang et al. 2019).

Micro-remediation of electronic waste is characterized as the utilization of microorganisms to reduce the contaminants to non-toxic or mild structures in the environment by the metabolization process of organisms (Ferronato and Torretta 2019). There is an assortment of actions taken to be put in charge of the throwing away of such constituents and substances the most effective of them being the biological ways (Pietrelli et al. 2019). We are going to focus on these biological ways to crumble the electronic waste, which leads to an amicable environment as shown in Fig. 2.

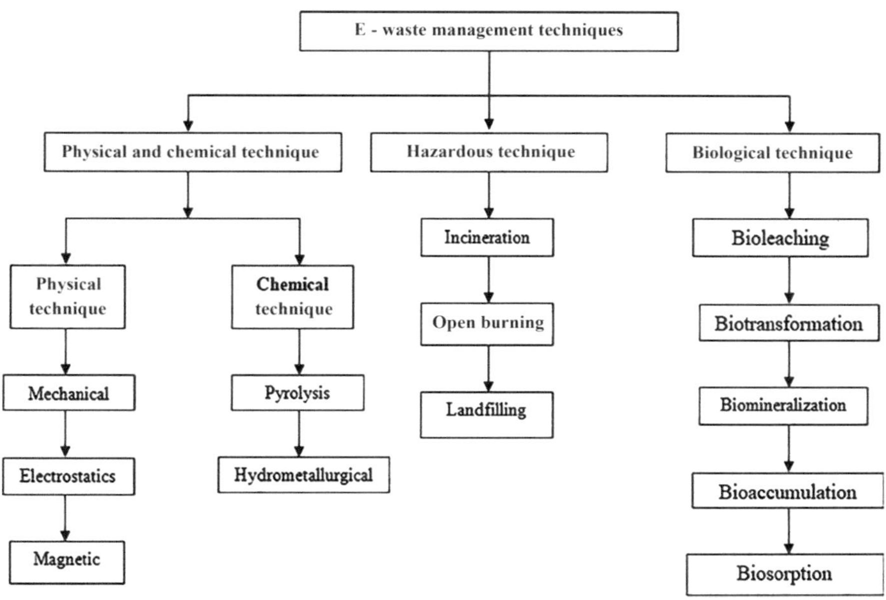

Fig. 2 Classification of E-waste management techniques

Bioleaching

Microorganisms break down metals into microscopic structures in bioleaching. Leaching is classified into two categories: direct and indirect. Direct leaching uses microorganism-molded acids to oxidize strong toxic metals, and spinning them into particles that get dissolved (Gunjal et al. 2020). Under metal strain, indirect filtering metal uses oxidizing microorganisms. Bleached anion mixes metal ions, releasing uncharged metal ions into the aqueous medium (Tichy et al. 1998). Bioleaching is typically used for copper extraction. Copper cyanide (CuCN) blocks from CuLps dissociated from bioleaching alcohol (Narayanasamy et al. 2017). Laptop lithium-ion batteries (LIBs) are bioleached in two steps by means of a blend of *Acidithiobacillus ferrooxidans* and *thiooxidans* (Heydarian et al. 2018). A new bioleaching method extracts metals from electronic waste. Few researchers examined the acidophilic bioleaching of Cu and Ni from E-waste. The sample size for various types of E-waste was less than 100 microns (Arshadi and Yaghmaei 2020). *Acidithiobacillus ferrooxidans* were adapted to the E-waste sample's 15 g/L concentration (Arshadi and Yaghmaei 2020; Arslan 2021). Electric Potential increases with a pH decrease. Cu and Ni were extracted 99 percent on day 11 and 98 percent on day 14 with the maximum bacterial population (Arshadi and Yaghmaei 2020). Field emission scanning electron microscopy (FE-SEM) shows bioleaching roughens the sample surface (Yamanea et al. 2018). It may show that bioleaching degrades the sample surface (Arshadi and Yaghmaei 2020). In bioleaching studies, pure *A. ferrooxidans* culture recovered copper (96%) and zinc (92%) but not nickel (55%) (Arslan 2021).

Microorganisms were used to remove heavy metals, including calcium, copper, gold, iron, manganese, osmium, platinum, tantalum, and zinc from waste printed circuit boards (WPCBs) (Mejía-Rodríguez et al. 2018). Adjusting temperature, pH, stirring speed, and extract density helps bacteria proliferate and dissolve metals (Yamanea et al. 2018). All consortia showed favorable results after adjusting for successive reseeding (Mejía-Rodríguez et al. 2018).

Citric acid improved Cu bioleaching (Diaz-Martinez et al. 2019). Physicochemical aspects like metallic species, quantity and type of electronic waste, pH, and temperature affect the chemical and biological leaching of printed circuit board (PCB) from cellular phones (Diaz-Martinez et al. 2019). Finally, the microbes efficiently bioleached metals from PCBs at ambient temperature without the need for agitation, resulting in reduced energy use compared to previous trials.

Biomineralization

Biomineralization pairs contaminated metal ions with microbe-produced anions or ligands as precipitation. Biomineralization occurs when microbes chemically modify an environment to precipitate minerals (Stocks-Fischer et al. 1999). The studies are done on biomineralization for secondary gold nanoparticles. Retrieval of gold from ESM was likely improved by ultrasound-assisted sulfuric acid and hydrogen peroxide mixture pretreatment to remove maximum competing copper, ultrasound-induced enhanced growth rate of *Chromobacterium* sp. for increased production of cyanide, as well as bioleaching for improved gold recovery. Microbes like *Pseudomonas, Bacillus, Achromobacter, Flavobacterium,* and *Acinetobacter* are sensitive to environmental contamination, while others like *Citrobacter koseri* and *Serratia ficaria* are capable of degrading spilled gasoline, diesel, crude oil, and oil-based paints, as well as additional lethal polycyclic aromatic hydrocarbons (PAHs).

Additionally, white-rot fungus has been effective in eradicating crude oil spills, plastics, radioactive waste, and electronic waste (Abatenh et al. 2017). The usage of CRISPR/Cas9 technology to genetically engineer organisms for detecting and cleaning environmental contaminants represents a novel approach in biological science. This technique enables for precise gene editing, including cutting, insertion, deletion, and recombination, making it more efficient than earlier approaches owing to developments in gene editing vectors and process optimization (Etaware 2021).

Biotransformation

Due to destructive metals, the erosion state is changed by electron expansion or deduction, resulting in precise compound properties (Lovley and Coates 1997).

There are two biotransformation backdrops.

- Direct enzymatic reduction groups multivalent toxic metallic ions along with free electrons from catalysts outside the cell.
- Indirect reduction of multivalent toxic metal particles in subsurface and sedimentary electrons by metal or sulfate reducing microscopic organisms is another possibility. The addition or removal of electrons changes the oxidation state of dangerous metals, changing their chemical properties (Prakash et al. 2012).

Biotransformation has two methods. Direct enzymatic reduction reduces multivalently dangerous metal ions by taking electrons from non-cell enzymes (Chaudhary et al. 2021). In sedimentary and underground settings, metal-reducing or sulfate-reducing bacteria may indirectly decrease and immobilize multivalently dangerous metal ions (Tabak et al. 2005).

Bioaccumulation

It involves the amalgamation of toxins within a live creature that move in cell biomass, cell growth, and cell concentration. Bioaccumulation demands a high metabolism. The cell cytoplasm combines responses for natural contaminants to create diverse mixes, but metals are buried (Demirba 2001).

Researchers analyzed waste luminophores from compact fluorescence lamps (CFLs) and fluorescent lamps (FLs) containing rare earth elements (REEs) quantitatively and qualitatively (Olayinka-Olagunju et al. 2021). Two times more REEs were found in CFL luminophores than in FL luminophores (Čížková et al. 2021). These outcomes propose that *Galdieria phlegrea* may bioaccumulate REEs from E-waste at low pH. Landfill contaminants may promote adipogenesis and obesity. Heavy metals may cause hepatotoxicity and nephrotoxicity due to strong correlations between liver and kidney metal concentrations and clinical markers of cellular damage (Gbadebo et al. 2021). This might provide the basis for eco-friendly REE bioaccumulation. Utilize low-cost waste glycerol as a carbon source for the purpose of cultivating microorganisms via mixotrophic or heterotrophic growth in order to facilitate large-scale production (Čížková et al. 2021).

The source of novel brominated flame retardant (NBFR) emissions affects their dispersion in aquatic environments and species (Chen et al. 2020). Some NBFR concentrations in coastal sediments from source locations may exceed the PNEC, and sediment NBFR levels are rising worldwide. Physicochemical properties affect NBFRs' environmental behaviors, destinies, and bioaccumulation potentials (Hou et al. 2021).

Biosorption

Biosorption can effectively settle essential pollutants outside of cellular association without requiring high metabolism, targeting ionized harmful metals within a container's soluble contaminants (Volesky and Holan 1995). Biosorption also enhances the extraction of precious metals and REE from mixed solutions, hence increasing the yield of valuable elements (Tunali and Yenigun 2021). Additional study is required to develop effective techniques for isolating algae and extracting the sorbed ions in order to get these elements (Tunali and Yenigun 2021).

Various reaction-based kinetic models have revealed that the uneven absorption of Hg (II) by different biosorbents, among these various biosorbents, banana peels were observed as the most capable choice for treating the water (Fabre et al. 2021). Diffusion models suggest that film diffusion primarily controls mercury sorption in banana peels, coffee waste, potato peels, and water hyacinth. The biosorption process is initially governed by the film diffusion method, followed by intraparticle diffusion (Ibrahim et al. 2016).

In addition to, this several microorganisms are also used for the remediation purpose of E-waste, which are listed in Table 4.

Alternative Methods for Managing Electronic Waste

Efficient management and regulation of E-waste are essential for the establishment of a sustainable society. As technological advancements continue, the creation of electronic devices leads to an inevitable increase in E-waste. Researchers stressed on this fact, asserting that "today's electronics will be future's electronic waste" (Nunes et al. 2021).

Table 4 E-waste treatment by biological technique uses microorganisms (Tabak et al. 2005; Prakash et al. 2012; Atkinson et al. 1998; Ahalya et al. 2003; Francis 2006; Hou et al. 2006; Juwarkar and Yadav 2010; Srinath et al. 2001)

S. no.	Biological technique	Source of microorganism	Name of microorganism	Removal of toxic metal
1	Bioleaching	Bacteria	*Acidithiobacillus thioxidans*	Arsenic, lead
2	Bioleaching	Bacteria	*Micrococcus roseus*	Cadmium
3	Bioleaching	Bacteria	*Thiobacillus ferrooxidans*	Arsenic, lead
4	Bioleaching	Fungus	*Aspergillus fumigatus*	Arsenic
5	Biomineralization	Bacteria	*Bacillus fusiformis*	Lead
6	Biomineralization	Bacteria	*Cupriavidus metallidurans*	Cadmium
7	Biomineralization	Bacteria	*Desulfotomaculum auripigmentum*	Arsenic
8	Biomineralization	Fungus	*Aspergillus flavus*	Lead
9	Biotransformation	Bacteria	*Anaeromyxobacter* sp.	Uranium
10	Biotransformation	Bacteria	*Clostridium sphenoides*	Uranium
11	Biotransformation	Fungus	*Fusarium oxysporum*	Cadmium
12	Biotransformation	Fungus	*Rhizopus oryzae*	Chromium
13	Bioaccumulation	Bacteria	*Bacillus circulans*	Chromium
14	Bioaccumulation	Bacteria	*Bacillus megaterium*	Chromium
15	Bioaccumulation	Bacteria	*Deinococcus radiodurans*	Uranium
16	Bioaccumulation	Fungus	*Aspergillus niger*	Chromium, lead
17	Biosorption	Bacteria	*Bacillus sphaericus*	Chromium
18	Biosorption	Bacteria	*Myxococcus xanthus*	Uranium
19	Biosorption	Bacteria	*Streptoverticillium cinnamoneum*	Lead
20	Biosorption	Fungus	*Saccharomyces cerevisiae*	Cadmium

E-waste consists of discarded electronic devices and their components that often contain hazardous and toxic constituents such as heavy metals, flame retardants, and various other toxic substances. Improper E-waste management can result in severe health issues and environmental impacts. Traditional methods of E-waste disposal, such as landfilling and incineration, are not sustainable due to risks of soil and water contamination and the emission of toxic fumes. Therefore, more effective and sustainable strategies are essentially required.

Researchers are increasingly turning to biological methods to recover valuable materials from E-waste. Bioleaching, bioaccumulation, biosorption, biomineralization and biotransformation techniques are being investigated for its potential to provide environmentally friendly and cost-effective solutions. Baniasadi et al. (2019) highlighted that these methods offer a more feasible strategy compared to conventional techniques.

In addition to these methods, genetically modified organisms (GMOs) can be a capable tool to optimize the effectiveness of E-waste management. Scientists are investigating the potential of GMOs to degrade E-waste more effectively. Ismail and Hanafiah 2020 pointed out that biological methods are eco-friendly and can be efficient in managing E-waste. Genetically modified microorganisms can be engineered to have enhanced capabilities for breaking down complex electronic waste components (Zeng et al. 2017).

The utilization of GMOs in E-waste management could revolutionize how we handle this growing problem. These organisms may be engineered to selectively target certain pollutants and can transform them into less detrimental substances. GMOs have the potential to be genetically manipulated to synthesize enzymes capable of decomposing plastics or metals found in E-waste, hence facilitating their extraction and recycling processes. Awasthi et al. (2018) discuss the potential of GMOs to transform E-waste into a liquid form, facilitating the recovery and reuse of valuable materials.

The genetically engineered microorganisms (GEMs) can also be a promising tool for the degradation of various components and metals generated through E-waste. The in silico study can be an effective tools and techniques for the construction of GEMs that help in the degradation of E-waste through in vitro methods (Awasthi et al. 2019a).

Despite the promising potential of biological methods, many microorganisms with potential for E-waste degradation remain undiscovered and unexplored. Ilankoon et al. (2018) noted the vast diversity of microorganisms with untapped potential for E-waste management. Continued research is essential to identify and harness these microorganisms' capabilities.

Other biological techniques, such as vermi-remediation and rhizofiltration, offer additional avenues for E-waste management. Vermi-remediation uses earthworms to process and detoxify waste. Earthworms are natural bioaccumulators, capable of transforming contaminants within their tissues. Researchers indicated that earthworms can biodegrade or biotransform hazardous substances into non-hazardous forms, helping detoxify E-waste and enhancing soil fertility through nutrient-rich vermicompost production (Sinha et al. 2010).

Rhizofiltration is a method that involves the use of plant roots to uptake and accumulates high levels of heavy metals from water that has been polluted. Plants with vast root systems have the capability to absorb significant amounts of metals, which are then collected and subjected to processing in order to extract the metals. This technology is very efficient in treating leachates from E-waste and avoiding the dispersion of pollutants in the environment.

Integrating these biological methods into E-waste management strategies offers a sustainable way out of the growing E-waste problem. Leveraging the natural processes of microorganisms and plants can help in the recovery of valuable components for reuse from E-waste and can also reduce the environmental shock. The eco-friendly nature of these methods makes them a feasible substitute for out-of-date E-waste management techniques, which often involve harmful chemicals and processes (Ikhlayel 2017).

Proper treatment and management of E-waste is essential for defending human well-being and the environment. Biological methods, including bioleaching, bioaccumulation, biosorption, biomineralization, biotransformation, utilization of GMOs, and the utilization and construction of GEMs can provide viable strategies for sustainable handling of electronic trash. These techniques are environmentally friendly, cost-effective, and capable of transforming hazardous E-waste into valuable resources. Ongoing research and innovation in this domain are essential for the progress of more efficient and effective technologies for the E-waste managment. By adopting these biological approaches, we can mitigate and treat the adverse effects of E-waste and help in promoting a clean and highly sustainable future.

Conclusion

The rapid advancement of electronic technologies has resulted in a substantial rise in E-waste, which presents substantial risk to both humans and the living world. The improper disposal and handling of E-waste help to release toxic and harmful compounds that have the potential to pollute air, soil, and water, which lead to severe health and ecological consequences. Addressing this growing problem necessitates effective and sustainable waste management solutions. Among the various methods, biological techniques have emerged as a promising and environmentally friendly approach for E-waste treatment, recovery, and management.

Microorganisms are at the forefront of these biological methods, serving as invaluable tools for transforming hazardous E-waste into less harmful or non-hazardous forms. Techniques such as bioleaching, bioaccumulation, biosorption, biomineralization, biotransformation, and chemisorption of metals enhanced by microbes, influence the natural metabolic processes of microorganisms to recover valuable metals and detoxify harmful substances.

Bioleaching, for instance, utilizes bacteria and fungi to solubilize metals from E-waste, making them more readily recoverable and reusable. Bioaccumulation sees microorganisms absorbing and concentrating toxic elements within their cells. Biosorption involves the binding of heavy metals to microbial biomass, facilitating their removal from the environment. Biomineralization results in the formation of stable mineral forms of metals, reducing their mobility and bioavailability, while biotransformation alters the chemical structure of hazardous compounds, rendering them less toxic. Finally, microbes are used to enhance chemisorption of metals, where microbial activity combines with chemical processes and enhances the adsorption of metals onto surfaces, aiding in their recovery.

The potential of genetically engineered microorganisms to degrade E-waste further expands the scope of biological treatment. These engineered strains can be tailored to target specific contaminants more efficiently, enhancing the overall effectiveness of solutions for managing electronic waste.

The advantages of these biological techniques are manifold. They are very cost-effective and feasible as well as environmentally friendly, minimizing secondary pollution. By incorporating these methods into E-waste treatment and management practices, we can mitigate the detrimental impacts of E-waste on human beings as well as on the ecosystem. These biological techniques can pave the way for a more accountable and sustainable approach to technological progress towards the treatment of E-waste.

References

Abatenh E, Gizaw B, Tsegaye Z (2017) Application of microorganisms in bioremediation review. J Environ Microbiol 1(1):02–09

Ahalya N, Ramachandra TV, Kanamadi RD (2003) Biosorption of heavy metals. Res J Chem Environ 7:71–78

Ahirwar R, Tripathi AK (2021) E-waste management: a review of recycling process, environmental and occupational health hazards, and potential solutions. Environ Nanotechnol Monit Manag 15:100409

Arshadi M, Yaghmaei S (2020) Bioleaching of basic metals from electronic waste PCB's. J Min Mech Eng 1:41–50

Arslan V (2021) Bacterial leaching of copper, zinc, nickel and aluminium from discarded printed circuit boards using acidophilic bacteria. J Mater Cycles Waste Manag 23:1–11

Atkinson BW, Bux F, Kasan HC (1998) Considerations for application of biosorption technology to remediate metal-contaminated industrial effluents. Water 24(2):129–135

Awasthi G, Kumari A, Pant AB et al (2018) *In silico* identification and construction of microbial gene clusters associated with biodegradation of undesired toxic materials. Microb Pathog 114: 340–343

Awasthi AK, Hasan M, Mishra YK et al (2019) Environmentally sound system for E-waste, biotechnological perspectives. Curr Res Biotechnol 1:58–64

Awasthi G, Yadav R, Srivastava P (2019a) De-novo hybrid protein design for biodegradation of organophosphate pesticides. Microbiol Biotechnol Lett 47(2):278–288

Balde CP, Wang F, Wong J et al (2015) The global e-waste monitor-2014, United Nations University. IAS-SCYCLE, Bonn

Baniasadi M, Vakilchap F, BahalooHoreh N et al (2019) Advances in bioleaching as a sustainable method for metal recovery from e-waste. J Ind Eng Chem 76:75–90

Bastiaan C, Zoeteman J, Krikke HR et al (2010) Handling WEEE waste flows, on the effectiveness of producer responsibility in a globalizing world. Int J Adv Manuf Technol 47:415–453

Bathurst PA, McMichael AJ, Wigg NR et al (1992) Environmental exposure to lead and children's intelligence at the age of seven years, The Port Pirie cohort study. N Engl J Med 327:1279–1284

Chaudhary A, Akram AM, Ahmad QUA et al (2021) Optimized biotransformation of acid-treated water melon peel hydrolyzate into ethanol. Braz J Biol 83:e253009

Chen Y, Guo M, Liu R et al (2020) Effects of novel brominated flame retardants and metabolites on cytotoxicity in human umbilical vein endothelial cells. Chemosphere 253:126653

Čížková M, Mezricky P, Mezricky D et al (2021) Bioaccumulation of rare earth elements from waste luminophores in the red algae, Galdieriaphlegrea. Waste Biomass Valorization 12(6): 3137–3146

Demirba A (2001) Heavy metal bioaccumulation by mushrooms from artificially fortified soils. Food Chem 74(3):293–301

Diaz-Martinez ME, Argumedo-Delira R, Sanchez-Viveros G et al (2019) Microbial bioleaching of Ag, Au and Cu from printed circuit boards of mobile phones. Curr Microbiol 76:536–544

Etaware PM (2021) Biomineralization of toxicants, recalcitrant and radioactive wastes in the environment using genetically modified organisms. J Agric Res Adv 3(2):28–36

Fabre E, Vale C, Pereira E et al (2021) Sustainable water treatment: use of agricultural and industrial wastes to remove mercury by biosorption. Water Air Soil Pollut 232(7):1–17

Ferronato N, Torretta V (2019) Waste mismanagement in developing countries: a review of global issues. Int J Environ Res Public Health 16(6):1060

Francis AJ (2006) Biotransformation of uranium and other actinides in radioactive wastes. J Alloys Compd 271:78–84

Gbadebo AM, Alabi OA, Alimba CG et al (2021) Metal bioaccumulation, cytogenetic and clinico-biochemical alterations in *Rattus norvegicus* exposed in situ to a municipal solid waste landfill in Lagos, Nigeria. Biol Trace Elem Res 200:1–16

Ghulam ST, Abushammala H (2023) Challenges and opportunities in the management of electronic waste and its impact on human health and environment. Sustainability 15(3):1837

Gollakota AR, Gautam S, Shu CM (2020) Inconsistencies of e-waste management in developing nations-facts and plausible solutions. J Environ Manag 261:110234

Grace PK, Sundar PS, Balaji D et al (2020) Sustainable electronic-waste management: implications on environmental and human health. In: Khan A, Inamuddin, Asiri AM (eds) E-waste recycling and management: present scenarios and environmental issues. Springer International Publishing, Germany, pp 201–218

Gunjal A, Waghmode M, Patil N et al (2020) Environmental management of e-waste by biological process. In: Sangetha J, Thangadurai D, Islam S (eds) Beneficial microbes for sustainable agriculture and environmental management. Apple Academic Press, New York, pp 139–155

Heydarian A, Mousavi SM, Vakilchap F et al (2018) Application of a mixed culture of adapted acidophilic bacteria in two-step bioleaching of spent lithium-ion laptop batteries. J Power Sources 378:19–30

Hou B, Zhu K, Lu J et al (2006) Research on bioremediation of petroleum contaminated soils and its perspectives. Sichuan Environ 6(24):131–140

Hou R, Lin L, Li H et al (2021) Occurrence, bioaccumulation, fate, and risk assessment of novel brominated flame retardants (NBFRs) in aquatic environments. Water Res 198:117168

Ibrahim WM, Hassan AF, Azab YA (2016) Biosorption of toxic heavy metals from aqueous solution by *Ulva lactuca* activated carbon. J Basic Appl Sci 3:241–249

Ikhlayel M (2017) Environmental impacts and benefits of state-of-the-art technologies for E-waste management. Waste Manag 68:458–474

Ikhlayel M (2018) An integrated approach to establish e-waste management systems for developing countries. J Clean Prod 170:119–130

Ilankoon IMSK, Ghorbani Y, Chong MN et al (2018) E-waste in the international context – a review of trade flows, regulations, hazards, waste management strategies and technologies for value recovery. Waste Manag 82:258–275

Islam A, Ahmed T, Awual MR et al (2020) Advances in sustainable approaches to recover metals from e-waste-a review. J Clean Prod 244:118815

Ismail H, Hanafiah MM (2020) A review of sustainable e-waste generation and management: present and future perspectives. J Environ Manag 264:110495

Jang YC, Townsend TG (2003) Leaching of lead from computer printed wire boards and cathode-ray tubes by municipal solid waste landfill leachates. Environ Sci Technol 37(20):4778–4784

Jiang B, Adebayo A, Jiad J et al (2019) Impacts of heavy metals and soil properties at a Nigerian e-waste site on soil microbial community. J Hazard Mater 362:187–195

Juwarkar AA, Yadav SK (2010) Bioaccumulation and biotransformation of heavy metals. In: Fulekar MH (ed) Bioremediation technology. Springer, Dordrecht, pp 266–284

Kumar A (2019) Exploring young adults'e-waste recycling behaviour using an extended theory of planned behaviour model: a cross-cultural study. Resour Conserv Recycl 141:378–389

Lovley DR, Coates JD (1997) Bioremediation of metal contamination. Curr Opin Biotech 8(3): 285–289

Mejía-Rodríguez BJ, Bossio-Cerpa LV, Albis-Arrieta AR, et al (2018) Recovery of heavy metals from waste printed circuit boards through microbiological leaching, using consortia of acidophilic chemolithotrophic bacteria. In: 7th International workshop advances in cleaner production Barranquilla, Colombia 21–22, pp 1–10

Mohammed A, Oyeleke SB, Ndamitso MM et al (2013) The impact of electronic waste disposal and possible microbial and plant control. Int J Eng Sci 2(10):35–42

Narayanasamy M, Dhanasekaran D, Thajuddin N (2017) Biological recovery of metals from electronic waste polluted environment using microorganisms. In: Bagyaraj DJ, Jamaluddin (eds) Microbes for restoration of degraded ecosystems. New India Publishing Agency, New Delhi, pp 149–172

Nunes IC, Kohlbeck E, Beuren FH et al (2021) Life cycle analysis of electronic products for a product-service system. J Clean Prod 314:127926

Olayinka-Olagunju JO, Dosumu AA, Olatunji-Ojo AM (2021) Bioaccumulation of heavy metals in pelagic and benthic fishes of Ogbese River, Ondo State, South-Western Nigeria. Water Air Soil Pollut 232(2):1–19

Pietrelli L, Ferro S, Vocciante M (2019) Eco-friendly and cost-effective strategies for metals recovery from printed circuit boards. Renew Sust Energ Rev 112:317–323

Prakash A, Satyanarayana T, Johri BN (2012) Microorganisms in environmental management, microbes, and environment. In: Satyanarayana T, Johri BN, Prakash A (eds) Microorganisms in environmental management. Springer, Dordrecht, New York 49(2)

Puckett J, Byster L, Westervelt S, et al (2002) Exporting harm, the high-tech trashing of Asia. Seattle, Basal Action Network. Available via www.ban.org. Accessed on 10 Mar 2024

Rautela R, Arya S, Vishwakarma S et al (2021) E-waste management and its effects on the environment and human health. Sci Total Environ 773:145623

Saha L, Kumar V, Tiwari J et al (2021) Electronic waste and their leachates impact on human health and environment: global ecological threat and management. Environ Technol Innov 24:102049

Sharma P, Fulekar MH, Pathak BE (2012) Waste- a challenge for tomorrow. Res J Recent Sci 1(3): 86–93

Singh N, Duan H, Yin F et al (2018) Characterizing the materials composition and recovery potential from waste mobile phones: a comparative evaluation of cellular and smart phones. ACS Sustain Chem Eng 6(10):13016–13024

Sinha RK, Agarwal S, Chauhan K et al (2010) Vermiculture technology, reviving the dreams of sir Charles Darwin for the scientific use of earthworms in sustainable development programs. Technol Invest 3:155–172

Srinath T, Verma T, Ramteke PW et al (2001) Chromium (VI) biosorption and bioaccumulation by chromate-resistant bacteria. Chemosphere 48(4):427–435

Stocks-Fischer S, Galinat JK, Bang SS (1999) Microbiological precipitation of CaCO3. Soil Biol Biochem 31(11):1563–1571

Tabak HH, Lens P, Van Hullebusch ED et al (2005) Developments in bioremediation of soils and sediments polluted with metals and radionuclides. Microbial processes and mechanisms affecting bioremediation of metal contamination and influencing metal toxicity and transport. Rev Environ Sci Biotechnol 4(3):115–156

Tichy R, Rulkens WH, Grotenhuis JTC et al (1998) Bioleaching of metals from soils or sediments. Water Sci Technol 37(8):119–127

Tunali M, Yenigun O (2021) Biosorption of Ag^{+} and Nd^{3+} from single-and multi-metal solutions (Ag^{+}, Nd^{3+}, and Au^{3+}) by using living and dried microalgae. J Mater Cycles Waste Manag 23(2):764–777

Volesky B, Holan ZR (1995) Biosorption of heavy metals. Biotechnol Prog 11(3):235–250

Widmer R, Oswald HK, Sinha DK et al (2005) Global perspectives on E-waste. Environ Impact Assess Rev 25:436–458

Withanage SV, Habib K (2021) Life cycle assessment and material flow analysis: two under-utilized tools for informing e-waste management. Sustainability 13(14):7939

Yamanea LH, Moraesa VT, Tenórioa JAS et al (2018) Influence of bacterial adaptation on copper bioleaching from printed circuit boards. Adv Biotechnol Microbiol 9(3):555761

Zeng X, Yang C, Chiang JF et al (2017) Innovating e-waste management: from macroscopic to microscopic scales. Sci Total Environ 575:1–5

The Environmental Impact on Antimicrobial Resistance: A Threat to Biodiversity

Farhan Ahmad Khan

Abstract

Antimicrobial resistance (AMR) is an escalating worldwide menace that presents substantial dangers to human health, biodiversity, and ecosystems. This chapter is devoted to the investigation of the environmental factors of AMR, with an emphasis on how pollutants and human activities affect the emergence and dissemination of resistance. Explaining how AMR develops in the environment, it focuses on the contribution of pharmaceutical production, farming practices, fish farming, water treatment, and human behavior in promoting resistance. AMR is also shown to have environmental consequences in this chapter, such as the influence on ecosystems and biodiversity. These are some of the major issues related to the management of AMR and some of the possible solutions that include regulation, technology, and global partnerships. The last section of the chapter focuses on the direction of further research and the use of the One Health approach and sustainable measures to combat the threat of AMR. This chapter is like a comprehensive analysis which offers an understanding of the relationship between the environment and AMR and can be used as a guide for solving this significant problem using the cooperation of various disciplines and policies.

Keywords

Pollutants · Anthropogenic activities · Endocrine disruptors · Pharmaceutical waste

F. A. Khan (✉)
Department of Pharmacology, Jawaharlal Nehru Medical College and Hospital, Aligarh Muslim University, Aligarh, Uttar Pradesh, India
e-mail: fakhan2@myamu.ac.in

S. Raisuddin et al. (eds.), *Environmental Toxicants and Lifestyle Diseases*,
https://doi.org/10.1007/978-981-95-0575-3_9

Introduction

Antimicrobial resistance (AMR) is one of the most important global public health concerns that now affects human, animal, and environmental health. Even though antibiotic abuse and overuse in humans and animals is a naturally occurring evolutionary phenomenon, it nonetheless poses a serious threat to the development and spread of AMR. AMR affects global health, financial sustainability, food security and sustainability, environmental welfare, and socioeconomic development.

AMR is a pressing and critical issue in contemporary healthcare, with far-reaching consequences for worldwide health, food security, and environmental sustainability. The resilience of bacteria to antimicrobial therapies poses a significant risk to the progress made in medicine over the past decades, making it progressively more challenging, if not impossible, to treat common infections effectively. The environment is crucial in the process of AMR formation and spread, since it serves as a source and pathway for resistant microorganisms and antimicrobial contaminants (Bertagnolio et al. 2023). Gaining insight into the environmental factors contributing to AMR is crucial to formulating comprehensive approaches to address this worldwide public health concern.

The Scope of the AMR Problem

The World Health Organization (WHO) has recognized AMR as one of the foremost worldwide public health perils confronting humanity. According to estimates, if no action is taken, AMR might result in 10 million fatalities per year and incur a cost of up to $100 trillion for the global economy by 2050 (World Health Organization report 2019). The growth of antibiotic-resistant strains is mostly caused by the excessive and inappropriate use of antibiotics in humans, animals, and crops, contributing to the spread of AMR.

Environmental Reservoirs and Pathways

Environmental reservoirs refer to natural or man-made sources where microorganisms or other agents can survive and potentially spread. Pathways, on the other hand, are the routes by which these agents might be transmitted or transferred.

The environment functions as both a storage and a route for the dissemination of AMR. Soil and water bodies contain a wide variety of microorganisms, including bacteria that possess genes for resistance. These settings can be polluted with antimicrobial agents and microorganisms that are resistant to them due to different human activities. Examples of sources that introduce substantial amounts of antibiotics and resistant microorganisms into the environment include agricultural runoff, wastewater effluents, and discharges from pharmaceutical manufacturing (Kummerer 2009; Larsson et al. 2007).

Human Activities and AMR Dissemination

Human activities play a key role in spreading AMR in the environment. Notable individuals who have made significant contributions include:

1. In the field of pharmaceutical manufacture, the wastewater discharged from pharmaceutical factories frequently contains significant concentrations of active medicinal substances. These waste substances can flow into bodies of water, leading to concentrated areas of pollution from antimicrobial substances that favor the growth of bacteria that are resistant to them (Larsson et al. 2007).
2. Agriculture: The utilization of antibiotics in livestock farming and agriculture to enhance growth and preventing disease results in the buildup of antimicrobial residues in soil and water. These remnants facilitate the choice and dissemination of genes that confer resistance among bacteria in the environment (Martinez 2009).
3. Wastewater Treatment: Wastewater treatment plants (WWTPs) play a crucial role in preventing the discharge of antibiotics and bacteria that are resistant to them into the environment. Nevertheless, numerous WWTPs lack the necessary capabilities to efficiently eliminate these pollutants, leading to their discharge into rivers, lakes, and coastal regions (Martinez 2009).
4. Human waste: Human waste, comprising expelled medications and resistant microorganisms, represents a substantial contributor to environmental AMR. Insufficient sewage systems and inadequate sanitation facilities might discharge these pollutants into natural water bodies, hence promoting the dissemination of resistance (Kummerer 2009).

Methods of Antimicrobial Resistance (AMR) Transmission

Various processes contribute to the dissemination of AMR in the environment:

1. Horizontal Gene Transfer (HGT) is the main process via which resistance genes are disseminated across bacteria. Mobile genetic elements, such as plasmids, transposons, and integrons, promote the transfer of resistance genes between different species and genera, hence facilitating the spread of AMR (Allen et al. 2010).
2. Biofilms are intricate communities of bacteria that attach to surfaces and are prevalent in both natural and artificial settings. Biofilms create an ideal environment for HGT and act as focal points for the gathering and dissemination of resistance genes (Davies and Davies 2010).

Environmental Impact of AMR

The environmental consequences of AMR are significant and complex. Antimicrobial pollution has the potential to disturb microbial communities, which are essential for the process of nutrient cycling and the overall functioning of ecosystems. Reducing the variety of microorganisms can lead to a series of negative impacts on larger ecological systems. Antibiotics present in aquatic habitats can have an impact on the well-being and viability of fish and other aquatic creatures. Antimicrobial residues present in soil can have detrimental effects on soil health, plant growth, and terrestrial fauna (Baquero et al. 2008).

Addressing the Environmental Dimensions of AMR

To tackle the environmental aspects of AMR, a comprehensive strategy is needed that involves collaboration among scientists, policymakers, and industrial stakeholders. Efficient monitoring and surveillance of AMR in the environment are crucial for comprehending its extent and influence. This involves creating standardized techniques for identifying and measuring resistance genes and antimicrobial residues in different environmental samples. Furthermore, robust regulatory frameworks are necessary to restrict the discharge of antimicrobials into the environment. The policies should prioritize the reduction of antibiotic usage in agriculture, enhancement of wastewater treatment procedures, and implementation of strict limits on pharmaceutical industry effluents (Rizzo et al. 2013; World Health Organization 2015).

Ultimately, effectively addressing AMR necessitates a comprehensive strategy that encompasses its environmental aspects. Future research should prioritize comprehending the intricate interplay between human activities and environmental AMR, devising efficient measures to mitigate its impact, and promoting international cooperation. Collaboration between policymakers, scientists, and industrial stakeholders is essential to protect public health and maintain the integrity of ecosystems.

Environmental Drivers of AMR

AMR is caused by a range of environmental conditions that promote the transmission and long-term existence of bacteria that are resistant to antimicrobial agents and the remnants of these agents. Gaining a comprehensive understanding of these factors is essential for formulating effective methods to reduce AMR. This section explores the primary environmental factors that contribute to AMR, such as pharmaceutical production and waste, agricultural methods, wastewater treatment and discharge, human waste and sewage, animal farming, and endocrine-disrupting substances.

Pharmaceutical Manufacturing and Waste

Pharmaceutical manufacture is a major contributor to the contamination of antimicrobial substances. Wastewater from manufacturing plants frequently contains elevated levels of active pharmaceutical substances, which are released into aquatic environments and promote the emergence and spread of antibiotic-resistant bacteria. Research has indicated that pharmaceutical wastewater can include antibiotic concentrations of several milligrams per liter, which is considerably greater than the amounts detected in municipal wastewater (Kummerer 2009). The absence of strict restrictions and efficient treatment methods worsens the problem.

Agricultural Practices

AMR is significantly exacerbated by the utilization of antibiotics in livestock production and agriculture. The utilization of antibiotics is widespread in order to stimulate growth and mitigate the occurrence of illnesses in animals, resulting in the buildup of antimicrobial remnants in soil and water. These remnants facilitate the choice and dissemination of genes that confer resistance among bacteria in the environment. The utilization of manure from treated animals as fertilizer exposes both residues and resistant bacteria into the environment, hence intensifying the dissemination of antimicrobial resistance (Martinez 2009).

Wastewater Treatment and Effluents

Wastewater treatment facilities (WWTPs) play a crucial role in preventing the discharge of antibiotics and microorganisms that are resistant to them into the environment. Nevertheless, numerous WWTPs lack the necessary capabilities to efficiently eliminate these pollutants. Research has demonstrated that even highly developed treatment methods may not entirely eradicate antibiotics and resistance genes, resulting in their discharge into rivers, lakes, and coastal regions. Antibiotics included in treated wastewater can promote the growth of bacteria that are resistant to them, leading to their dissemination throughout the water system (Rizzo et al. 2013).

Human Waste and Sewage

Human waste, which contains both excreted antibiotics and bacteria that are resistant to antibiotics, is a significant factor contributing to the spread of antimicrobial resistance in the environment. Insufficient sanitation facilities and sewage systems might discharge these pollutants into natural water bodies, hence promoting the dissemination of resistance. The inadequate disposal of unused antibiotics and the existence of medicines in human waste emphasize the necessity for efficient waste management and public awareness efforts (Kummerer 2009).

Animal Husbandry

Animal husbandry techniques, especially in the context of large-scale industrial farming, provide a substantial contribution to environmental AMR. Excessive utilization of antibiotics in animal feed and inadequate waste management methods lead to the discharge of antimicrobial residues and microorganisms that are resistant to treatment into the environment. These practices result in the pollution of soil, water, and crops, facilitating the transmission of resistance genes across the food chain and the environment (Marshall and Levy 2011).

Endocrine Disruptors

Endocrine disruptors are substances that disturb the hormonal systems of organisms. Several antimicrobial drugs, especially those employed in veterinary care, have the ability to function as endocrine disruptors. These compounds can significantly impact the reproductive, developmental, and immune systems of wildlife, resulting in changes to microbial ecosystems and the emergence of resistant strains. Endocrine disruptors in the environment can worsen the spread of AMR by disturbing natural microbial ecosystems and encouraging the survival of bacteria that are resistant to antibiotics (Sumpter and Jobling 2013).

Environmental Transmission Pathways

AMR can be transmitted through many environmental routes, including as water, soil, and air. Water bodies that are contaminated can operate as reservoirs and pathways for bacteria that are resistant to treatment and residues of antimicrobial substances. Soil contaminated with animal dung or wastewater effluents can include bacteria that are resistant to antibiotics and can be transmitted through agricultural activities. Resistant bacteria can be transmitted through the air via dust particles and aerosols that are produced by agricultural operations and wastewater treatment plants (Allen et al. 2010).

Impact of Climate Change

Climate change is a developing element that is affecting the spread of AMR. Fluctuations in temperature, patterns of rainfall, and occurrences of severe weather might impact the ability of resistant bacteria to survive, spread, and be transmitted in the environment. For instance, elevated temperatures can stimulate the proliferation of bacteria and the breakdown of antibiotics, while intense rainfall and flooding can amplify the discharge of antimicrobial residues into bodies of water. Comprehending the influence of climate change on AMR is crucial for formulating adaptive approaches to alleviate its consequences (MacFadden et al. 2018).

Mechanisms of AMR Spread in the Environment

The proliferation of AMR in the environment is propelled by intricate mechanisms that encompass the interplay of microbial populations, the transmission of resistance genes, and the impact of human activities. This section examines the main ways in which AMR spreads in the environment. These include horizontal gene transfer, the influence of antimicrobial pollutants, the creation of biofilms, and the involvement of mobile genetic elements.

Horizontal Gene Transfer (HGT)

Horizontal gene transfer (HGT) is a crucial process that enables the rapid dissemination of AMR. Horizontal gene transfer refers to the process by which genetic material is horizontally transferred between different organisms, rather than being passed down vertically from parent to offspring. The process is aided by mobile genetic elements, such as plasmids, transposons, and integrons, which have the ability to transfer resistance genes between different species and genera. HGT, also known as horizontal gene transfer, can occur through three primary mechanisms: transformation, transduction, and conjugation (Frost et al. 2005).

1. Transformation: This process entails the absorption of extracellular DNA by a bacterial cell. This DNA may come from deceased or ruptured cells and may contain genes that confer resistance. Upon entering the cell, the foreign DNA can be incorporated into the bacterial genome, resulting in the acquisition of novel resistance characteristics (Thomas and Nielsen 2005).
2. Transduction: This process involves the transfer of genetic material between bacteria through the action of bacteriophages, which are viruses that infect bacteria. During the course of infection, bacteriophages have the potential to inadvertently encapsulate bacterial DNA, which may include genes that confer resistance, and transmit it to different bacterial hosts (DePaepe et al. 2014).
3. Conjugation: This refers to the process of DNA transfer between bacterial cells through direct physical contact. Conjugation frequently entails the utilization of plasmids, which are circular DNA molecules capable of harboring numerous resistance genes. Plasmids have the ability to transfer between several bacterial species, facilitating the fast dissemination of resistance throughout microbial communities (Heuer and Smalla 2007).

Selective Pressure from Antimicrobial Pollutants

The presence of antimicrobial contaminants in the environment exerts a significant influence on the development and spread of AMR. Antibiotics in soil, water, or sediments provide an environment conducive to the survival and proliferation of only resistant bacteria. This phenomenon of selective pressure leads to an augmentation in the prevalence of resistance genes within microbial populations (Kummerer 2009).

1. Environmental persistence of antibiotics: Antibiotics exhibit environmental persistence, as they possess stability and can endure in the environment for prolonged durations. Their existence in soil and water exerts a constant selective force, fostering the survival of strains that are resistant (Martinez 2008).
2. Antibiotic residues in agriculture: The utilization of antibiotics in agriculture results in the buildup of antimicrobial remnants in soil and water. These remnants stimulate the growth of bacteria that are resistant to antibiotics in agricultural environments, and these bacteria can then spread to the wider surroundings through the flow of water and the use of irrigation (Marshall and Levy 2011).
3. Wastewater and effluents: Residual antibiotics are frequently present in effluents originating from wastewater treatment plants, pharmaceutical production facilities, and hospitals. These waste substances discharge antibiotics into natural water sources, causing a selective influence on microbial communities in water and facilitating the proliferation of resistance (Rizzo et al. 2013).

Formation of Biofilms

Biofilms are organized populations of microorganisms that are adhered to surfaces and enclosed within a self-generated extracellular matrix. Biofilms are commonly seen in both natural and manmade environments and have a substantial impact on the long-term existence and transmission of antimicrobial resistance (Flemming and Wingender 2010). Biofilms create a favorable environment for HGT and act as focal points for the gathering and dissemination of resistance genes.

1. Enhanced resistance: Biofilm bacteria demonstrate heightened resistance to antibiotics in comparison to planktonic (free-living) bacteria. The extracellular matrix and the close proximity of cells within biofilms establish a defensive setting that restricts the infiltration of antibiotics and amplifies horizontal gene transfer (Stewart and Costerton 2001).
2. Hotspots for gene exchange: Biofilms serve as optimal locations for horizontal gene transfer, facilitating the exchange of genes. The compact microbial populations and the existence of movable genetic components, such as plasmids and transposons, enable the transfer of resistance genes inside biofilms (Madsen et al. 2012).

3. Environmental reservoirs: Biofilms found in natural habitats including riverbeds, sediments, and plant roots, can operate as storage areas for bacteria that are resistant to certain substances. Biofilms have the ability to release bacteria that are resistant to treatment into the environment, hence facilitating the spread of antimicrobial resistance (Stalder et al. 2012).

Role of Mobile Genetic Elements

Mobile genetic elements (MGEs) are segments of genetic material capable of transferring between genomes and are essential for facilitating the dissemination of resistance genes. Mobilizable genetic elements encompass plasmids, transposons, integrons, and insertion sequences (Frost et al. 2005).

1. Plasmids: These refer to extrachromosomal DNA molecules capable of harbouring numerous resistance genes. Conjugation facilitates the transfer of resistance genes between bacteria, allowing resistance to proliferate among many species and conditions (Norman et al. 2009).
2. Transposons: Sometimes known as "jumping genes," these are specific DNA sequences that have the ability to relocate themselves within the genome. They have the ability to transport resistance genes and transfer them between chromosomes, plasmids, and other MGEs, hence promoting the dissemination of antimicrobial resistance (Heuer and Smalla 2007).
3. Integrons: Genetic elements that have the ability to capture and assimilate gene cassettes, which may include those that encode antibiotic resistance. They have a crucial function in the organization and spreading of resistance genes throughout microbial communities (Cambray et al. 2010).

Environmental Transmission Pathways

AMR is transmitted in the environment through multiple channels, including water, soil, air, and food.

1. Waterborne transmission: Water sources that are contaminated, such as rivers, lakes, and coastal areas, operate as reservoirs and pathways for bacteria that are resistant to treatment and residues of antimicrobial substances. Aquatic ecosystems are especially susceptible to pollution caused by agricultural runoff, wastewater discharges, and the release of pharmaceuticals (Pruden et al. 2006).
2. Soil and sediment: Serve as storage areas for both resilient bacteria and antibiotic residues. The application of manure and biosolids in agricultural techniques introduces antibiotics and resistance genes into the soil. These pollutants have the ability to endure and disperse through the process of irrigation and the flow of water (Heuer et al. 2011).

3. Airborne transmission: This can occur through the dispersion of dust particles and aerosols that are produced by agricultural activities, wastewater treatment plants, and composting facilities. Aerobic bacteria have the ability to disperse across large distances, thereby disseminating genes that confer resistance in many habitats (Finley et al. 2013).
4. Food chain: An additional route for the spread of AMR. Contaminated water and soil can transmit bacteria that are resistant to antibiotics into crops and livestock. Human exposure to resistant bacteria can occur through the consumption of contaminated food products, which in turn contributes to the spread of antimicrobial resistance (Woolhouse et al. 2015).

Environmental Impact of AMR

The environmental consequences of AMR are significant and complex, impacting ecosystems, biodiversity, and human health. This section examines several ways in which AMR impacts the environment, encompassing the pollution of natural resources, the disturbance of microbial populations, the ramifications for wildlife, and the wider ecological repercussions.

Contamination of Natural Resources

The existence of antimicrobial residues and germs that are resistant to treatment in natural resources such as soil, water, and air is a major cause for concern. These toxins arise from several sources, such as agricultural activities, pharmaceutical production, and wastewater treatment plants (Martinez 2009).

1. Soil contamination: This occurs when antibiotics are used in agriculture, namely in animal and agricultural production, resulting in the buildup of antimicrobial residues in the soil. Manure and biosolids used as fertilizers frequently include antibiotics, which can endure in the soil and exert selection pressure on bacteria, leading to the development of antibiotic-resistant strains (Heuer et al. 2011).
2. Water pollution: Water bodies serve as significant repositories for drug residues and bacteria that are resistant to antibiotics. Runoff from agricultural areas, discharge from wastewater treatment plants, and effluents from pharmaceutical enterprises are sources of pollution that contaminate rivers, lakes, and groundwater (Kummerer 2009). This pollution not only impacts marine organisms but also presents hazards to human well-being when consuming polluted water (Pruden A et al. 2006).
3. Airborne dissemination: Refers to the spread of antibiotics and resistant bacteria through the air, usually carried by dust particles and aerosols. This is particularly common in agricultural environments and while handling animal manure. These airborne pollutants have the ability to disperse across vast distances, hence

facilitating the spread of AMR in the surrounding environment (Finley et al. 2013).

Disruption of Microbial Communities

The introduction of antibiotics into the environment causes disturbances in natural microbial populations, resulting in imbalances that can have significant ecological ramifications.

1. Alteration of soil microbiota: Soil microbes have a vital function in the process of nutrient cycling, decomposition of organic materials, and the well-being of plants. Contamination with antibiotics can modify the makeup and operation of soil microbiota, leading to a decrease in the variety of microorganisms and affecting the fertility of the soil (Wellington et al. 2013).
2. Impact on aquatic microorganisms: Aquatic ecosystems depend on a harmonious microbial community for vital functions like nutrient cycling and water purification. The presence of antibiotic residues in water can disturb these communities, resulting in reduced water quality and compromised ecosystem functions (Rizzo et al. 2013).
3. Biofilm formation: Biofilms, consisting of microorganisms adhered to surfaces, can serve as a habitat for germs that are resistant to treatment. Biofilms can develop on both natural and artificial surfaces such as rocks, plant roots, and water pipelines. This can result in the long-term presence and spread of AMR, making it more difficult to control microbial contamination (Flemming and Wingender 2010).

Implications for Wildlife

The proliferation of AMR in the environment has both direct and indirect impacts on wildlife. Animals can serve as reservoirs and vectors for bacteria that are resistant to antibiotics, which helps to spread AMR between different environments and species.

1. Exposure through contaminated resources: Wildlife can come into contact with antibiotics and resistant bacteria through water, soil, and food sources that have been contaminated. For instance, fish and aquatic invertebrates residing in contaminated water bodies have the ability to amass antibiotics and resistant bacteria. These substances can subsequently be transmitted via the food chain (Martinez et al. 2009).
2. Impact on health and biodiversity: Antibiotic exposure can damage the health of wildlife populations, making them more susceptible to infections. In addition, the dissemination of AMR might diminish biodiversity by promoting the proliferation of resistant strains at the expense of susceptible ones, resulting in alterations

in the composition of species and the functioning of ecosystems (Arnold et al. 2016).

3. Interplay with human activities: Wildlife frequently engages with human surroundings, including urban areas, agricultural regions, and trash disposal sites. These interactions can promote the transmission of antibiotic-resistant bacteria among wildlife, domestic animals, and humans, establishing a cycle of dissemination of antimicrobial resistance (Franklin et al. 2016).

Broader Ecological Consequences

The dissemination of AMR in the environment has wider ecological ramifications that go beyond direct health effects. These encompass the disturbance of ecosystem functions, difficulties in waste management, and the possibility of worldwide spread.

1. Ecosystem services: Robust microbial communities are crucial for ecosystem services such as the efficient cycling of nutrients, the maintenance of soil fertility, and the effective filtration of water. The antibiotic pollution can disturb these communities, leading to impaired services that have negative effects on agricultural output, water quality, and overall ecosystem health (Kummerer 2009).
2. Waste management challenges: The existence of antibiotic residues and resistant bacteria in waste products adds complexity to waste management procedures. The treatment of agricultural runoff, wastewater, and industrial effluents is becoming increasingly difficult, necessitating the use of modern technology and management strategies to control the emergence of antimicrobial resistance (Kümmerer et al. 2018).
3. Global dissemination: AMR is a worldwide problem, as bacteria that are resistant to treatment and the genes responsible for this resistance can move across national boundaries through different routes. The global spread of AMR is facilitated by international trade, travel, and migratory animals. Therefore, it is crucial to have coordinated global initiatives to tackle this urgent problem (Woolhouse et al. 2015).

Key Challenges and Solutions

The problem of AMR poses numerous substantial obstacles that encompass scientific, regulatory, and societal spheres. To tackle these difficulties, a comprehensive approach involving all stakeholders is necessary. This section explores the primary obstacles linked to AMR in the environment and suggests viable strategies to successfully address these problems.

Key Challenges

1. *Environmental Contamination and Spread*
   - Dispersed pollution sources: Antibiotics are introduced into the environment through several dispersed sources, such as agricultural runoff, wastewater treatment plants, and pharmaceutical manufacturing. Controlling these dispersed sources is intricate because of the extensive and fluctuating characteristics of pollution (Martinez 2009).
   - Persistence and bioaccumulation: Antibiotics and bacteria that are resistant to them can endure in the environment for prolonged durations and accumulate in the food chain, presenting enduring ecological and health hazards (Heuer et al. 2011).
2. *Detection and Monitoring*
   - Absence of standardized techniques: The absence of standardized techniques hinders the detection and quantification of antibiotic residues and resistance genes in environmental matrices. This impedes the efficient surveillance and juxtaposition of data across diverse research and geographies (Kummerer 2009).
   - Resource constrictions: This might hinder monitoring efforts due to limitations in financing, experience, and technical infrastructure, especially in low- and middle-income countries (Pruden et al. 2006).
3. *Regulatory and Policy Gaps*
   - Insufficient regulations: Current regulations may not sufficiently cover the environmental aspects of AMR. A significant number of countries do not have comprehensive regulations in place to effectively manage the use and disposal of antibiotics, especially in the fields of agriculture and aquaculture (Finley et al. 2013).
   - Challenges in enforcement: Despite the presence of regulations, enforcement might be ineffective due to limited resources, lack of political determination, and difficulties in monitoring compliance (Wellington et al. 2013).
4. *Public Awareness and Behavior*
   - Limited awareness: The general public has a limited understanding of the environmental consequences of antibiotic usage and the dangers of AMR. This leads to insufficient backing for essential policy measures and conscientious behaviors (Rizzo et al. 2013).
   - Antibiotic misuse: The improper and excessive use of antibiotics in human medicine, agriculture, and aquaculture significantly contribute to the development of antimicrobial resistance (AMR). The incorrect use of medical resources is frequently motivated by a lack of knowledge, financial constraints, and insufficient availability of suitable healthcare services (Flemming and Wingender 2010).

Possible Resolutions

1. *Improved Waste Management*
   - Advanced treatment technologies: Incorporating state-of-the-art wastewater treatment technologies, such as membrane filtration, activated carbon adsorption, and advanced oxidation processes, can effectively decrease the presence of antibiotics and resistant bacteria in effluents (Martinez et al. 2009).
   - Decentralized treatment solutions: In order to address the waste management needs of rural and isolated areas, the implementation of decentralized treatment systems, such as built wetlands and anaerobic digesters, can be highly successful. These systems are capable of treating both agricultural and domestic waste, resulting in a significant reduction in environmental contamination (Arnold et al. 2016).
2. *Improved Monitoring and Surveillance*
   - Standardization of Methods: The development and implementation of standardized methods for sampling, detecting, and measuring antibiotics and resistance genes are essential for efficient monitoring and accurate data comparability (Franklin et al. 2016).
   - Global surveillance networks: Creating worldwide networks for monitoring environmental AMR might enable the exchange of data, analysis of trends, and coordinated actions in response to emergent risks (Kummerer 2009).
3. *Strengthening Regulatory Frameworks*
   - Comprehensive legislation: Developing comprehensive legislation is crucial to address all forms of antibiotic pollution, encompassing agricultural, aquaculture, and industrial discharge. Robust enforcement procedures should support this legislation (Kümmerer et al. 2018).
   - International collaboration: Collaborating internationally and harmonizing policies can effectively address the cross-border implications of AMR. The World Health Organization (WHO) has established initiatives such as the Global Action Plan on AMR to give a structured approach for addressing this issue (Woolhouse et al. 2015).
4. *Raising Public Awareness*
   - Education campaigns: Public educational initiatives can enhance public knowledge of the environmental consequences of antibiotics and the significance of responsible utilization. The campaigns should focus on reaching several stakeholders, such as farmers, healthcare practitioners, and the general public (World Health Organization 2015).
   - Community engagement: Involving communities in the process of monitoring and managing the use and disposal of antibiotics can cultivate a feeling of accountability and endorsement for policy actions. Community-driven projects can be especially impactful in rural and resource-constrained environments (Centre for Disease Control and Prevention 2019).
5. *Promoting Research and Innovation*
   - Exploring alternatives: Promoting scientific investigations into substitutes for antibiotics, including as probiotics, bacteriophages, and vaccinations, can

decrease the dependence on antibiotics in both agriculture and medicine (O'Neill 2016).

- Advancing novel technologies: The development of cutting-edge technology to identify, track, and reduce AMR in the environment can significantly improve our capacity to address this problem with greater efficiency. For instance, the utilization of biosensors and mobile laboratories helps expedite the process of doing quick and efficient on-site testing (Pei et al. 2006).

Conclusion and future directions

AMR poses a substantial and escalating danger to worldwide public health, with severe environmental and biodiversity consequences. The complex interaction between environmental factors and the dissemination of AMR highlights the need for a comprehensive strategy that includes scientific investigation, policy formulation, public education, and global cooperation. This chapter has emphasized the environmental factors that contribute to AMR, how resistance spreads across different ecosystems, and the significant environmental consequences that arise from this phenomenon.

Summarizing the Challenges

AMR is worsened by a range of environmental variables, such as pollution caused by pharmaceutical manufacture, inappropriate disposal of antibiotics, agricultural practices, and the use of antibiotics in animal farming. These methods deliver substantial amounts of antimicrobial drugs into natural environments, which then exert selective pressure on microbial populations, resulting in the spread of resistant strains (Kummerer 2009). The enduring presence of antibiotics and resistant microorganisms in the environment presents enduring hazards to both human and ecological well-being.

Comprehensive Strategies for Mitigation

Tackling antimicrobial resistance (AMR) necessitates a comprehensive approach that encompasses:

Strengthening Regulatory Frameworks

It is crucial to develop and enforce strict laws to effectively manage the discharge of antibiotics into the environment from different sources. This encompasses more stringent regulations for the production of pharmaceuticals, the control of agricultural runoff, and the management of garbage (Finley et al. 2013).

Advanced Wastewater Treatment

This involves the utilization of cutting-edge technologies in wastewater treatment facilities to efficiently eliminate antibiotics and antibiotic-resistant bacteria prior to the discharge of treated wastewater into natural aquatic ecosystems (Rizzo et al. 2013).

Implementing Comprehensive Global Surveillance and Monitoring Systems

For effectively monitoring the dissemination of AMR in the environment. It is crucial to have standardized procedures in place for monitoring antibiotic residues and resistance genes to effectively address this issue (Pruden et al. 2006).

Research and Innovation

Facilitating investigations into alternative therapies and preventive strategies, such as the advancement of vaccinations and the use of bacteriophages. The implementation of advanced diagnostic tools and treatment procedures can have a substantial impact on the management of AMR as well.

Public Awareness and Education

Increasing knowledge and understanding among the general public, healthcare professionals, and farmers about the negative environmental consequences of misusing antibiotics, and encouraging the responsible use of antibiotics. Education initiatives and community participation have a vital role in altering behaviors and endorsing policy actions (World Health Organization 2015).

Promoting International Collaboration

Facilitating global cooperation and alignment of policy to tackle the cross-border implications of AMR. Effective collaboration is crucial for the exchange of knowledge, resources, and optimal methods among nations (Centers for Disease Control and Prevention 2019).

Prospects for the Future

To effectively address AMR, it is necessary to focus attention on certain specific areas going forward. Transitioning to sustainable agricultural practices that decrease dependence on antibiotics is crucial for the long-term viability of agriculture. These methods encompass integrated pest management, organic farming, and the utilization of alternative therapies (Franklin et al. 2016).

Implementing a One Health strategy that acknowledges the interdependence of human, animal, and environmental health. Collaborative endeavors across these sectors can result in more efficient approaches for controlling AMR (Arnold et al. 2016). Technological Advancements: Ongoing allocation of resources towards the creation of novel technologies to identify, track, and reduce AMR in the environment. This includes the utilization of biosensors, portable laboratories, and sophisticated analytical methods (O'Neill 2016). Policy and Governance: Enhancing the policy frameworks and governance structures at both national and international levels to guarantee the efficient implementation and enforcement of rules designed to manage AMR (Kümmerer et al. 2018).

Promoting interdisciplinary research entails integrating knowledge from several fields, such as microbiology, environmental science, public health, and social sciences, to find comprehensive solutions for the AMR epidemic (Woolhouse et al. 2015).

Ultimately, combating AMR is a multifaceted and continuous endeavor that requires collaborative actions from every segment of society. Through a thorough comprehension of the environmental aspects of AMR and the implementation of comprehensive policies, we may effectively reduce its impact and protect the well-being of the public and ecosystems for future generations.

Acknowledgments I would like to express my heartfelt gratitude to all those who contributed to the development of this chapter. I am especially grateful to Dr. Sreemedha Choudhury, Resident, Department of Pharmacology, JNMC, Aligarh Muslim University for her dedicated efforts in shaping the content and ensuring the accuracy of the chapter.

References

Allen HK, Donato J, Wang HH, Cloud-Hansen KA, Davies J, Handelsman J (2010) Call of the wild: antibiotic resistance genes in natural environments. Nat Rev Microbiol 8(4):251–259

Arnold KE, Williams NJ, Bennett M (2016) The role of wildlife in the dissemination of antimicrobial-resistant. Biol Lett 12(8):20160137

Baquero F, Martínez JL, Cantón R (2008) Antibiotics and antibiotic resistance in water environments. Curr Opin Biotechnol 19(3):260–265

Bertagnolio S, Suthar AB, Tosas O et al (2023) Antimicrobial resistance: strengthening surveillance for public health action. PLoS Med 20(7):e1004265

Cambray G, Guerout AM, Mazel D (2010) Integrons. Annu Rev Genet 44:141–166

Centers for Disease Control and Prevention (2019) Antibiotic resistance threats in the United States, 2019. Retrieved from https://www.cdc.gov/drugresistance/biggest-threats.html. Accessed on 12 Apr 2024

Davies J, Davies D (2010) Origins and evolution of antibiotic resistance. Microbiol Mol Biol Rev 74(3):417–433
DePaepe M, Leclerc M, Tinsley CR et al (2014) Bacteriophages: an underestimated role in human and animal health? Front Cell Infect Microbiol 4:39
Finley RL, Collignon P, Larsson DG et al (2013) The scourge of antibiotic resistance: the important role of the environment. Clin Infect Dis 57(5):704–710
Flemming HC, Wingender J (2010) The biofilm matrix. Nat Rev Microbiol 8(9):623–633
Franklin AM, Aga DS, Cytryn E et al (2016) Antibiotics in agroecosystems: introduction to the special section. J Environ Qual 45:377–393
Frost LS, Leplae R, Summers AO et al (2005) Mobile genetic elements: the agents of open source evolution. Nat Rev Microbiol 3(9):722–732
Heuer H, Smalla K (2007) Manure and sulfadiazine synergistically increased bacterial antibiotic resistance in soil over at least two months. Environ Microbiol 9(3):657–666
Heuer H, Schmitt H, Smalla K (2011) Antibiotic resistance gene spread due to manure application on agricultural fields. Curr Opin Microbiol 14(3):236–243
Kummerer K (2009) Antibiotics in the aquatic environment: a review- part I. Chemosphere 75(4): 417–434
Kümmerer K, Dionysiou DD, Olsson O et al (2018) A path to clean water. Science 361(6399): 222–224
Larsson DG, de Pedro C, Paxeus N (2007) Effluent from drug manufactures contains extremely high levels of pharmaceuticals. J Hazard Mater 148(3):751–755
MacFadden DR, McGough SF, Fisman D et al (2018) Antibiotic resistance increases with local temperature. Nat Clim Chang 8(6):510–514
Madsen JS, Burmølle M, Hansen LH et al (2012) The interconnection between biofilm formation and horizontal gene transfer. FEMS Immunol Med Microbiol 65(2):183–195
Marshall BM, Levy SB (2011) Food animals and antimicrobials: impacts on human health. Clin Microbiol Rev 24(4):718–733
Martinez JL (2008) Antibiotics and antibiotic resistance genes in natural environments. Science 321(5887):365–367
Martinez JL (2009) Environmental pollution by antibiotics and by antibiotic resistance determinants. Environ Pollut 157(11):2893–2902
Martinez JL, Fajardo A, Garmendia L et al (2009) A global view of antibiotic resistance. FEMS Microbiol Rev 33(1):44–65
Norman A, Hansen LH, Sørensen SJ (2009) Conjugative plasmids: vessels of the communal gene pool. Philos Trans R Soc Lond Ser B Biol Sci 364(1527):2275–2289
O'Neill J (2016) Tackling drug-resistant infections globally: final report and recommendations. HM Government and Wellcome Trust Review Antimicrobial Resistance, London. Retrieved from https://amr-review.org/. Accessed on 12 Apr 2024
Pei R, Kim SC, Carlson KH et al (2006) Effect of river landscape on the sediment concentrations of antibiotics and corresponding antibiotic resistance genes (ARG). Water Res 40(12):2427–2435
Pruden A, Pei R, Storteboom H et al (2006) Antibiotic resistance genes as emerging contaminants: studies in northern Colorado. Environ Sci Technol 40(23):7445–7450
Rizzo L, Manaia C, Merlin C et al (2013) Urban wastewater treatment plants as hotspots for antibiotic-resistant bacteria and genes spread into the environment: a review. Sci Total Environ 447:345–360
Stalder T, Barraud O, Casellas M et al (2012) Integron involvement in environmental spread of antibiotic resistance. Front Microbiol 3:119
Stewart PS, Costerton JW (2001) Antibiotic resistance of bacteria in biofilms. Lancet 358(9276): 135–138
Sumpter JP, Jobling S (2013) The occurrence, causes, and consequences of estrogens in the aquatic environment. Environ Toxicol Chem 22(2):221–231
Thomas CM, Nielsen KM (2005) Mechanisms of, and barriers to, horizontal gene transfer between bacteria. Nat Rev Microbiol 3(9):711–721

Wellington EM, Boxall AB, Cross P et al (2013) The role of the natural environment in the emergence of antibiotic resistance in gram-negative bacteria. Lancet Infect Dis 13(2):155–165
Woolhouse M, Ward M, van Bunnik B et al (2015) Antimicrobial resistance in humans, livestock, and the wider environment. Philos Trans R Soc Lond Ser B Biol Sci 370(1670):20140083
World Health Organization (2015) Global action plan on antimicrobial resistance. Retrieved from https://www.who.int/antimicrobial-resistance/global-action-plan/en/. Accessed on 10 Apr 2024
World Health Organization (2019) Ten threats to global health in 2019. Retrieved from https://www.who.int/news-room/spotlight/ten-threats-to-global-health-in-2019. Accessed on 10 Apr 2024

Role of Perfluoroalkyl Substances (PFAS) on the Reproductive System

Suramya, Mehjbeen Javed, Shaesta Shahid, and Sheikh Raisuddin

Abstract

Perfluoroalkyl substances (PFAS) are a class of man-made chemicals, popular in industrial applications for their heat, oil, and water-resistant properties. A wide range of applications includes common consumer products such as raincoats, non-stick cookware, waterproof clothing, electrical cables, carpets resistant to stains, pizza boxes, and fire-floating foams. In this chapter, the physiochemical properties, and bioaccumulation potential have been discussed. The later part of this chapter is devoted to the adverse effects on the reproductive system of the male and female populations. Studies from epidemiological and experimental animals have also been included. Recent studies on the mechanisms leading to reproductive dysfunction have been highlighted. Despite the current limitations in data, it is anticipated that this chapter could provide valuable understanding for future research in deciphering the role of PFAS in life style diseases.

Keywords

Perfluoroalkyl substances · Leydig cells dysfunction · Reproductive abnormalities · Ovarian dysfunction

Suramya · S. Shahid
Molecular Toxicology Laboratory, Department of Medical Elementology and Toxicology, School of Chemical and Life Sciences, Jamia Hamdard, New Delhi, India
e-mail: suramya_sch@jamiahamdard.ac.in; shaestashahid_sch@jamiahamdard.ac.in

M. Javed · S. Raisuddin (✉)
Department of Medical Elementology and Toxicology, Jamia Hamdard, School of Chemical and Life Sciences, New Delhi, Delhi, India
e-mail: mehjabeenjaved_sch@jamiahamdard.ac.in; sraisuddin@jamiahamdard.ac.in

S. Raisuddin et al. (eds.), *Environmental Toxicants and Lifestyle Diseases*,
https://doi.org/10.1007/978-981-95-0575-3_10

Introduction

The world is impending towards low fertility. It has drawn alarming concern among researchers worldwide. The causative agents remain elusive. There are enough pieces of evidence that suggest exposure to environmental contaminants is one of the major problems directly affecting reproductive health among the young population. According to a study published in The Lancet, birth rates in industrial countries with high economic activity have been below the replacement level. The global fertility rate (GFR) has been reduced to half over the years, which could lead to an inverted population pyramid, increasing the number of older people than the working-age populations. Recent scientific studies have investigated that exposure to certain environmental contaminants leads to reproductive-related disorders. Among the major emerging contaminants, perfluoroalkyl substances (PFAS) have been extensively used in industrial applications due to their heat, oil, and water-resistance properties making them useful in our daily lives (Skutlarek et al. 2006). It is found in thousands of products including water-resistant clothing, oil-resistant cookware, electric cables, food packaging, paints, and firefighting foams (Wee and Aris 2023). PFAS was discovered accidentally in the 1940s; however, it was much later that the world got aware of its potential to induce toxic effects. These chemicals are obnoxious guests that overstay inside the body. This may be because of their strong and stable Carbon-Flourine (C-F) bond that resists degradation (Brunn et al. 2023). Hence, they got the nickname "Forever Chemicals". Once entered into the body they don't leave. Now the question arises what does PFAS do to your body? The answer is nothing good. PFAS has been detected in the blood and serum of human and wildlife populations worldwide.

PFAS exposure is ubiquitous. There is overwhelming evidence that links exposure to these substances (both long- and short-chain, depending on the carbon chain length) with the endocrine system of the body, hence classified under endocrine-disrupting chemicals (EDC). According to the Environmental Protection Agency (EPA), EDCs are the "Exogenous agents that interrupts with synthesis, secretion, transport, metabolism, binding action, or elimination of natural blood-borne hormones that are found in the body and also accountable for homeostasis, reproduction, and developmental processes" (Diamanti-Kandarakis et al. 2009). EDCs are a group of chemicals that exhibit hormone-like properties, mimicking fatty acids, hence disrupting the developmental and biological system of the body by interrupting the hormone system. Several researchers have evaluated the potential role of PFAS in altering cholesterol levels (Morgan et al. 2023), and disturbances in hormone production (Shelly et al. 2019), disrupting thyroid levels, risk of diabetes (Roth and Petriello 2022), insulin resistance (Schillemans et al. 2021), decreasing the immune response (Von Holst et al. 2021) increasing the risk of birth defects, risk of premature ovarian insufficiency (POI) (Zhang et al. 2018), delayed puberty (Pinney et al. 2023), and neurological disorders (Brown-Leung and Cannon 2022). However, the exact mechanism of action of several PFAS is still unclear. Once entered into the body it primarily binds to serum albumin, which is a major transporter protein and hence travels throughout the body. The binding of PFAS compounds to various biological components such as proteins and membranes has been explored to elucidate the mechanism of toxic effects exerted by such chemicals (Zhao et al. 2023). Long-chain PFAS, i.e., PFOS (perfluorooctane sulfonate) and PFOA (perfluorooctanoic acid) were extensively used in industries as surfactants for several decades. Their persistence in the human body (half-lives) raises concern all over the globe. Consecutively, short-chain PFAS was introduced as a replacement and comparatively safer alternative for long-chain PFAS. But the studies show controversial results (Wang et al. 2014; Brendel et al. 2018) making short-chain PFAS an equally worrisome alternative and a matter of concern among researchers and scientists.

Negative effects of long- and short-chain PFAS depend not only on the route, duration of exposure and dose administered but also on the species, gender, and age of the exposed population. The inconsistency in results needs to be analyzed thoroughly, increasing the opportunities for future research.

Physiochemical Properties

There is a whole class of compounds that falls under the tree of PFAS. Nearly, 4700 anthropogenic chemicals lie under this tree. Chemicals differ from carbon chain length and functional groups exhibits unique properties. Structurally, most hydrogen atoms are replaced by fluorine atoms making a highly stable and strong bond between C-F which resists degradation (Baker and Knappe 2021). PFAS has both a hydrophobic (water-repelling) and oleophobic (oil-repelling) nature making it more complicated to study and likable among industrial applications. Some short-chain PFAS are soluble in water, leading to surface and groundwater contamination via leaching from contaminated soils worldwide. Table 1 consists of some known PFAS with their chemical formula and half-lives according to Agency for Toxic Substances and Disease Registry (ASTDR).

Source, Exposure, and Health Hazards

PFAS have been majorly utilized in manufacturing processes such as non-stick cookware, raincoats, stain-resistant carpets, fire-floating foams, cosmetics, and dental floss. They reach the water sources via industrial discharge, contaminating the surface water and groundwater (Skutlarek et al. 2006). Dietary exposure, mainly via drinking water is the primary route of exposure (Roth et al. 2020) also consumption of contaminated fish and other animals involved in the food chain greatly affects the exposure. The exposure sources are illustrated in Fig. 1. Studies have presented occupational exposure occurs due to inhalation of dust particles of PFAS compounds also affected by dermal contact.

Table 1 Some popularly known PFAS along with abbreviations, formulas, chemical structures, and half-lives in humans (ASTDR Perfluoalkyl PFAS ToxGuide)

Group	Chemical name	Abbreviation	Formula	Half-Lives in humans
Perfluoro-alkane sulfonic acids	Perfluorobutane sulfonic acid	PFBS	$C_4HF_9O_3S$	28 days
	Perfluorohexane sulfonic acid	PFHxS	$C_6HF_{13}O_3S$	4.7–35 years
	Perfluoroheptane sulfonic acid	PFHpS	$C_7HF_{15}O_3S$	1.46 years
	Perfluorooctane sulfonic acid	PFOS	$C_8HF_{17}O_3S$	3.5–27 years
Perfluoro-alkane-carboxylic acid	Perfluorobutanoic acid	PFBA	$C_3F_7CO_2H$	3–4 days
	Perfluorohexanoic acid	PFHxA	$C_6HF_{11}O_2$	<2 years
	Perfluorohepatnoic acid	PFHpA	$C_7HF_{13}O_2$	14–49 days
	Perfluorooctanoic acid	PFOA	$C_8HF_{15}O_2$	2.1–10.1 years

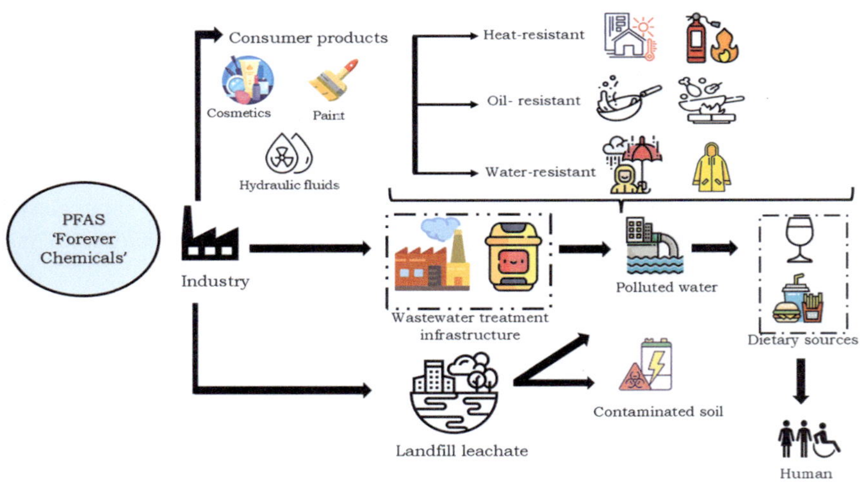

Fig. 1 Illustrates sources of exposure to perfluoroalkyl substances

How Does PFAS Affect Male Reproductive Health?

Male infertility is a growing problem in the high middle, and middle sociodemographic index (SDI) region (Huang et al. 2023). Epidemiological and experimental studies have found an association between PFAS exposure and reproductive abnormalities (Lopez-Espinosa et al. 2016; Raymer et al. 2012), showing a negative correlation between PFOS and PFOA in serum and testosterone levels (Joensen et al. 2013). Testes produce the male reproductive hormones, that is, testosterone, a main androgen facilitates growth, development and maintenance of male reproductive systems and produces sperm. Present epidemiological studies aligned with the PFAS exposure and male infertility in various populations. In a study, an association between male reproductive hormone levels shows lower testosterone levels, and higher estradiol (Zhang et al. 2014) with legacy PFAS raises concerns over continuous exposure to such chemicals. The reduction in semen quality (Toft et al. 2012) across the world is a prominent issue in the reproductive wellness of the male population. Studies have also found testicular cancers, as triggered after PFAS exposure (Seyyedsalehi and Boffetta 2023). Furthermore, studies also suggested that PFOA negatively affects Leydig cells, and Sertoli cells (Lu et al. 2019) which are responsible for hormonal support and provides nourishment for sperm production. In vitro, studies about PFOA also depicted that it negatively affects human spermatogenesis and delays puberty in children (Lopez-Espinosa et al. 2016) concerns about its potential effect. Table 2 shows the list of epidemiological and experimental studies that associated exposure to PFAS with the risk of male reproductive disorders. However, the outcomes greatly depend on the dose, duration, and type of PFAS analyzed. The studies performed do not always focus on environmentally relevant concentrations, and due to the variation ranges of its half-life that exist between species, the interpretation of the outcome demands a deeper insight into the real scenario. Serum PFAS association could largely link to decreased sperm count and concentration (Vested et al. 2013), and decreased sperm motility (Song et al. 2018). The action mechanism of PFAS is majorly endocrine-disrupting that shows the ability of these chemicals to interfere with functioning of hypothalamus-pituitary-gonadal (HPG) axis, affecting hormonal signaling pathways.

Epidemiological as well as animal studies have shown fluctuations in hormones. The inverse relationship between testosterone or elevated estrogen levels shows the disparity in testosterone/estrogen ratio linking to the reduced enzymatic activity of aromatase inhibitors responsible for converting testosterone to estrogen. Furthermore, testis, leydig cells, and sertoli cells work together to maintain a healthy spermatozoon. Oxidative stress is another major mechanistic factor when activated can damage sperm DNA, proteins, lipids, etc., eventually leading to infertility. Some results have been linked with enhanced apoptosis and reduction in the proliferation of germ cells (Qu et al. 2016). However, the outcomes of the studies largely depend upon the type of PFAS used and species-specific consequences. Hence, these studies reflected a wide range of doses, time of exposure, and target species, yet consistently indicate disruption in hormones. They highlighted alterations in testosterone levels and disturbed steroidogenic pathways, suggesting the negative effects are conserved across different experimental conditions.

Table 2 Summary of the previous studies investigating the effect of PFAS on male reproductive parameters

S. No.	PFAS exposure	Dose/duration/time of exposure	Species/age/design	Outcomes	References
1	PFOS PFOA	500, 1000, 2000, 3000 (μM), (in vitro capacitation period, lasted 4 hours)	Boars	Sperm capacitation was affected and sperm mortality was reported	Ortiz-Sánchez et al. (2022)
2	PFOS	(0, 2, 20, and 200 μg/L), 21 days	Zebrafish	In testis, expression of FSHR, LHR, and AR was downregulated	Bao et al. (2020)
3	Ammonium PFOA (APFO)	25 mg/kg/day APFO for 14 days.	Male CD rats (11–12 weeks)	Negatively affects Leydig cells, inhibits the release of testosterone	Biegel et al. (1995)
4	Perfluorodecanoic acid (PFDA)	20–80 mg/kg, IP, 7 days	Mice	Decreased plasma testosterone, decreased plasma androgen concentrations, hypophagia	Bookstaff et al. (1990)
5	PFNA	0.2 and 0.5 mg/kg, Orally administered (90 days)	Sprague-Dawley rats, postnatal day 25 to adulthood postnatal day114	Non-uniform degenerative changes in the seminiferous tubules, adverse effects on testicular expression of steroidogenic markers, serum levels of cholesterol and testosterone, sperm parameters and litter size	Singh and Singh (2018)
6	PFOA	0.31, 1.25, 5, and 20 mg/kg/day daily for 28 days by gavage	Male BALB/c mice (6–8 weeks)	Reduction in hormonal levels such as testosterone and progesterone Disrupted spermatogenesis Reduced sperm count	Zhang et al. (2014)
7	PFOA & PFOS	0, 25 or 50 mg/kg/day PFOA ($n = 12$ per group) for 9 consecutive days	Adult Sprague-Dawley rats	The expression levels of specific genes of Leydig cells and protein level downregulated inhibit the development of Leydig cells	Lu et al. (2019)
8	PFOA & PFOS	0.001, 0.01, and 0.1 mmol/L PFOS or PFOA for 48 h	*Caenorhabditis elegans*	Damage to germ cells negatively affects spermatogenesis	Yin et al. (2020)

9	PFOS	0, 0.5, and 10 mg/kg/day PFOS oral gavage, 5 weeks	Adult mice	Alterations in testicular ER expression, decreased proliferation and increased apoptosis of germ cells	Qu et al. (2016)
10	PFOS & PFOA	Adulthood	Cross-sectional study of 247 men	Negative correlation between sex hormone binding globulin, testosterone level, and serum concentration	Joensen et al. (2012)
11	PFOA & PFOS	In utero	169 participants, median age 20 cross-sectional	Low sperm concentration and reduced sperm quality	Vested et al. (2013)
12	PFOA, PFOS, PFHxS, PFNA, PFAA	Adulthood	Male patients cross-sectional	Increased sperm DNA damage, chromosomal unbalance	Governini et al. (2015)
13	PFOA	Adulthood	Participants ($n = 32,254$) reported 2507 validated cancers (21 different cancer types)	Prevalence of testicular cancer	Barry et al. (2013)
14	PFOA, PFOS, PFHxS, PFHNA	Adulthood	588 partners of pregnant women	Increased sperm morphology abnormality	Toft et al. (2012)

How Does PFAS Affect Female Reproductive Health?

Young children and women are greatly influenced by the exposure to PFAS. The number of studies examining the link between female reproductive disorders and EDCs has reportedly increased over time. The female reproductive system consists of ovaries, a uterus, fallopian tubes, and vagina. A complete endocrine system from the hypothalamus-pituitary axis to the ovaries plays a major role in the proper functioning of reproductive organs. Any sort of abnormalities in this system leads to infertility-related issues such as uterine disorder, endometriosis, premature ovarian insufficiency (POI), and polycystic ovarian syndrome. PFAS exposure can also affect lactation and mammary gland development. Previous experimental studies are highlighted in Table 3 concerning various aspects of female fertility and its plausible link with exposure to potential PFAS. The estrogenic and anti-androgenic effects of PFAS in the prenatal and postnatal period and may negatively affect the female reproductive health during adolescence and later stages.

Epidemiological studies are crucial in identifying chemical exposure and risk to human health. The real-world scenario often depicts exposure to different mixtures making it more relevant than studies on animal models. Previous studies indicated exposure to PFOA, and PFOS is linked with polycystic ovarian syndrome (PCOS) in China (Wang et al. 2019), the UK (Heffernan et al. 2018), and the USA (Vagi et al. 2014). Observational studies on PFAS have shown an association of PFAS with breast cancer (Li et al. 2022). PFAS accumulation in ovarian fluids and follicular fluids makes it prone to cause several diseases associated with ovarian, uterine tissues and plausible mechanism mainly involves hormonal disruption which led to disturb a complete HPG axis in the body.

Ovary Ovary is the site of folliculogenesis and facilitates the maturation of oocyte, also plays an essential role in steroidogenesis. Experimental studies have a piece of evidence suggesting PFAS exposure links to depleted ovarian reserve (Hallberg et al. 2019) and also inhibition of steroidogenic enzyme activities (Lin et al. 2023), and premature ovarian insufficiency (Zhang et al. 2018). Altered oocyte development at the earliest stages of folliculogenesis was observed (Domínguez et al. 2016). A histopathological study on adult female mice exposed to 0.1 mg PFOS/kg/day for 4 months by oral administration had a reduced number of preovulatory follicles and an increased number of atretic follicles, prolonged disastrous, hormonal disturbances (Feng et al. 2017). Furthermore, upon exposure to neonates to 0.1 and 1 mg PFOA/kg/day, primordial follicles were significantly reduced (Du et al. 2019). Another crucial role of the ovary is steroidogenesis. The coordination of theca cells and granulosa cells responsible for the secretion of reproductive hormones, i.e., androgen, luteinizing hormone, and follicle-stimulating hormone. PFAS exposure may result in imbalances between these hormones leading to reproductive abnormalities (Yang et al. 2009).

Uterus Exposure to PFAS led to structural and functional alterations in the uterus often linked to hormonal regulations. Epidemiological studies suggest an association

Table 3 Summary of previous studies investigating the effect of PFAS on female reproductive parameters

S. No.	PFAS exposure	Dose/duration	Species/age	Outcomes	References.
1	PFOA	0.01 mg/kg, 0.02, oral gavage, 24 hours	Female offspring CD-1 mice, Post natal day (PND)-18-20	Increased uterine weight Histopathological changes in the uterus, cervix and vagina	Dixon et al. (2012)
2	PFOS and PFOA	0.1, 1, 10 mg/kg, injected, PND 1–5 or PND 26–30	Sprague-Dawley rats, PND 1-5/ PND 26-30	Elevated E_2 and LH levels, Irregular estrous cyclicity, decreased number of secondary follicles, growing follicles, atretic follicles, and reduced corpus luteum	Du et al. (2019)
3	PFHxS	5, 50 mg/kg/day 42 days, Intragastric administration	ICR mice, 8-weeks old	Decreased ovarian weight, Prolonged estrous cyclicity Decreased number of secondary follicles, antral follicles, and CL	Yin et al. (2021)
4	PFNA	0.1, 10 mg/mL, In vitro, 22 h	Bovine oocytes (in vitro)	Impaired oocyte developmental competence, Alterations to lipid accumulation in blastocysts	Hallberg et al. (2019)
5	F53B	1.5, 3, 6, 112 mg/L, 6–132 hpf, Via tank water	Zebrafish (adult)	Delayed hatchings Increased birth defects Reduced survival rates	Shi et al. (2017)
6	PFHxA	20, 100, 500 mg/kg Gavage, 90 days	Crl:CD (SD) rats, young adult	No alterations in estrous cyclicity No reproductive and developmental effects	Loveless et al. (2009)
7	PFOA	0.1–10 mg/kg BW, 29 days	Balb/c and C57BL/6 mice. peripubertal period (21 through 50 days of age)	Delayed vaginal opening, inhibition of mammary gland and uterine development	Yang et al. (2009)
8	PFOA, PFOS, undecafluoro-2-methyl-3-oxahexanoic acid, and PFBS	15 days, oral administration	Female CD-1 mice, gestational expsoure	Disturbed estrous cycle, increased mRNA transcripts for steroidogenic enzymes and fatty acid synthesis-related genes, and increased collagen content in the ovary	Clark et al. (2024)

(continued)

Table 3 (continued)

S. No.	PFAS exposure	Dose/duration	Species/age	Outcomes	References.
9	PFNA, PFDA, PFHxS	Cohort study	Odense Child Cohort (OCC). Pregnant women, January 2010 and December 2012	Maternal PFAS exposure was associated with reduced AGD in girls, lower birth weight among boys, and higher birth weight among girls	Lind et al. (2017)
10	PFHxS, PFDoA PFTrDA,	Shanghai-Minhang Birth Cohort Study (S-MBCS)	Pregnant women at 12–16 gestational weeks (362 mother-infant pairs)	Increased AGD, in female neonates	Li et al. (2022)
11	Mixture PFAS	Case-control study	From 2014 to 2016, 366 women with PCOS-related infertility and 577 control participants without PCOS	High serum concentration in PCOS patients, increased prevalence of PCOS in women of reproductive age	Zhan et al. (2023)
12	Mixture PFAS	Cohort study	Singapore Preconception Study of Long-Term Maternal and Child Outcomes (S-PRESTO), 1032 women aged from 18 to 45 years	Higher PFAS exposures may be associated with decreased fertility in women. Reduction in the likelihood of clinical pregnancy and live birth	Cohen et al. (2023)

between PFAS exposure to the progression of uterus leiomyoma or fibroids (Hammarstrand et al. 2021), irregular menstrual cycle (Zhou et al. 2017), endometriosis-related infertility (Campbell et al. 2016). The studies of fibroids linked with EDC are still evolving. Nevertheless, PFAS-dependent alterations in the uterus were observed with direct and indirect exposure in mammalian studies such as. PFAS dependant gravimetric changes in rodents were also observed (Dixon et al. 2012). A detailed study by Nie et al. (2024) evaluated uterine development during puberty in rats, resultant uterine dysplasia was observed induced by ROS-mediated apoptosis. Further research is required to understand the mechanistic action of PFAS through which it contributes to uterine disorders. The disrupted lining of uterus may compromise implantation leading to infertility.

Hormonal disruption PFAS can alter hormones by imitating fatty acids leading to menstrual cycle irregularity, due to alterations in hormones crucial for reproductive health such as progesterone and estrogen. Similarly, the PFAS exposure is also linked to disruption of thyroid hormone, leading to various metabolic issues such as dyslipidemia and hypo or hyper-thyroidism. It can also obstruct with ovulation process leading to infertility. Furthermore, the exposure to PFAS if left untreated, or continuous to a longer duration, could link to risk of miscarriages, premature birth and developmental disorders.

The mechanism through which PFAS negatively affects the female reproductive system is mainly via endocrine disruption. It can interfere with the hormone signaling pathway led to various reproductive anomalies during puberty and lead to the progression of several diseases. Another possible reason is linked with the elevated oxidative stress which leads to damage to ovarian cells responsible for the reduction in oocyte quality, and risk of infertility. Alteration in mRNA expression is also another plausible impact of PFAS-dependant toxicity. Recently, a study evaluated the effect of PFOA exposure on kisspeptin signaling and found the downregulation of the kisspeptin 1 receptor (*Kiss1r*) (Zhang et al. 2020). Since Kisspeptin signaling is associated with the hypothalamus and plays a very crucial for the ovarian cycle, any changes will lead to dysregulation of the ovarian cycle and disrupt the hormone signaling. However, the precise mechanism needs more exploration of associated pathways both at cellular and molecular level to understand the potential of such chemicals to cause health related abnormalities.

Conclusion

Perfluoroalkyl substances have emerged as persistent contaminants responsible for various deteriorating effects on the reproductive systems of both males and females. Given the mounting evidence of its widespread presence and its detection in serum, plasma, follicular fluids, breast milk, etc., it has raised concern among the populations. In recent times, various approaches to phase out the PFAS from water bodies is the main area of research. Researchers are facing many challenges due to its extreme chemical stability, complex structure, and high solubility. Also, traditional water remediation methods are not effective on PFAS removal. Therefore, advance methods such as ion exchange resins or activated carbon filtration are needed, but are economically not friendly and also their byproducts are harmful. The reduction in production of PFAS products and limited human exposure becomes safer option as the initial level. Despite the policies and regulatory efforts, PFAS are still continuously used in industries such as firefighting foams and non-stick cookware due to unregulated laws in various countries. Moreover, several nations do not have extensive monitoring of PFAS, because of this many persistence and contaminations often neglected. Strict policies must be imposed worldwide to ban the PFAS products and other alternative bioremediation approaches must be introduced to eliminate PFAS from the environment.

To get the deeper insights into PFAS- induced toxicity, further studies should investigate the dose-response relationships of PFAS family compounds individually and also in mixture form, considering lifestyle factors, sex-specific vulnerabilities, and thorough mechanistic investigation. International efforts with affirmative approaches, including awareness of public, innovative researches and rigid regulations is obligatory to lighten the enduring consequences and to safeguard succeeding generations.

References

Baker ES, Knappe DRU (2021) Per- and polyfluoroalkyl substances (PFAS)-contaminants of emerging concern. Anal Bioanal Chem 414(3):1187–1188

Bao M, Zheng S, Liu C et al (2020) Perfluorooctane sulfonate exposure alters sexual behaviors and transcriptions of genes in hypothalamic–pituitary–gonadal–liver axis of male zebrafish (*Danio rerio*). Environ Pollut 267:115585

Barry V, Winquist A, Steenland K (2013) Perfluorooctanoic acid (PFOA) exposures and incident cancers among adults living near a chemical plant. Environ Health Perspect 121(11–12): 1313–1318

Biegel LB, Liu RC, Hurtt ME et al (1995) Effects of ammonium perfluorooctanoate on leydig cell function: in vitro, in vivo, and ex vivo studies. Toxicol Appl Pharmacol 134(1):18–25

Bookstaff RC, Moore RW, Ingall GB et al (1990) Androgenic deficiency in male rats treated with perfluorodecanoic acid. Toxicol Appl Pharmacol 104(2):322–333

Brendel S, Fetter É, Staude C et al (2018) Short-chain perfluoroalkyl acids: environmental concerns and a regulatory strategy under REACH. Environ Sci Eur 30(1):9

Brown-Leung JM, Cannon JR (2022) Neurotransmission targets of per- and polyfluoroalkyl substance neurotoxicity: mechanisms and potential implications for adverse neurological outcomes. Chem Res Toxicol 35(8):1312–1333

Brunn H, Arnold G, Körner W et al (2023) PFAS: forever chemicals-persistent, bioaccumulative and mobile. Reviewing the status and the need for their phase out and remediation of contaminated sites. Environ Sci Eur 35:30

Campbell S, Raza M, Pollack AZ (2016) Perfluoroalkyl substances and endometriosis in US women in NHANES 2003–2006. Reprod Toxicol 65:230–235

Clark KL, George JW, Davis JS (2024) Adolescent exposure to a mixture of per- and polyfluoroalkyl substances (PFAS) depletes the ovarian reserve, increases ovarian fibrosis, and alters the Hippo pathway in adult female mice. Toxicol Sci 202(1):36–49

Cohen NJ, Yao M, Midya V et al (2023) Exposure to perfluoroalkyl substances and women's fertility outcomes in a Singaporean population-based preconception cohort. Sci Total Environ 873:162267

Diamanti-Kandarakis E, Bourguignon J, Giudice LC et al (2009) Endocrine-disrupting chemicals: an endocrine society scientific statement. Endocr Rev 30(4):293–342

Dixon D, Reed CE, Moore AB et al (2012) Histopathologic changes in the uterus, cervix and vagina of immature CD-1 mice exposed to low doses of perfluorooctanoic acid (PFOA) in a uterotrophic assay. Reprod Toxicol 33(4):506–512

Domínguez A, Salazar Z, Arenas E et al (2016) Effect of perfluorooctane sulfonate on viability, maturation and gap junctional intercellular communication of porcine oocytes in vitro. Toxicol In Vitro 35:93–99

Du G, Hu J, Huang Z et al (2019) Neonatal and juvenile exposure to perfluorooctanoate (PFOA) and perfluorooctane sulfonate (PFOS): advance puberty onset and kisspeptin system disturbance in female rats. Ecotoxicol Environ Saf 167:412–421

Feng X, Cao X, Zhao S et al (2017) Exposure of pregnant mice to perfluorobutanesulfonate causes hypothyroxinemia and developmental abnormalities in female offspring. Toxicol Sci 155(2): 409–419

Governini L, Guerranti C, De Leo V et al (2015) Chromosomal aneuploidies and DNA fragmentation of human spermatozoa from patients exposed to perfluorinated compounds. Andrologia 47: 1012–1019

Hallberg I, Kjellgren J, Persson S et al (2019) Perfluorononanoic acid (PFNA) alters lipid accumulation in bovine blastocysts after oocyte exposure during in vitro maturation. Reprod Toxicol 84: 1–8

Hammarstrand S, Jakobsson K, Andersson E et al (2021) Perfluoroalkyl substances (PFAS) in drinking water and risk for polycystic ovarian syndrome, uterine leiomyoma, and endometriosis: a Swedish cohort study. Environ Int 157:106819

Heffernan A, Cunningham TK, Drage DS et al (2018) Perfluorinated alkyl acids in the serum and follicular fluid of UK women with and without polycystic ovarian syndrome undergoing fertility treatment and associations with hormonal and metabolic parameters. Int J Hyg Environ Health 221(7):1068–1075

Huang B, Wang Z, Kong Y et al (2023) Global, regional and national burden of male infertility in 204 countries and territories between 1990 and 2019: an analysis of global burden of disease study. BMC Public Health 23(1):2195

Joensen UN, Veyrand B, Antignac J et al (2012) PFOS (perfluorooctanesulfonate) in serum is negatively associated with testosterone levels, but not with semen quality, in healthy men. Hum Reprod 28(3):599–608

Joensen UN, Veyrand B, Antignac J et al (2013) PFOS (perfluorooctanesulfonate) in serum is negatively associated with testosterone levels, but not with semen quality, in healthy men. Hum Reprod 28(3):599–608

Li X, Song F, Liu X et al (2022) Perfluoroalkyl substances (PFASs) as risk factors for breast cancer: a case-control study in Chinese population. Environ Health 21(1):83

Lin C, Nian M, Xie M et al (2023) Associations of per- and polyfluoroalkyl substances and alternatives with reproductive hormones in women of childbearing age. Int J Hyg Environ Health 250:114158

Lind DV, Priskorn L, Lassen TH (2017) Prenatal exposure to perfluoroalkyl substances and anogenital distance at 3 months of age in a Danish mother-child cohort. Reprod Toxicol 68: 200–206

Lopez-Espinosa M, Mondal D, Armstrong BG et al (2016) Perfluoroalkyl substances, sex hormones, and insulin-like growth factor-1 at 6–9 years of age: a cross-sectional analysis within the c8 health project. Environ Health Perspect 124(8):1269–1275

Loveless SE, Slezak B, Serex T et al (2009) Toxicological evaluation of sodium perfluorohexanoate. Toxicology 264(1–2):32–44

Lu H, Zhang H, Gao J et al (2019) Effects of perfluorooctanoic acid on stem Leydig cell functions in the rat. Environ Pollut 250:206–215

Morgan S, Mottaleb MA, Kraemer MP et al (2023) Effect of lifestyle-based lipid lowering interventions on the relationship between circulating levels of per-and polyfluoroalkyl substances and serum cholesterol. Environ Toxicol Pharmacol 98:104062

Nie P, Lan Y, You T et al (2024) F-53B mediated ROS affects uterine development in rats during puberty by inducing apoptosis. Ecotoxicol Environ Saf 277:116399

Ortiz-Sánchez PB, Roa-Espitia AL, Fierro R et al (2022) Perfluorooctane sulfonate and perfluorooctanoic acid induce plasma membrane dysfunction in boar spermatozoa during in vitro capacitation. Reprod Toxicol 110:85–96

Pinney SM, Fassler CS, Windham GC et al (2023) Exposure to perfluoroalkyl substances and associations with pubertal onset and serum reproductive hormones in a longitudinal study of young girls in Greater Cincinnati and the San Francisco Bay area. Environ Health Perspect 131(9):97009

Qu J, Lu C, Xu C et al (2016) Perfluorooctane sulfonate-induced testicular toxicity and differential testicular expression of estrogen receptor in male mice. Environ Toxicol Pharmacol 45:150–157

Raymer JH, Michael LC, Studabaker WB et al (2012) Concentrations of perfluorooctane sulfonate (PFOS) and perfluorooctanoate (PFOA) and their associations with human semen quality measurements. Reprod Toxicol 33(4):419–427

Roth K, Petriello MC (2022) Exposure to per- and polyfluoroalkyl substances (PFAS) and type 2 diabetes risk. Front Endocrinol 13:965384

Roth K, Imran Z, Liu W et al (2020) Diet as an exposure source and mediator of per- and polyfluoroalkyl substance (PFAS) toxicity. Front Toxicol 2:601149

Schillemans T, Shi L, Donat-Vargas C et al (2021) Plasma metabolites associated with exposure to perfluoroalkyl substances and risk of type 2 diabetes – a nested case-control study. Environ Int 146:106180

Seyyedsalehi MS, Boffetta P (2023) Per- and poly-fluoroalkyl substances (PFAS) exposure and Risk of kidney, liver, and testicular cancers: a systematic review and meta-analysis. Med Lav 114(5):e2023040

Shelly C, Grandjean P, Oulhote Y et al (2019) Early life exposures to perfluoroalkyl substances in relation to adipokine hormone levels at birth and during childhood. J Clin Endocrinol Metab 104(11):5338–5348

Shi G, Cui Q, Pan Y et al (2017) 6:2 Chlorinated polyfluorinated ether sulfonate, a PFOS alternative, induces embryotoxicity and disrupts cardiac development in zebrafish embryos. Aquat Toxicol 185:67–75

Singh S, Singh SK (2018) Chronic exposure to perfluorononanoic acid impairs spermatogenesis, steroidogenesis and fertility in male mice. J Appl Toxicol 39(3):420–431

Skutlarek D, Exner M, Farber H (2006) Perfluorinated surfactants in surface and drinking waters. Environ Sci Pollut Res Int 13:299–307

Song X, Tang S, Zhu H et al (2018) Biomonitoring PFAAs in blood and semen samples: investigation of a potential link between pfaas exposure and semen mobility in China. Environ Int 113: 50–54

Toft G, Jönsson B, Lindh C et al (2012) Exposure to perfluorinated compounds and human semen quality in arctic and European populations. Hum Reprod 27(8):2532–2540

Vagi SJ, Azziz-Baumgartner E, Sjödin A et al (2014) Exploring the potential association between brominated diphenyl ethers, polychlorinated biphenyls, organochlorine pesticides, perfluorinated compounds, phthalates, and bisphenol a in polycystic ovary syndrome: a case–control study. BMC Endocr Disord 14:86

Vested A, Ramlau-Hansen CH, Olsen SF et al (2013) Associations of in utero exposure to perfluorinated alkyl acids with human semen quality and reproductive hormones in adult men. Environ Health Perspect 121(4):453–458

Von Holst H, Nayak P, Dembek Z et al (2021) Perfluoroalkyl substances exposure and immunity, allergic response, infection, and asthma in children: review of epidemiologic studies. Heliyon 7(10):e08160

Wang Z, Cousins IT, Scheringer M et al (2014) Global emission inventories for C4-C14 perfluoroalkyl carboxylic acid (PFCA) homologues from 1951 to 2030, Part I: production and emissions from quantifiable sources. Environ Int 70:62–75

Wang W, Zhou W, Wu S et al (2019) Perfluoroalkyl substances exposure and risk of polycystic ovarian syndrome related infertility in Chinese women. Environ Pollut 247:824–831

Wee SY, Aris AZ (2023) Revisiting the "forever chemicals", PFOA and PFOS exposure in drinking water. NPJ Clean Water 6:57

Yang C, Tan YS, Harkema JR et al (2009) Differential effects of peripubertal exposure to perfluorooctanoic acid on mammary gland development in C57Bl/6 and Balb/c mouse strains. Reprod Toxicol 27(3–4):299–306

Yin J, Jian Z, Zhu G et al (2020) Male reproductive toxicity involved in spermatogenesis induced by perfluorooctane sulfonate and perfluorooctanoic acid in *Caenorhabditis elegans*. Environ Sci Pollut Res 28(2):1443–1453

Yin X, Di T, Cao X et al (2021) Chronic exposure to perfluorohexane sulfonate leads to a reproduction deficit by suppressing hypothalamic kisspeptin expression in mice. J Ovarian Res 14(1):141

Zhan W, Qiu W, Ao Y et al (2023) Environmental exposure to emerging alternatives of per- and polyfluoroalkyl substances and polycystic ovarian syndrome in women diagnosed with infertility: a mixture analysis. Environ Health Perspect 131(5):57001

Zhang H, Lu Y, Luo B et al (2014) Proteomic analysis of mouse testis reveals perfluorooctanoic acid-induced reproductive dysfunction via direct disturbance of testicular steroidogenic machinery. J Proteome Res 13(7):3370–3385

Zhang S, Tan R, Pan R et al (2018) Association of perfluoroalkyl and polyfluoroalkyl substances with premature ovarian insufficiency in Chinese women. J Clin Endocrinol Metab 103(7): 2543–2551

Zhang Y, Cao X, Chen L et al (2020) Exposure of female mice to perfluorooctanoic acid suppresses hypothalamic kisspeptin-reproductive endocrine system through enhanced hepatic fibroblast growth factor 21 synthesis, leading to ovulation failure and prolonged dioestrus. J Neuroendocrinol 32(5):e12848

Zhao L, Teng M, Zhao X et al (2023) Insight into the binding model of per- and polyfluoroalkyl substances to proteins and membranes. Environ Int 175:107951

Zhou W, Zhang L, Tong C et al (2017) Plasma perfluoroalkyl and polyfluoroalkyl substances concentration and menstrual cycle characteristics in preconception women. Environ Health Perspect 125(6):067012

Environmental Toxicants and Lifestyle Diseases

Toxic Effect of Nanoparticles on the Gut Microbiota

Aryan Shukla, Abhishek Kumar Mishra, Nayanika Tiwari, and Garima Awasthi

Abstract

This chapter provides a comprehensive analysis of the effects of nanoparticles on gut microbiota, emphasizing their interactions and potential repercussions of their application in medical and environmental contexts. It begins with an overview of nanoparticle characteristics, such as size, shape, and surface properties, which dictate their biological interactions and technological applications. Then it delves into the critical roles of various gut microbiota and their interactions with nanoparticles, highlighting how these particles can influence health by altering microbial communities and functions. The application of nanoparticles in drug delivery systems is discussed, highlighting their potential in enhancing treatment efficacy, while also considering their toxicity and the challenges posed by their interaction with biological systems. Emerging concerns about nanoparticle safety, their environmental impact, and their regulations form a significant part of the discussion, outlining the current understanding and gaps in research. This chapter proposes microbiota-specific protocols to evaluate the safe application of nanoparticles, stressing the need for a balanced approach that maximizes therapeutic benefits while minimizing risks. This synergy aims to guide future research and regulatory efforts, ensuring that nanoparticle technologies are developed and used responsibly with a clear understanding of their interactions with the gut microbiota and the broader ecosystem.

Keywords

Nanoparticles · Gut microbiota · Gastrointestinal tract

A. Shukla · A. K. Mishra · N. Tiwari · G. Awasthi (✉)
Amity Institute of Biotechnology, Amity University Uttar Pradesh, Lucknow Campus, Lucknow, Uttar Pradesh, India
e-mail: gawasthi@lko.amity.edu

S. Raisuddin et al. (eds.), *Environmental Toxicants and Lifestyle Diseases*,
https://doi.org/10.1007/978-981-95-0575-3_11

Introduction

Nanoparticles, tiny entities smaller than 100 nm in at least one dimension, are designed to leverage their unique size-related characteristics and capabilities, which are markedly distinct from their larger, bulk forms. This distinction is primarily due to their enhanced surface-to-volume ratio and quantum mechanical effects. Nanoparticles can be synthesized from diverse materials, such as metals, metal oxides, carbon, and polymers, and they exhibit various forms, including spheres, rods, tubes, and quantum dots.

There is a broad spectrum of nanoparticle types, each distinguished by specific features and potential uses. For example, silver nanoparticles (Ag NPs) are celebrated for their germ-fighting abilities and find applications across a wide range of products, from electronic gadgets to healthcare instruments (Ahamed et al. 2010). Titanium dioxide nanoparticles (TiO_2 NPs), known for their white pigment properties, are commonly incorporated into food additives and sunscreen lotions (Baranowska-Wójcik et al. 2019). Nanoparticles derived from carbon, such as fullerenes and carbon nanotubes, are prized for their electrical conductivity and structural strength, making them valuable in the fields of material science and electronic engineering (Braydich-Stolle et al. 2005). Moreover, nanoparticles made from other metal oxides, such as zinc oxide (ZnO) and copper oxide (CuO), find extensive usage across a variety of industrial and consumer goods (Karlsson et al. 2008).

There is a growing trend towards eco-friendly nanoparticle synthesis, especially plant-based methods. Utilizing plant extracts for the reduction and stabilization of nanoparticle formation presents an economical and green alternative to conventional production methods. Nanoparticles developed through these biological processes show great promise across various sectors due to their compatibility with biological systems and their tiny size (Singh et al. 2023). The technique of creating nanoparticles with plant extracts stands out for its environmental advantages over traditional chemical synthesis. It employs sustainable biological materials and offers superior control over the nanoparticles' features, which enhances their medicinal properties. This approach is in line with the goals of sustainable and environmentally conscious manufacturing (Singh et al. 2023).

Moreover, the integration of ionic liquids (ILs) in nanoparticle production represents a notable innovation in the realm of green chemistry. ILs facilitate a greener and more efficient process for nanoparticle creation, enabling precise adjustments to aspects like surface design and particle size distribution. This strategy is especially pertinent in fields like healthcare and biotechnology, supporting the ethos of green chemistry (Feder-Kubis et al. 2024). However, despite the environmental advantages of green synthesis methods, they encounter challenges related to scale-up, efficiency, and economic viability. Traditional production techniques, despite their environmental drawbacks, tend to achieve better yields and are more suited for mass production. Ongoing research aims to find a middle ground that reconciles efficiency, affordability, and environmental friendliness in nanoparticle production. Efforts are concentrated on refining eco-friendly synthesis approaches and discovering new sustainable resources. The push towards green chemistry in the synthesis of nanoparticles marks an essential move towards ecological stewardship. Nonetheless, the quest to harmonize efficiency, cost, and environmental considerations continues to be a critical focus for future exploration and innovation.

Application of Nanoparticles in Current Clinical Practices/Sources of Exposure

Applications are illustrated in Fig. 1.

- Revolutionizing biologics delivery with lipid nanoparticles (LNPs): LNPs are at the forefront of medical innovation, particularly in their role of transporting therapeutic agents directly into cells while maintaining low toxicity levels.

Fig. 1 Applications of nanoparticles in medicine

Their small size, compatibility with biological systems, and ability to break down naturally make them ideal for the delivery of genes and proteins. LNPs stand out for their effectiveness in gene and protein therapy, offering new avenues for treatment (John et al. 2024).

- Tackling neurodegenerative diseases with nanoparticles: In the fight against neurodegenerative conditions such as Alzheimer's, Parkinson's, and Huntington's diseases, nanoparticles are proving to be invaluable. Their capacity to cross the blood–brain barrier allows for targeted drug delivery. Success has been seen in reducing Alzheimer's amyloid-beta buildup and neurotoxicity, improving dopamine levels in Parkinson's, and introducing novel treatments for Huntington's disease (Fatima and Siddique 2024).
- Harnessing natural products for enhanced nano-drug delivery: Merging the worlds of natural products and nanotechnology marks a significant leap in pharmaceutical sciences. This fusion aims to maximize the therapeutic benefits of natural compounds through nanotechnology, improving the precision, delivery, and effectiveness of drugs (Lv et al. 2024).
- Advancing cancer immunotherapy with nanoparticles: In cancer immunotherapy, nanoparticles are instrumental in achieving targeted drug delivery and controlled release, minimizing side effects. This synergy between drug delivery systems and immunotherapy promises to elevate both the efficacy and safety of cancer treatments (Lv et al. 2024).
- Innovations in dental implantology through Nanotechnology: The application of nanotechnology in dental implantology is enhancing drug delivery, offering better therapeutic outcomes, controlled release, antimicrobial benefits, and improved integration with bone (Hakim et al. 2024). These advancements promise improved healing and long-term success in dental implants.
- Nano-drug delivery systems in oral cancer therapy: For oral cancer treatment, nano-drug delivery systems, including liposomes and solid lipid nanoparticles, are showing great potential. These systems are designed for controlled drug release and targeted therapy, aiming for treatments that are both more effective and less harmful (Kapoor et al. 2024).
- LNPs: A gateway to improved biological drug delivery: Lipid nanoparticles are becoming increasingly crucial for the delivery of biological drugs, thanks to their biocompatibility and protective qualities. They play a pivotal role in maintaining the stability and solubility of sensitive biological agents, paving the way for advancements in drug delivery (Kapoor et al. 2024).

The rapid advancement of nanoparticle-based drug delivery systems is opening new possibilities in medicines shown in Fig. 1. Their integration with natural products, biologics, and targeted disease treatments, such as for cancer and neurodegenerative disorders, underscores their transformative potential. Nevertheless, despite this progress, challenges in targeting precision, solubility, and controlled drug release remain critical areas for ongoing research and development.

Overview of Human Microbiota

The human microbiota encompasses a complex and diverse array of microorganisms residing within and on our bodies, playing an essential role in sustaining health and systemic balance. This intricate community, composed of commensal, symbiotic, and some pathogenic organisms, is pivotal for numerous bodily functions, as shown in Table 1. It aids digestion, wards off disease-causing pathogens, activates immune responses, and even exerts influence on the central nervous system (Sieow et al. 2019).

Particularly noteworthy is the gut microbiota, critical for its involvement in metabolic processes such as the breakdown of dietary polysaccharides for energy (Turnbaugh et al. 2008). Unlike the varying species present, the gut microbiota's overall composition is relatively stable across different individuals, suggesting overlapping functionalities that support one another (Bauer and Thiele 2018). The initial establishment of the human microbiota, often inherited from the mother, underscores the significance of early microbial colonization in human health (Ramos et al. 2021). Beyond the gastrointestinal tract, other microbiota, such as those inhabiting the vaginal area, play distinct roles in pregnancy and may affect the newborn's microbial composition (Kervinen et al. 2019). The skin microbiota, too, is subject to change due to factors, including antimicrobial peptides, which modify its composition (Rademacher et al. 2021).

A balanced microbial composition is vital for normal physiological operations, whereas imbalances, or dysbiosis, can lead to various health issues (Li et al. 2021). Dysbiosis can be triggered by multiple factors, such as environmental pollution, dietary choices, lifestyle changes, and even conditions like space travel, all of which

Table 1 Composition and functions of human microbiota

S. no.	Microbiota type	Primary location	Key functions	Notable microorganisms
1	Gut microbiota	Gastrointestinal tract	Digestion, immune activation, pathogen defense	*Bacteroides*, *Firmicutes*
2	Urogenital microbiota	Urinary and the genital tract	Maintaining a healthy pH and preventing infections, such as urinary tract infections and bacterial vaginosis.	*Lactobacillus reuteri*, *L. rhamnosus*, *L. acidophilus*, *L. delbrueckii*, and *Bifidobacterium longum*
3	Breast milk microbiota	In human milk	Influences the construction of the infant gut microbiota	*Lactobacillus reuteri* and *Bifidobacterium breve*
4	Vaginal microbiota	Vaginal area	Pregnancy health, newborn microbial colonization	*Lactobacillus* species
5	Skin microbiota	Skin surface	Protection against pathogens, immune response	*Staphylococcus*, *Corynebacterium*

may alter the microbiota's makeup and functionality (Liu et al. 2020a, b). Understanding the human microbiota's composition and functions is crucial for researching dysbiosis, and prevalent diseases, and exploring new therapeutic strategies (Belizário and Napolitano 2015).

The significance of the human, especially the gut microbiota, in health maintenance, disease prevention, and the facilitation of various physiological functions is undeniable. Ongoing research into the human microbiota's composition, roles, and interactions continues to shed light on its critical contribution to overall health.

Emerging Concerns and Research Gaps

A study that delves into how nanomaterials (NMs) might adversely affect human health and the environment, indicating a need for a nuanced view of nanotoxicology that includes negative and controversial findings (Kumah et al. 2023). Although nanoparticles have many beneficial uses, some can pose risks under specific conditions. These revelations underscore the importance of comprehensive safety assessments and risk analyses in the development and use of nanomaterials. As the field of nanotoxicology progresses, ongoing studies aim to decipher the intricate dynamics between nanoparticles and biological entities, essential for the safe application of nanotechnology in health and environmental arenas.

Properties of Metal Nanoparticles

Physical and Chemical Properties

Metal nanoparticles (NPs) possess distinctive physical and chemical attributes that set them apart from their larger, bulk-material counterparts, largely due to their nanoscale dimensions. These attributes encompass particle size, shape, surface charge, hydrophobicity, rigidity, and the presence of surface-targeting ligands, all of which can be finely tuned to influence immune reactions and enhance medical treatments (Ke et al. 2019). Some of these properties are listed in Table 2.

Nanoparticles typically measure between 1 nm and 100 nm, granting them unique optical qualities, such as varied coloration stemming from absorption in the visible spectrum. Their reactivity, strength, and additional properties are closely linked to their size, shape, and structural makeup (Khan et al. 2019). The surface energy of metal NPs, crucial for processes such as nucleation and growth, is affected by their size. Studies using computational chemistry have found that the size-related variance in surface energy is minimal for certain shapes of metal nanoparticles, such as gold or silver icosahedra and cuboctahedra (Molleman and Hiemstra 2018). The shape of metal nanoparticles is a vital determinant of their properties. Various forms, including spherical, rod-shaped, or cubic, can influence surface energy and reactivity differently. By integrating atomic and thermodynamic perspectives, the influence

Table 2 Properties of metal nanoparticles

S. no.	Property	Description	Influence on applications
1	Size	1–100 nm, unique optical qualities (e.g., varied coloration due to visible spectrum absorption)	Enhanced reactivity, strength, and specialized optical properties
2	Shape	Spherical, rod-shaped, cubic, etc., affects surface energy and reactivity	Different shapes influence catalytic efficiency and reactivity
3	Surface charge	Stability and utility are influenced by the surface charge, which can be adjusted with organic molecules	Modulates interactions with biological systems and chemical processes
4	Hydrophobicity	Degree of water repellence, adjusted by surface functionalization	Affects dispersion in solvents and interaction with biological membranes
5	Rigidity	Structural firmness, influenced by the material composition and size	Impacts durability in various applications, including medical implants
6	Surface-targeting ligands	Organic molecules (e.g., mercapto derivatives) attach to surfaces, forming interfacial bonds	Introduces unique optical and electronic properties
7	Catalytic behavior	Structural and electronic characteristics adapt during chemical reactions	Enhances performance in heterogeneous catalysis
8	In situ characterization	Observing structural and chemical transformations during reactions	Provides insights for optimizing catalytic processes

of shape on the surface energy of metal nanoparticles has been comprehensively explained (Molleman and Hiemstra 2018).

The stability and utility of metal nanoparticles are significantly influenced by their surface charge and the way their surfaces are functionalized. Attaching organic molecules, such as mercapto derivatives, to the surface is an effective strategy for adjusting nanoparticle properties. The formation of interfacial bonds between metal and sulfur, as well as covalent bonds with carbon and nitrogen, can introduce unique optical and electronic characteristics, attributed to the delocalization of charges within the particle (Hu et al. 2016).

Additionally, the behavior of metal nanoparticles in catalytic processes is of considerable interest. Their structural and electronic characteristics can adapt during chemical reactions, a fundamental aspect of heterogeneous catalysis. The application of in situ characterization methods has enabled the observation of these nanoparticles as they undergo structural and chemical transformations during activation and catalytic reactions (Zhou et al. 2018). This evolving understanding of metal nanoparticles opens new pathways for their application in areas ranging from medicine to industrial catalysis.

Structure and Chemical Formulae of NPs

In the scholarly discussions surrounding metal nanoparticles, their characterization predominantly focuses on their metallic attributes rather than any ionic states. This preference stems from the transformation of metal ions undergo during nanoparticle formation, where they transition to a metallic state instead of remaining as free ions. Nanoparticles are essentially clusters of atoms or molecules, and this aggregation allows them to manifest distinct properties that are not observed in isolated ions in a solution. The idea that nanoparticles might exist in ionic forms could arise from a misconception regarding the role of ions in nanoparticle synthesis. Metal ions serve as the initial ingredients in the creation of nanoparticles; they are typically present in metal salts, used as precursors in various chemical synthesis routes. Through reduction reactions, these ions are converted into metal nanoparticles, which then exhibit characteristics pertinent to their metallic, not ionic, forms. Their individual properties are listed in Table 3.

Table 3 Properties of individual metal nanoparticles

S. no.	Metal NPs	Properties
1	CuO	Copper oxide nanoparticles (CuO) have a monoclinic crystal structure with a coordination number of four. This means that each copper atom is linked to four oxygen atoms in a square planar configuration in the (110) plane. CuO is a p-type semiconductor with a narrow energy gap of 1.2–2.4 eV
2	Cu_2O	Copper Oxide is also called as cuprous oxide, an inorganic compound with the chemical formula Cu_2O. It is covalent in nature. Copper(I) oxide crystallizes in a cubic structure. It is easily reduced by hydrogen when heated. It undergoes disproportionation in acid solutions producing copper (II) ions and copper. When the cupric oxide is gently heated with metallic copper, it is converted into cuprous oxide. It acts as a good corrosion resistance, due to reactions at the surface between the copper and the oxygen in air to give a thin protective oxide layer
3	$Cu(OH)_2$	The diameter of $Cu(OH)_2$ nanowires ranges from 33 nm to 46 nm and the length ranges from 400 nm to 2000 nm. $Cu(OH)_2$ is the most toxic compound against E. vermiculata. It's used as a broad-spectrum foliar fungicide on fruits, vegetables, and ornamentals.
4	Fe_2O_3	Iron oxide nanoparticles (Fe_2O_3) are particles with diameters between 1 nm and 100 nm. The two main forms are magnetite and maghemite.
5	$Fe(OH)_3$	$Fe(OH)_3$, or iron (III) hydroxide, is a chemical compound made of iron, hydrogen, and oxygen. $Fe(OH)_3$ nanoparticles are inexpensive and environmentally friendly. They can be prepared by increasing the pH of solutions of cetyltrimethylammonium bromide and Fe(III), which has been complexed with salicylate at pH 3. $Fe(OH)_3$ nanoparticles can also be prepared using an aqueous extract of Acacia catechu leaves as a green reducing and stabilizing agent. $Fe(OH)_3$ nanoparticles can be used as a precursor to generate monodisperse iron oxide nanoparticles. These nanoparticles can be organic-soluble and ferromagnetic, with rod-like and superparamagnetic dot-like shapes and sizes ranging from 12 nm to 25 nm. Iron oxide nanoparticles are widely used in various biomedical applications, including molecular imaging, hyperthermia treatment, protein separation, and drug delivery.

(continued)

Table 3 (continued)

S. no.	Metal NPs	Properties
6	MgO	Magnesium oxide nanoparticles (MgONPs) are non-toxic, antibacterial agents that can be metabolized and absorbed by the body. They have many applications, including: -Lightweight, heat-insulating, and sound-proofing materials, Refractory fiber board, Metallic ceramics, Silicon steel sheet, High-grade ceramic materials, Adhesives, Additives, Crucibles and Smelters
7	Mg $(OH)_2$	Magnesium hydroxide ($Mg(OH)_2$) nanoparticles (NPs) are white powder particles with low toxicity and many applications: It is used in antacids and laxatives because of their low cost and low risk of adverse effects. $Mg(OH)_2$ NPs have antibacterial, antifungal, and protozoal effects. In industrial use as a non-hazardous alkali to neutralize acidic wastewaters, as a flame retardant in various polymer compositions, and as a precursor for magnesium oxide refractory ceramics. It is used in the bierock method of building artificial reefs.
8	Zn $(OH)_2$	Zinc hydroxide nanoparticles ($Zn(OH)_2$) can be synthesized by precipitation and heated in a diol solution to produce zinc oxide (ZnO) nanoparticles. The size and shape of the particles affect how light is scattered, so zinc oxide nanoparticles can be gray or other colors. ZnO nanoparticles have gained interest in the agricultural and food industries for their ability to kill or reduce the activity of microorganisms. They are also used in many industrial products, such as rubber, paint, coating, and cosmetics.
9	FeO	Iron(II) oxide, or ferrous oxide, is an inorganic compound with the formula FeO. It is a black powder that is sometimes confused with rust, which is hydrated iron(III) oxide.
10	Fe $(OH)_2$	Iron oxide nanoparticles are particles with diameters between 1 nm and 100 nm. They are made up of maghemite (γ-Fe_2O_3) and/or magnetite (Fe_3O_4) particles. Iron oxide nanoparticles have many applications, including: Biomedical: As carriers for drug delivery and imaging contrast agents, as gene carriers for gene therapy, and as therapeutic agents for hyperthermia-based cancer treatments Magnetic data storage: Because of their superparamagnetic properties Biosensing: Because of their specific magnetic behaviors and properties.
11	NiO	Nickel oxide (NiO) nanoparticles are transition metal oxides with a cubic lattice structure. They have a wide range of applications, including: Battery electrodes, Photo-electron devices, Ion storage materials, Gas sensors, Magnetic materials, Thermo electric materials, Catalysts, Fuel cells, Dye-sensitized photocathodes and Electrochromic films
12	Ni $(OH)_2$	Nickel (II) hydroxide ($Ni(OH)_2$) nanoparticles are used in nickel–cadmium batteries, nickel-metal hydride batteries, and rechargeable batteries. They are also important components of the surface layers that form on nickel metal and nickel-based alloys. $Ni(OH)_2$ nanoparticles have a critical size of 7.9 nm, which is where they have the best electrochemical performance. They can be synthesized using hydrothermal and ultrasonication methods.
13	Cu_2O	Cuprous oxide (Cu_2O) nanoparticles are spherical and have a characteristic surface plasmon band at 650 nm. They are used in many applications due to their unique magnetic, electronic, and optical properties. Some of their applications include: catalysis, solar cells, batteries, gas sensors, CO oxidation, photodegradation, and photocatalyst.

Solubility and Aggregation Behavior

The solubility and aggregation behaviors of metal NPs are pivotal in determining their environmental impact, availability for biological systems, and utility in pollution remediation efforts. Two significant studies shed light on these aspects for various metal nanoparticles, offering valuable insights into their interactions with environmental factors. The first investigation delved into the aggregation dynamics of ferric iron oxyhydroxide nanoparticles and their efficacy in adsorbing heavy metals. It was discovered that the aggregation process, influenced by variables like pH levels, ionic strength, and temperature changes through drying or freezing, significantly alters the aggregate structure. This change, in turn, affects the nanoparticles' capability to capture and retain metal ions, such as copper. Although aggregation tended to diminish the overall copper adsorption capacity, it simultaneously increased metal retention due to physical constraints and the emergence of stronger binding sites, potentially enhancing the nanoparticles' ability to trap metal ion pollutants (Gilbert et al. 2009).

The second research endeavor focused on the concurrent dissolution and aggregation behaviors of metal sulfide nanoparticles, particularly those of nickel sulfide (NiS), copper sulfide (CuS), and cobalt sulfide (CoS), assessed through single-particle inductively coupled plasma mass spectrometry (spICP-MS). This study highlighted that metal release kinetics encompassing both dissolution and disaggregation vary substantially under different environmental settings. Release rates were observed to be quickest under highly oxidizing conditions, decelerating in nearly neutral and acidic environments. Factors such as solution pH significantly influenced the dissolution kinetics and aggregation states, thereby affecting the overall metal release rates. Furthermore, the study established that the release rates of various metal sulfides aligned with predictions based on their crystalline structure, solubility products, and surface area metrics, with the release rate order identified as CoS > NiS > CuS. This indicates that the inherent characteristics of the nanoparticles critically dictate their solubility and aggregation tendencies (Mansor et al. 2022).

Collectively, these studies underscore the intricate relationship between the environmental settings and the fundamental properties of metal nanoparticles. This relationship plays a crucial role in modulating the nanoparticles' solubility and aggregation patterns, which, in turn, influence their environmental ramifications.

Negative Interaction Mechanisms with Biological Systems

The toxicity associated with nanoparticles is a multifaceted issue influenced by several characteristics, including their size, shape, chemical makeup, surface charge, and state of agglomeration. Research indicates that nanoparticles have the potential to cause oxidative stress, inflammation, DNA damage, and even lead to the death of cells (Ahamed et al. 2010; Karlsson et al. 2008; Mansor et al. 2022; Hu and Gao 2010). Notably, silver nanoparticles have demonstrated toxicity toward mammalian cells, with adverse impacts observed in organs such as the skin, liver, lungs, brain, and reproductive systems (Ahamed et al. 2010). Titanium dioxide nanoparticles, too, have been identified as triggers for oxidative stress and genotoxicity (Baranowska-Wójcik et al. 2019). With the rising usage of nanoparticles, concerns about their toxicity and the implications for human health and environmental safety have intensified. Nanoparticles have various pathways to enter the human body via inhalation, ingestion, or through the skin and have been observed to accumulate in organs including the lungs (Mansor et al. 2022), liver (Baranowska-Wójcik et al. 2019), and brain (Hu and Gao 2010). Their minuscule size enables them to evade the body's natural defenses, posing a risk of adverse health outcomes.

The central nervous system (CNS) is particularly vulnerable to nanoparticles, which can traverse the blood–brain barrier and interact with neuronal cells, potentially leading to neurotoxic consequences (Hu and Gao 2010). The reproductive system is also at risk, with evidence suggesting nanoparticles can adversely affect fertility in both sexes and hinder fetal development in animal studies (Brohi et al. 2017).

Given these potential hazards, there's an urgent call for thorough investigations into the long-term impacts of nanoparticle exposure on both human health and the environment. Developing strategies to limit exposure and mitigate the harmful effects of nanoparticles is essential for ensuring the safe application of nanotechnologies (Sajid et al. 2015).

Reactivity Inside Biological Systems

A focus on gold nanoparticles (AuNPs) has shed light on their molecular impact, underscoring their importance in fields such as nanomedicine, bioimaging, and photothermal therapy. Despite their beneficial applications, AuNPs' interactions with biological entities can induce toxic effects, influenced by a mix of inherent and external factors. Grasping these factors is pivotal in minimizing the risks tied to the use of AuNPs in biomedical endeavors (Nguyen and Falagan-Lotsch 2023).

In the realm of mRNA nanomedicine, a piece in Advanced Drug Delivery Reviews, 2023 delves into the intricacies of targeted mRNA delivery via nanotechnology. This research underlines the complex interactions nanomedicines have with biological tissues and fluids upon administration, involving a dynamic engagement with biomacromolecules. These interactions are central to comprehending the therapeutic effectiveness and safety of nanomedicines, particularly their influence on the human microbiome (Saiding et al. 2023). In another review addressing human and environmental concerns, a thorough scoping review examined a wide array of nanoparticles, spanning metallic (oxides, pure metals) to carbon-based (fullerenes, carbon nanotubes, graphene) types (Kumah et al. 2023). This review underscored the critical need to understand the repercussions of nanoparticles on human health and the environment. It proposed diverse biomarkers and models for gauging human and environmental exposure to nanoparticles, drawing on studies from various global locations to emphasize the universal relevance of nanotoxicology research.

Mechanisms of Nanoparticle Toxicity

The scrutinized research delves into the cellular and molecular mechanisms underpinning nanoparticle-induced toxicity. Nanoparticles exhibit complex interactions with cells and tissues, instigating a spectrum of toxicological outcomes. Deciphering these interactions is crucial for evaluating the hazards of nanoparticle exposure and for the development of safer nanomaterials. The effects of nanoparticles have been investigated across multiple biological systems, including the respiratory, nervous, endocrine, immune, and reproductive systems, along with their carcinogenic potential (Xuan et al. 2023). Studies have pinpointed specific mechanisms of toxicity, such as reactive oxygen species (ROS) buildup, mitochondrial disruption, inflammatory responses, apoptosis, DNA damage, alterations in the cell cycle, and epigenetic changes (Deng et al. 2023). Furthermore, the biological transport and physiological responses of plants to nanoparticles have been studied, revealing impacts like growth inhibition, structural harm, oxidative stress, metabolic dysfunctions, cellular death, and genetic damage (Kang et al. 2023). The mechanobiological effects of nanoparticles on cells have also been explored, offering insights into the underlying processes of nanoparticle toxicity and their possible applications in drug delivery and tissue engineering.

Oxidative Stress Induction

The capacity of metal nanoparticles to trigger oxidative stress is a central mechanism behind their toxicity. Oxidative stress occurs when there's a disruption in the balance between the production of ROS and the biological system's ability to neutralize these reactive compounds or repair the damage they cause. An overload of ROS can damage proteins, cell membranes, and DNA, leading to a slew of negative outcomes such as inflammation, cell death through apoptosis, and other cellular damage. Research across various studies indicates that metal nanoparticles can spark the production of ROS, with the extent of this induction varying based on their size, shape, surface characteristics, composition, solubility, and how they aggregate. Examples include ambient ultrafine particles and cationic polystyrene nanospheres, which have been linked to cellular ROS production, a reduction in glutathione (GSH) levels, and toxic oxidative stress resulting in mitochondrial and organellar damage (Xia et al. 2006). Zinc oxide (ZnO) nanoparticles, similarly, have been identified as catalysts for cellular toxicity, leading to ROS generation, oxidative injury, and inflammation (Xia et al. 2008). Copper oxide nanoparticles have been connected to oxidative stress, as well as toxicity and cell death induced by endoplasmic reticulum (ER) stress in liver tissues and cells (Liu et al. 2020a, b).

However, the adverse effects of nanoparticles are not limited to oxidative stress alone but can also include physical interference with cellular structures (Shvedova et al. 2012). For instance, silver nanoparticles are known to cause oxidative stress, apoptosis, and autophagy in mouse embryonic fibroblasts (Lee et al. 2014). Gold nanoparticles, though less extensively researched for their toxicity, have been observed to induce oxidative stress and autophagy in human lung fibroblasts (Li et al. 2010).

It's critical to recognize that not all metal oxide nanoparticles are detrimental. Cerium oxide (CeO_2) nanoparticles, for example, have been shown to reduce ROS production and bolster cellular defenses against oxidative stress (Xia et al. 2008). This underscores the complexity of nanoparticle-induced toxicity and emphasizes the importance of a thorough understanding of the diverse factors influencing their biological interactions. The toxicity associated with metal nanoparticles is intricately linked to their propensity to induce oxidative stress, eliciting a spectrum of biological responses, from cytotoxic to cytoprotective effects. Grasping the specific characteristics of nanoparticles that drive oxidative stress is essential for devising ways to minimize their harmful impacts (Xia et al. 2006; Xia et al. 2008; Liu et al. 2020a, b; Shvedova et al. 2012; Lee et al. 2014; Li et al. 2010).

Alteration of Microbial Membrane Integrity

The dissemination of metal nanoparticles into our bodies and environment has sparked considerable concern regarding their adverse effects, especially on microbial ecosystems. These nanoparticles can compromise the integrity of microbial membranes, leading to significant repercussions for microbial communities and the environmental processes they underpin. Such alterations in membrane integrity can diminish microbial biomass, enzymatic activities, and biodiversity, which are critical for maintaining soil fertility and overall ecosystem health.

Research has indicated that nanoparticles, such as TiO_2 and ZnO, can adversely affect soil bacterial communities by decreasing microbial biomass and diversity and altering community compositions (Ge et al. 2011). The introduction of metal-based nanoparticles into soil is linked with reduced microbial biomass and enzymatic activities, negatively impacting the composition of microbial communities, including bacteria, fungi, and yeasts. This decline poses potential risks to human health, agricultural productivity, and soil characteristics (Rajput et al. 2018).

In agricultural settings, especially in soils treated with biosolids, various metal nanoparticles have been shown to influence soil enzyme activities and microbial community structures. For instance, silver nanoparticles (AgNPs) were found to inhibit specific enzymes, whereas titanium dioxide (TiO_2NPs) nanoparticles had minimal or slightly adverse effects on enzymatic activity. Elevated levels of metal ions or AgNPs led to shifts in the microbial community composition, whereas nTiO_2 resulted in negligible changes (Asadishad et al. 2018).

Metal nanoparticles also affect aquatic microbial ecosystems vital for decomposing plant debris in water bodies. Studies have shown that nano copper oxide and nano silver more significantly inhibit bacterial biomass than fungal biomass and decrease fungal sporulation rates, inducing shifts in fungal and bacterial community structures (Pradhan et al. 2011). Furthermore, the interference of metal-based nanoparticles, such as ZnONPs, TiO_2NPs, and AgNPs, with bacterial quorum sensing (QS) systems has been documented. These nanoparticles can disrupt QS by affecting signal reception and response or autoinducer biosynthesis, with impacts varying based on nanoparticle characteristics and external conditions like UV irradiation (Gómez-Gómez et al. 2019). The toxicity of metal and metal oxide nanoparticles to microbial systems hinges on factors like the nanoparticles' composition, size, shape, and surface coating. Environmental conditions can modify the surface coating, potentially altering the nanoparticles' microbial toxicity (Suresh et al. 2013). The uncontrolled release of metal-based nanoparticles into terrestrial ecosystems, particularly agricultural lands, threatens the diversity of beneficial microbial populations, such as soil bacteria and fungi (Ameen et al. 2021). Studies involving nanostructured TiO_2 nanoparticles have shown that they can significantly compromise microbial membranes, even under environmental conditions mirroring those in natural water bodies. The damage was more pronounced in free-living cells than in biofilm-encapsulated cells, highlighting the protective effect of biofilms against TiO_2 nanoparticle exposure (Battin et al. 2009). Assessments using fluorescent-based assays have revealed that TiO_2 nanoparticles impact the membrane integrity of Gram-negative bacteria at elevated concentrations, while ZnO nanoparticles tend to aggregate and precipitate, showing minimal impact on cell membrane integrity (Mileyeva-Biebesheimer et al. 2010).

Lastly, employing heavy metal nanoparticles in membrane bioreactors (MBRs) as an antifouling strategy may affect MBR functionality, including the performance of aerobic and anaerobic biological processes and microbial communities involved in nutrient cycling and anaerobic digestion (Cheng and Hong 2020). This underscores the intricate balance between leveraging nanoparticle advantages and managing their environmental and microbial impacts.

Impact of Metal Nanoparticles on Human Microbiota

In vitro ***studies*** The impact of nanoparticles on human microbiota is a rapidly evolving area of research, whereas nitro studies reveal complex interactions and varied effects based on the type and concentration of nanoparticles involved. Studies have demonstrated that AgNPs exhibit concentration-dependent toxicity, being more toxic to bacteria than to human enterocytes. This suggests that AgNPs might selectively target gut microbiota, potentially disrupting its balance, which could have implications for gut health (Braydich-Stolle et al. 2005).

Research focusing on metallic nanoparticles like Cu, Zn, CuO, and ZnO has found that these particles can reduce the diversity and abundance of antibiotic resistance genes (ARGs) in the microbiota. This effect is primarily due to the inhibition of bacterial growth and the induction of oxidative stress, which could have both beneficial and detrimental consequences (Su et al. 2019).

Comparative studies of the effects of different nanoparticles on the gut microbiota indicate material-specific impacts. For instance, titanium dioxide nanoparticles (TiO_2 NPs) are shown to have a lesser direct inhibitory effect on human gut microbiota than AgNPs, suggesting that the material composition of nanoparticles plays a critical role in their microbial interactions (Agans et al. 2019). Exposure to nano-ZnO has led to concentration-dependent decreases in the production of short-chain fatty acids, altering the composition and diversity of the gut microbiota as well as the gut resistome. This underscores the complex and potentially harmful effects of nanoparticles on gut health (Zhang et al. 2021). While non-biodegradable nanoparticles have been shown to induce apoptosis in human intestinal epithelium models, a mechanism involving hydrogen peroxide indicates a bystander effect that might progressively compromise the intestinal barrier, posing significant long-term health risks (Thubagere and Reinhard 2010).

In vivo *studies* Nanoparticles have been shown to affect the populations of various microbes in the gut, where certain types enhance the growth of beneficial bacteria and reduce the number of harmful ones, as observed in the intestinal microbiota of silkworms (Li et al. 2019). Interestingly, selenium nanoparticles (Se NPs) have demonstrated a capability to reduce the abundance of certain pathogens in poultry microbiota without significantly disturbing the overall microbial community. These points to a potential for targeted microbial interventions using nanoparticles (Gangadoo et al. 2019). Environmental exposure to nanoparticles, such as $nTiO_2$, nZnO, and nSe in zebrafish, has affected the alpha diversity and composition of gut microbiota across different developmental stages. Such findings indicate that nanoparticles can influence microbial communities in various biological contexts, not just in humans but across species (Chen et al. 2021). Another review discusses the influence of gut microbiota on the effectiveness of nanomaterials used in cancer diagnosis, imaging, and theranostics. It underscores the potential of nanotechnology in targeting and manipulating gut microbiota to enhance cancer treatment, while also stressing the importance of understanding the complex interactions between NMs, gut microbiota, and the host for effective application in cancer prevention and treatment (Li et al. 2023).

Human study The interplay of AgNPs, gut microbiota, and probiotics was investigated (Zorraquín-Peña et al. 2020) by studying the effects of non-lethal amounts of AgNPs on the human gut microbiota. This was relevant to real-life exposure via food, pharmaceuticals, and consumer items. The study used in vitro models using human fecal matter to investigate the effects of AgNP exposure. The results revealed that although the core bacterial population remained unchanged, there were considerable alterations seen in uncommon species. Furthermore, there was a modification in the ratio of Firmicutes to Bacteroidetes (F/B), suggesting alterations in the microbial makeup. The research demonstrated that the inclusion of the probiotic strain *Bacillus subtilis* (*B. subtilis* or BS) effectively alleviated the adverse impacts of AgNPs, resulting in a reduction of the F/B ratio and the preservation of crucial metabolic pathways. Although functional disparities were seen in cultures treated with AgNPs, these disparities were not present in samples treated with a combination of AgNPs and *Bacillus subtilis* (Zorraquín-Peña et al. 2020). Crucially, the presence of AgNPs did not have a substantial effect on the total diversity of the gut microbiota or its metabolic and toxic characteristics. This study highlights the possible preventative effect of probiotics on safeguarding the gut flora against changes caused by AgNP. The usage of AgNPs in consumer items is extensive, and their consumption is a prevalent means of exposure, highlighting the need to comprehend their impact on gastrointestinal well-being.

A collective review of 22 in vitro and 46 in vivo studies concludes that engineered nanomaterials (ENMs) such as carbon nanotubes, titanium dioxide, and zinc oxide are also implicated in affecting gut microbiota (Pietroiusti et al. 2016). This interaction can lead to clinical disorders, including colitis, obesity, and immunological dysfunctions, highlighting the broad implications of nanoparticle exposure on health. These studies underscore the significant and diverse impacts of nanoparticles on human microbiota. While silver nanoparticles show high toxicity to bacteria, potentially beneficial for altering harmful bacterial populations, other engineered nanomaterials could negatively affect the microbiota, leading to health disorders. Additionally, the interaction of nanoparticles with antibiotic-resistance genes and their influence on key microbial processes like short-chain fatty acid production highlights the potential health risks associated with nanoparticle exposure. Nevertheless, some nanoparticles, such as selenium, offer a more nuanced interaction, targeting specific pathogens without broadly disrupting microbiota balance. These findings collectively emphasize the need for further research to fully understand the material-specific effects of nanoparticles on human health and the microbiota. The research suggests that while nanomaterials offer promising therapeutic benefits, especially in cancer therapy, there is a need for careful consideration of their potential negative impacts on the human microbiome. Understanding the nano-bio interactions is crucial for optimizing the benefits and minimizing the risks associated with the use of nanomaterials in medical treatments.

The current state of research in nano-bio interactions and microbiome impact is evolving, with an increasing understanding of the complex dynamics between nanomaterials and the human microbiome. This knowledge is pivotal for advancing the field of nanomedicine, particularly in ensuring the safe and effective use of nanoparticles in therapeutic applications.

Regulatory and Safety Considerations

There is an ongoing effort among regulatory bodies to standardize assessment practices for nanomaterials, although there remains some uncertainty in the regulations and safety standards. Efforts to mitigate the toxicity of nanoparticles are ongoing, but complete safety has not yet been achieved. Regulatory frameworks aimed at addressing the potential dangers of nanomaterials to human health and the environment have been updated, with revised OECD guidelines incorporating further risk assessment investigations. In the workplace, risk management frameworks for engineered nanoparticles focus on established industrial hygiene practices and health monitoring to handle potential risks.

The current regulatory frameworks for nanoparticles are primarily focused on identifying and managing potential risks to both human health and the environment. These regulations are informed by ongoing research into the toxicological impacts of nanoparticles and aim to implement protective measures such as comprehensive risk assessments, safe handling protocols, and health surveillance. These guidelines are under continuous review and adaptation to incorporate new scientific findings about nanoparticle effects. A key element in determining their potential toxicity is the chemical stability of nanoparticles, underscoring the necessity of understanding their physicochemical characteristics in regulatory contexts.

Risk assessment models Nanomedicines leverage engineered nanomaterials to enhance medical therapies, but their minuscule size and distinct properties also bring potential toxicological concerns. Implementing thorough risk assessment models and strategies is critical to ensuring their safe application in healthcare. The physicochemical attributes of nanomaterials, such as size, shape, and surface features, play pivotal roles in determining their biological interactions and potential toxicity. It is recommended to employ an extensive data collection strategy that includes these parameters for effective hazard identification (Oberdörster et al. 2005). The complex relationship between particle characteristics and toxicity highlights the necessity of nuanced risk assessments for regulating nanomaterials in consumer products. This underscores the limitation of making broad toxicological generalizations based solely on size (Tsuji et al. 2006).

An adapted testing strategy for nanomaterials used in medical applications involves modifying traditional methods to consider the unique properties of nanomaterials. This strategy should encompass a variety of toxicity tests and cell models, and account for nanoparticle features and experimental conditions (Dusinska et al. 2015). To gauge toxicity, in vitro assays are instrumental in pinpointing specific biological pathways and understanding cellular-level effects, while in vivo assays provide broader insights into impacts on entire organisms. A layered testing approach that incorporates both in vitro and in vivo assays is advocated for a holistic assessment (Oberdörster et al. 2005). A future sophisticated testing approach for assessing nanomaterial safety includes tiered in vitro tests, computational toxicology, high-throughput screening, and omics-based techniques. This method aims to bridge in vitro toxicity findings with in vivo risk evaluations (Chen et al. 2018). Key tenets of nanotoxicology, such as dosage, dose rate, and biokinetics, are crucial in evaluating the safety of engineered nanomaterials. An exhaustive characterization of nanoparticles, especially surface properties, is essential to differentiate between their beneficial and harmful effects (Oberdörster 2010).

The safety evaluation of nanomedicines demands a comprehensive approach that acknowledges the distinct physicochemical properties of nanomaterials. Risk assessment models must incorporate both in vitro and in vivo testing methods, recognize the complexity of nanoparticle behaviors, and employ advanced analytical technologies. These models are shaped by nanotoxicology principles, which highlight the importance of thorough material characterization, understanding dose-response relationships, and biokinetics to mitigate potential toxic effects and optimize therapeutic outcomes.

Conclusion

The necessity for microbiota-specific protocols in the study and application of nanoparticles is critical due to their intricate interactions with the human microbiota, as highlighted in this chapter and summarized in Fig. 2. These interactions can significantly affect health outcomes in diverse and sometimes contradictory ways. Establishing tailored protocols is essential for accurately evaluating these effects, maximizing therapeutic benefits, and reducing potential risks associated with nanoparticle exposure. Below are essential elements that these protocols should encompass (Fig. 2).

- Microbial Diversity and Function Assessment: Protocols need to incorporate techniques to examine shifts in microbiota diversity and functionality. This should include sequencing methods like 16S rRNA and metagenomic sequencing to detect taxonomic changes and functional metagenomics to explore alterations in microbial gene expression and metabolic processes. Such measures are crucial

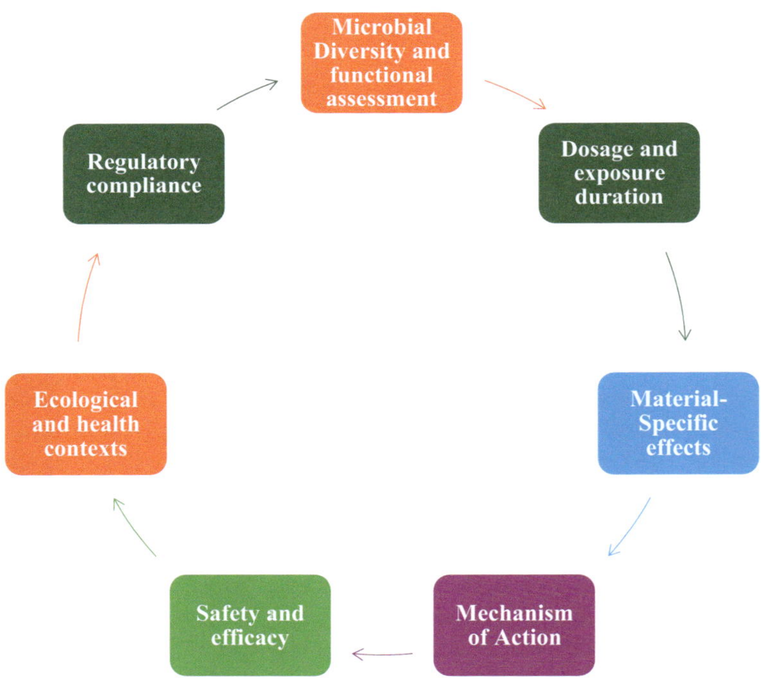

Fig. 2 Essential elements of safety assessment protocols of metal nanoparticles

to grasp how nanoparticles influence the structure and behaviour of microbial communities in the body.

- Dosage and Exposure Duration: Protocols must establish standardized dosing and exposure timelines that reflect realistic usage scenarios to effectively mimic the impact of nanoparticles. This should account for both short-term and prolonged exposures to evaluate immediate and enduring effects on the microbiota.
- Material-Specific Effects: Different nanoparticle materials interact variably with the microbiota, necessitating protocols that address the distinct impacts of various nanoparticle types, such as metallic, carbon-based, or polymer-based nanoparticles. Understanding these interactions helps determine which nanoparticles may disrupt microbial communities under specific conditions.
- Mechanisms of Action: It is vital to decipher the mechanisms by which nanoparticles influence microbiota. Protocols should include methods to investigate processes like oxidative stress induction, disruption of membrane integrity, and modulation of immune responses. Insights from these studies will assist in designing nanoparticles that minimize negative interactions with the microbiota.
- Safety and Efficacy: Evaluating both the safety and therapeutic efficacy of nanoparticles on microbiota is necessary. Protocols should assess the potential toxic effects, such as harm to beneficial bacteria or promotion of antibiotic resistance genes, alongside potential advantages like the reduction of pathogenic bacteria.
- Ecological and Health Contexts: Protocols should reflect the ecological and health contexts, considering variations in microbiota across different body sites and among individuals, as well as the health status of subjects, which may influence microbiota composition. This ensures that research findings apply to a broad population and can be translated into practical medical or environmental interventions.
- Regulatory Compliance: Protocols must adhere to current regulatory standards regarding nanoparticle use, ensuring that all research adheres to ethical guidelines and legal stipulations. This includes compliance with regulations from authoritative bodies such as the FDA or EMA concerning the application of nanomaterials in consumer goods and medical treatments.

By addressing these key areas, microbiota-specific protocols will effectively characterize the complex interactions between nanoparticles and microbiota, guiding the safe and efficacious application of nanotechnologies in various domains. These protocols are fundamental in advancing our understanding of nano-bio interactions and ensuring that the deployment of nanotechnologies does not negatively impact human health or the environment.

References

Agans RT, Gordon A, Hussain S et al (2019) Titanium dioxide nanoparticles elicit lower direct inhibitory effect on human gut microbiota than silver nanoparticles. Toxicol Sci 172(2): 411–416

Ahamed M, Alsalhi MS, Siddiqui MK (2010) Silver nanoparticle applications and human health. Clin Chim Acta 411(23-24):1841–1848

Ameen F, Alsamhary K, Alabdullatif JA et al (2021) A review on metal-based nanoparticles and their toxicity to beneficial soil bacteria and fungi. Ecotoxicol Environ Saf 213:112027

Asadishad B, Chahal S, Akbari A et al (2018) Amendment of agricultural soil with metal nanoparticles: effects on soil enzyme activity and microbial community composition. Environ Sci Technol 52(4):1908–1918

Baranowska-Wójcik E, Szwajgier D, Oleszczuk P et al (2019) Effects of titanium dioxide nanoparticles exposure on human health – a review. Biol Trace Elem Res 193:118–129

Battin TJ, Kammer FV, Weilhartner A et al (2009) Nanostructured TiO2: transport behavior and effects on aquatic microbial communities under environmental conditions. Environ Sci Technol 43(21):8098–8104

Bauer E, Thiele I (2018) From network analysis to functional metabolic modeling of the human gut microbiota. mSystems 3(3):110–128

Belizário JE, Napolitano M (2015) Human microbiomes and their roles in dysbiosis, common diseases, and novel therapeutic approaches. Front Microbiol 6:1050

Braydich-Stolle L, Hussain S, Schlager JJ et al (2005) In vitro cytotoxicity of nanoparticles in mammalian germline stem cells. Toxicol Sci 88(2):412–419

Brohi RD, Wang L, Talpur HS et al (2017) Toxicity of nanoparticles on the reproductive system in animal models: a review. Front Pharmacol 8:606

Chen R, Qiao J, Bai R et al (2018) Intelligent testing strategy and analytical techniques for the safety assessment of nanomaterials. Anal Bioanal Chem 410(26):6051–6066

Chen P, Huang J, Rao L et al (2021) Environmental effects of nanoparticles on the ecological succession of gut microbiota across zebrafish development. Sci Total Environ 806:150963

Cheng H, Hong PY (2020) Nanoparticles applied in membrane bioreactors: potential impact on reactor performance and microbial communities. In: Contaminants of emerging concern in water and wastewater. Elsevier, pp 207–236

Deng R, Zhu Y, Wu X et al (2023) Toxicity and mechanisms of engineered nanoparticles in animals with established allergic asthma. Int J Nanomedicine 18:3489–3508

Dusinska M, Boland S, Saunders M et al (2015) Towards an alternative testing strategy for nanomaterials used in nanomedicine: lessons from NanoTEST. Nanotoxicology 9(1):118–132

Fatima J, Siddique YH (2024) Application of nanocomposites and nanoparticles in treating neurodegenerative disorders. CNS Neurol Disord Drug Targets 23(10):1217–1233

Feder-Kubis J, Wirwis A, Policht M et al (2024) Principles and practice of greener ionic liquid-nanoparticles biosystem. Green Chem 26:3072–3124

Gangadoo S, Bauer BW, Bajagai YS et al (2019) In vitro growth of gut microbiota with selenium nanoparticles. Anim Nutr 5(4):424–431

Ge Y, Schimel JP, Holden PA (2011) Evidence for negative effects of TiO_2 and ZnO nanoparticles on soil bacterial communities. Environ Sci Technol 45(4):1659–1664

Gilbert B, Ono RK, Ching KA et al (2009) The effects of nanoparticle aggregation processes on aggregate structure and metal uptake. J Colloid Interface Sci 339(2):285–295

Gómez-Gómez B, Arregui L, Serrano S et al (2019) Unravelling mechanisms of bacterial quorum sensing disruption by metal-based nanoparticles. Sci Total Environ 696:133869

Hakim LK, Yari A, Nikparto N et al (2024) The current applications of nano and biomaterials in drug delivery of dental implant. BMC Oral Health 24(1):126

Hu YL, Gao JQ (2010) Potential neurotoxicity of nanoparticles. Int J Pharm 394(1-2):115–121

Hu P, Chen L, Kang X et al (2016) Surface functionalization of metal nanoparticles by conjugated metal-ligand interfacial bonds: impacts on intraparticle charge transfer. Acc Chem Res 49(10): 2251–2260

John R, Monpara J, Swaminathan S et al (2024) Chemistry and art of developing lipid nanoparticles for biologics delivery: focus on development and scale-up. Pharmaceutics 16(1):131

Kang ME, Weng Y, Liu Y et al (2023) A review on the toxicity mechanisms and potential risks of engineered nanoparticles to plants. Rev Environ Contam Toxicol 261(1):5

Kapoor D, Chilkapalli SC, Prajapati BG et al (2024) The astonishing accomplishment of biological drug delivery using lipid nanoparticles: an ubiquitous review. Curr Pharm Biotechnol 25(15): 1952–1968

Karlsson HL, Cronholm P, Gustafsson J et al (2008) Copper oxide nanoparticles are highly toxic: a comparison between metal oxide nanoparticles and carbon nanotubes. Chem Res Toxicol 21(9): 1726–1732

Ke X, Howard GP, Tang H et al (2019) Physical and chemical profiles of nanoparticles for lymphatic targeting. Adv Drug Deliv Rev 151–152:72–93

Kervinen K, Kalliala I, Glazer-Livson S et al (2019) Vaginal microbiota in pregnancy: role in induction of labor and seeding the neonate's microbiota? J Biosci 44:116

Khan I, Saeed K, Khan I (2019) Nanoparticles: properties, applications and toxicities. Arab J Chem 12(7):908–931

Kumah EA, Fopa RD, Harati S et al (2023) Human and environmental impacts of nanoparticles: a scoping review of the current literature. BMC Public Health 23(1):1059

Lee YH, Cheng FY, Chiu HW et al (2014) Cytotoxicity, oxidative stress, apoptosis and the autophagic effects of silver nanoparticles in mouse embryonic fibroblasts. Biomaterials 35(16):4706–4715

Li JJ, Hartono D, Ong CN et al (2010) Autophagy and oxidative stress associated with gold nanoparticles. Biomaterials 31(23):5996–6003

Li M, Li F, Lu Z et al (2019) Effects of TiO2 nanoparticles on intestinal microbial composition of silkworm, *Bombyx mori*. Sci Total Environ 704:135273

Li W, Deng X, Chen T (2021) Exploring the modulatory effects of gut microbiota in anti-cancer therapy. Front Oncol 11:644454

Li X, Wei H, Qi J et al (2023) Interactions of nanomaterials with gut microbiota and their applications in cancer therapy. Sensors 23(9):4428

Liu H, Lai W, Liu X et al (2020a) Exposure to copper oxide nanoparticles triggers oxidative stress and endoplasmic reticulum (ER)-stress induced toxicology and apoptosis in male rat liver and BRL-3A cell. J Hazard Mater 401:123349

Liu Z, Luo G, Du R et al (2020b) Effects of spaceflight on the composition and function of the human gut microbiota. Gut Microbes 11(4):807–819

Lv Y, Li W, Liao W et al (2024) Nano-drug delivery systems based on natural products. Int J Nanomedicine 19:541–569

Mansor M, Alarcon H, Xu J et al (2022) Simultaneous insight into dissolution and aggregation of metal sulfide nanoparticles through single-particle inductively coupled plasma mass spectrometry. ACS Earth Space Chem 6(3):541–550

Mileyeva-Biebesheimer ON, Zaky A, Gruden CL (2010) Assessing the impact of titanium dioxide and zinc oxide nanoparticles on bacteria using a fluorescent-based cell membrane integrity assay. Environ Eng Sci 27:329–335

Molleman B, Hiemstra T (2018) Size and shape dependency of the surface energy of metallic nanoparticles: unifying the atomic and thermodynamic approaches. Phys Chem Chem Phys 20(31):20575–20587

Nguyen NH, Falagan-Lotsch P (2023) Mechanistic insights into the biological effects of engineered nanomaterials: a focus on gold nanoparticles. Int J Mol Sci 24(4):4109

Oberdörster G (2010) Safety assessment for nanotechnology and nanomedicine: concepts of nanotoxicology. J Intern Med 267(1):89–105

Oberdörster G, Maynard A, Donaldson K et al (2005) Principles for characterizing the potential human health effects from exposure to nanomaterials: elements of a screening strategy. Part Fibre Toxicol 2:1–35
Pietroiusti A, Magrini A, Campagnolo L (2016) New frontiers in nanotoxicology: gut microbiota/ microbiome-mediated effects of engineered nanomaterials. Toxicol Appl Pharmacol 299:90–95
Pradhan A, Seena S, Pascoal C et al (2011) Can metal nanoparticles be a threat to microbial decomposers of plant litter in streams? Microb Ecol 62:58–68
Rademacher F, Gläser R, Harder J (2021) Antimicrobial peptides and proteins: interaction with the skin microbiota. Exp Dermatol 30(10):1496–1508
Rajput VD, Minkina T, Sushkova S et al (2018) Effect of nanoparticles on crops and soil microbial communities. J Soils Sediments 18:2179–2187
Ramos C, Calus M, Schokker D (2021) Persistence of functional microbiota composition across generations. Sci Rep 11(1):19007
Saiding Q, Zhang Z, Chen S et al (2023) Nano-bio interactions in mRNA nanomedicine: challenges and opportunities for targeted mRNA delivery. Adv Drug Deliv Rev 203:115116
Sajid M, Ilyas M, Basheer C et al (2015) Impact of nanoparticles on human and environment: review of toxicity factors, exposures, control strategies, and future prospects. Environ Sci Pollut Res Int 22:4122–4143
Shvedova AA, Pietroiusti A, Fadeel B et al (2012) Mechanisms of carbon nanotube-induced toxicity: focus on oxidative stress. Toxicol Appl Pharmacol 261(2):121–133
Sieow B, Nurminen T, Ling H et al (2019) Meta-omics- and metabolic modeling-assisted deciphering of human microbiota metabolism. Biotechnol J 14(9):1800445
Singh H, Desimone MF, Pandya S et al (2023) Revisiting the green synthesis of nanoparticles: uncovering influences of plant extracts as reducing agents for enhanced synthesis efficiency and its biomedical applications. Int J Nanomedicine 18:4727–4750
Su Y, Wu D, Xia H et al (2019) Metallic nanoparticles induced antibiotic resistance genes attenuation of leachate culturable microbiota: the combined roles of growth inhibition, ion dissolution and oxidative stress. Environ Int 128:407–416
Suresh AK, Pelletier DA, Doktycz MJ (2013) Relating nanomaterial properties and microbial toxicity. Nanoscale 5(2):463–474
Thubagere A, Reinhard BM (2010) Nanoparticle-induced apoptosis propagates through hydrogen-peroxide-mediated bystander killing: insights from a human intestinal epithelium in vitro model. ACS Nano 4(7):3611–3622
Tsuji J, Maynard A, Howard P et al (2006) Research strategies for safety evaluation of nanomaterials, part IV: risk assessment of nanoparticles. Toxicol Sci 89(1):42–50
Turnbaugh PJ, Hamady M, Yatsunenko T et al (2008) A core gut microbiome in obese and lean twins. Nature 457(7228):480–484
Xia T, Kovochich M, Brant J et al (2006) Comparison of the abilities of ambient and manufactured nanoparticles to induce cellular toxicity according to an oxidative stress paradigm. Nano Lett 6(8):1794–1807
Xia T, Kovochich M, Liong M et al (2008) Comparison of the mechanism of toxicity of zinc oxide and cerium oxide nanoparticles based on dissolution and oxidative stress properties. ACS Nano 2(10):2121–2134
Xuan L, Ju Z, Skonieczna M et al (2023) Nanoparticles-induced potential toxicity on human health: applications, toxicity mechanisms, and evaluation models. MedComm 4(4):e327
Zhang T, Zhu X, Guo J et al (2021) Toxicity assessment of nano-ZnO exposure on the human intestinal microbiome, metabolic functions, and resistome using an in vitro colon simulator. Environ Sci Technol 55(10):6884–6896
Zhou Y, Jin C, Li Y et al (2018) Dynamic behavior of metal nanoparticles for catalysis. NanoToday 20:101–120
Zorraquín-Peña I, Cueva C, Bartolomé B et al (2020) Silver nanoparticles against foodborne bacteria: effects at intestinal level and health limitations. Microorganisms 8(1):132

The Nexus of Chemicals and Cancer: Environmental Toxicants and Lifestyle Effects Unveiled

Adela Perolla

Abstract

Cancer is a complex disease defined as an uncontrolled cellular growth, primarily caused by genetic mutations interfering with normal cell cycle control, leading to the development and progression of tumors. The rise in global cancer cases is strongly associated with the aging population. By 2050, the United States is projected to experience a nearly 50% rise in cancer cases, largely due to population growth and aging. While cancer incidence and mortality rates are decreasing in industrialized Western countries, there is a rising trend in economically transitioning and less developed regions, driven by unhealthy lifestyles including smoking, lack of physical activity, and consumption of high-calorie diets, which are significant contributors to rising cancer rates, especially in developing nations. The decline in natural selection over the past 150 years has led to an accumulation of harmful genetic mutations, further increasing cancer risks. Exposure to toxic substances also plays a significant role in the development of cancer. The tumor microenvironment (TME) is vital in cancer progression, with its interactions influencing treatment outcomes and prognosis. Environmental toxicants continue to present significant risks, particularly for vulnerable populations, emphasizing the need for comprehensive strategies to mitigate these risks and improve cancer prevention and treatment.

A. Perolla (✉)
Faculty of Medicine, University of Medicine, Tirana, Albania

Department of Internal Medicine, Service of Hematology, University Hospital Centre Mother Teresa, Tirana, Albania

S. Raisuddin et al. (eds.), *Environmental Toxicants and Lifestyle Diseases*,
https://doi.org/10.1007/978-981-95-0575-3_12

Keywords

Carcinogenesis · Environmental toxicants · Lifestyle diseases · Epigenetics · Inflammation

Introduction

Cancer is a multifaceted and complex disease characterized by the uncontrolled growth and spread of abnormal cells. It develops due to various genetic mutations in oncogenes and tumor suppressor genes, which result in unchecked cell division and the formation of tumors, usually requiring up to six mutations to advance through stages of tumor development (Torpy et al. 2010).

The conversion from healthy cells to cancerous cells is linked to changes in many processes (Huang and Zhou 2021), such as deoxyribonucleic acid (DNA) repair systems, cell proliferation rates, cell-cycle control, and apoptosis. Distinct chromosomal abnormalities are linked to various types of cancer, helping in their identification and treatment (Diddens 2023; Hagan and Sharrocks 2002). Therefore, we are facing a disease of uncontrolled proliferation by transformed cells subject to natural selection, modernizing the definition to include genetic and epigenetic changes that lead to lethal phenotypes (Brown et al. 2023).

The progression of cancer is greatly influenced by the interactions between the tumor and its surrounding microenvironment. Recent advancements in cancer research indicate that integrating immune-related factors into clinical stratification systems could improve the accuracy of prognosis and predictions of treatment, as fundamentally a genetic disease (Galon and Bruni 2020).

While cancer incidence and mortality rates are decreasing in the US and other developed Western countries, they are on the rise in economically transitioning and less developed regions, driven by unhealthy behaviors and infectious agents (Weir et al. 2021). By 2035, it is expected that around 14 million new cancer cases will emerge globally, accounting for nearly 60% of the total cancer burden. The largest increases are anticipated in regions such as China, the Middle East, and Northern Africa, with less developed areas projected to see a 144% rise in new cases, compared to a 54% increase in more developed regions (Pilleron et al. 2019). The increasing prevalence of Western lifestyles, characterized by habits such as smoking, sedentary behavior, and the eating of high-calorie meals, is a significant factor in the escalating rates of cancer in developing nations (Jemal et al. 2010). The increase in cancer incidence worldwide can be attributed to the accumulation of mutant alleles in human populations due to a decrease in natural selection over the past 150 years. Countries with limited opportunities for natural selection have significantly higher average cancer incidence rates compared to those with greater opportunities (You and Henneberg 2018).

Exposure to environmental and chemical carcinogens can result in genetic mutations and alterations in gene expression patterns, leading to the development of cancer. Chemical carcinogens promote cancer by altering and activating proto-oncogenes while deactivating tumor suppressor genes. The overall risk of cancer is determined by the combined effect of these agents on various genetic and epigenetic processes (Harris 1991). The adaptability of cancer cells to their microenvironment, including low oxygen levels and high acidity, is crucial for their invasion and proliferation (Fang et al. 2008).

Tumors adapt to unfavorable environments through genetic or non-genetic changes, metabolic signal alteration, or reconfiguration of their environment through migration to other organs (Kim et al. 2022). The tumor microenvironment is vital in cancer progression, facilitating cancer cell survival, local invasion, and metastasis, and offering potential targets for therapeutic intervention (Rasheed and Koyyala 2020).

Definition and Sources of Environmental Toxicants

Environmental toxicants are detrimental substances found in the environment that can have negative impacts on living species, including humans, animals, and plants. These toxicants can originate from various sources and persist in the environment, leading to the accumulation of these substances in organisms and causing long-lasting health consequences. They are synthetic organic compounds that are resistant to metabolism and have a long-lasting presence in the environment. They accumulate in the food chain, causing harm to fish, wildlife, and humans (Younglai et al. 2007). Environmental toxicology studies the origins, detrimental effects, and strategies to mitigate environmental pollution, which has an impact on both human health and ecosystems (US EPA 2013).

Environmental contaminants encompass many harmful elements, including heavy metals, pesticides, industrial chemicals, and biological agents such as mycotoxins and bacterial toxins. The presence of these substances in the environment poses a significant challenge to public health and underscores the need for comprehensive regulation and monitoring (Younglai et al. 2007; US-EPA 2013; Vaz 2021).

Environmental toxicants have significant and lasting impacts on health, including an elevated risk of cancer, worsening of infectious diseases, and accelerated aging due to stem cell dysfunction. These effects are often mediated by epigenetic changes and oxidative stress. Effective regulation and public health policies are crucial to reduce these risks and safeguard future generations. These toxicants can induce long-term health effects by causing changes in DNA methylation and histone modifications, resulting in chronic diseases and being passed down to future generations (Barouki et al. 2018).

Environmental toxicants, such as dioxins and polychlorinated biphenyls (PCBs), can harm human development and reproduction due to their persistent presence in the environment and their harmful effects on the body (Younglai et al. 2007; Vaz 2021). Additionally, substances like BPA, phthalates, pesticides, and persistent contaminants can negatively impact mammalian reproduction through mechanisms passed down through multiple generations (Rebuzzini et al. 2022). Heavy metals (such as lead and cadmium) and pesticides are significant pollutants that pose severe risks to human health, including damage to the nervous system, cancer-causing properties, and disruption of hormone function. These pollutants can build up in agricultural soil and plants, entering the food chain and affecting human well-being (Alengebawy et al. 2021).

A systematic review conducted in 2015 investigated the effects of environmental toxicants on the aging process of stem cells. It revealed that exposure to pesticides, tobacco smoke, radiation, and heavy metals can induce oxidative stress and lead to epigenetic changes. These changes have a negative effect on germ cells and embryos, resulting in a decline in regenerative capacity and an increased risk of age-related diseases (Hodjat et al. 2015; Younglai et al. 2007).

Exposure to environmental toxicants can have negative effects on the development and reproduction of humans, impacting both male and female fertility. Toxicants like polychlorinated biphenyls and dioxins can contribute to these adverse effects by persisting in the environment and affecting how they work (Barouki et al. 2018; Hodjat et al. 2015). Additionally, these toxicants can also lead to various non-communicable diseases and contribute to the ageing process by interfering with the function of stem cells (Alengebawy et al. 2021; Sarah et al. 2022).

Environmental toxicants accumulate in organisms over time through the food chain, resulting in long-term health effects and potential transgenerational consequences. The ability of these toxicants to persist and accumulate in soil and plants presents significant risks to the environment and human health (Rebuzzini et al. 2022; Alengebawy et al. 2021). They exert their effects through different cellular and molecular mechanisms, such as oxidative stress, DNA methylation, and histone modification, which can cause alterations in cell function and the ageing process (Hodjat et al. 2015; Worley and Parker 2019).

The Impact of Environmental Toxicants on the Development of Cancer

Environmental pollutants are increasingly recognized as significant contributors to the development of cancer. These harmful substances, known as "toxicants," encompass a diverse array of chemicals present in food, water, air, and consumer goods. They have the capacity to disrupt hormone pathways, modify the tumor microenvironment, and induce genetic and epigenetic alterations, all of which heighten the likelihood of cancer.

In a systematic review published in 2018 was carried out the contribution of environmental chemicals to the development of breast cancer. They discovered a connection between a higher risk of breast cancer and substances like DDT, dioxins, perfluorooctanesulfonamide (PFOSA), and air pollution. Exposure that takes place during crucial periods for breast development increases the significance of this risk. According to the study, these substances influence the development of cancer by altering the epigenetics of genes that suppress cancer, such as the global hypomethylation and hypermethylation of tumor suppressor genes (Rodgers et al. 2018; Park 2015).

Exposure to environmental toxicants is a significant risk factor for cancer, influencing both genetic and epigenetic mechanisms. These toxicants can disrupt gene expression without altering DNA sequences, promoting carcinogenesis by reprogramming cells into cancer stem cells. These chemicals affect molecular targets involved in inflammation, such as cyclooxygenase, nuclear factor kappa B, and cytokines. They can increase the number of methyl groups in the promoter CpG island, turn off genes that stop tumors from growing, make the genome less stable, and start inflammatory pathways like Signal Transducer and Activator of Transcription 3 (STAT3) and Nuclear Factor kappa-light-chain-enhancer of activated B cells (NF-κB) (Panahi et al. 2016). Toxic chemicals in the environment, like bisphenol A, atrazine, and phthalates, can throw off the balance between pro- and anti-inflammatory responses. This can raise the risk of cancer by turning on NFκB, which then strongly activates Interleukin-6 (IL-6), which then turns on the STAT3 protein. Both cancer cells and surrounding inflammatory immune cells produce excessive and continuous levels of IL-6 and other cytokines, leading to chronic STAT3 activation. STAT3 influences gene transcription, promoting cell proliferation, resulting in resistance to apoptosis, angiogenesis development, immune evasion and invasion, and finally giving rise to metastasis, which are all key characteristics of cancer (Thompson et al. 2015). During studying the molecular mechanisms that connect environmental toxicants to cancer development was found that polyphenols from plants, such as epigallocatechin-gallate, resveratrol, curcumin, and anthocyanins, may help lower the risk of cancer and lessen the harmful effects of environmental toxicants (Lagoa et al. 2022).

An article published in the International Journal of Environmental Research and Public Health, in 2019 clearly specify that the risk of cancer may increase due to the concurrent presence of several conditions: excessive nitrate intake from drinking water, leading to increased nitrite production and blood hypoxia; natural and therapeutic radiation exposure, which can elevate plasma nitric oxide (NO) levels; and iodine and/or selenium deficiency, as well as exposure to air pollution, particularly in volcanic areas (Fiore et al. 2019).

Compounds like nickel chloride, Bisphenol A (BPA), and methylmercury disrupt the interactions within the tumor microenvironment, facilitating tumor growth and metastasis. They induce preneoplastic changes in cells such as stromal cells, fibroblasts, and endothelial cells, and disrupt both innate and adaptive immune functions, affecting antigen-presenting cells and B and T lymphocytes. By promoting angiogenesis, they ensure cancer cells receive the necessary nutrients and oxygen. Additionally, by inducing chronic inflammation, which fuels the growth and evolution of neoplastic cells through a continuous supply of pro-inflammatory cytokines and growth factors, in combined actions, creates a supportive environment for tumor development and progression (Fiore et al. 2019; Casey et al. 2015).

Long-lasting organic pollutants and endocrine disruptors contribute to the spread of breast cancer and reduce the effectiveness of chemotherapy by promoting the transition from epithelial to mesenchymal cells and the development of cancer stem cell traits (Koual et al. 2020). Epithelial-to-mesenchymal transition (EMT), refers to the process where epithelial cells transform into cells with mesenchymal traits, enhancing their capacity to migrate and invade surrounding tissues (Lv et al. 2021). This allows primary tumor cells to migrate and infiltrate other areas. Cancer stem cells can also undergo the EMT process again, perpetuating a cycle of invasion and metastasis (Moustakas and De Herreros 2017). Exposure to small amounts of common pollutants can cause damage to DNA, changes in gene activity, and increased cell invasion, all of which contribute to resistance to chemotherapy (Abdel-Rahman et al. 2018; Lagunas-Rangel et al. 2022).

Numerous studies have examined the role of environmental and occupational exposures in the development of blood cancers, such as leukemias and lymphomas. Key mechanisms involve genetic and epigenetic alterations, such as overall reduction in DNA methylation and excessive methylation of genes that suppress tumor growth (Park 2015; Panahi et al. 2016).

Farmers and industrial workers exposed to asbestos, aromatic hydrocarbons, fertilizers, mineral oils, pesticides, and radiation have a significantly increased risk of hematological malignancies. In 2020, a cross-sectional descriptive study explored the presence of toxic environmental factors in patients diagnosed with hematological neoplasms. The results showed that non-Hodgkin lymphoma was the most frequent diagnosis (25.5%), with a higher incidence among individuals aged 61–80 years (50%), farmers (31%), and those from rural areas (57%). A notable 64.4% of patients had been exposed to toxins, mainly pesticides, with an average exposure duration of 14 years, and the last contact was 9.2 years prior. Protective measures were rarely used, and 90% of participants were unaware of safety protocols (Favier Rodríguez et al. 2023). In addition, another case-control study, which included 620 cases of hematological malignancies, found that exposure to asbestos, aromatic hydrocarbons, fertilizers, mineral oils, pesticides, and radiation is significantly linked to an increased risk of these malignancies (Pasqualetti et al. 1991).

Occupational exposure to benzene and aromatic amines is closely linked to a higher risk of developing hematological cancers. Benzene, in particular, has a well-established connection to acute myeloid leukemia (AML) and non-Hodgkin's lymphoma, with the risk increasing alongside higher exposure levels and longer durations of exposure. Physical and chemical agents, such as therapeutic alkylating agents, topoisomerase II inhibitors, and ionizing radiation, primarily induce AML by causing DNA damage that leads to gene or chromosomal mutations. In contrast, biological agents and certain immunosuppressive chemicals tend to cause lymphoid neoplasms by interfering with immune function. While benzene is strongly associated with AML, there is also emerging but limited evidence of its potential link to lymphoid neoplasms. Ethylene oxide and 1,3-butadiene are primarily associated with lymphoid cancers, while the relationship between formaldehyde and leukemia, particularly AML, remains controversial but potentially significant. All the environmental agents reviewed exhibit genotoxic properties, leading to gene mutations, chromosomal alterations, and/or micronuclei formation in vivo (Eastmond et al. 2014).

In China, a large-scale study involving 74,828 workers found that workers exposed to benzene levels below 10 parts per million (ppm) had a 2.2% increased risk of developing any blood cancer and a 3.2% increased risk of acute nonlymphocytic leukemia and related myelodysplastic syndromes (Hayes et al. 1997).

Additionally, individuals with more than a decade of benzene exposure face a 4.2 times greater risk of developing non-Hodgkin's lymphoma, especially if the exposure occurred at least 10 years before diagnosis. The timing of exposure is a critical factor in determining the risk of specific diseases. However, a systematic review conducted in 2021 did not find strong evidence connecting exposure to hazardous substances, such as benzene, with myeloproliferative neoplasms (MPNs) (Allahverdi et al. 2021).

Benzene, ethylene oxide, and 1,3-butadiene cause hematological cancers by damaging genes and altering chromosomes through genotoxic mechanisms (Pasqualetti et al. 1991; Eastmond et al. 2014). In the early stages of cancer development caused by environmental toxins, changes in DNA methylation and histone modifications play a crucial role (Park 2015; Panahi et al. 2016). Environmental toxins can disrupt immune function, leading to an imbalance between pro- and anti-inflammatory responses, which contributes to the formation of tumors. Additionally, these toxins can disrupt the migration, polarization, differentiation, and immune-stimulatory functions of macrophages, potentially leading to inflammatory disorders and neoplastic diseases. Toxicants also affect the polarization and function of macrophages, promoting the invasiveness of tumors and suppressing the immune system (Sepand et al. 2021). Furthermore, research has linked airborne pollutants, specifically volatile organic compounds (VOCs), to increased occurrences of blood cancer (Clapp et al. 2008).

Lifestyle

Lifestyle refers to the daily routines, behaviors, and habits that shape how individuals or groups live, encompassing factors such as diet, physical activity, sleep patterns, stress management, and social interactions. It is the way individuals express their values and beliefs through their day-to-day actions, including both tangible factors such as demographics and intangible ones like personal preferences (Ahmad 2020).

Recognizing the substantial influence of lifestyle on both physical and mental well-being is essential. Unfavorable lifestyles can result in a range of health problems, including metabolic disorders, cardiovascular ailments, and mental health conditions (Brivio et al. 2023). These behaviors are influenced by factors such as geography, economics, politics, culture, and religion. Lifestyle encompasses people's unique way of life in a specific region and time, including their daily activities, work roles, recreational habits, and dietary practices (Definition of LIFESTYLE 2024; https://www.merriam-webster.com/dictionary/lifestyle). According to the World Health Organization (WHO), 60% of health and quality of life factors are associated with lifestyle. Unhealthy lifestyles, characterized by poor diet, lack of exercise, substance abuse, and improper technology use, are considered "at-risk groups." These lifestyles can lead to various health issues, including metabolic diseases, cardiovascular conditions, hypertension, and mental health disorders (Ahmad 2020).

Nutrition and Body Mass Index (BMI)

Urbanization and modernization have caused a shift in dietary patterns towards foods that are high in calories but low in nutrients. This has led to an increase in obesity rates worldwide.

Obesity is defined as having a body-mass index (BMI) of 30 kg/m^2 or higher and is now a major global health concern, surpassing undernutrition and infectious diseases in terms of its impact (Kopelman 2000). The main factors contributing to the rise in obesity are genetic factors, the increased availability of high-calorie foods, and a decrease in physical activity due to modern lifestyles. Obesity plays a significant role in the development of chronic diseases such as diabetes, coronary heart disease, certain types of cancer, and respiratory disorders.

An epidemiological analysis conducted by Ziraba et al. (2009) examined demographic and health surveys from seven sub-Saharan African countries. The study focused on the trends in overweight and obesity among urban women from 1992 to 2005. The results revealed a substantial 35% increase in overweight and obesity among urban women during this period. Interestingly, the rise was more prominent among the poorest women, with a 50% increase, compared to the richest women who only experienced a 7% increase. Non-educated and primary-educated women saw a significant 45–50% increase, while women with secondary education or higher actually experienced a 10% decrease. The prevalence of overweight and obesity increased by approximately 5% each year. Although the rate of change did not differ significantly between poor and rich women, it was notably higher among non-educated women (Ziraba et al. 2009).

Obesity is caused by a combination of factors: a rise in the consumption of foods that lack nutrients, are high in sugar, and contain saturated fats, together with a decrease in physical activity. This, in turn, leads to the development of major chronic diseases related to diet (Kumar et al. 2016).

Economic growth leads to an increase in the availability of processed foods and a decrease in physical activity due to more sedentary jobs and lifestyles. Urban areas often lack access to fresh and healthy foods as well as recreational facilities. The design of the urban environment promotes inactive lifestyles and the consumption of unhealthy foods, which increases the risk of obesity (Jaime et al. 2011).

Studies conducted in São Paulo and Detroit demonstrate the significant impact of local food environments on diet and physical activity. For instance, living in a "food desert" like Detroit is associated with a higher BMI. However, this association becomes less significant when considering factors such as demographics, disease status, shopping and eating behaviors, dietary intakes, and knowledge about diet

(Budzynska et al. 2013; Dake et al. 2016). Obesity is promoted by limited access to healthy food options and exposure to certain environments. In low-income urban areas, there is a high consumption of energy-dense foods and low levels of physical activity, which worsen obesity rates (Johnson et al. 2021; Ntandou et al. 2009). Obesity significantly increases the risk of hypertension, coronary heart disease, and stroke. Urban populations have higher rates of these conditions, mainly due to dietary and lifestyle factors (Kopelman 2000; Kumar et al. 2016).

A high BMI is strongly linked to type 2 diabetes. Urban areas, where people have sedentary lifestyles and poor eating habits, have higher rates of diabetes. The rate of obesity among adults in rural areas of the United States is notably higher than in urban areas. Specifically, the obesity rate in rural areas is 39.6%, while in urban areas it is 33.4% (Befort et al. 2012). Obesity is associated with an elevated risk of multiple types of cancer, such as esophageal, colorectal, postmenopausal breast, endometrial, and kidney cancer (Fitzgibbon and Stolley 2010). This connection is attributed to the inflammatory and hormonal changes that occur with excessive body fat (Kopelman 2000). Increased consumption of processed and red meats is linked to a higher risk of colorectal cancer (Key et al. 2004).

A research, using both mouse and drosophila models in 2018 to investigate the molecular links between diet, obesity, type II diabetes, and cancer, demonstrated that utilizing diet and cancer models in Drosophila can rapidly enhance our understanding of the shared molecular mechanisms that connect diet and diet-related metabolic disorders to a poor cancer prognosis (Warr et al. 2018).

A comprehensive analysis of 37 studies conducted between 1990 and 2017 reveals a complex relationship between high sugar consumption and cancer risk. While excessive sugar intake can lead to disruptions in insulin-glucose regulation, oxidative stress, inflammation, and increased body fat, the overall findings from epidemiological studies are not definitive. Most of these studies did not establish a significant connection between total sugar, sucrose, or fructose intake and cancer risk. However, some research indicated a 60–200% increased risk of cancer associated with high consumption of added sugars and sugary beverages, with a specific link to an elevated risk of breast cancer associated with sugary drink consumption (Makarem et al. 2018; Debras et al. 2020).

A study published in 2011 examined the relationship between different types of dietary sugars and the risk of developing 24 various cancers, drawing on data from 435,674 participants aged 50–71 from the NIH-AARP Diet and Health Study, tracked over a 7.2-year period. Dietary sugar intake was measured using a 124-item food frequency questionnaire (FFQ). The analysis identified 29,099 cancer cases in men and 13,355 in women. The study found a positive correlation between added sugars and esophageal adenocarcinoma, added fructose and small intestine cancer, and all sugars studied with pleural cancer. Interestingly, in women, an inverse association was observed between all types of sugars and ovarian cancer. No significant links were found between dietary sugars and the risk of colorectal or other major cancers. The authors note that potential measurement errors in the FFQ-reported dietary sugars could have influenced the results, highlighting the necessity for further research, particularly regarding the significant associations found with rare cancers (Nöthlings et al. 2007). While other studies have not found a significant correlation, it is necessary to do additional research to explore the potential detrimental impact of added sugars and sugary beverages on the risk of developing cancer.

Recent studies indicate a potential link between elevated blood sugar and insulin levels and the risk of pancreatic cancer, possibly influenced by high sugar consumption, which raises insulin demand and reduces insulin sensitivity. Diets with a high glycemic load, as well as the intake of fructose and sucrose, have been connected to a higher risk of various cancers, particularly pancreatic cancer, especially among individuals who are overweight or obese. A 2006 Swedish study involving 77,797 participants aged 45–83 examined the association between added sugar intake and pancreatic cancer. Over an average follow-up of 7.2 years, 131 new cases of pancreatic cancer were documented. The study found significant links between the intake of added sugars, soft drinks, and sweetened fruit soups or stewed fruit and an increased risk of pancreatic cancer, with the highest consumption groups showing hazard ratios of 1.69 for added sugar, 1.93 for soft drinks, and 1.51 for sweetened fruit soups or stewed fruit. These results suggest that high sugar consumption may elevate the risk of pancreatic cancer, highlighting the need for dietary changes to reduce this risk (Larsson et al. 2006).

While excessive sugar intake is widely recognized as a key risk factor for obesity, type 2 diabetes, and cardiovascular diseases, its link to cancer is less well-defined. A study from the French NutriNet-Santé cohort (2009–2019) involving 101,279 participants explored the connection between sugar intake and the risk of overall, breast, and prostate cancer. Using 24-h dietary records, the study found that higher total sugar consumption was associated with an increased overall cancer risk, particularly significant in breast cancer. The study also identified a link between increased cancer risk and the intake of added sugars, free sugars, sucrose, and sugars from milk-based desserts, dairy products, and sugary beverages (Tasevska et al. 2012).

A study involving 102,865 adults discovered that consuming higher amounts of artificial sweeteners, specifically aspartame and acesulfame-K, is associated with a greater likelihood of developing cancer. The research, which took into account various factors that could affect the results, found significant connections between the overall risk of cancer and the specific risks of breast and obesity-related cancers. Although there are some limitations to the study, such as potential bias in participant selection and other factors that were not fully accounted for, these findings indicate that health authorities should reconsider the safety of these commonly used food additives (Debras et al. 2022).

Drug Addiction

The consumption of alcohol, tobacco, opioids, and illicit drugs greatly heightens the likelihood of getting different forms of cancer. The cancer-causing properties of these substances and their influence on the advancement of cancer have been extensively demonstrated in multiple research studies.

Alcohol Use and the Risk of Cancer Development

Alcohol consumption is a well-documented risk factor for several types of cancer, including those of the head and neck, esophagus, liver, colon, rectum, and breast. Even minimal alcohol intake is associated with an increased cancer risk, with no safe level of consumption identified (Moussas and Papadopoulou 2017). Epidemiological studies demonstrate a strong link between the amount of alcohol consumed and cancer mortality. Additionally, substance abuse, particularly with opioids and cannabis, may contribute to cancer development and negatively impact the mental health and quality of life of cancer patients. The American Society of Clinical Oncology (ASCO) can play a key role in promoting public education and policy efforts to reduce cancer risk (LoConte et al. 2018; Rehm et al. 2019).

Research indicates that both moderate and heavy alcohol consumption can increase the risk of various cancers. High levels of alcohol intake (≥4 drinks per day) significantly raise the likelihood of developing cancers such as oral and pharyngeal cancer, esophageal squamous cell carcinoma, laryngeal cancer, colorectal cancer, breast cancer, and pancreatic cancer. Even moderate alcohol consumption, such as 1–2 drinks per day, is linked to an elevated risk of those cancers. Furthermore, low levels of alcohol consumption (≤1 drink per day) can increase the risk of oral and pharyngeal cancer and esophageal squamous cell carcinoma by about 20–30% (Pelucchi et al. 2011).

Meta-analyses have revealed that alcohol intake significantly raises the likelihood of developing malignancies in the oral cavity, pharynx, esophagus, and larynx. There is also a statistically significant increase in the risk of cancers in the stomach, colon, rectum, liver, female breast, and ovaries (Bagnardi et al. 2001).

Approximately 4% of all cancer cases worldwide are attributed to alcohol consumption, which raises the risk for several types of cancer, including those of the upper aerodigestive tract, liver, colorectum, and breast. This review emphasizes the epidemiological and mechanistic evidence linking alcohol to cancer. Ethanol, the primary component in alcoholic beverages, contributes to cancer development through multiple pathways, some of which remain incompletely understood. Acetaldehyde, a metabolite of ethanol, induces inflammation and oxidative stress, leading to lipid peroxidation and alteration of DNA methylation, interfering with its synthesis and repair (Rumgay et al. 2021).

Another large-scale study involving over 1.2 million middle-aged women found a clear link between increased alcohol consumption and a higher risk of cancers in various parts of the body, including the oral cavity, pharynx, esophagus, rectum, liver, and breast. The study indicated that even moderate alcohol consumption in women significantly raises the risk of certain cancers, leading to an excess of about 15 cancer cases per 1000 women by age 75. The risk was notably higher among those who also smoked (Allen et al. 2009).

The evidence strongly supports that even modest alcohol consumption increases the likelihood of developing several types of cancer. Implementing public health initiatives to reduce alcohol consumption could significantly decrease the incidence of alcohol-related cancers.

The Correlation Between Tobacco Consumption and the Likelihood of Developing Cancer

Tobacco use is a leading cause of various cancers, as highlighted by recent in-depth analyses and assessments. Tobacco smoke contains over 250 harmful chemicals, many of which are carcinogenic and contribute to the development of cancers such as lung, mouth, throat, bladder, and pancreatic cancer. These chemicals have the potential to damage DNA, induce mutations, and promote the growth of cancer cells through multiple mechanisms (Moussas and Papadopoulou 2017). Tobacco smoking accounts for nearly 87% of lung cancer deaths, with male smokers facing approximately 23 times the risk of developing lung cancer compared to non-smokers, and female smokers facing a risk 13 times higher (Moussas and Papadopoulou 2017).

The rise in e-cigarette use, particularly among younger individuals, has sparked concerns about their potential health effects. A survey was conducted with 4695 current and former e-cigarette users, with a median age of 34, revealed that 78.9% of participants experienced an adverse event within 6 h of using an e-cigarette, with headaches, anxiety, and coughing being the most common issues, and notably influenced by product characteristics and user behaviors, and identified sociodemographic disparities, with higher rates of adverse events reported among Hispanic participants and those with at least a college education (Cuomo et al. 2024).

Smoking is responsible for at least 30% of all cancer-related deaths, with 87% of lung cancer deaths in men and 70% in women being attributed to smoking (O'Keeffe et al. 2018). Lung cancer is the primary cause of cancer-related deaths for both sexes. Furthermore, tobacco significantly increases the risk of oral and pharyngeal cancers, which, despite accounting for only 1–2% of all cancers, have high mortality rates (Bassiony et al. 2015). A comprehensive analysis of multiple reviews confirms the strong link between the use of smokeless tobacco (SLT) and oral cancer. The estimated risks for oral cancer due to SLT use vary significantly, with regional differences and product-specific risks (Asthana et al. 2019).

In 2021, a meta-analysis demonstrated that exposure to secondhand smoke (SHS) significantly raises the risk of developing oral cancer. The analysis synthesized data from five studies, involving a total of 1179 cases and 5798 controls. Among the 3452 individuals exposed to SHS and 3525 who were not, the overall odds ratio (OR) for oral cancer was 1.51 (95% CI 1.20 to 1.91, $p = 0.0004$), indicating a marked increase in risk. The findings suggested a clear link between SHS exposure and the elevated risk of oral cancer, particularly with prolonged exposure (Mariano et al. 2022).

In October 2017, a comprehensive analysis examined the relationship between oral cancer and tobacco use in the Middle East and North Africa (MENA) region, with a particular focus on shammah, a form of smokeless tobacco. The study found that individuals who use shammah face a substantially higher risk of developing oral cancer compared to non-users. Specifically, shammah users were nearly 39 times more likely to develop oral cancer (Quadri et al. 2019).

While lung cancer is well known to be related to smoking, other cancers such as those affecting the oral cavity, larynx, esophagus, pancreas, kidneys, bladder, stomach, and cervix are also linked to tobacco use. Additionally, passive smoking increases the risk of lung cancer and maternal smoking during pregnancy is associated with a higher likelihood of childhood cancers like Wilms tumor, acute lymphoblastic leukemia, and non-Hodgkin's lymphoma. Researchers are working to identify high-risk groups to reduce cancer incidence and mortality. However, preventing smoking altogether is deemed more effective, with estimates suggesting that 30% of all cancer cases could be prevented this way (Schuster-Kolbe and Rauchen 1994). A meta-analysis confirmed the causal association of smoking with various cancers, including the nasal cavities, paranasal sinuses, nasopharynx, stomach, liver, kidney, uterine cervix, and myeloid leukemia (Schuster-Kolbe and Rauchen 1994). Current, former, and ever smokers have increased risks of 1.07, 1.08, and 1.09, respectively, compared to never smokers. The risk increases with smoking intensity and duration in a linear fashion with both the intensity and duration of smoking. Specifically, the relative risk (RR) for smoking 20 cigarettes per day is 1.12, and it increases to 1.26 for 40 cigarettes per day. Similarly, the relative risk for 20 years of smoking is 1.05, rising to 1.11 for 40 years of smoking (Scala et al. 2023).

The relationship between cigarette smoking and prostate cancer is complex. While smoking is associated with a lower incidence of prostate cancer, it increases the risk of death from the disease. A systematic review and meta-analysis in 2023 examined the incidence of prostate cancer and the risk of death. The analysis included 15 studies and found that smoking is inversely related to the incidence of prostate cancer, with a relative risk of 0.84. However, within the review, five cohort studies revealed that current smokers have a significantly higher risk of dying from prostate cancer, with a relative risk of 1.42, indicating a 42% increased risk compared to non-smokers (Al-Fayez and El-Metwally 2023).

A systematic review and meta-analysis of epidemiological studies up to September 2018 examined the link between cigarette smoking and colorectal cancer (CRC) risk, analyzing data from 188 studies. The results showed that current smokers had a relative risk (RR) of 1.14, and former smokers had an RR of 1.17, compared to never smokers. The risk of CRC increased with higher smoking intensity and longer duration. Notably, those who had quit smoking for over 25 years had a significantly lower risk than current smokers. Smoking was strongly associated with CRCs characterized by high CpG island methylator phenotype, BRAF mutation, or high microsatellite instability, but not with Kirsten rat sarcoma (KRAS) or tumor suppressor p53 (TP53) mutations. The study concluded that smoking increases CRC risk in a dose-dependent manner, and quitting smoking significantly reduces this risk, especially for CRCs associated with microsatellite instability, highlighting the importance of smoking cessation in CRC prevention (Botteri et al. 2020).

Tobacco smoke comprises more than 250 detrimental compounds, a significant number of which are recognized as carcinogens. These chemicals have the potential to cause DNA damage, trigger mutations, and facilitate the progression of cancer through several mechanisms (Kachhap et al. 2011).

Smoking during cancer treatment negatively impacts treatment efficacy and increases toxicity. For example, smoking during radiotherapy is associated with higher risks of locoregional recurrence and radiation-induced toxicity. In a meta-analysis investigating the impact of smoking on cancer treatment efficacy and toxicity across various cancer types, findings revealed that smoking during radiotherapy or chemoradiotherapy is linked to worse outcomes and a higher risk of toxicity. Specifically, smoking increased the risk of locoregional recurrence and worsened disease-free survival, with pooled hazard ratios (HRs) of 1.56 for radiotherapy and 4.28 for chemoradiotherapy. Additionally, smoking heightened the risk of radiation-induced toxicity.

In lung cancer patients treated with EGFR tyrosine kinase inhibitors (EGFR-TKIs), smoking was associated with poorer progression-free survival (HR: 1.43). Conversely, smoking was linked to better progression-free survival in patients receiving checkpoint inhibitors (HR: 0.70). No significant connection was observed between smoking and the efficacy or toxicity of chemotherapy, though the evidence was considered low in certainty. This meta-analysis highlights the detrimental impact of smoking on cancer treatments like radiotherapy and EGFR-TKIs, while also suggesting a possible beneficial effect with checkpoint inhibitors (Kachhap et al. 2011).

Opioids and Illicit Drugs

Opioids and specific illicit substances have been recognized as factors that can cause cancer, with the potential to interfere with the normal functioning of the immune system, which may impact cancer development. However, research findings remain inconclusive, as some studies indicate that opioids like morphine could either stimulate or inhibit tumor growth, depending on the dosage. Similarly, cannabis consumption has been linked to cancer, but its precise role is unclear and may be influenced by concurrent use of tobacco and alcohol (Moussas and Papadopoulou 2017).

A systematic review and meta-analysis suggested a potential link between opioid use- both prescription and illicit- and an increased risk of pancreatic cancer (PC). Covering the period from 1999 to 2016, the analysis employed mixed-effects regression models to examine the relationship between opioid death rates and PC incidence. The findings showed that both PC rates and opioid death rates rose over time at the national and state levels. Among other factors, only obesity prevalence exhibited a similar upward trend. Significantly, a state's opioid death rate from 4 years earlier was a strong predictor of both the initial incidence of PC and the annual change in PC rates (Barlass et al. 2021).

Opium use, in particular, is increasingly recognized as a potential risk factor for cancer. A comprehensive study review found that opium use is linked to an increased risk of cancers in the esophagus, stomach, larynx, lung, and urinary bladder. Given that approximately 16.5 million people worldwide use opiates, with four million using raw opium, these findings are particularly significant (Kamangar et al. 2014).

Opioids can induce immunosuppression through central mechanisms, which are mediated by the MOP (mu opioid peptide) receptor. These mechanisms particularly affect areas such as the periaqueductal gray matter. Additionally, peripheral mechanisms involve MOP receptors on immune cells, which reduce the activity of macrophages and impair the function of B-cells. The study found that different opioids have varying degrees of immunosuppressive effects. Morphine and fentanyl have a significant potential for immunosuppression, while buprenorphine and tramadol have minimal effects, making them potentially safer in terms of the immune system (Zajączkowska et al. 2018). A study conducted on 300 advanced cancer patients to examine the relationship between tobacco use, symptom expression, and alcohol and illicit drug use showed that among the patients, 40% had never smoked, 49% were former smokers, and 11% were current smokers. There was no significant difference in opioid dosage based on smoking status, suggesting that a smoking history in advanced cancer patients may indicate an increased likelihood of opioid misuse (Kim et al. 2016).

Opioids are a highly prevalent class of illicit drugs. People with opioid use disorder have a high risk of all-cause and overdose-specific mortality, with modifiable risk factors like HIV and injection drug use predicting mortality risk. This was demonstrated with a systematic review and meta-analysis analyzing 92 studies encompassing 101 cohorts from various regions. Mortality rates were notably higher in low/middle-income countries, among HIV-positive individuals, and injection drug users (Bahji et al. 2020).

A comprehensive analysis of the safety of opioids in breast cancer treatment revealed conflicting evidence regarding their influence on cancer progression. Although preclinical studies indicated a potential association between opioid usage and disease advancement, clinical studies failed to validate these findings (Lucia et al. 2021).

An investigation was conducted to examine the connection between cannabis use and the initiation of opioid use, as well as the subsequent development of opioid use disorders, dependence, or abuse. Six studies from the United States, Australia, and New Zealand, involving a total of 102,461 participants, were included in the analysis. The results indicated that individuals who use cannabis are significantly more likely to start using opioids and to develop opioid use disorders, abuse, or dependence compared to those who do not use cannabis (Wilson et al. 2022).

Maternal Substance Use

Research indicates that maternal use of tobacco, alcohol, and illicit drugs during pregnancy is associated with an elevated risk of certain cancers in their children, particularly leukemias and fibrosarcomas (Auger et al. 2019; Wimberly et al. 2024; Sasazuki et al. 2012). A study conducted in Quebec, involving a large cohort of 785,438 newborns, revealed that children exposed to these substances through maternal use had higher cancer rates compared to those who were not exposed. Specifically, maternal use of illicit drugs was linked to an increased risk of acute lymphoblastic leukemia and fibrosarcoma in children, while tobacco use by mothers was associated with a heightened risk of acute myeloid leukemia and fibrosarcoma (Auger et al. 2019).

Another research study, which examined survey data from 3145 families with children diagnosed with cancer, found that the use of illegal drugs during pregnancy was associated with an increased incidence of intracranial embryonal tumors, medulloblastoma, supratentorial primitive neuroectodermal tumors (PNET), and retinoblastoma. Additionally, moderate to heavy alcohol consumption during pregnancy was strongly linked to a higher risk of non-Hodgkin lymphoma. However, no significant connection was identified between prenatal smoking and any specific subtype of childhood cancer (Wimberly et al. 2024).

Physical Activity

Adopting a combination of healthy lifestyle factors, such as not smoking, moderate alcohol consumption, limiting salt-preserved foods, staying physically active, and maintaining a healthy body mass index, can significantly lower the risk of cancer. In an epidemiological study involving 5576 Japanese men and women aged 45–74 years, researchers explored the impact of these lifestyle factors on cancer risk. The findings revealed that greater adherence to healthy lifestyle factors led to a significant reduction in cancer risk for both men and women. For men, the risk decreased by 14% for each additional healthy factor and for women, the risk reduction was 9% per additional factor. The study also noted that the reduction in cancer risk was particularly significant among elderly women, underscoring the importance of maintaining a healthy lifestyle to reduce cancer incidence in this group (Sasazuki et al. 2012).

The European Prospective Investigation into Cancer and Nutrition (EPIC) cohort study assessed the impact of a healthy lifestyle on cancer risk among 391,158 participants and found that each point increase in the HLIS was associated with a 5% reduction in overall cancer risk for men and a 4% reduction for women. Those with higher HLIS scores showed significantly lower cancer risks compared to those with lower scores: 28% for men and 24% for women across all cancers, 41% and 33% for alcohol-related cancers, 49% and 46% for tobacco-related cancers, 41% and 26% for obesity-related cancers, and 21% for female reproductive cancers. These findings highlight the importance of adopting healthier lifestyle practices in reducing cancer risk, particularly for men (Petrova et al. 2021).

Epidemiological evidence increasingly suggests that physical activity plays a beneficial role in lowering cancer risk (Colberg et al. 2016). Furthermore, physical activity also reduces the time it takes for stool to pass through the body and the duration of contact between pathogens and the gastrointestinal mucosa, which may lower the risk of developing cancer. The recognition of skeletal muscle as an endocrine organ has further clarified these mechanisms (Damato et al. 2021).

General guidelines recommend at least 150 min of moderate or 75 min of vigorous physical activity weekly (Damato et al. 2021). The World Cancer Research Fund (WCRF) further suggests that increasing this to 150–300 min or 7–15 MET hours per week can be particularly effective in lowering cancer risk (Physical activity and cancer risk 2022; https://www.wcrf.org/diet-activity-and-cancer/risk-factors/physi cal-activity-and-cancer-risk).

Combined Lifestyle Factors

Adhering to multiple healthy lifestyle behaviors, such as not smoking, moderate alcohol consumption, maintaining a healthy body and an equilibrated diet, and performing regular physical activity, can significantly reduce cancer risk. Studies have shown that individuals who follow these combined healthy behaviors have a lower incidence of cancer than those who do not (Sasazuki et al. 2012).

It was estimated that 16% of new colorectal cancer cases (22% in men and 11% in women) were linked to the failure to follow all five healthy lifestyle behaviors. The study concludes that following a combination of these healthy lifestyle factors is linked to a reduced incidence of CRC in European populations (Petrova et al. 2021).

Genetic Interactions

Genetic interactions in cancer risk can be greatly influenced by lifestyle choices, with research suggesting that a healthy lifestyle may reduce the genetic predisposition to certain cancers. A prospective cohort study using data from the UK Biobank, which included 195,822 White European participants aged 37–73 with no prior cancer history, examined the relationship between lifestyle, genetic risk, and the incidence of 13 cancer types. While no significant interactions were observed on the multiplicative scale, additive interactions were noted for colorectal, breast, pancreatic, lung, and bladder cancers. This suggests that following a healthy lifestyle has a more significant impact on reducing cancer risk for those with higher genetic susceptibility (Physical activity and cancer risk 2022, https://www.wcrf.org/diet-activity-and-cancer/risk-factors/physical-activity-and-cancer-risk).

Conclusion

Lifestyle choices significantly impact the risk of cancer development, with key factors including obesity, smoking, alcohol mad illicit drug consumption, diet, and physical activity. Smoking is a major contributor to the development of numerous cancers, with lung cancer being one of the most commonly associated. Similarly, alcohol consumption elevates the risk of cancers in areas such as the head and neck, esophagus, liver, breast, and colon. Poor dietary habits and obesity are linked to colorectal, breast, endometrial, and other cancers. Regular physical activity is protective, while a sedentary lifestyle contributes to higher cancer risk. Embracing a combination of healthy habits—including avoiding smoking, consuming alcohol in moderation, maintaining a healthy weight, and engaging in regular physical activity—can significantly lower the risk of cancer. For those who have survived cancer, making these lifestyle choices not only enhances prognosis and quality of life but also decreases the likelihood of recurrence. Even those with genetic predispositions to cancer can benefit from these healthy behaviors, which play a crucial role in public health strategies for cancer prevention and management.

References

Abdel-Rahman WM, Aravind SR, Nair VA et al (2018) PO-001 environmental toxins can transform mammary cells and induce chemo-resistance in breast cancer via multiple mechanisms. ESMO Open 3:A21

Ahmad I (2020) Lifestyle and health education. Gomal J Med Sci 17(3):63–64

Alengebawy A, Abdelkhalek ST, Qureshi SR et al (2021) Heavy metals and pesticides toxicity in agricultural soil and plants: ecological risks and human health implications. Toxics 9(3):42

Al-Fayez S, El-Metwally A (2023) Cigarette smoking and prostate cancer: a systematic review and meta-analysis of prospective cohort studies. Tob Induc Dis 21:1–12

Allahverdi N, Yassin M, Ibrahim M (2021) Environmental factors, lifestyle risk factors, and host characteristics associated with Philadelphia negative myeloproliferative neoplasm: a systematic review. Cancer Control 28:107327482110468

Allen NE, Beral V, Casabonne D et al (2009) Moderate alcohol intake and cancer incidence in women. J Natl Cancer Inst 101(5):296–305

Asthana S, Vohra P, Labani S (2019) Association of smokeless tobacco with oral cancer: a review of systematic reviews. Tob Prev Cessat 5:34

Auger N, Goudie C, Low N et al (2019) Maternal use of illicit drugs, tobacco or alcohol and the risk of childhood cancer before 6 years of age. Drug Alcohol Depend 200:133–138

Bagnardi V, Blangiardo M, La Vecchia C et al (2001) Alcohol consumption and the risk of cancer: a meta-analysis. Alcohol Res Health 25(4):263–270

Bahji A, Cheng B, Gray S et al (2020) Mortality among people with opioid use disorder: a systematic review and meta-analysis. J Addict Med 14(4):e118–e132

Barlass U, Deshmukh A, Beck T et al (2021) Opioid use as a potential risk factor for pancreatic cancer in the United States: an analysis of state and national level databases. PLoS One 16(1): e0244285

Barouki R, Melén E, Herceg Z et al (2018) Epigenetics as a mechanism linking developmental exposures to long-term toxicity. Environ Int 114:77–86

Bassiony M, Aqil M, Khalili M et al (2015) Tobacco consumption and oral, pharyngeal and lung cancers. Open Cancer J 8(1):1–11

Befort CA, Nazir N, Perri MG (2012) Prevalence of obesity among adults from rural and urban areas of the United States: findings from NHANES (2005–2008). J Rural Health 28(4):392–397

Botteri E, Borroni E, Sloan EK et al (2020) Smoking and colorectal cancer risk, overall and by molecular subtypes: a meta-analysis. Am J Gastroenterol 115(12):1940–1949

Brivio F, Viganò A, Paterna A et al (2023) Narrative review and analysis of the use of 'lifestyle' in health psychology. Int J Environ Res Public Health 20(5):4427

Brown JS, Amend SR, Austin RH et al (2023) Updating the definition of cancer. Mol Cancer Res 21(11):1142–1147

Budzynska K, West P, Savoy-Moore RT et al (2013) A food desert in Detroit: associations with food shopping and eating behaviours, dietary intakes and obesity. Public Health Nutr 16(12): 2114–2123

Casey SC, Vaccari M, Al-Mulla F et al (2015) The effect of environmental chemicals on the tumor microenvironment. Carcinogenesis 36(Suppl 1):S160–S183
Clapp RW, Jacobs MM, Loechler EL (2008) Environmental and occupational causes of cancer: new evidence 2005–2007. Rev Environ Health 23(1):1–37
Colberg SR, Sigal RJ, Yardley JE et al (2016) Physical activity/exercise and diabetes: a position statement of the American Diabetes Association. Diabetes Care 39(11):2065–2079
Cuomo RE, Purushothaman VL, Mackey TK et al (2024) Rates of adverse events and related risk factors following e-cigarette use. J Public Health 46(2):e230–e239
Dake F, Thompson A, Ng S et al (2016) The local food environment and body mass index among the urban poor in Accra, Ghana. J Urban Health 93:438–455
Damato TM, Oliveira CB, Franco MR et al (2021) Characteristics associated with people with chronic low back pain meeting physical activity guidelines and recommendations for sedentary behavior: a cross-sectional study. J Manip Physiol Ther 44(5):378–388
Debras C, Chazelas E, Srour B et al (2020) Total and added sugar intakes, sugar types, and cancer risk: results from the prospective NutriNet-Santé cohort. Am J Clin Nutr 112(5): 1267–1279
Debras C, Chazelas E, Srour B et al (2022) Artificial sweeteners and cancer risk: results from the NutriNet-Santé population-based cohort study. PLoS Med 19(3):e1003950
Definition of LIFESTYLE. Merriam-webster.com. Available at: https://www.merriam-webster.com/dictionary/lifestyle. Accessed 27 July 2024
Diddens I (2023) The role of cytogenetics in health and research. Res J Zool 5:2
Eastmond DA, Keshava N, Sonawane B (2014) Lymphohematopoietic cancers induced by chemicals and other agents and their implications for risk evaluation: an overview. Mutat Res Rev Mutat Res 761:40–64
Fang JS, Gillies RD, Gatenby RA (2008) Adaptation to hypoxia and acidosis in carcinogenesis and tumor progression. Semin Cancer Biol 18(5):330–337
Favier Rodríguez RM, Rodríguez Matos G, Chibas Muñoz EE (2023) Exploring links between toxic-environmental factors and hematologic malignancies: considerations for data-driven health decision making. Data Metadata 2:3
Fiore M, Oliveri Conti G, Caltabiano R et al (2019) Role of emerging environmental risk factors in thyroid cancer: a brief review. Int J Environ Res Public Health 16(7):1185
Fitzgibbon ML, Stolley MR (2010) Diet and cancer. In: Holland JC, Breitbart WS, Jacobsen PB, Marguerite SL, Matthew JL, McCorkle R (eds) Psycho-oncology, 2nd edn. Oxford University Press, Oxford, UK, pp 22–27
Galon J, Bruni D (2020) Tumor immunology and tumor evolution: intertwined histories. Immunity 52(1):55–81
Hagan I, Sharrocks AD (2002) Understanding cancer: from the gene to the organism: conference on genes and cancer. EMBO Rep 3(5):415–419
Harris CC (1991) Chemical and physical carcinogenesis: advances and perspectives for the 1990s. Cancer Res 51(18 Suppl):5023s–5044s
Hayes RB, Yin SN, Dosemeci M et al (1997) Benzene and the dose-related incidence of hematologic neoplasms in China. J Natl Cancer Inst 89(14):1065–1071
Hodjat M, Rezvanfar MA, Abdollahi M (2015) A systematic review on the role of environmental toxicants in stem cells aging. Food Chem Toxicol 86:298–308
Huang R, Zhou PK (2021) DNA damage repair: historical perspectives, mechanistic pathways and clinical translation for targeted cancer therapy. Sig Transduct Target Ther 6(1):254
Jaime PC, Duran AC, Sarti FM et al (2011) Investigating environmental determinants of diet, physical activity, and overweight among adults in Sao Paulo. Brazil J Urban Health 88(3): 567–581
Jemal A, Center MM, DeSantis C et al (2010) Global patterns of cancer incidence and mortality rates and trends. Cancer Epidemiol Biomarkers Prev 19(8):1893–1907

Johnson AR, Arasu S, Singstock M et al (2021) Obesity among the urban poor: evidence from a community-based study among adults residing in an underprivileged area of Bengaluru city, India. Indian J Health Sci Biomed Res 14(3):348–355
Kachhap S, Keshamouni VG, Qian DZ et al (2011) Cigarette smoke and cancer. J. Oncologia 2011: 1–2
Kamangar F, Shakeri R, Malekzadeh R et al (2014) Opium use: an emerging risk factor for cancer? Lancet Oncol 15(2):e69–e77
Key TJ, Schatzkin A, Willett WC et al (2004) Diet, nutrition and the prevention of cancer. Public Health Nutr 7(1a):187–200
Kim YJ, Dev R, Reddy A et al (2016) Association between tobacco use, symptom expression, and alcohol and illicit drug use in advanced cancer patients. J Pain Symptom Manag 51(4):762–768
Kim SJ, Khadka D, Seo JH (2022) Interplay between solid tumors and tumor microenvironment. Front Immunol 13:882718
Kopelman PG (2000) Obesity as a medical problem. Nature 404(6778):635–643
Koual M, Tomkiewicz C, Cano-Sancho G et al (2020) Environmental chemicals, breast cancer progression, and drug resistance. Environ Health 19(1):117
Kumar CA, Lekha BS, Poojitha G et al (2016) Obesity-related hypertension and awareness of its complications among the people. Paripex Indian J Res 5(10):96–98
Lagoa R, Marques-da-Silva D, Diniz M et al (2022) Molecular mechanisms linking environmental toxicants to cancer development: significance for protective interventions with polyphenols. Semin Cancer Biol 80:118–144
Lagunas-Rangel FA, Liu W, Schiöth HB (2022) Can exposure to environmental pollutants be associated with less effective chemotherapy in cancer patients? Int J Environ Res Public Health 19(4):2064
Larsson SC, Bergkvist L, Wolk A (2006) Consumption of sugar and sugar-sweetened foods and the risk of pancreatic cancer in a prospective study. Am J Clin Nutr 84(5):1171–1176
LoConte NK, Brewster AM, Kaur JS et al (2018) Alcohol and cancer: a statement of the American Society of Clinical Oncology. J Clin Oncol 36(1):83–93
Lucia M, Luca T, Federica DP et al (2021) Opioids and breast cancer recurrence: a systematic review. Cancer 13(21):5499
Lv J, Zhang S, Liu Y et al (2021) NR2F1-AS1/miR-190a/PHLDB2 induces the epithelial-mesenchymal transformation process in gastric cancer by promoting phosphorylation of AKT3. Front Cell Dev Biol 9:688949
Makarem N, Bandera EV, Nicholson JM et al (2018) Consumption of sugars, sugary foods, and sugary beverages in relation to cancer risk: a systematic review of longitudinal studies. Annu Rev Nutr 38(1):17–39
Mariano LC, Warnakulasuriya S, Straif K et al (2022) Secondhand smoke exposure and oral cancer risk: a systematic review and meta-analysis. Tob Control 31(5):597–607
Moussas GI, Papadopoulou AG (2017) Substance abuse and cancer. Psychiatrike 28(3):234–241
Moustakas A, de Herreros AG (2017) Epithelial–mesenchymal transition in cancer. Mol Oncol 11(7):715–717
Nöthlings U, Hoffmann K, Bergmann MM et al (2007) Fitting portion sizes in a self-administered food frequency questionnaire. J Nutr 137(12):2781–2786
Ntandou G, Delisle H, Agueh V et al (2009) Abdominal obesity explains the positive rural-urban gradient in the prevalence of the metabolic syndrome in Benin. West Africa Nutr Res 29(3): 180–189
O'Keeffe LM, Taylor G, Huxley RR et al (2018) Smoking as a risk factor for lung cancer in women and men: a systematic review and meta-analysis. BMJ Open 8(10):e021611
Panahi Y, Beiraghdar F, Amirhamzeh A et al (2016) Environmental toxicant exposure and cancer: the role of epigenetic changes and protection by phytochemicals. Curr Pharm Des 22(2): 130–140
Park HL (2015) Environmental toxicants, epigenetics, and cancer. Mol Integrative Toxicol, pp 131–154

Pasqualetti P, Casale R, Colantonio D et al (1991) Occupational risk for hematological malignancies. Am J Hematol 38(2):147–149
Pelucchi C, Tramacere I, Boffetta P et al (2011) Alcohol consumption and cancer risk. Nutr Cancer 63(7):983–990
Petrova D, Molina-Montes E, Amiano P et al (2021) Evidence update on the relationship between diet and the most common cancers from the European prospective investigation into cancer and nutrition (EPIC) study: a systematic review. Nutrition 13(10):3582
Physical activity and cancer risk (2022) WCRF International. Available at: https://www.wcrf.org/diet-activity-and-cancer/risk-factors/physical-activity-and-cancer-risk
Pilleron S, Gnangnon F, Noronha V et al (2019) Global cancer incidence in older adults, 2012 and 2035: a population-based study. Int J Cancer 144(1):49–58
Quadri MFA, Tadakamadla SK, John T (2019) Smokeless tobacco and oral cancer in the Middle East and North Africa: a systematic review and meta-analysis. Tob Induc Dis 17:56
Rasheed AA, Koyyala VPB (2020) The tumor microenvironment. Indian J Med Paediatr Oncol 41(2):213–214
Rebuzzini P, Fabozzi G, Cimadomo D et al (2022) Multi- and transgenerational effects of environmental toxicants on mammalian reproduction. Cells 11(19):3163
Rehm J, Soerjomataram I, Ferreira-Borges C et al (2019) Does alcohol use affect cancer risk? Curr Nutr Rep 8(3):222–229
Rodgers KM, Udesky JO, Rudel RA et al (2018) Environmental chemicals and breast cancer: an updated review of epidemiological literature informed by biological mechanisms. Environ Res 160:152–182
Rumgay H, Murphy N, Ferrari P et al (2021) Alcohol and cancer: epidemiology and biological mechanisms. Nutrients 13(9):3173
Sarah RH, Shariful MI, Zamiur MR et al (2022) Pivotal role of environmental toxicants on developing infectious diseases in LMICs: a protocol for a systematic review and meta-analysis. BMJ Open 12(7):e058927
Sasazuki S, Inoue M, Iwasaki M et al (2012) Combined impact of five lifestyle factors and subsequent risk of cancer: the Japan Public Health Center Study. Prev Med 54(2):112–116
Scala M, Bosetti C, Bagnardi V et al (2023) Dose-response relationships between cigarette smoking and breast cancer risk: a systematic review and meta-analysis. J Epidemiol 33(12):640–648
Schuster-Kolbe J, Rauchen LH (1994) Smoking and cancer. Wien Med Wochenschr 144(22–23): 540–544
Sepand MR, Bigdelou B, Ho JQ et al (2021) The role of macrophage polarization and function in environmental toxicant-induced cancers. Environ Res 196:110933
Tasevska N, Park Y, Jiao L et al (2012) Sugars in diet and risk of cancer in the NIH-AARP Diet and Health Study. Int J Cancer 130(1):159–169
Thompson PA, Khatami M, Baglole CJ et al (2015) Environmental immune disruptors, inflammation and cancer risk. Carcinogenesis 36:S232–S253
Torpy JM, Lynm C, Glass RM (2010) JAMA patient page. Cancer: the basics. JAMA 303(11):1108
US EPA OA (2013) U.S. Environmental Protection Agency. Available at: http://epa.gov
Vaz KK (2021) Polychlorinated biphenyls and heavy metals: source of emission, harmful effects and prevention. J Sci Res 65(08):27–32
Warr CG, Shaw KH, Azim A et al (2018) Using mouse and drosophila models to investigate the mechanistic links between diet, obesity, type II diabetes, and cancer. Int J Mol Sci 19(12):4110
Weir HK, Thompson TD, Stewart SL et al (2021) Cancer incidence projections in the United States between 2015 and 2050. Prev Chronic Dis 18:210006
Wilson J, Mills K, Freeman TP et al (2022) Weeding out the truth: a systematic review and meta-analysis on the transition from cannabis use to opioid use and opioid use disorders, abuse or dependence. Addiction 117(2):284–298
Wimberly CE, Gulrajani NB, Russ JB et al (2024) Maternal prenatal use of alcohol, tobacco, and illicit drugs and associations with childhood cancer subtypes. Cancer Epidemiol Biomarkers Prev 33(3):347–354

Worley JR, Parker GC (2019) Effects of environmental stressors on stem cells. World J Stem Cells 11(9):565–577
You W, Henneberg M (2018) Cancer incidence increasing globally: the role of relaxed natural selection. Evol Appl 11(2):140–152
Younglai E, Wu Y, Foster W (2007) Reproductive toxicology of environmental toxicants: emerging issues and concerns. Curr Pharm Des 13(29):3005–3019
Zajączkowska R, Leppert W, Mika J et al (2018) Perioperative immunosuppression and risk of cancer progression: the impact of opioids on pain management. Pain Res Manag 2018:1–8
Ziraba A, Fotso J, Ochako R (2009) Overweight and obesity in urban Africa: a problem of the rich or the poor. BMC Public Health 9:465–465

Environmental Chemical-Induced Heat Shock Proteins in Toxicology: Past, Present, and Future Trends

Ashutosh Pandey, Hitesh Singh Chouhan, Debapratim Kar Chowdhuri, and Anurag Sharma

Abstract

Heat shock proteins (HSPs) have been implicated in maintaining cellular homeostasis and survival. Xenobiotics pose stress to the exposed organisms at different levels of organization. To protect cellular homeostasis, organisms exert defense responses. One of the major cellular defense response pathways includes HSP. A plethora of studies have demonstrated the altered expression of HSP in different model systems against xenobiotic exposure. This chapter is intended to highlight the contribution/implication of HSP in xenobiotics-induced stress by reviewing the relevant toxicological studies in the past two decades. HSPs act as molecular chaperones, vital in maintaining cellular homeostasis under normal and pathophysiological conditions. A wide range of xenobiotics induces HSP expression in the exposed organisms. HSPs are implicated in toxicology as early indicators and modulators of xenobiotic-induced stress. This chapter has also discussed the

A. Pandey (✉)
Department of Molecular & Human Genetics, Baylor College of Medicine, Houston, TX, USA
e-mail: ashutosh.pandey@bcm.edu

H. S. Chouhan
National Institute of Neurological Disorders and Stroke, National Institutes of Health, Bethesda, MD, USA

D. K. Chowdhuri
Embryotoxicology Laboratory, CSIR-Indian Institute of Toxicology Research (CSIR-IITR), Lucknow, Uttar Pradesh, India

A. Sharma (✉)
Department of Environmental Health and Toxicology, Nitte (Deemed to be University), Nitte University Centre for Science Education and Research (NUCSER), Mangaluru, Karnataka, India
e-mail: anuragsharma@nitte.edu.in

S. Raisuddin et al. (eds.), *Environmental Toxicants and Lifestyle Diseases*,
https://doi.org/10.1007/978-981-95-0575-3_13

implications/role of HSPs in maintaining cellular homeostasis and survival against xenobiotic stress. Relevant cited studies in this chapter are from the last two decades.

Keywords

Cellular stress · Cytotoxicity · Environmental chemicals · Heat shock proteins · Stress genes · Xenobiotics

Introduction

The consistent release of pollutants into the environment has led to contaminating all essential abiotic resources, including air, water, and soil. These pollutants, including agrochemicals, industrial effluents, household waste materials, and other anthropogenic compounds, are collectively called xenobiotics. Exposure to these xenobiotics poses a risk to living organisms in different ecosystems, including aquatic, terrestrial, and aerial. Occupationally and non-occupationally, humans are continuously being exposed to these xenobiotics. Living organisms, including humans, experience stress at their cellular or sub-organismal level against these environmental pollutants/xenobiotics, depending on the length of exposure, either acute (short-term) or chronic (long-term). To survive against xenobiotic stress, living organisms exert stress responses at their cellular and organismal level. Because of the differences in multicellular body organization, organisms show a variety of organismal or sub-organismal responses; however, among the responses, the cellular stress response pathways are highly conserved across the taxa. One of the highly conserved stress responses, marked by the increased expression of heat shock genes (HSGs) against unfavorable conditions, including thermal stress, is called the heat shock response (HSR). HSGs translate into the heat shock proteins (HSPs), which play molecular chaperone functions to maintain cellular homeostasis and have been used in toxicological studies as early biomarkers and modulators of cellular toxicity induced by xenobiotics. This chapter is intended to accentuate the significance of HSP as biomarkers as well as the component of cellular adaptation response against xenobiotic stress by covering the toxicological studies during the last two decades.

Heat Shock Proteins

Sixty years ago, Italian researcher Ferruccio Ritossa discovered the HSR in *Drosophila* as a response against elevated temperature and a pesticide, 2,4-dinitro-o-cresol (DNP) (Ritossa 1962). Almost a decade after discovering HSR, Tissiers and colleagues found that HSR results in synthesizing specific proteins called HSP or stress proteins (Tissieres et al. 1974). Although heat shock response is comprehensively studied in *Drosophila*, other organisms, including prokaryotes, yeast, invertebrates, and vertebrates, including humans, exhibit a similar response at their cellular level (Sikora and Grzesiuk 2007; Talorete et al. 2008; Verghese et al. 2012). Based on the molecular weight range and sequence homology, HSPs are classified into five major families, i.e., the HSP100 family, the HSP90 family, the HSP70 family, the HSP60 family, and the small HSP family. The HSP100 family includes the HSP with a molecular weight between 100 and 104 kDa, which are present in yeast, bacteria, and plants but absent in animals including human (Kedzierska-Mieszkowska and Zolkiewski 2021; Mishra and Grover 2016) and essential for normal growth conditions and cellular survival under thermal stress (Queitsch et al. 2000). HSP100 chaperones function to dissociate the protein aggregation under heat stress (Parsell et al. 1994). These proteins form hexamers, and their chaperone functions are ATP-dependent (Bukau et al. 2006). The HSP90 family includes HSP with a molecular weight range of 82–90 kDa and is present in all kingdoms of the organism except archaea (Chen et al. 2006). These proteins are also ATP-dependent and involved in indispensable cellular activities, including protein folding and degradation, cell cycle, and cell death (Birbo et al. 2021; Hoter et al. 2018). The HSP70 family includes the HSP with a molecular weight between 68 and 75 kDa and is highly evolutionarily conserved among the HSP families across the taxa. The HSP60 family comprises proteins of a molecular weight range from 58 to 65 kDa, facilitating the folding and trafficking of nascent proteins. HSP60 family proteins were earlier thought to function in mitochondria (Ostermann et al. 1989). Later, studies evidenced their presence in the cytosol, cell membrane, and extracellular space (Meng et al. 2018). These proteins are also ATP-dependent and involved in cell death and migration (Ostermann et al. 1989). HSP with molecular weight between 15 and 30 kDa is called small HSP (sHSP). These proteins are ATP-independent and function in protein quality control and signal transduction (de Los Reyes and Casas-Tinto 2022).

A transcription regulator regulates the HSP expression called the heat shock factor (HSF). *Drosophila* has one HSF (Clos et al. 1990), and in mammals, there are four HSFs, namely, HSF1, HSF2, HSF3, and HSF4 (Akerfelt et al. 2010). Under normal conditions, the heat shock factor (HSF) remains suppressed due to its binding with HSP. Upon heat shock/thermal stress, HSF is dissociated from HSP, undergoes trimerization and phosphorylation, and subsequently translocates to the nucleus where it binds to a *cis*-acting sequence, heat shock element (HSE) (Pelham 1982) to induce HSP expression (Fig. 1). The HSE is also highly conserved across the species (Guertin et al. 2010). Induction of HSP is not limited to exposure to elevated temperature. Since the discovery of HSP, many studies have demonstrated the induction of HSP against exposure to xenobiotics, including agrochemicals, effluents, heavy metals, and nanoparticles (Abdel-Latif et al. 2021; Doganlar and Doganlar 2015; Gupta et al. 2010; Pandey et al. 2013; Sharma et al. 2012). Among the HSP families, the proteins belonging to the HSP70 family have been demonstrated to be induced against several stress environments, including high temperature and xenobiotic exposure, and have been established as early indicators of cellular stress (Moreira-de-Sousa et al. 2018). Several studies have evidenced the protective roles of HSP against xenobiotic-induced cellular damage. Figure 2

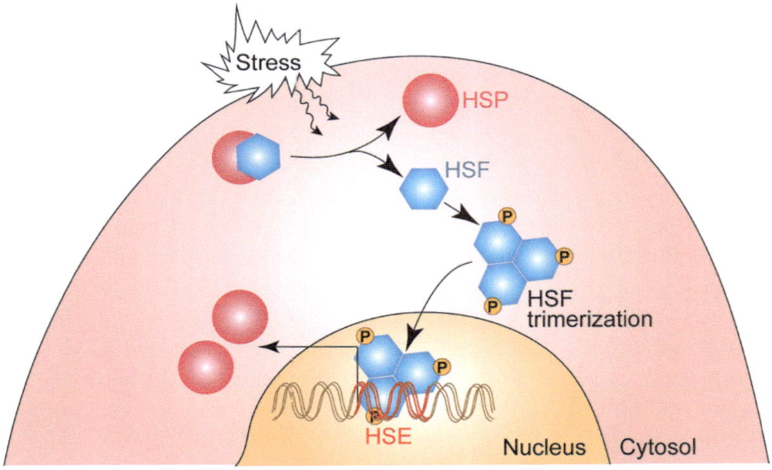

Fig. 1 Schematic representation of the induction of heat shock protein (HSP). In normal conditions, HSP prevents the activation of heat shock factor (HSF). Upon stress, HSF gets detached from the HSP and undergoes phosphorylation and trimerization. Activated HSF then enters the nucleus and binds to the heat shock element (HSE) to induce HSP transcription

represents a cartoon showing inducers/stimulators, functions, and implications of HSP in toxicology.

Xenobiotic Stress: Impacts from Cellular to Organismal Level in Exposed Organisms

The term xenobiotics is implied for the compounds/substances not formed in the organisms and considers different chemicals, including agrochemicals, drugs, food additives, industrial waste, and several other environmental pollutants. Humans are estimated to be exposed to approximately 3×10^6 xenobiotics during their entire lives, and with growing advancements in lifestyle and technologies, this number would go on to a higher level (Idle and Gonzalez 2007).

Depending on the route of exposure, these xenobiotics can reach to the cells of different organs of an exposed organism. An encounter of these chemicals with the cells leads to cellular stress by inducing redox imbalance, damage to cellular components or constituents, and subsequent cell death. In addition, certain chemicals can exert specific inhibition on cellular physiology. Oxidative stress has been commonly associated with organisms exposed to xenobiotics. Consequences of oxidative stress may range from minimal reversible short-term disruption in cellular physiology to irreversible cellular damage or cell death. Oxidative stress may exert effects from the cellular to the organismal level, which depends on the magnitude of chemical exposure in terms of the concentration and duration of exposure. Developmental delay, structural anomaly, behavioral changes, or decreased lifespan and

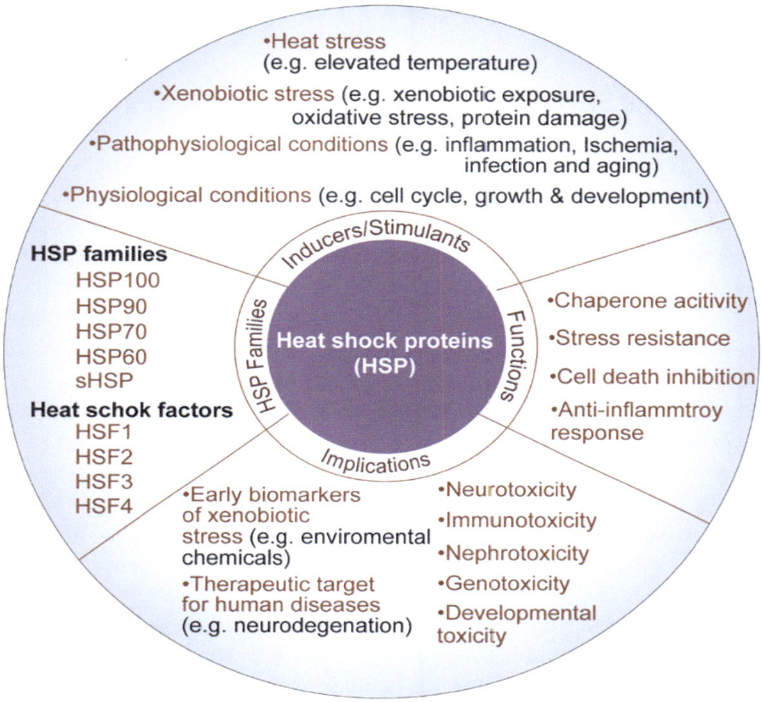

Fig. 2 Cartoon represents information about families, inducers, functions, and implications of HSPs. HSP families include HSP100, HSP90, HSP70, HSP60 and small HSP. Heat shock factors (HSF1, HSF2, HSF3, and HSF4) are induced by heat stress, xenobiotic stress, and pathophysiological and physiological conditions. Functions are chaperone activity, stress resistance, and anti-inflammatory response. Induction of HSPs is implicated as an early biomarker of xenobiotic stress

health span may infer the latter. Exposure to many xenobiotics, including pesticides or heavy metals, leads to stress through pathophysiological conditions such as the onset of neurodegenerative disorders and immunological disease. Both acute and chronic exposure to xenobiotics exerts stress on different organ systems through the redox imbalance, leading to oxidative stress, categorized as neurotoxicity, immunotoxicity, nephrotoxicity, genotoxicity, and developmental toxicity.

Several biological responses/pathways act on different steps during cellular stress to maintain cellular physiology and survival. HSPs are one of the majorly involved components of cellular homeostasis that function during physiological conditions as well as in response to pathophysiological and xenobiotic stress. The following sections will cover the significance of HSP in toxicology as early/first-tier indicators and modulators of xenobiotics-induced toxicity.

Implication of HSP in Xenobiotic Stress: A Perspective in Toxicology

HSP as Sentinel Species Against Xenobiotic-Induced Stress

Xenobiotics prevalence in the environment puts the health of living organisms at risk, necessitating the assessment of adversities/toxicity. Xenobiotics induce stress at the cellular level of the exposed organisms; therefore, targeting the gene expression associated with sensitive stress response pathways would provide essential readouts of early stress induced by xenobiotics. The biomarker is one of the prime essentialities to assess the exposure to any xenobiotics as it can raise a biological alarm before the adverse effects become irreversible. It can be defined as an adaptive biological response, a change in biochemical, molecular, or cellular physiology, or organismal behavior. Simple and reliable detection, sensitivity, and specificity are the important criteria for biomarkers (Moreau et al. 2008). At the current state of knowledge, numerous biomarkers have been projected and established against specific types of xenobiotic exposure and environmental risk assessment.

The very first evidence of induction of HSR in *Drosophila* against high temperature and DNP exposure by Ritossa (1962) has been well-stretched in toxicological studies by using HSP as an early indicator/stress biomarker against several xenobiotics. Since HSPs are involved in cellular homeostasis and repair of proteins, their induction (i.e., heat shock response) across the taxa is conserved, and their expression is targeted as a readout of cellular stress in different organisms (both aquatic and terrestrial) against exposure to xenobiotics such as toxins, chemicals, pesticides, heavy metals, radiations, oxidants, free radicals, viruses and other stressors (Gupta et al. 2010; Moreira-de-Sousa et al. 2018). Three decades ago, a study on a euryhaline rotifer, *Brachionus plicatilis,* reported more than four-fold upregulation in the expression of HSP58 against exposure to copper sulphate (Cochrane et al. 1991). This was probably the first toxicological study that evidenced HSP as a stress biomarker. A report by Sander and Martin (1993) showed the increased level of HSP60 and HSP70 in fish and mussels and suggested that HSP might provide a quantitative readout of an adverse environmental situation in the exposed organism. With the line of quantification of xenobiotic exposure severity, genetic engineering technology utilized the HSP gene to make transgenic models harboring HSP gene promoter sequence fused with the coding sequence of different reporter genes, e.g., a bacterial gene *LacZ* (encodes for β-galactosidase) and a green fluorescent protein-expressing gene (GFP). Under different stress conditions, HSP promoters would induce these reporter proteins' expression, which can serve as a qualitative or quantitative readout of cellular stress. In this context, the different transgenic models were developed using *Drosophila*, *Caenorhabditis elegans*, Zebrafish, yeast, mice, and cell lines (Krone et al. 2005; Mutwakil et al. 1997; Nazir et al. 2003; Nisamedtinov et al. 2008; Wirth et al. 2002).

Many studies have utilized expression level as an early indicator or a cellular stress biomarker in the last two decades. To name a few, Xu et al. (2022) reported that anthranilic diamide insecticide chlorantraniliprole exposure upregulated the expression of *HSP90*, *HSP70*, *HSP68*, and *HSP20* in armyworm, *Spodotera frugiperda*. Another study showed that the combined exposure of copper and norfloxacin altered the expression of *HSP* in *C.elegans* (Liu et al. 2022a). Exposure to a nematicide, fluopyram, altered the expression of *HSP* (*HSP-16.1* and *HSP16.2*) in *C.elegans* (Liu et al. 2022b). Exposure to an endocrine disruptor, nonylphenol, altered the expression of *HSP27* in *Drosophila* (Dwivedi et al. 2022). Benzo[α] pyrene induced the expression of six *HSPs* (*HSP22–1*, *HSP22–2*, *HSP40A*, *HSP60*, *HSP70*, and *HSP90*) in the exposed clams, *Ruditapes philippinarum* (Gao et al. 2022). Among the HSP families, HSP70 has been extensively utilized as a first-tier indicator of cellular stress against xenobiotic exposure (Gupta et al. 2010; Moreira-de-Sousa et al. 2018).

Modulatory Role of HSP in Xenobiotic Stress

Increased ROS and oxidative stress and subsequent induction of cell death are central to the xenobiotic-induced cellular stress in the exposed organisms. HSP can interpose oxidative stress-mediated cellular damage at different steps depending upon the diverse cellular functions. The chaperone activity of HSP (mainly HSP90 and HSP70) can prevent protein misfolding and aggregation and direct the damaged proteins to the degradation pathway. This would preserve the protein homeostasis. At the later stage, when the severity of xenobiotic stress leads to the induction of apoptosis, HSP (such as HSP90, HSP70, and HSP27) function as a negative apoptosis regulator. Collectively, increased HSP level buffers cellular homeostasis and survival during xenobiotic stress. Figure 3 represents a schematic of the preventive HSP's function against oxidative damage induced by xenobiotic exposure. The following sections will highlight the modulatory role of HSP in xenobiotic-induced toxicity in different organ systems.

Protective Role of HSP in Xenobiotics-Induced Neurotoxicity

Neurons are attributed to transmitting and processing the signals, eventually controlling individuals' sensory, integrative, and motor functions. Xenobiotic exposure hampers neuron function or exhibits neuronal cell death, resulting in neuropathological abnormalities (Rock and Patisaul 2018). Dysregulated protein homeostasis is key to initiating neurodegenerative disease (Webster et al. 2019). In general, HSP functions in protein folding, trafficking, and degradation of misfolded proteins and is also involved in neurotransmitter release (Dukay et al. 2019). A protective role for HSP has been demonstrated in xenobiotic-induced neuronal damage. The HSP70 efficacy was tested against rotenone (one of the model-neurotoxicants) induced inflammation in human neuroblastoma (SH-SY5Y cells) (Li et al. 2019). This study indicated increased oxidative stress and modulated NF-κB, STAT3 levels, and apoptosis in rotenone-treated SH-SY5Y cells were rescued with the HSP70 supplementation. A study on the *Drosophila* model showed that paraquat

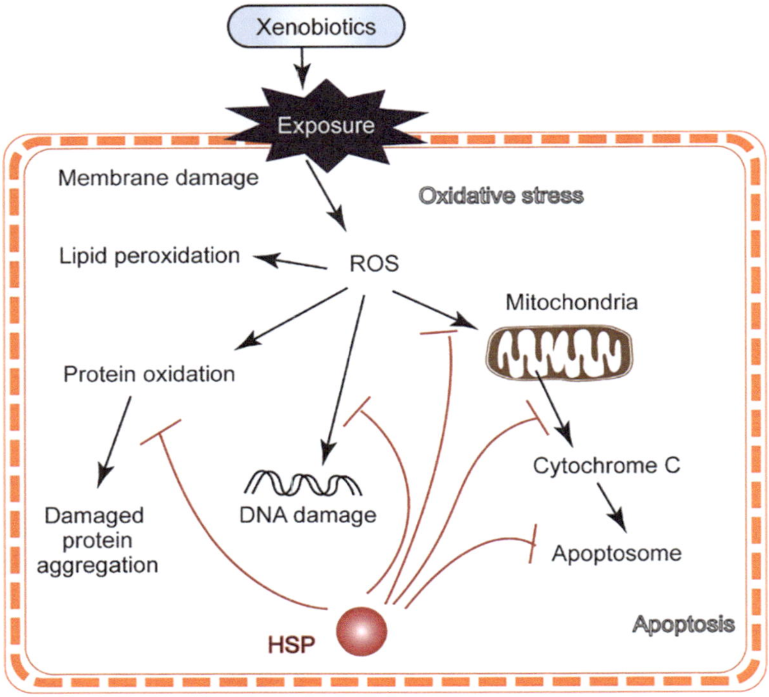

Fig. 3 Schematic representation of xenobiotic induced cellular stress. Xenobiotics can cause membrane damage, reactive oxygen species (ROS) generation, lipid peroxidation, and protein oxidation leading to oxidative stress and subsequent cell death pathway activation. Heat shock protein can prevent cellular stress at the indicated level (red arch lines)

(a model-neurotoxicant) exposure induces ROS generation and JNK-mediated dopaminergic (DA) neuronal cell death, which was mitigated by the *HSP70* overexpression in dopaminergic cells (Shukla et al. 2014). Damage in DA is also induced by another neurotoxicant, 1-methyl-4-pheny-1,2,3,6-tetrahydropyridine (MPTP). A study on the mice model showed that geldanamycin's pretreatment rescues against MPTP-induced dopaminergic neurotoxicity (Shen et al. 2005). Interestingly, this study observed HSP70 induction in the geldanamycin-treated group, which suggests a shielding function of HSP70 in neuronal health against xenobiotic exposure.

The protective role of HSP is also demonstrated against heavy metal-induced toxicity. Manganese is one of the naturally occurring heavy metals. Accumulation of manganese in the brain leads to neurotoxicity. Avila et al. (2016) observed that *HSP70* mutants of *C. elegans* were more susceptible to manganese-induced neuronal injury. This study found increased protein oxidation and dopaminergic neurodegeneration and decreased organism survival in *HSP70* mutants of *C.elegans* compared to wild-type worms. Arsenic is another naturally occurring metalloid known to cause neurotoxicity in humans. In a study, arsenic trioxide (As_2O_3) exposure to *Gallus gallus* (chicken) resulted in histological changes in the brain tissue, accompanied by decreased glutathione peroxidase and catalase activity, increased malondialdehyde levels, and increased HSP60 and HSP70 mRNA/protein levels (Zhao et al. 2017). This indicates that increased HSP expression might be a protective response against As_2O_3-induced neuronal injury. Increased amyloid-beta (Aβ) and the presence of hyperphosphorylated tau are linked to the progression of Alzheimer's disease (Nisbet and Gotz 2018). A highly toxic heavy metal, cadmium, increases Aβ levels and tau protein phosphorylation in the exposed SN56 cholinergic neuronal cells by downregulating *HSP90*, *HSP70,* and *HSP27* (Moyano et al. 2018). Collectively, these studies indicate the HSP's function in protecting the neurons against xenobiotic stress.

Small HSPs have also been documented for their neuroprotective efficacy in the invertebrate and mammalian models. Targeted overexpression of *HSP26* and *HSP27* in the neuronal cell confers resistance to the *Drosophila* model against paraquat-induced oxidative stress (Liao et al. 2008). Ubiquitous *HSP27* overexpression partially rescues the neuronal cell death in dichlorvos (an organophosphate pesticide) exposed *Drosophila* (Pandey et al. 2016). Alpha B-crystallin (αBC) belongs to the small HSP family and has been revealed to play a neuroprotective function in rats against permethrin (an insecticide) induced neuronal cell death (Guvenc et al. 2013). The amyloid fibrils form α-Synuclein is a well-documented hallmark of Parkinson's disease (Breydo et al. 2012). Fan et al. (2006) found that overexpression of *HDJ-1* (human homologs of *HSP40*) in SK-N-SH cells could suppress α-synuclein accumulation by increasing protein ubiquitination upon proteasome activity 1-methyl-4-phenylpyridinium ion (MPP^+) exposure. These studies establish a neuroprotective function role of small HSPs.

Modulation of Nephrotoxicity by HSP

Excretion, detoxification, regulation of extracellular fluid, and electrolyte homeostasis are vital functions of the kidney. However, more blood flow and the presence of metabolic enzymes make the kidney susceptible to pollutants (Vervaet et al. 2017). The toxicants could severely affect kidney function by increasing oxidative damage, cell-cycle arrest, inflammation, DNA damage, and apoptosis (Salama et al. 2021). HSPs are constitutively expressed in the kidney and involved in various functions, including protein folding, transport, and degradation (Nayak Rao 2016). The cytoprotective role of different HSPs has also been documented during acute or chronic kidney injury (Nayak Rao 2016). Cisplatin is an alkylating agent, and its exposure to rats decreased *HSP72* expression, accompanied by acute renal injury (Zhou et al. 2003). Further, HSP72 induction through uranyl acetate pretreatment increased the Bcl-2/Bax ratio and reduced apoptosis in cisplatin-exposed organisms. Similarly, HSP60 was shown to attenuate cisplatin-mediated renal toxicity by suppressing Bax-induced apoptosis in renal proximal HK-2 tubular cells against dimethylthiourea (Tsuji et al. 2009). HSP27 was reported to protect the rat renal proximal tubular cells survival against uranyl acetate toxicity (Fujigaki et al. 2010). Using the Drosophila renal system, which comprises nephrocytes and Malpighian tubules, Saini et al. (2023) observed that overexpression of *HSP27* reduced the cadmium-induced oxidative stress and restored the function of Malpighian tubule, i.e., efflux activity and fluid secretion rate. These studies in diverse model systems conclude the modulatory function of HSPs in nephrotoxicity.

Protective Role of HSP in Xenobiotics-Induced Genotoxicity

Genetic instability is one of the major factors in initiating many diseases. A certain class of xenobiotics, such as heavy metals, selectively cause genotoxicity, i.e., DNA damage in the exposed model system (Mishra et al. 2011; Singh et al. 2011). These xenobiotics can interact with the DNA directly (DNA adduct) or indirectly (free radical formation), which leads to single- or double-strand DNA break. Studies have demonstrated the protective role of HSP in xenobiotic-induced genotoxicity. Xiao et al. (2002) reported the association of higher levels of HSP70 with DNA damage protection in lymphocytes in coke-oven emission-exposed workers. Another study observed an inverse relationship between the HSP70 level and banzo[a] pyrene-induced DNA damage (Gao et al. 2004). Increased HSP70 level showed a protective effect against banzo[a] pyrene-induced DNA damage in genotoxicity in HEL culture cells. Doulias et al. (2007) reported that the increased HSP70 level attenuates H_2O_2-induced nuclear single-strand DNA breaks by maintaining cellular iron homeostasis and stabilizing the lysosomal membrane in the exposed HeLa cells. In addition, Yang et al. (2009) identified increased co-localization between enhanced HSP70 and XPA and XPG nucleotide excision repair genes in the nuclei of banzo[a]pyrene-exposed 16-HPE1 cells. They suggested that this co-localization might be involved in correctly positioning DNA repair machinery around DNA lesions to foster the repair process. Dong et al. (2013) reported that the increased HSP70 level in response to estrogen rescues oxidative stress-mediated adversities and DNA damage in the $A\beta_{25-35}$ fragment-exposed SH-SY5Y cell line. Increased expression of *HSP70* in *Drosophila* has been demonstrated to provide protection against Cr (VI)-induced DNA damage by promoting de novo synthesis of GSH and modulation of cellular redox impairment (Chaouhan et al. 2021).

HSP Modulates Xenobiotics-Induced Hepatotoxicity

The liver is the primary organ for xenobiotic metabolism. The liver bio-transforms the toxicants through phase I and II metabolic reactions and makes the toxicant more water-soluble. In this process, the formation of free radicals causes the liver to be susceptible to toxicity. In physiological conditions, HSP help in liver metabolism (Archer et al. 2018) and its regeneration (Wolf et al. 2014), whereas in the xenobiotic-stressed situation, they play a cytoprotective role. For instance, liver damage has been documented due to carbon tetrachloride (CCL_4) exposure (Ullah et al. 2020). A study by Song et al. (2007) reported that CCL_4 exposure to *HSP70* knockout mice showed more alanine transaminase levels (ALT), neutrophilic infiltration, necrosis, and a slower recovery rate than the wild-type animals, thus suggesting a functional role of HSP70 in CCL_4 -induced liver injury. Following affirmation of the cytoprotective role of HSP, Escobar Mdel et al. (2009) demonstrated the activation of HSP70 in cadmium-exposed HepG2 cells. The study concluded that exposure to cadmium activates mitogen-activated protein kinases (MAPKs) signaling cascade. The MAPK activates oxidative stress-sensing transcription factor 'activator protein-1 (AP-1)', promoting HSP70 induction. In a study, the exposure of 2,3,7,8-tetrachlorodibenzo-p-dioxin to rats resulted in increased levels of HSP (*HSP27*, *HSP105*, and *HSP90*) in the liver, which was supposed to be a part of the protective mechanism (Kim et al. 2012).

Protective Role of HSP in Xenobiotics-Induced Developmental and Reproductive Toxicity

Structural abnormalities, growth retardation, and functional impairment are the common developmental effects of exposure to xenobiotics (Ullah et al. 2020; Song et al. 2007; Escobar Mdel et al. 2009; Kim et al. 2012), while altered fecundity, fertility, and sperm function, defective spermatogenesis and oogenesis, and reproductive hormonal imbalance are the key reproductive impairment indices (Siracusa et al. 2018; Tiwari et al. 2011; Wen et al. 2017). Studies have reported the protective roles of HSP on toxicant-induced developmental/reproductive toxicity. For instance, Dwivedi et al. (2022) observed a reduced and delayed *Drosophila* larva development when exposed to an endocrine disruptor, nonylphenol, along with reduced *HSP27* mRNA, increased oxidative stress, and cell death in the larval gut of the exposed organism. Subsequent experiments found that overexpression of *HSP27* in the organism's gut lessened the nonylphenol-induced developmental toxicity. The study concluded that nonylphenol-induced ROS in the gut promotes systemic toxicity in the organism, whereas *HSP27* overexpression in the gut reduced oxidative stress through Thioredoxin Reductase (TrxR) mediated regenerative pathway. Mukhopadhyay et al. (2003a, b) reported that failure of *HSP70* induction in the testes and the accessory glands of male flies resulted highly sensitivity to reproductive toxicity against argemone oil and chrome plating industry effluents. Anbara et al. (2020) observed an altered sperm quality and histomorphometric changes upon aspartame exposure to male mice. Increased oxidative stress and reduced *HSP70–2* expression might be possible reasons for these aspartame-associated reproductive indices.

HSP in Nanotoxicology

The advancement of nanotechnology and the widespread use of nanoparticles in various applications, including food and medicine, have made nanoparticles one of the potential health hazards. In the last two decades, several research groups have examined the risk of nanoparticles using different model systems, and "nanotoxicology" has emerged as a distinguished aspect of toxicology (Demir 2021). Having at least one dimension in the nanoscale, nanoparticles get access to the cellular system and impose the stress either by a physical hindrance to the organelles, including membranes, or through direct, reactive oxygen species generation. Both conditions can lead to proteotoxicity and, therefore, can trigger the induction of HSP (Fig. 4). Several studies (both in vivo and in vitro) have evidenced the utility of HSP as an early indicator of nanoparticle-induced cellular stress. Temiz

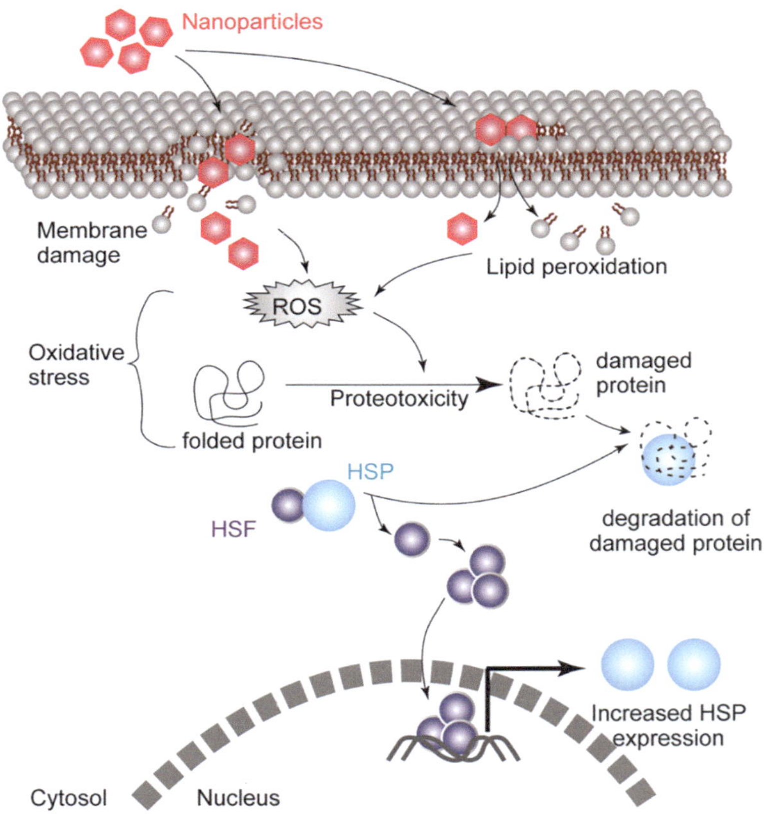

Fig. 4 Schematic representation of nanoparticles induced cellular toxicity. Nanoparticles can cause membrane damage and lipid peroxidation through direct physical interaction or through reactive oxygen species (ROS) generation, which can further lead to proteotoxicity and subsequently induce the expression of HSP. HSF Heat shock factor

and Kargin (2021) reported that aluminum oxide nanoparticle-exposed Nile fish, *Oreochromis niloticus* exhibit increased redox imbalance, decreased antioxidant enzymes, and increased HSP70 level. It was shown that Nile fish exposed to copper oxide nanoparticles showed increased expression of oxidative stress-related genes and HSP70 (Abdel-Latif et al. 2021). Silver nanoparticles exposed third instar *Drosophila* larvae showed increased oxidative stress and upregulation of *HSP70* expression (Ahamed et al. 2010). Silver nanoparticle exposure to human cell lines (A549 and HepG2) induced *HSPA1A* expression and oxidative stress (Xin et al. 2015). Iron oxide nanoparticle-exposed male mice showed oxidative stress and increased expression of HSP60, 70, and 90 (Sundarraj et al. 2017). Amorphous silica nanoparticles induced redox imbalance and upregulated the expression of *HSP70* and *HSP22* in exposed third-instar *Drosophila* larvae (Pandey et al. 2013). Copper nanoparticles exposed puffer fish, *Takifugu fasciatus,* exhibited increased expression of *HSP70* and *HSP90* (Wang et al. 2019). A study on fertilized chicken embryos showed the altered expression of small HSP (*HSPB1, HSPB5, HSPB8,* and *HSPB9*) against iron oxide nanoparticle exposure. This study suggested the protective role of small HSP during embryogenesis and urged for HSP expression profiling in nanotoxicity assessment (Basaki et al. 2021). Considering the sensitivity of HSP induction against nanoparticle exposure, additional approaches are warranted to dissect the underlying mechanisms of HSP modulation, as these findings can fine-tune the potential health risk due to the use of nanomaterials in food and medicines.

Prolonged/Long-Term Xenobiotic Stress: HSP and Hormesis

Considering the persistence of xenobiotics in the environment, the concept of prolonged exposure attains a more realistic approach to toxicity assessment. Depending on the strength of chemical exposure (in terms of concentration), prolonged exposure to xenobiotics may exert different outcomes toward organismal health. Prolonged exposure to relatively higher concentrations of xenobiotics may decrease the lifespan and health span, while mild stress, in the long run, may increase the lifespan and health span. Latter is termed as hormesis, in which long-term mild stress is beneficial to the exposed organism. Hormesis is also termed as adaptive homeostasis or preconditioning (Berry and Lopez-Martinez 2020). Hormesis is seen against different stimulators, including oxygen, temperature, radiation, and xenobiotics (heavy metals, pesticides, and water disinfectant by-products). Studies demonstrating hormesis in different model systems revealed that prolonged mild stress uses mild ROS generation/oxidative stress to evoke cellular adaptive responses, including antioxidant defense response, which protects the organism from the stress exerted by other stressors (Fig. 5). Studies have evidenced the involvement of HSP in hormesis. Hormesis achieved through elevated temperature and radiation was reported to be associated with upregulated *HSP70* and *HSP90* in Drosophila and Cimex (Benoit et al. 2009; Zhikrevetskaya et al. 2015). Damelin et al. (2000) reported the upregulation of *HSP70* in mouse cells against a low dose of cadmium upon its prolonged exposure.

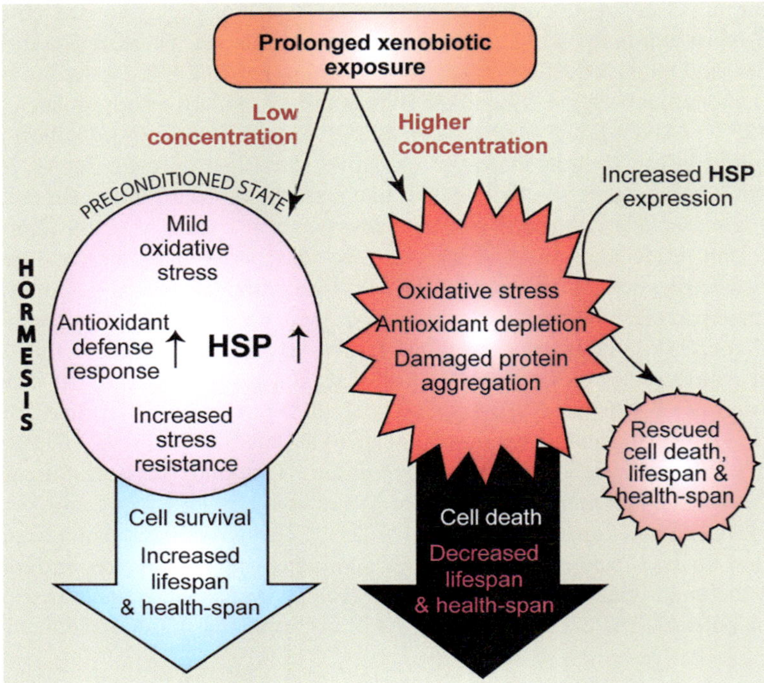

Fig. 5 Schematic representation of prolonged exposure to low concentrations of xenobiotics. Prolonged exposure to a very low concentration of xenobiotics may result in beneficial effects through increased levels of HSP and antioxidant capacity, leading to enhanced cell survival, and life- and health span. Exposure to a high concentration of xenobiotics results in oxidative stress, antioxidant depletion, and reduced life and health span

Prolonged exposure to a low water disinfectant by-product dichloroacetic acid concentration was reported to increase lifespan and health span associated with upregulated *HSP27* in *Drosophila* (Pandey et al. 2014). On the other hand, prolonged exposure to relatively high concentrations of environmental chemicals decreases the lifespan and health span of the organism. Prolonged dichlorvos exposure was reported to reduce lifespan and health-span associated with neuronal cell death, while overexpression of *HSP27* protected against these adversities (Pandey et al. 2016). In the context of long-term exposure, a multigeneration exposure approach is relevant to insecticidal resistance in insects. The effect of multigenerational exposure to an industrial pollutant, benzene, was studied by Sharma et al. (2018). The study observed that organisms become resistant to benzene toxicity in subsequent generations, accompanied by altered *HSP* levels (*HSP60*, *HSP70*, *HSP26*, and *HSP27)* and altered activities of cytochrome P450 and phase I, II, and III enzymes. The study further suggested that the altered *HSP* levels from generation to generation might be part of the organism's adaptive process towards benzene resistance. Table 1 enlists the studies showing modulation of HSP against

Table 1 Studies reported the involvement of HSP in prolonged/long-term mild xenobiotic stress

Toxicants/xenobiotics		Stress proteins	Stress protein level	Model system	Cellular/molecular functions	References
Prolonged exposure with low concentration	Cadmium	HSP70	Increased	McCoy CRM mouse cells	Enhanced cellular antioxidant defense response	Damelin et al. (2000)
	Dichloroacetic acid (DCA)	HSP27	Increased	*Drosophila melanogaster* (adult flies)	Increase lifespan and health-span of the exposed organism	Pandey et al. (2014)
	Dichlorvos	HSP27	Increased			Pandey et al. (2016)
Multigeneration exposure	Benzene	HSP26, 27, 60 and 70	Increased		Development of insecticidal resistance in insects	Sharma et al. (2018)

long-term/prolonged exposure to xenobiotics. Taken together, HSP are involved in cellular as well as organismal survival against prolonged exposure to xenobiotics. The cellular adaptation, as stated above, also confers an improved lifespan. This phenomenon is called hormesis. The adaptive condition/state resulting from long-term exposure to mild stress is called a preconditioned state. However, exposure to high concentrations leads to cell death and decreased lifespan. Genetic or pharmacological induction of HSP can rescue the adverse effects, including cell death and reduced lifespan.

Limitations

Studies in the last two decades have demonstrated /suggested the candidature/ the use of HSP as an early indicator of cellular stress against various toxicants. However, several concerns have also been raised, such as:

1. Toxicants have different chemical properties, and the exposure route may be different. Therefore, a particular type of HSP might not be a universal bioindicator.
2. The HSPs are evolutionary conserved. Nevertheless, the expression of HSP can differ in various animal model systems upon similar toxicant exposure.
3. Studies have observed the downregulation of HSP upon xenobiotic exposure (Dwivedi et al. 2022; Uwizeyimana et al. 2018; Rhee et al. 2009). Whether the downregulation of HSP be considered a cellular stress indicator?
4. HSPs are highly susceptible to stress. Therefore, its rapid and transient induction creates a suspicion of its specificity.
5. A study reported that the ecdysone receptor mediates the *HSP27* induction upon nonylphenol exposure (Dwivedi et al. 2022). This raises whether *HSP* induction upon toxicant exposure is a direct or indirect cellular process.

Conclusions

Most xenobiotics induce cellular stress by disturbing cellular redox balance, which further alters the protein homeostasis at different levels, including protein folding, tracking, and degradation. The cellular antioxidant response is elevated to cope with the oxidative stress upon redox imbalance. Acting as molecular chaperones, HSP responds systemically to maintain protein homeostasis. Upon exposure to xenobiotics, expression of HSP is upregulated to ameliorate proteotoxicity; therefore, these proteins contribute to the cellular defense response against xenobiotic stress. In addition to the studies establishing HSP as a sentinel species against xenobiotic exposure, several studies have emphasized the protective role of these proteins against xenobiotic-induced cellular stress in the last two decades. Hence, the protective efficacy of the HSP against the cellular adversities induced by xenobiotic exposure provides a strong base for using these as therapeutic targets. Given the conserved heat shock response across the species and the modulation of HSP in different model systems at their cellular and organismal levels, further mechanistic studies are prompted. Insights from these studies can be implicated towards utilizing HSP modulation as a therapeutic target against the adversities/disease-induced upon xenobiotic exposure.

Acknowledgments AS thankfully acknowledges the financial support by the Nitte Research Grant (NU/DR/NUFR1/NUCSER/2019–20/01) and Science and Engineering Research Board (ECR/2016/001863).

Disclosure of Interests All authors declare they have no conflicts of interest.

References

Abdel-Latif HMR, Dawood MAO, Mahmoud SF et al (2021) Copper oxide nanoparticles alter serum biochemical indices, induce histopathological alterations, and modulate transcription of cytokines, HSP70, and oxidative stress genes in *Oreochromis niloticus*. Animals (Basel) 11(3): 652

Ahamed M, Posgai R, Gorey TJ et al (2010) Silver nanoparticles induced heat shock protein 70, oxidative stress and apoptosis in *Drosophila melanogaster*. Toxicol Appl Pharmacol 242(3):263–269

Akerfelt M, Morimoto RI, Sistonen L (2010) Heat shock factors: integrators of cell stress, development and lifespan. Nat Rev Mol Cell Biol 11(8):545–555

Anbara H, Sheibani MT, Razi M (2020) Long-term effect of aspartame on male reproductive system: evidence for testicular histomorphometrics, hsp70-2 protein expression and biochemical status. Int J Fertil Steril 14(2):91–101

Archer AE, Rogers RS, Von Schulze AT et al (2018) Heat shock protein 72 regulates hepatic lipid accumulation. Am J Phys Regul Integr Comp Phys 315(4):R696–R707

Avila DS, Benedetto A, Au C et al (2016) Involvement of heat shock proteins on Mn-induced toxicity in *Caenorhabditis elegans*. BMC Pharmacol Toxicol 17(1):54

Basaki M, Keykavusi K, Sahraiy N et al (2021) Small heat shock protein's gene expression response to iron oxide nanoparticles in the brain. Biol Trace Elem Res 200(4):1791–1798

Benoit JB, Lopez-Martinez G, Teets NM et al (2009) Responses of the bed bug, *Cimex lectularius*, to temperature extremes and dehydration: levels of tolerance, rapid cold hardening and expression of heat shock proteins. Med Vet Entomol 23(4):418–425

Berry R, Lopez-Martinez G (2020) A dose of experimental hormesis: when mild stress protects and improves animal performance. Comp Biochem Physiol A Mol Integr Physiol 242:110658

Birbo B, Madu EE, Madu CO et al (2021) Role of HSP90 in cancer. Int J Mol Sci 22(19):10317

Breydo L, Wu JW, Uversky VN (2012) Alpha-synuclein misfolding and Parkinson's disease. Biochim Biophys Acta 1822(2):261–285

Bukau B, Weissman J, Horwich A (2006) Molecular chaperones and protein quality control. Cell 125(3):443–451

Chaouhan HS, Jha RR, Patel DK et al (2021) Cr(VI)-induced DNA damage is lessened by the modulation of hsp70 via increased GSH de novo synthesis in *Drosophila melanogaster*. J Biochem Mol Toxicol 35(8):e22819

Chen B, Zhong D, Monteiro A (2006) Comparative genomics and evolution of the HSP90 family of genes across all kingdoms of organisms. BMC Genomics 7:156

Clos J, Westwood JT, Becker PB et al (1990) Molecular cloning and expression of a hexameric Drosophila heat shock factor subject to negative regulation. Cell 63(5):1085–1097

Cochrane BJ, Irby RB, Snell TW (1991) Effects of copper and tributyltin on stress protein abundance in the rotifer Brachionus plicatilis. Comp Biochem Physiol C Comp Pharmacol Toxicol 98(2–3):385–390

Damelin LH, Vokes S, Whitcutt JM et al (2000) Hormesis: a stress response in cells exposed to low levels of heavy metals. Hum Exp Toxicol 19(7):420–430

Demir E (2021) A review on nanotoxicity and nanogenotoxicity of different shapes of nanomaterials. J Appl Toxicol 41(1):118–147

Doganlar O, Doganlar ZB (2015) Effects of a mixture of volatile organic compounds on total DNA and gene expression of heat shock proteins in *Drosophila melanogaster*. Arch Environ Contam Toxicol 68(2):395–404

Dong Y, Wang Y, Yu H et al (2013) Involvement of heat shock protein 70 in the DNA protective effect from estrogen. Am J Alzheimers Dis Other Dement 28(3):269–277

Doulias PT, Kotoglou P, Tenopoulou M et al (2007) Involvement of heat shock protein-70 in the mechanism of hydrogen peroxide-induced DNA damage: the role of lysosomes and iron. Free Radic Biol Med 42(4):567–577

Dukay B, Csoboz B, Toth ME (2019) Heat-shock proteins in neuroinflammation. Front Pharmacol 10:920

Dwivedi S, D'Souza LC, Shetty NG et al (2022) Hsp27, a potential EcR target, protects nonylphenol-induced cellular and organismal toxicity in *Drosophila melanogaster*. Environ Pollut 293:118484

Escobar Mdel C, Souza V, Bucio L et al (2009) MAPK activation is involved in cadmium-induced Hsp70 expression in HepG2 cells. Toxicol Mech Methods 19(8):503–509

Fan GH, Zhou HY, Yang H et al (2006) Heat shock proteins reduce alpha-synuclein aggregation induced by MPP+ in SK-N-SH cells. FEBS Lett 580(13):3091–3098

Fujigaki Y, Sun Y, Fujikura T et al (2010) Immunohistochemical study of heat shock protein 27 with respect to survival and regeneration of proximal tubular cells after uranyl acetate-induced acute tubular injury in rats. Ren Fail 32(1):119–125

Gao YJ, Xiao CF, Chen S et al (2004) In vitro study on role of Hsp70 expression in DNA damage of human embryonic lung cells exposed to Benzo[a]pyrene. Biomed Environ Sci 17(2):144–152

Gao Z, Yao L, Pan L (2022) Gene expression and functional analysis of different heat shock protein (HSPs) in *Ruditapes philippinarum* under BaP stress. Comp Biochem Physiol C Toxicol Pharmacol 251:109194

Guertin MJ, Petesch SJ, Zobeck KL et al (2010) Drosophila heat shock system as a general model to investigate transcriptional regulation. Cold Spring Harb Symp Quant Biol 75:1–9

Gupta SC, Sharma A, Mishra M et al (2010) Heat shock proteins in toxicology: how close and how far? Life Sci 86(11–12):377–384

Guvenc D, Kabak YB, Atmaca E et al (2013) Biological significance of the overexpression of HSP70 and alpha B-crystallin in rat substantia nigra exposed to different doses of permethrin. Arh Hig Rada Toksikol 64(1):47–55

Hoter A, El-Sabban ME, Naim HY (2018) The HSP90 family: structure, regulation, function, and implications in health and disease. Int J Mol Sci 19(9):2560

Idle JR, Gonzalez FJ (2007) Metabolomics. Cell Metab 6(5):348–351

Kedzierska-Mieszkowska S, Zolkiewski M (2021) Hsp100 molecular chaperone ClpB and its role in virulence of bacterial pathogens. Int J Mol Sci 22(10):5319

Kim HS, Park SY, Yoo KY et al (2012) Induction of heat shock proteins and antioxidant enzymes in 2,3,7,8-TCDD-induced hepatotoxicity in rats. Korean J Physiol Pharmacol 16(6):469–476

Krone PH, Blechinger SR, Evans TG et al (2005) Use of fish liver PLHC-1 cells and zebrafish embryos in cytotoxicity assays. Methods 35(2):176–187

Li H, Yang J, Wang Y et al (2019) Neuroprotective effects of increasing levels of HSP70 against neuroinflammation in Parkinson's disease model by inhibition of NF-kappaB and STAT3. Life Sci 234:116747

Liao PC, Lin HY, Yuh CH et al (2008) The effect of neuronal expression of heat shock proteins 26 and 27 on lifespan, neurodegeneration, and apoptosis in Drosophila. Biochem Biophys Res Commun 376(4):637–641

Liu L, He S, Tang M et al (2022a) Pseudo toxicity abatement effect of norfloxacin and copper combined exposure on *Caenorhabditis elegans*. Chemosphere 287:132019

Liu Y, Zhang W, Wang Y et al (2022b) Oxidative stress, intestinal damage, and cell apoptosis: toxicity induced by fluopyram in *Caenorhabditis elegans*. Chemosphere 286:131830

de Los Reyes T, Casas-Tinto S (2022) Neural functions of small heat shock proteins. Neural Regen Res 17(3):512–515

Meng Q, Li BX, Xiao X (2018) Toward developing chemical modulators of Hsp60 as potential therapeutics. Front Mol Biosci 5:35

Mishra RC, Grover A (2016) ClpB/Hsp100 proteins and heat stress tolerance in plants. Crit Rev Biotechnol 36(5):862–874

Mishra M, Sharma A, Negi MP et al (2011) Tracing the tracks of genotoxicity by trivalent and hexavalent chromium in Drosophila melanogaster. Mutat Res 722(1):44–51

Moreau X, Saez G, Thiery A et al (2008) ELISA detection of multixenobiotic resistance transporter induction in indigenous freshwater Chironomidae larvae (Diptera): a biomarker calibration step for in situ monitoring of xenobiotic exposure. Environ Pollut 153(3):574–581

Moreira-de-Sousa C, de Souza RB, Fontanetti CS (2018) HSP70 as a biomarker: an excellent tool in environmental contamination analysis – a review. Water Air Soil Pollut 229:264

Moyano P, Garcia JM, Lobo M et al (2018) Cadmium alters heat shock protein pathways in SN56 cholinergic neurons, leading to Abeta and phosphorylated tau protein generation and cell death. Food Chem Toxicol 121:297–308

Mukhopadhyay I, Saxena DK, Bajpai VK et al (2003a) Argemone oil induced cellular damage in the reproductive tissues of transgenic *Drosophila melanogaster*: protective role of 70 kDa heat shock protein. J Biochem Mol Toxicol 17(4):223–234

Mukhopadhyay I, Saxena DK, Chowdhuri DK (2003b) Hazardous effects of effluent from the chrome plating industry: 70 kDa heat shock protein expression as a marker of cellular damage in transgenic *Drosophila melanogaster* (hsp70-lacZ). Environ Health Perspect 111(16): 1926–1932

Mutwakil MH, Reader JP, Holdich DM et al (1997) Use of stress-inducible transgenic nematodes as biomarkers of heavy metal pollution in water samples from an English river system. Arch Environ Contam Toxicol 32(2):146–153

Nayak Rao S (2016) The role of heat shock proteins in kidney disease. J Transl Int Med 4(3): 114–117

Nazir A, Saxena DK, Kar Chowdhuri D (2003) Induction of hsp70 in transgenic Drosophila: biomarker of exposure against phthalimide group of chemicals. Biochim Biophys Acta 1621(2):218–225

Nisamedtinov I, Lindsey GG, Karreman R et al (2008) The response of the yeast *Saccharomyces cerevisiae* to sudden vs. gradual changes in environmental stress monitored by expression of the stress response protein Hsp12p. FEMS Yeast Res 8(6):829–838

Nisbet RM, Gotz J (2018) Amyloid-beta and tau in Alzheimer's disease: novel pathomechanisms and non-pharmacological treatment strategies. J Alzheimer's Dis 64(s1):S517–S527

Ostermann J, Horwich AL, Neupert W et al (1989) Protein folding in mitochondria requires complex formation with hsp60 and ATP hydrolysis. Nature 341(6238):125–130

Pandey A, Chandra S, Chauhan LK et al (2013) Cellular internalization and stress response of ingested amorphous silica nanoparticles in the midgut of *Drosophila melanogaster*. Biochim Biophys Acta 1830(1):2256–2266

Pandey A, Vimal D, Chandra S et al (2014) Long-term dietary exposure to low concentration of dichloroacetic acid promoted longevity and attenuated cellular and functional declines in aged *Drosophila melanogaster*. Age (Dordr) 36(3):9628

Pandey A, Saini S, Khatoon R et al (2016) Overexpression of hsp27 rescued neuronal cell death and reduction in life- and health-span in *Drosophila melanogaster* against prolonged exposure to dichlorvos. Mol Neurobiol 53(5):3179–3193

Parsell DA, Kowal AS, Singer MA et al (1994) Protein disaggregation mediated by heat-shock protein Hsp104. Nature 372(6505):475–478

Pelham HR (1982) A regulatory upstream promoter element in the Drosophila HSP70 heat-shock gene. Cell 30(2):517–528

Queitsch C, Hong SW, Vierling E et al (2000) Heat shock protein 101 plays a crucial role in thermotolerance in Arabidopsis. Plant Cell 12(4):479–492

Rhee JS, Raisuddin S, Lee KW et al (2009) Heat shock protein (Hsp) gene responses of the intertidal copepod *Tigriopus japonicus* to environmental toxicants. Comp Biochem Physiol C Toxicol Pharmacol 149(1):104–112

Ritossa F (1962) A new puffing pattern induced by temperature shock and DNP in drosophila. Experientia 18:571–573

Rock KD, Patisaul HB (2018) Environmental mechanisms of neurodevelopmental toxicity. Curr Environ Health Rep 5(1):145–157

Saini S, Rani L, Shukla N et al (2023) Hsp27 over expression protect against cadmium induced nephrotoxicity in *Drosophila melanogaster*. Comp Biochem Physiol C Toxicol Pharmacol 273: 109716

Salama SA, Mohamadin AM, Abdel-Bakky MS (2021) Arctigenin alleviates cadmium-induced nephrotoxicity: targeting endoplasmic reticulum stress, Nrf2 signaling, and the associated inflammatory response. Life Sci 287:120121

Sander BM, Martin LS (1993) Stress proteins as biomarkers of contaminant exposure in archived environmental samples. Sci Total Environ 1:459–470

Sharma A, Mishra M, Shukla AK et al (2012) Organochlorine pesticide, endosulfan induced cellular and organismal response in *Drosophila melanogaster*. J Hazard Mater 221-222:275–287

Sharma D, Singh MP, Vimal D et al (2018) Benzene induced resistance in exposed *Drosophila melanogaster*: outcome of improved detoxification and gene modulation. Chemosphere 201: 144–158

Shen HY, He JC, Wang Y et al (2005) Geldanamycin induces heat shock protein 70 and protects against MPTP-induced dopaminergic neurotoxicity in mice. J Biol Chem 280(48):39962–39969

Shukla AK, Pragya P, Chaouhan HS et al (2014) Heat shock protein-70 (Hsp-70) suppresses paraquat-induced neurodegeneration by inhibiting JNK and caspase-3 activation in Drosophila model of Parkinson's disease. PLoS One 9(6):e98886

Sikora A, Grzesiuk E (2007) Heat shock response in gastrointestinal tract. J Physiol Pharmacol 58(3):43–62

Singh MP, Mishra M, Sharma A et al (2011) Genotoxicity and apoptosis in *Drosophila melanogaster* exposed to benzene, toluene and xylene: attenuation by quercetin and curcumin. Toxicol Appl Pharmacol 253(1):14–30

Siracusa JS, Yin L, Measel E et al (2018) Effects of bisphenol A and its analogs on reproductive health: a mini review. Reprod Toxicol 79:96–123

Song JY, Li L, Ahn JB et al (2007) Acute liver toxicity by carbon tetrachloride in HSP70 knock out mice. Exp Toxicol Pathol 59(1):29–34

Sundarraj K, Raghunath A, Panneerselvam L et al (2017) Iron oxide nanoparticles modulate heat shock proteins and organ specific markers expression in mice male accessory organs. Toxicol Appl Pharmacol 317:12–24

Talorete T, Limam A, Kawano M et al (2008) Stress response of mammalian cells incubated with landfill leachate. Environ Toxicol Chem 27(5):1084–1092

Temiz O, Kargin F (2021) Toxicological impacts on antioxidant responses, stress protein, and genotoxicity parameters of aluminum oxide nanoparticles in the liver of *Oreochromis niloticus*. Biol Trace Elem Res 200(3):1339–1346

Tissieres A, Mitchell HK, Tracy UM (1974) Protein synthesis in salivary glands of Drosophila melanogaster: relation to chromosome puffs. J Mol Biol 84(3):389–398

Tiwari AK, Pragya P, Ravi Ram K et al (2011) Environmental chemical mediated male reproductive toxicity: *Drosophila melanogaster* as an alternate animal model. Theriogenol 76(2): 197–216

Tsuji T, Kato A, Yasuda H et al (2009) The dimethylthiourea-induced attenuation of cisplatin nephrotoxicity is associated with the augmented induction of heat shock proteins. Toxicol Appl Pharmacol 234(2):202–208

Ullah H, Khan A, Baig MW et al (2020) Poncirin attenuates CCL4-induced liver injury through inhibition of oxidative stress and inflammatory cytokines in mice. BMC Complement Med Ther 20(1):115

Uwizeyimana H, Wang M, Chen W et al (2018) Ecotoxicological effects of binary mixtures of siduron and Cd on mRNA expression in the earthworm *Eisenia fetida*. Sci Total Environ 610-611:657–665

Verghese J, Abrams J, Wang Y et al (2012) Biology of the heat shock response and protein chaperones: budding yeast (Saccharomyces cerevisiae) as a model system. Microbiol Mol Biol Rev 76(2):115–158

Vervaet BA, D'Haese PC, Verhulst A (2017) Environmental toxin-induced acute kidney injury. Clin Kidney J 10(6):747–758

Wang T, Wen X, Hu Y et al (2019) Copper nanoparticles induced oxidation stress, cell apoptosis and immune response in the liver of juvenile *Takifugu fasciatus*. Fish Shellfish Immunol 84: 648–655

Webster JM, Darling AL, Uversky VN et al (2019) Small heat shock proteins, big impact on protein aggregation in neurodegenerative disease. Front Pharmacol 10:1047

Wen HJ, Chen CC, Wu MT et al (2017) Phthalate exposure and reproductive hormones and sex-hormone binding globulin before puberty – Phthalate contaminated-foodstuff episode in Taiwan. PLoS One 12(4):e0175536

Wirth D, Christians E, Munaut C et al (2002) Differential heat shock gene hsp70-1 response to toxicants revealed by in vivo study of lungs in transgenic mice. Cell Stress Chaperones 7(4): 387–395

Wolf JH, Bhatti TR, Fouraschen S et al (2014) Heat shock protein 70 is required for optimal liver regeneration after partial hepatectomy in mice. Liver Transpl 20(3):376–385

Xiao C, Chen S, Li J et al (2002) Association of HSP70 and genotoxic damage in lymphocytes of workers exposed to coke-oven emission. Cell Stress Chaperones 7(4):396–402

Xin L, Wang J, Wu Y et al (2015) Increased oxidative stress and activated heat shock proteins in human cell lines by silver nanoparticles. Hum Exp Toxicol 34(3):315–323

Xu L, Zhao J, Xu D et al (2022) Application of transcriptomic analysis to unveil the toxicity mechanisms of fall armyworm response after exposure to sublethal chlorantraniliprole. Ecotoxicol Environ Saf 230:113145

Yang J, Liu X, Niu P et al (2009) Correlations and co-localizations of Hsp70 with XPA, XPG in human bronchial epithelia cells exposed to benzo[a]pyrene. Toxicology 265(1–2):10–14

Zhao P, Guo Y, Zhang W et al (2017) Neurotoxicity induced by arsenic in *Gallus Gallus*: regulation of oxidative stress and heat shock protein response. Chemosphere 166:238–245

Zhikrevetskaya S, Peregudova D, Danilov A et al (2015) Effect of low doses (5-40 cGy) of gamma-irradiation on lifespan and stress-related genes expression profile in *Drosophila melanogaster*. PLoS One 10(8):e0133840

Zhou H, Kato A, Yasuda H et al (2003) The induction of heat shock protein-72 attenuates cisplatin-induced acute renal failure in rats. Pflugers Arch 446(1):116–124

Role of Environmental Toxicants and Lifestyle Diseases in Infertility

David Chukwuma Nwikwe

Abstract

Environmental toxicants and lifestyle diseases are critical factors that can significantly impact both male and female reproductive health. Environmental toxicants, such as endocrine-disrupting chemicals, heavy metals, smoking, radiation, pesticides, and pollutants, and lifestyle diseases, age, and sexually transmitted diseases, have been comprehensively discussed in this chapter to be some of the causative agents of infertility. It has been extensively discussed that one in six individuals are victim of infertility, with approximately ten million couples projected to suffer the same fate by 2025. However, developing countries are the victims of this disease. It was also discussed that the prevalence of this disease is increasing globally with the upsurge in the use of assisted reproductive technologies. Therefore, factors like healthy lifestyle, avoidance of delayed parenthood, regular screening, awareness campaign, effective policy, and research are some of the mitigation strategies. While treatment measures like the use of herbs, traditional practices, counseling, and sensitization also have promise.

Keywords

Environmental toxicants · Heavy metals · Infertility · Lifestyle diseases · Reproductive health · Toxins

D. C. Nwikwe (✉)
Department of Chemical Sciences (Biochemistry Unit), Faculty of Science, Kings University, Odeomu, Osun State, Nigeria
e-mail: dc.nwikwe@kingsuniversity.edu.ng

S. Raisuddin et al. (eds.), *Environmental Toxicants and Lifestyle Diseases*,
https://doi.org/10.1007/978-981-95-0575-3_14

Introduction

Infertility is the inability of a couple (or individuals of reproductive age) to conceive after 1 year or more of regular unprotected sexual intercourse. It is a disease of either the male or female reproductive system, globally affecting millions of couples, and poses a substantial public health challenge with an alarming rate. Despite this disease being caused by various factors, developing research has provided detailed reports on the role of environmental toxicants and lifestyle diseases in its pathogenesis (World Health Organization 2023; Lee et al. 2024). Primary infertility characterizes couples who are unable to conceive their first child despite consistent efforts without contraceptives, that is, the inability to achieve pregnancy. Its prevalence can be expressed mathematically as the number of infertile women in marriage against the total number of fertile and infertile women in marriage. Secondary infertility characterizes a situation in which a couple that has been in a marriage for at least 5 years since their last live childbirth finds it difficult for another live child, despite not using contraceptives, that is, the ability to achieve at least one prior conception. Its prevalence can be expressed mathematically as the number of infertile women in marriage against the total number of fertile and infertile women in marriage (Kundu et al. 2023). It is worth noting that parameters for women in a fertile marriage for each expression vary during the process of any survey.

The prevalence of infertility varies among countries and localities. For instance, fertility rate among Koreans was reported to have fallen from 0.72 in 2022 to 0.68 in 2024 per woman (Lee and Kim 2024; Lee et al. 2024), while that of Brazil was from 1.682 in 2022 to 1.657 in 2024 per woman. Similarly, fertility rate of Nigeria dropped from 5.212 in 2021 to 5.144 in 2022, and from 5.076 in 2023 to 5.009 in 2024 (Macrotrends 2024). The World Health Organization (WHO) is alarmed that the rates of infertility are on the rise; 1 in 6 persons are affected worldwide. This implies that by 2025, about ten million couples will have infertility challenges (World Health Organization 2023; Clinical Trials Arena 2024). Stats of infertility for most couples revealed that infertility is high in developing countries, with South and Central Africa scoring highest (Kundu et al. 2023). Recently, the prevalence of this disease is increasing with the upsurge in the use of assisted reproductive technologies (ART) (Sharma et al. 2023).

Infertility is classically caused by several factors, such as environmental toxicants, lifestyle, or even genetic or physiological factors, responsible for a significant upsurge in progression of infertility in the last 50 years (Sharma et al. 2023). It may also be caused by a simple ailment that may meddle with ovulation, be harmful to the fallopian tubes, or even cause hormonal complexities (Bala et al. 2021). Worldwide, it has been estimated that 8–12% of couples are suffering from infertility. It is estimated that around 48 million couples and 186 million persons who are in their reproductive age are infertile; half of these couples are present in sub-Saharan Africa and South Asia (Sharma and Shrivastava 2022; Biggs et al. 2023). Although infertility is a reproductive health issue of both genders, reports have it that females suffer the consequences more than the males by enduring substantial societal humiliations in several African norms (Kissiwaa and Fouché

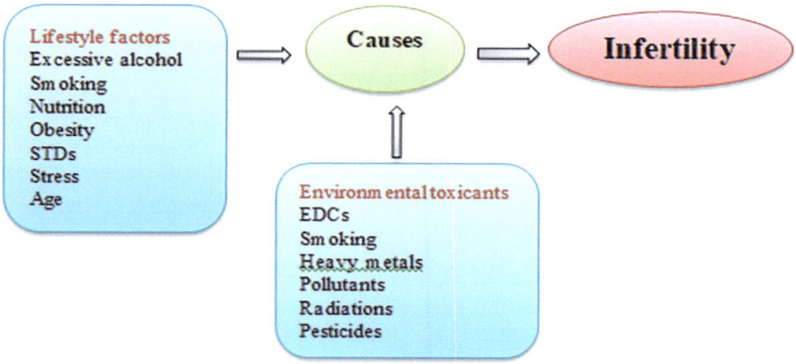

Fig. 1 Correlation between lifestyle and environmental toxicants in the progression of infertility

2024). Irrespective of gender, it is a polygenic disorder, which includes various genes, interactions between the genes, and correlations between the genes, its environment, and lifestyle, affecting how infertility develops (Sharma et al. 2023). Environmental toxicants such as industrial chemicals, heavy metals, radiations, pollutions, and pesticides, including lifestyle diseases such as diabetes, age, stress, hypertension, poor nutrition, obesity, and sexually transmitted diseases (STDs), have been reported to play critical roles in influencing male and female reproductive health (Fig. 1).

The diseases associated with infertility combines endometriosis, pelvic fiery disorder, polycystic ovarian syndrome (PCOS), uterine fibroid tumor, premature ovarian distress, and natural hazards. Although female infertility consists of ovulation irregularities, tubal blockage, uterine disorder, chemical awkwardness, age-related components, and past tubal ligation, male infertility consists of premature spermatozoa (Zulfiqar and Khan 2024). Although there are no known symptoms of infertility, the leading symptom is the inability to get pregnant. Nevertheless, in some cases, some females with this disease may face the challenges of irregular monthly periods or even no period, as the case may be. While for males, symptoms such as hormonal instability causing abnormal hair growth or sexual function are frequently observed.

Environmental toxicants are "ever-present" around us, and their harmful effects on our health have gathered increasing attention. Industrial substances such as endocrine-disrupting chemicals (EDCs), which include polychlorinated biphenyls (PCBs), bisphenol A (BPA), and phthalates, have been connected to disrupting hormonal regulation, impairing reproductive performance, and contributing to infertility (Kawa et al. 2021; Ghosh et al. 2022). Exposure to heavy metals like mercury, lead, and cadmium has been linked with reduced quality of sperm in men, with increased risks of conditions such as endometriosis and PCOS in the female gender (Dar et al. 2023). In males, phthalates, BPA, heavy metals, pesticides or herbicides, air pollutants, and electromagnetic frequencies exposures through computers and phones, have been linked with sperm quality, volume, viability, and motility (Biggs et al. 2023). These toxins can cause infertility in four ways such as endocrine disruption, damage to the male and female reproductive systems, and impaired fetal viability (Pizzorno 2018).

Lifestyle diseases such as diabetes, stress, poor diet, obesity, hypertension, and cardiovascular diseases are gradually attaining epidemic proportions worldwide, thereby impacting reproductive health. It is connected to the way an individual lives. That is, habits that constrain a person from activity and cause that individual towards a sedentary lifestyle, which is capable of causing numerous health complications and life-threatening consequences. Economic advancement and globalization are two major factors that have accompanied an increase in global lifestyle diseases. In 2018, WHO estimated the global incidence of mortality caused by lifestyle diseases to be approximately 71% (41 million) (Balwan and Kour 2021). Diabetes (types 1 and 2) interrupts glucose metabolism, thereby causing complications such as menstrual fluctuations in females and erectile dysfunction in males, eventually affecting fertility (Pizzorno 2018). Similarly, hypertension and cardiovascular diseases are known to impair blood flow to reproductive organs, thereby contributing to ovulatory dysfunction, diminished ovarian reserve, male factor, tubal factor, erectile dysfunction, unexplained infertility, and multifactorial etiologies (Jain and Singh 2024), further exacerbating infertility. Since our lifestyle routines are pivotal in defining the risk of developing lifestyle diseases, it is imperative to consider healthy habits and a decent way of life in order to mitigate an individual's risk of developing lifestyle diseases.

Environmental Toxicants

Environmental toxicants comprise a wide range of chemicals present in our environment, including industrial chemicals, radiations, pesticides, heavy metals, and pollutants. These deleterious substances are capable of disrupting normal reproductive functions by interfering with hormonal regulation, damaging reproductive organs, thereby affecting the quality of gametes. These toxins impact humans throughout their lifespan (including prenatally) and can pose deleterious effects ranging from ovulatory dysfunction to increasing risk of cancer to altered sperm quality. It has been reported in the United States that pregnant women got exposed to more than 43 different potential chemical toxins, according to the National Health and Nutrition Survey of 2003 to 2004 (Jain and Singh 2024). Some common environmental toxicants associated with infertility include:

Endocrine-Disrupting Chemicals (EDCs)

Polychlorinated biphenyls (PCBs), Bisphenol A (BPA), and phthalates are common EDCs that mimic or interfere with human hormones, thereby disrupting reproductive processes. These chemicals are known to be exogenous, and when exposed in utero or puberty may cause infertility, thereby predisposing to the onset of several diseases that impact the reproductive function, such as diabetes, endocrine cancer, and obesity (Kundu et al. 2023).

The sources of EDCs vary. While some are from industries, e.g., industrial chemicals such as pesticides, lubricants, and flame-retardants, some are household consumables such as tea, lavender oils, phthalates, and BPA products, some are soy products from plant-derived phytoestrogens, and others are estradiol and diethylstilbesterol, which are pharmaceutical products. Some of these toxins have been implicated in the disruption of the endocrine system by altering endogenous hormone production or participating in the peripheral distribution of endogenous hormones. Several molecular pathways have been proposed on how this can be attained, but the nuclear receptor pathway is well-known. This works by the binding of EDCs to the hormone receptors of androgens, estrogens, progestins, and thyroid glands. This chemistry either disrupts or augments their cellular activity. Furthermore, some mechanisms have proposed the possibility of prenatal and early exposure causing epigenetic effects, thereby predisposing the victims to numerous diseases such as obesity, rare cancers, and diabetes (Jain and Singh 2024). EDCs have also been associated with reproductive development conditions, PCOS, ovarian dysfunction, and subfertility in females. Bisphenol A in animal experiments has been associated with uterine polyps, vaginal adenosis, ovarian cysts, and impaired implantation in in vitro fertilization (IVF) (Pizzorno 2018). Females with PCOS were reported to have high serum BPA with increasing insulin resistance and hyperandrogenism linked with increased BPA concentrations (Kandaraki et al. 2011). Epidemiological studies have shown that BPA negatively impacts quality of sperm (Li et al. 2011). People undergoing IVF usually record a longer period of trying to conceive before pregnancy is achieved as well as worse fertilization rates when exposed to PCBs. But when exposed to phthalates, have been linked to lower antral follicles with mature oocytes (Jain and Singh 2024). Dichlorodiphenyltrichloroethane (DDT) has been linked with decreased concentration of spermatozoa, stable morphology, and motility in semen analyses (De Jager et al. 2006). Although EDCs are well known to affect reproductive systems, it also affect some other parts of the endocrine system capable of affecting fertility, such as thyroid, transport, and metabolism abnormalities, which may be linked with hypothyroidism (Oulhote et al. 2016).

Smoking

Smoking is known to be a leading environmental toxin that causes preventable deaths and morbidity linked with stroke (Adeloye et al. 2019), which influences fertility. It is a significant environmental toxin, with about 17.8% adult smokers in the United States (Jamal et al. 2014). Toxic metals, such as cadmium and cotinine, present in the follicular fluid of the ovaries, impact oocyte quality, thereby causing infertility. This cotinine level is also elevated in nonsmoking females who live with smokers (Jain and Singh 2024). Meanwhile, research in smoking males revealed diminished sperm concentration and motility, while animal experiments revealed poor ability of spermatozoa to bind to the zona pellucida to enable fertilization (Pasqualotto et al. 2008). Soares et al. (2007) in a meta-analysis comprising seven research works stated that smokers undergoing assisted reproduction (such as IVF) need two times as many IVF cycles compared with nonsmokers before conception. Surprisingly, the rates of pre-term births were reported to have dropped by 25% when Ireland restricted smoking in workplaces (Pizzorno 2018).

Heavy Metals

Cases of exposure to heavy metals such as cadmium, arsenic, lead, and mercury have been associated with menstrual abnormalities, diminished sperm quality, and impaired fertility in females and males. Exposure to these toxins can emanate from our natural environment, including housing or work environments, for example, mercury and arsenic through food and water supply, and lead through paint (Jain and Singh 2024). It is a misnomer to believe that lead exposure can only be achieved through lead paints; lead exposure can also be attained through cosmetics, water from lead pipes, herbal supplements, and construction sites. The level of lead contamination can be determined from 'high risk' individuals using venous blood levels. A value >15.47 parts per billion (ppb) is linked with a two-fold risk of infertility compared to those below the threshold (Lei et al. 2015). Spontaneous abortion and pre-term birth have been implicated in increased exposure to lead toxin (Jelliffe-Pawlowski et al. 2006). High levels of lead toxin in blood can be managed during the pre-conception stage by simply identifying and removing the toxins or through chelation in severe situations (Jain and Singh 2024).

Consumption of mercury-containing diets, e.g., predatory fish, is a major route of mercury exposure. This toxin enters the predatory fish through industrial wastes that find its way into the sea mainly through human activities. Although it can be contacted through other minor means such as the use of toning creams and gold mining, diet remains the ultimate source (Chan 2011). Its toxicity during pregnancy has been related to neurodevelopmental disorders in progeny, and also linked with infertility in women (Maeda et al. 2019). High levels of mercury toxin can be managed during the pre-conception stage by simply identifying and removing the toxins or through chelation in severe situations (Jain and Singh 2024).

Exposure to cadmium can be made possible from batteries (especially rechargeable ones) and cadmium-containing plastics and paints. This toxin can find its way into our diet from the soil, thereby exposing the food (wheat, rice, shellfish, and certain vegetables) to cadmium. Research conducted on infertile couples undertaking IVF revealed a decrease in oocyte fertilization and implantation rates for those with high cadmium concentrations (Bloom et al. 2012). High levels of cadmium can be managed during the pre-conception phase by simply identifying and eliminating the toxins or by chelation in severe situations (Jain and Singh 2024). Conversely, exposure to arsenic is often encountered through groundwater, industrial effluents, and certain pesticides. This toxin has also been implicated in male infertility (Wang et al. 2016), though it can be easily removed from the blood as its mode of the treatment. Despite the numerous hazardous impacts of these toxicants, there is no available potent remedy for their treatment.

Pollutants

Pollutants, especially air pollutants, including sources such as polycyclic aromatic hydrocarbons (PAHs), particulate matter (PM), and volatile organic compounds (VOCs), have been associated with adverse reproductive complications, including infertility and pregnancy challenges. Exposure to these endocrine disruptors can be linked with numerous health effects, such as cardiovascular and respiratory diseases, while other studies have linked it with having negative impact on fertility. One study conducted on men exposed to air pollution revealed a lower X to Y sperm chromosome ratio with abnormal morphology of 51.78%. Furthermore, exposure to this pollutant has a negative impact on sperm aneuploidy, sperm concentration, DNA damage, and sperm motility (Wieczorek et al. 2024).

These pollutants, especially air-borne pollutants, have been linked with about seven million mortalities yearly, and this colossal loss documented yearly is more than the combined figure from communicable diseases such as HIV/AIDS, tuberculosis, and malaria (Sokan-Adeaga et al. 2023). Most of these losses are recorded in developing nations due to innumerable factors such as urbanization, deforestation, industrialization, and poverty. There are evidences linking air pollutants with female fertility. A study by Gaskins et al. (2019) linked particulate matters ($PM_{2.5}$) to decrease ovarian reserve. Other studies on long-term exposure to SO_2 or PM were linked with lower antral-follicle count, pregnancy loss, and birth weight. Lead, copper, and diesel exhaust are common air pollutants implicated in endocrine activity, which may affect female reproductive system. These toxicants disrupt the endocrine activity by exerting anti-estrogenic, estrogenic, and anti-androgenic activities, and some even interfere with the thyroid and impact metabolic disorders (such as insulin resistance obesity) which are related to infertility (Seli and Taylor 2023; Wieczorek et al. 2024).

However, be it acute or chronic exposure, the emanating health complications from pollutants depend on the nature of the chemical and the affected organ. More so, the degree of toxicity varies among organs, even for similar contaminants. Studies have shown that environmental and occupational sources are the main channels through which individuals are exposed to these harmful chemicals and their combinations. Although there has been no established simple law to protect or intervene against emanating reproductive diseases (e.g., infertility) associated with pollutant exposure, bioactive compounds in phytochemicals of medicinal plants may be a promising therapy (Dwivedi et al. 2022; Sokan-Adeaga et al. 2023).

Radiations

Advancements in science and technology in modern era has led to the production of numerous useful home and industrial appliances to aid human existence. As such, individuals in modern societies are exposed to various radiations in their daily activities, such as computer, television, mobile phones, occupational tools, including important medical devices (e.g., anticancer therapies, diagnostic imaging, and interventional radiology procedures) (Marci et al. 2018). These electromagnetic radiations, emitted by these devices (including radiations from transmission masts), may impact humans both genetically and biologically. Therefore, human exposure to extreme radiations and other deleterious substances may impact reproductive organs, causing infertility (Kesari et al. 2018).

Pesticides

Reproductive health may suffer harmful effects of various degrees when exposed to various pesticides. It is a harmful compound that can alter the male reproductive health, particularly semen, of exposed workers. This is because spouses of agricultural field officers directly using pesticides and hazardous chemicals like DDT have been reported to encounter a series of stillbirths, miscarriages, and even spontaneous abortions (Banerjee and Bhattacharya 2019). Organophosphorus pesticides remain one of the most commonly used pesticides and of consist of esters of phosphoric acid or thiophosphoric acid. They are known for interfering with the potentials of hormonal, examples include dimethoate, malathion, and parathion. These pesticides disrupt the functions of cholinesterase, regular cellular metabolic processes, decrease insulin production, and hinder mitochondrial activities, thereby prompting oxidative stress and impacting the endocrine and nervous systems (Karami-Mohajeri and Abdollahi 2011), including reproductive health.

Lifestyle Diseases

Lifestyle disease is a disease characterized by unhealthy living, such as a sedentary lifestyle, poor or unhealthy diet, smoking, stress, and excessive alcohol intake, which adversely affect the reproductive system by impairing reproductive organs, causing hormonal imbalance, and even increasing oxidative stress (Balwan and Kour 2021). Diseases like diabetes, obesity, and hypertension are globally reaching epidemic proportions, thereby exerting a significant impact on reproductive health. Lifestyle or environmental factors have been increasingly linked with male infertility. There has been an established link between quality of sperm and pregnancy outcomes with foods rich in soy, red meat, processed meats, full-fat dairy, and simple sugars (Biggs et al. 2023). Modern lifestyle and economic evolution have drifted human lifestyles towards sedentary routines with increasing intakes of processed foods and refined sugars excessively (Kumar and Kotwal 2023).

Impairment of sperm motility, sperm count, sperm morphology, and sexual functions has also been linked with adiposity, heavy smoking, alcohol dependence and use, and use of androgenic steroids (Sansone et al. 2018). Some common lifestyle factors that contribute to infertility include the following:

Excessive Alcohol Intake

Excessive alcohol intake has been linked with numerous health and social implications, including lifestyle diseases and infertility in humans. It can abruptly impair quality sperm production, disrupt hormonal regularity, and decrease ovarian functionality, thereby increasing the risk of infertility (Sansone et al. 2018; Lee et al. 2024). Drinking alcohol and smoking tobacco are avoidable and unhealthy lifestyles that are deleterious to health and may impact male fertility. For instance, the hypothalamic–pituitary axis, influenced by excessive alcohol intake, mostly impairs the male reproductive system (Ramgir and Abilash 2019). Ethanol has been implicated in animal studies to inhibit gonadotropin-releasing hormone (GnRH), a hormone released in the hypothalamus. This is possible by decreasing the amount of GTP-binding proteins of Rab family, which play significant roles in controlling membrane and protein trafficking. At the pituitary, alcohol intake causes diminished sperm quality, motility, volume, and viability. It lowers the levels of follicle-stimulating hormones (FSH), luteinizing hormones (LH), although its effect on the levels of testosterone is unclear (Alghobary and Mostafa 2022).

Obesity

The upsurge of obesity and overweight raises health concerns among reproductive individuals. Obesity is a health disease associated with insulin resistance, hormonal irregularities, and inflammation, which can impair ovulation in females and reduce the quality of sperm in males. Obesity and overweight are caused by the storage of excess fats, which impact the reproductive health negatively (Sharma et al. 2023). The incidence of obesity in an individual emanates from having a body weight that is more than 20% of the right age, while being overweight is having a body mass index (BMI) of 25 to 29.9. As done by other health risks, obesity and overweight have impacted human reproductive health, causing infertility in both genders (Sharma et al. 2013).

A study in Turkey revealed that most women trying to treat infertility were obese (Dag and Dilbaz 2015). Abnormal reproductive hormone levels were implicated in obese individuals. This disease comes with high levels of cholesterol, heart disease, diabetes, cancer, insomnia, gall bladder disease, and kidney failure (Baheerati and Devi 2018). While weight reduction in overweight and obese females has been linked with improving fertility, normalizing menstruation, and increased chance of spontaneous ovulation and conception in anovulation, high BMI values may increase chance of miscarriage, low chance of pregnancy, complications during pregnancy, and diminished wellbeing of fetus (Sustarsic et al. 2023; Lee et al. 2024). However, overweight in females can impact the treatment practicability of ovarian brokenness (Zulfiqar and Khan 2024).

Nutrition

Nutrients derived from diets play significant roles in the human reproductive system. Hence, eating healthy diets and good nutrition in reproductive age has been associated with improving fertility in both genders (Gaskins and Chavarro 2018). Change in diets (especially poor diets) may play a substantial role in fertility when used alone or in combination with fertility treatment. Several factors have been linked with impacting male and female fertility. Factors such as gynecological diseases, which cause female infertility, environmental factors like EDCs, and lifestyles such as stress, poor nutrition, and unbalanced diet were implicated (Silvestris et al. 2019). Nevertheless, certain diets need to be avoided to improve fertility challenges. They include trans-fats, excessive caffeine, excess alcohol intake, low-fat dairy, high-carbohydrate and sugar foods (especially soda), and processed foods and meats, while berries (rich in antioxidants and anti-inflammatory phytochemicals), citrus fruits (rich in vitamin C and folate) are essential as they aid in boosting both male and female fertility as well as aid in fetal growth.

Stress

Chronic stress has been reported to disrupt the hypothalamic-pituitary-adrenal (HPA) axis, thereby causing hormonal irregularities that affect reproductive functions in males and females (Zulfiqar and Khan 2024). Oxidative stress caused by an upsurge in reactive oxygen species (ROS) generation causes lipid peroxidation, hence damage to sperm DNA, thereby causing infertility. More so, alterations in sperm DNA through DNA adduct formation (especially involvement of PAHs) cause modifications in gene expression and DNA methylation regulations, thereby leading to male infertility (Sharma et al. 2023; Zulfiqar and Khan 2024).

Stress can be in the form of physical, social, or psychological, which is an integral factor of any society or inhabitant. Although there is no conclusive opinion on the interplay between stress and infertility and vice versa, it has been exclusively linked with some factors such as societal pressure, diagnosis, treatments, testing, treatment failures, unfulfilled wishes and desires, including fiscal cost. Despite infertility being an emotional issue for many couples, recommending counseling for them may be resourceful throughout the treatment period, which may play out to be the last viable help to their problem (Sharma et al. 2013; Sharma et al. 2023).

Smoking

Smokes from tobacco contain detrimental chemicals that are capable of damaging reproductive organs, reducing sperm count, quality, and motility, thereby increasing the risk of infertility and pregnancy difficulties. It has become an increasing trend among youths as it has been implicated to be one of the causes of male infertility. The negative effect of tobacco on sperm is that it results in low sperm count with high incidence of morphological anomalies, including defects in the tail, neck, and head of the sperm cell (Bundhun et al. 2019). It has been observed that males who smoke cigarettes have a low level of acrosin (protease found in sperm acrosome), attributed to reduced inducibility of the acrosome reaction. Another study revealed an increased seminal plasma cadmium level but a low activity of plasma Ca^{2+}-ATPase (ATP2B4) in the sperm cells of males who smoke, which was attributed to being responsible for the low sperm motility observed in these individuals. An enzyme responsible for cell-cycle regulation and DNA damage response, known as checkpoint kinase 1 (CHK1), has been linked with low potency in the sperm cells of smokers due to low levels of CHK1, which can cause reduced repair of sperm DNA (Cui et al. 2016).

Age

Age plays a crucial role in human reproductive health. In several societies where education (including other factors) is prioritized over early childbearing, many couples tend to delay childbearing for their career (Sharma et al. 2013). This lateness in childbearing has caused infertility in many marriages, which has been an increasing trend recently, as age plays an indispensable role in fertility. This is because female biological reproducibility ripens and decays with age, that is, fruitfulness is peak at ages 18–24, which begins to drop from 27 years and declines at 35 years (Zulfiqar and Khan 2024). In older females, fertility issue is mostly caused by the quality and reduced number of eggs or even reproductive health issues affecting fertility. Similarly, younger males less than 40 years are more fertile than those older than 40, which in most cases results in birth defects and genetic abnormalities in the offspring of the latter (Mayo Clinic 2023).

Sexually Transmitted Diseases (STDs)

Several known diseases contracted through sex (such as chlamydia and gonorrhea) have been linked with perpetual damage to the reproductive tract in men and women, thereby leading to infertility. It has been specifically implicated in those between 15 and 50 years. Prevention remains the most important measure of STDs because it causes acute infections and reproductive complications such as infertility (Sharma et al. 2023).

Most frequently, STDs (such as chlamydia and gonorrhea) are associated with pelvic inflammatory disease (PID), as this infection, which mostly arises from the lower vaginal tract, affects the fallopian tubes, uterus, and even the ovaries (El-Kharoubi 2023). It has been reported that 33.6% of females between the ages of 35–44 had experienced salpingitis, while about 16.1% had at least an encounter with a PID, and 10.7% experienced salpingitis without encountering any PID (Mansour 2023). As it is well known that adhesions, scarring, and partial or complete closure of the fallopian tubes are all possible outcomes of PID in females, thus could result in the fallopian tube's inner layer losing its ciliated epithelial cells, which would hinder ovum transit and raise the possibility of ectopic pregnancy and infertility by preventing eggs from being fertilized and also from traveling to the oviduct. Pelvic pain that persists in most cases may also be caused by adhesions. PID is typically misdiagnosed and can manifest as acute, chronic, or subclinical symptoms, and this may be slight or deceptive, thereby causing many females not to experience any symptoms. However, they realize this only when experiencing chronic pelvic pain or infertility (El-Kharoubi 2023; Mansour 2023).

Interaction Between Environmental Toxicants and Lifestyle Diseases

The lipid milieu of cells, including oxidative stress and antioxidant status, can be dictated by nutrients in diets. The nutritional status of an individual can modulate these parameters, which may be deleterious to biological processes, thereby stimulating the effects of environmental toxicants to cause dysfunction or even disease (Hennig et al. 2007). Infertility is a disease caused by genetics, environmental toxicants, and lifestyles. Environmental toxicants can impair DNA of sperm and eggs, disrupt hormonal stability, and harm reproductive cells. The hormonal stability, reproductive health, and the amount and quality of eggs and sperm can be adversely affected by life routines such as excessive alcohol intake, smoking, poor diet, age, and stress. Difficulty is usually encountered in predicting the rate at which infertility will advance from one person to another as a result of complexity in interactions between environmental toxicants and lifestyles, which vary among individuals (Sharma et al. 2023). Therefore, the interaction between lifestyle diseases and environmental toxicants can worsen health risks, including an impact on fertility.

Obesity and Endocrine Disruptive Compounds

EDCs (such as phthalates and BPA) have been linked to various metabolic conditions such as insulin resistance and obesity. Equally, obesity can change the body's metabolic processes and distribution of EDCs, thereby possibly causing toxicity and adverse health complications. How this works is that EDCs may disrupt the endocrine functions, thereby contributing to metabolic abnormalities and obesity. Conversely, obesity can impact metabolic processes and clearance of EDCs, thereby resulting in increased circulating levels and increased toxicity (Casals-Casas and Desvergne 2011).

Mitigation Strategies

Comprehensive strategies that will address both the individual and population-based factors are necessary to mitigate the combined effects of lifestyle diseases and environmental toxicants that can pose reproductive health risks. Therefore, the implementation of these strategies may be sufficient to minimize the combined effects of environmental toxicants and lifestyle diseases on reproductive health, thereby enhancing fertility preservation, boosting the chances of well-being of the populace of reproductive age.

Preventive Measures

One of the crucial steps towards mitigating the devastating effects of environmental toxicants and lifestyles on reproductive health is to implement proactive preventive measures to reduce exposure to these toxins and chemicals, such as regulating the use of harmful chemicals in consumables, improving the quality of air, and encouraging sustainable practices. Equally, it is of necessity to encourage lifestyle adjustments, including avoiding physical inactivity through regular physical activity (to reduce PCOS in females), embracing balanced diets (especially folate-rich diets like pea and broccoli), shunning tobacco smoking and excessive alcohol intake, abstinence from STDs (or by practicing safer sex techniques among those with multiple sex partners, e.g., use of contraceptives), treat STDs rapidly, sustaining a healthy weight, and avoid delaying parenthood.

Screening and Monitoring

It is important to incorporate screening for environmental exposures and unhealthy lifestyles into routine healthcare visits, especially for those planning to conceive. Equally, monitoring environmental quality and trends in lifestyle-related health behaviors to identify those at higher risk of infertility and implement targeted interventions.

Education and Awareness

It is crucial to raise awareness about the risks connected with environmental toxicants and lifestyle diseases on reproductive health through informative educational campaigns that target persons, healthcare givers, and policymakers. It is necessary to provide information on how lifestyle decisions, such as smoking cessation, regular exercise, and diet, can mitigate the adverse effects of environmental exposures on reproductive health systems.

Effective Policy and Regulation

To mitigate the effects of these harmful substances on reproductive health, it is pertinent to advocate for policies and regulations meant at eradicating environmental toxicants and limiting the use of hazardous chemicals in consumables. Necessity calls for the implementation of public health initiatives to encourage healthy lifestyles, such as urban planning strategies that will support healthy living, tobacco control guidelines, and improved access to balanced diets.

Fostering Multidisciplinary Collaboration

Another ingredient in mitigating the deleterious effects of hazardous substances in reproductive health is to foster collaboration among environmental researchers, healthcare experts, community organizations, policymakers, among other stakeholders to develop all-inclusive approaches to preserve fertility. It is also essential to encourage interdisciplinary research and practice that will integrate knowledge from various disciplines to address the complex relationship between environmental toxicants and lifestyle diseases affecting fertility.

Effective Research and Surveillance

It is pertinent to invest in effective and durable research to "X-ray" the interplays between environmental toxicants and lifestyle diseases on reproductive health, and identify effective mitigation strategies to curb it. Surveillance systems are also needed to monitor developments in environmental exposures and lifestyle-related health performances, to allow for appropriate intervention and policy implementation.

Treatment Measures

Since infertility is a disease that brings frustration, fear, anger, isolation, shame, guilt, confusion, hopelessness, etc., effective treatment measures will bring relief and hope. Infertility treatment in both male and female can be treated firstly by identifying the underlying etiology by health practitioners before necessary measures are adopted. Few of these measures include hormone therapy, timing of intercourse, intra-uterine insemination (IUI) to circumvent challenges with sperm/

cervical mucus interactions, IVF-embryo transfer, gamete intra-fallopian transfer (GIFT), or even intra-cytoplasmic sperm injection (ICSI) (Carson and Kallen 2021).

Herbal Therapy

Since time immemorial, the human race has been consistently using herbs or plant-based products to enhance sexual craving as well as to boost performance and the pleasure involved. The active ingredients and extracts of these plants have been resourceful in improving sexual enhancement and performance, thereby improving reproductive health (especially egg and spermatogenesis). Ipso facto, several drugs with herbal origin have been authenticated for their impact on reproductive health and fertility, thereby serving as a lead for identifying new chemical compounds for treating infertility caused by environmental toxins and lifestyles. Several parts of these plants have been adopted for use, such as leaves, seeds, flowers, and fruits. Some of these plants include *Matricaria chamomilla*, *Vitex Agnus-castus*, *Punica granatum*, *Piper guineense*, *Allium sativum*, *Persea americana*, and *Syzygium aromaticum* (Dwivedi et al. 2022; Nwikwe et al. 2023; Ugiomoh et al. 2023).

Traditional Practice

In some traditional settings, infertility is mostly believed to be caused by the female's past life, womb damage, and incompatibility between the couple examined. These traditional health practitioners usually believe that infertility is caused by a curse or witchcraft. This condition is usually treated through fasting, prayers, and sacrifices. It is also managed in some places by engaging in guidance and counseling, the use of herbal concoctions, and contraceptives.

Counseling

Victims of infertility are counseled to adopt relevant proactive measures to treat infertility. Health practitioners have resulted into counseling affected individuals into adopting measures such as lifestyle change (especially in obese and overweight women with PCOS who can gain from weight drop, exercising, and lifestyle change to restore ovulatory cycle and attain pregnancy) (Melo et al. 2015) and timing of the most fertile period for intercourse (usually the 6 days before normal ovulation). Females may gain from referral to a health practitioner comfortable with recommending pulsatile administration of gonadotropin-releasing hormone (GnRH) or gonadotropins with LH and LH activity to induce ovulation. Victims are counseled and advised to shun tobacco use, reduce alcohol intake, and target a BMI of less than 30 kg/m^2 to boost their rates of natural conception or use ART. Since obesity damages fertility and fertility treatment response (including IVF), the victims are advised to reduce weight before infertility treatment or conception (Lindsay and Vitrikas 2015).

Holistic Measure

Individual and community actions and change of policy level are paramount in the prevention of environmental toxin exposure. During antenatal counseling and pregnancy, reproductive healthcare givers often counsel patients observed to have reproductive health challenges on the need for exposure prevention. These caregivers can foil exposure by adopting suitable reporting and effective patient education. Any patient suspected to have been exposed to toxins or observed after screening should be referred to the Environmental Health Specialty Unit or occupational medicine programs (Jain and Singh 2024).

Holistically, treatment of toxicity such as exposure to heavy metals, EDCs, and smoking cigarette may be managed with cessation counseling to avoid these unhealthy habits and protect them from exposure during assessment for infertility. Studies have shown that avoidance of BPA-rich sources like canned foods may reduce BPA levels (Rudel et al. 2011). Such individuals may be advised to avoid processed foods and fast foods, as well as products containing EDCs like DES and BPA, and adopt diets rich in vegetables, fruits, and fibers. In fact, placing bans on EDC-containing products (e.g., infant formula and baby bottles) by regulatory agencies like the Food and Drug Administration (FDA) may be pivotal in reducing exposures and improving fertility (Jain and Singh 2024).

Educating the patients through counseling and education guides to minimize or prevent exposure is crucial. This involves creating awareness campaigns and sensitizing them on the right choice of living to prevent unhealthy living, such as exposure to toxins, poor diets, and unhealthy lifestyles. In females, they can adopt weight loss drugs, hormonal therapy (e.g., gonadotropins, bromocriptine), and chemical treatment (e.g., perganol) measures. This will proffer solutions to some of the challenges of infertility.

Conclusion and Future Directions

The rate at which a combination of environmental toxicants and lifestyle factors has impacted reproductive organs is alarming. Environmental toxicants, such as endocrine-disrupting chemicals (e.g., bisphenol A, phthalates, polychlorinated biphenyls), heavy metals (e.g., lead, cadmium, arsenic, mercury), smoking, radiations, and air pollutants (e.g., polycyclic aromatic hydrocarbons, particulate matter), and lifestyle diseases, such as obesity, excessive alcohol intake, poor nutrition, stress, smoking, age, and sexually transmitted diseases (e.g., gonorrhea and chlamydia), have been linked to adversely affect both male and female fertility. Exposure to these hazardous chemicals and lifestyle routines may negatively impact sperm aneuploidy, sperm concentration, DNA damage, and sperm motility, affecting male fertility. Some even disrupt the endocrine activity by exerting anti-estrogenic, estrogenic, and anti-androgenic activities, and even interfere with the thyroid, causing metabolic disorders. While some increase the chance of miscarriage, low chance of pregnancy, complications during pregnancy, and diminished well-being of fetus in females. Physical activity, controlled BMI, improved air quality, healthy diet, training and re-training of healthcare experts, avoiding smoking, effective environmental-friendly policy, awareness and sound education, law enforcement, are some of the strategies that can be adopted (especially by couples) to mitigate their effects on reproductive health. Finally, the treatment measures include IUI, hormone therapy, GIFT, timing of intercourse, IVF-ET, and ICSI through counseling, use of herbs, etc.

References

Adeloye D, Auta A, Fawibe A et al (2019) Current prevalence pattern of tobacco smoking in Nigeria: a systematic review and meta-analysis. BMC Public Health 19:1–14

Alghobary M, Mostafa TJA (2022) Addiction and human male fertility: a systematic review and a critical appraisal. Andrology 10:1073–1095

Baheerati M, Devi GR (2018) Obesity in relation to infertility. Res J Pharm Technol 11:3183–3185

Bala R, Singh V, Rajender S et al (2021) Environment, lifestyle, and female infertility. Reprod Sci 28:617–638

Balwan WK, Kour S (2021) Lifestyle diseases: the link between modern lifestyle and threat to public health. Saudi J Med Pharm Sci 7(4):179–184

Banerjee P, Bhattacharya J (2019) Impact of oxidative stress on infertility, with emphasis on infertility management strategies. Glob J Fertil Res 4:10–18

Biggs SN, Kennedy J, Lewis SL et al (2023) Lifestyle and environmental risk factors for unexplained male infertility: study protocol for Australian male infertility exposure (AMIE), a case control study. Reprod Health 20(1):32

Bloom MS, Fujimoto VY, Steuerwald AJ (2012) Background exposure to toxic metals in women adversely influences pregnancy during *in vitro* fertilization (IVF). Reprod Toxicol 34(3): 471–481

Bundhun PK, Janoo G, Bhurtu A et al (2019) Tobacco smoking and semen quality in infertile males: a systematic review and meta-analysis. BMC Public Health 19:1–11

Carson SA, Kallen AN (2021) Diagnosis and management of infertility: a review. JAMA 326(1): 65–76

Casals-Casas C, Desvergne B (2011) Endocrine disruptors: from endocrine to metabolic disruption. Annu Rev Physiol 73:135–162

Chan TY (2011) Inorganic mercury poisoning associated with skin-lightening cosmetic products. Clin Toxicol (Phila) 49(10):886–891

Clinical Trials Arena (2024) 2024 kicks off with a surge in clinical trials tackling women's infertility. https://www.clinicaltrialsarena.com/analyst-comment/2024-kicks-off-with-a-surge-in-clinical-trials-tackling-womens-infertility/?cf-view. Accessed 12 Apr 2024

Cui X, Jing X, Wu X et al (2016) Potential effect of smoking on semen quality through DNA damage and the down-regulation of CHK1 in sperm. Mol Med Rep 14:753–761

Dag ZÖ, Dilbaz B (2015) Impact of obesity on infertility in women. J Turk Ger Gynecol Assoc 16: 111

Dar BA, Yaqoob A, Mir GJ et al (2023) Effect of environmental contaminants on female reproductive health: a narrative review. BME Horizon 1(3):1–20

De Jager C, Farias P, Barraza-Villarreal A et al (2006) Reduced seminal parameters associated with environmental DDT exposure and p,p'-DDE concentrations in men in Chiapas, Mexico: a cross-sectional study. J Androl 27(1):16–27

Dwivedi S, Kushalan S, Paithankar JG et al (2022) Environmental toxicants, oxidative stress and health adversities: interventions of phytochemicals. J Pharm Pharmacol 74(4):516–536

El-Kharoubi AF (2023) Tubal pathologies and fertility outcomes: a review. Cureus 15(5):e38881

Gaskins AJ, Chavarro JE (2018) Diet and fertility: a review. Am J Obstet Gynecol 218:379–389

Gaskins AJ, Mínguez-Alarcón L, Fong KC et al (2019) Exposure to fine particulate matter and ovarian reserve among women from a fertility clinic. Epidemiology 30(4):486–491

Ghosh A, Tripathy A, Ghosh D (2022) Impact of endocrine disrupting chemicals (EDCs) on reproductive health of human. Proc Zool Soc 75(1):16–30

Hennig B, Ettinger AS, Jandacek RJ et al (2007) Using nutrition for intervention and prevention against environmental chemical toxicity and associated diseases. Environ Health Perspect 115(4):493–495

Jain M, Singh M (2024) Environmental toxins and infertility. In: StatPearls [Internet]. StatPearls Publishing, Treasure Island. Available via https://www.ncbi.nlm.nih.gov/books/NBK576379/. Accessed 10 Apr 2024

Jamal A, Agaku IT, O'Connor E (2014) Current cigarette smoking among adults-United States, 2005-2013. Morb Mortal Wkly Rep 63(47):1108–1112

Jelliffe-Pawlowski LL, Miles SQ, Courtney JG et al (2006) Effect of magnitude and timing of maternal pregnancy blood lead (Pb) levels on birth outcomes. J Perinatol 26(3):154–162

Kandaraki E, Chatzigeorgiou A, Livadas S et al (2011) Endocrine disruptors and polycystic ovary syndrome (PCOS): elevated serum levels of bisphenol A in women with PCOS. J Clin Endocrinol Metab 96(3):E480–E484

Karami-Mohajeri S, Abdollahi M (2011) Toxic influence of organophosphate, carbamate, and organochlorine pesticides on cellular metabolism of lipids, proteins, and carbohydrates: a systematic review. Hum Exp Toxicol 30(9):1119–1140

Kawa IA, Fatima Q, Mir SA et al (2021) Endocrine disrupting chemical bisphenol A and its potential effects on female health. Diabetes Metab Syndr 15(3):803–811

Kesari KK, Agarwal A, Henkel R (2018) Radiations and male fertility. Reprod Biol Endocrinol 16: 1–16

Kissiwaa AVM, Fouché N (2024) Ghanaian women's experiences of unsuccessful *in vitro* fertilisation treatment, unravelling their meanings: a Heideggerian hermeneutic phenomenological study. BMC Pregnancy Childbirth 24(1):212

Kumar S, Kotwal N (2023) Millets (Shrianna) and lifestyle diseases: a healing touch. Med J Armed Forces India 79(3):249–252

Kundu S, Ali B, Dhillon P (2023) Surging trends of infertility and its behavioural determinants in India. PLoS One 18(7):e0289096

Lee J, Kim C (2024) Asian Pacific: in South Korea, world's lowest fertility rate plunges again in 2023. https://www.reuters.com/world/asia-pacific/south-koreas-fertility-rate-dropped-fresh-record-low-2023-2024-02-28/#:~:text=South%20Korea%20has%20previously%20projected, rate%20of%200.55%20last%20year. Accessed 12 Apr 2024

Lee J, Choo CW, Moon KY et al (2024) Risk factors for infertility in Korean women. J Korean Med Sci 39(10):e85

Lei HL, Wei HJ, Ho HY et al (2015) Relationship between risk factors for infertility in women and lead, cadmium, and arsenic blood levels: a cross-sectional study from Taiwan. BMC Public Health 15:1220

Li DK, Zhou Z, Miao M et al (2011) Urine bisphenol-A (BPA) level in relation to semen quality. Fertil Steril 95(2):625–630

Lindsay TJ, Vitrikas KR (2015) Evaluation and treatment of infertility. Am Fam Physician 91(5): 308–314

Macrotrends (2024) India birth rate 1950–2024. www.macrotrends.net. Accessed 12 Apr 2024

Maeda E, Murata K, Kumazawa Y et al (2019) Associations of environmental exposures to methylmercury and selenium with female infertility: a case-control study. Environ Res 168: 357–363

Mansour HAEH (2023) Infertility diagnosis and management. Beni-Suef Univ J Basic Appl Sci 12(1):81

Marci R, Mallozzi M, Di Benedetto L et al (2018) Radiations and female fertility. Reprod Biol Endocrinol 16:1–12

Mayo Clinic (2023) Infertility. https://www.mayoclinic.org/diseases-conditions/infertility/symptoms-causes/syc-20354317#:~:text=These%20include%20 growths%20called%20uterine,an%20egg%20and%20sperm%20unite. Accessed 14 May 2024

Melo AS, Ferriani RA, Navarro PA (2015) Treatment of infertility in women with polycystic ovary syndrome: approach to clinical practice. Clinics 70(11):765–769

Nwikwe DC, Oladipo SO, Ikyo PM et al (2023) Antioxidant and anti-inflammatory potentials of methanol extract of *Syzygium aromaticum* L. flower buds on formalin-induced paw oedema in rats. Biokemistri 35(1):34597–34609

Oulhote Y, Chevrier J, Bouchard MF (2016) Exposure to polybrominated diphenyl ethers (pbdes) and hypothyroidism in Canadian women. J Clin Endocrinol Metab 101(2):590–598

Pasqualotto FF, Umezu FM, Salvador M (2008) Effect of cigarette smoking on antioxidant levels and presence of leukocytospermia in infertile men: a prospective study. Fertil Steril 90(2): 278–283

Pizzorno J (2018) Environmental toxins and infertility. Integr Med (Encinitas, Calif.) 17(2):8–11

Ramgir SS, Abilash V (2019) Impact of smoking and alcohol consumption on oxidative status in male infertility and sperm quality. Indian J Pharm Sci 81:933–945

Rudel RA, Gray JM, Engel CL et al (2011) Food packaging and bisphenol A and bis(2-ethyhexyl) phthalate exposure: findings from a dietary intervention. Environ Health Perspect 119(7): 914–920

Sansone A, Di Dato C, de Angelis C et al (2018) Smoke, alcohol and drug addiction and male fertility. Reprod Biol Endocrinol 16:1–11

Seli DA, Taylor HS (2023) The impact of air pollution and endocrine disruptors on reproduction and assisted reproduction. Curr Opin Gynecol Obstet 35(3):210–215

Sharma A, Shrivastava D (2022) Psychological problems related to infertility. Cureus 14(10): e30320

Sharma R, Biedenharn KR, Fedor JM et al (2013) Lifestyle factors and reproductive health: taking control of your fertility. Reprod Biol Endocrinol 11:1–15

Sharma M, Sharma B, Jamwal RS et al (2023) Role of genetics, environmental, and lifestyle factors responsible for infertility: a review. Fertil Steril Res 10(3):131–138

Silvestris E, Lovero D, Palmirotta R (2019) Nutrition and female fertility: an interdependent correlation. Front Endocrinol 10:346

Soares SR, Simon C, Remohí J et al (2007) Cigarette smoking affects uterine receptiveness. Hum Reprod 22(2):543–547

Sokan-Adeaga AA, Sokan-Adeaga MA, Sokan-Adeaga ED et al (2023) Environmental toxicants and health adversities: a review on interventions of phytochemicals. J Public Health Res 12(2): 1–20

Sustarsic A, Hadzic V, Meulenberg CJW et al (2023) The influence of lifestyle interventions and overweight on infertility: a systematic review, meta-analysis, and meta-regression of randomized controlled trials. Front Med 10:1264947

Ugiomoh IG, Agogbua JU, Okan-Richard A (2023) Plants used in the treatment of female infertility and other related health issues in Agbor, Ika south, Delta state. Nigeria World J Adv Res Rev 18(1):622–634

Wang X, Zhang J, Xu W et al (2016) Low-level environmental arsenic exposure correlates with unexplained male infertility risk. Sci Total Environ 571:307–313

Wieczorek K, Szczęsna D, Radwan M et al (2024) Exposure to air pollution and ovarian reserve parameters. Sci Rep 14(1):461

World Health Organization (2023) 1 in 6 people globally affected by infertility: WHO. www.who.int/news/item/04-04-2023-1-in-6-people-globally-affected-by-infertility. Accessed 04 Apr 2024

Zulfiqar N, Khan AS (2024) Study of female infertility. Biomed J Sci Tech Res 55(1):46656–46658

Contribution of Environmental Toxicants and Lifestyle in the Development of Cardiovascular Diseases

Sarah Patel, Haroon Habib Beigh, and Kudaibergen Osmonaliev

Abstract

Various environmental toxicants and lifestyle factors significantly predispose a host to the development of cardiovascular disease (CVD). Some of the main environmental toxicants implicated in CVD development include, but are not limited to, heavy metals, bisphenol A (BPA), and other chemicals such as pesticides. Unhealthy lifestyle habits such as sedentary behavior, smoking, alcohol consumption, and poor diet, paired with environmental toxicants, increase the risk of CVD. The environment, being a complex system, requires a combination of factors to be accounted for. The social environment greatly influences the natural environment, such as pollution in an industrial environment. This, in turn, may be related to an individual's socio-economic status and its place of residence. These distinct characteristics shape individual choices, thereby affecting the individual's outcome in the progression of CVD. Of utmost importance is an understanding of the outcome of the combined individual and environmental factors and the resultant risk of developing cardiovascular diseases. This, in turn, allows for alternate methods of prevention and treatment planning, in order to combat the ever-increasing burden of patients with cardiovascular diseases.

Keywords

Cardiovascular disease · Environmental toxicants · Lifestyle habits

S. Patel · H. H. Beigh (✉) · K. Osmonaliev (✉)
Faculty of Medicine, AlaToo International University, Bishkek, Kyrgyz Republic
e-mail: haroon.habibbeigh@alatoo.edu.kg; kudaibergen.osmonaliev@alatoo.edu.kg

S. Raisuddin et al. (eds.), *Environmental Toxicants and Lifestyle Diseases*,
https://doi.org/10.1007/978-981-95-0575-3_15

Introduction

The exact mechanism of action of environmental toxicants in the progress of cardiovascular disease (CVD) is not well defined. Hypertension, which acts as a predisposing factor for the cardiovascular diseases, has been linked to the chronic consumption of alcohol and smoking, as well as exposure to heavy metals such as lead, mercury, arsenic, and particulate matter (PM), among others. Environmental toxicants or lifestyle factors individually possess the ability to result in the cardiovascular diseases; if these factors are combined, it may further increase predisposition to the development of cardiovascular diseases (Fig. 1) (Bhatnagar 2017).

Environmental Toxicants and the Development of CVD

These are toxic chemicals that may be natural or artificial in origin and are introduced into the environment either by human activities or through natural events. The primary source of these compounds is the chemical industry (Habeeb et al. 2022). Short-term exposure to air pollution has been linked to myocardial infarction. Workers who are chronically exposed to styrene have a higher incidence rate of developing CVD. Studies show increases in CVD and mortality in workers who are exposed to diesel exhaust particles (DEP), as well as individuals exposed to urban particulate matter (PM). These xenobiotics cause immune cell disruption and inflammation, which can lead to the development of CVD (Habeeb et al. 2022; Bourdrel et al. 2017).

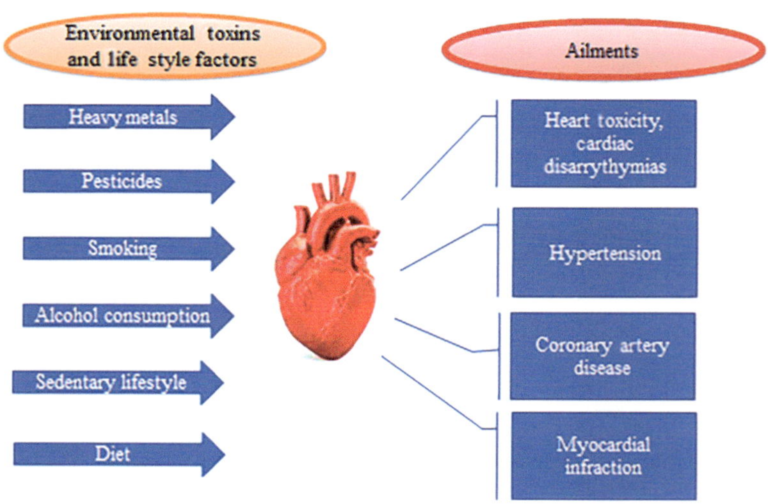

Fig. 1 Representation of contributing factors toward several cardiac diseases

Lead

A wide range of things we come into contact with on a daily basis contain lead (Pb). Pb is one of the constituents of airborne pollutants alongside cadmium (Cd), nickel (Ni), and zinc (Zn), among others. Tobacco smoke, known to contain over 7000 chemicals with almost 700 known to be carcinogenic and toxic, is an abundant reservoir of metal contaminants. There are multiple mechanisms by which these PM increase the risk of developing CVD by producing an environment more prone to the development of atherosclerosis, which collectively act toward creating a favorable environment for the development of myocardial infarction and death (Lamas et al. 2021). Increased use of lead since industrialization has led to a 1000-fold increase in the content of lead in bone. A multitude of sources of lead used in daily life, such as leaded gasoline, lead-based paint, tobacco smoke, electronic cigarettes, makeup, soil, food, and water, have led to this outcome, as well as the industrial use of lead for the production of goods (Pascual 2020).

Contaminants with a size of ≤10 μm and ≤2.5 μm is considered to be highly toxic as it can travel deep into the lungs and exert its effects by activating the local and systemic inflammatory response activating through neural receptors and thereafter moving straight to the bloodstream. Once in the bloodstream, it enters the red blood cells and remains there until the end of the cell's lifespan (Lamas et al. 2021). Lead exerts its force by three methods. First, it may indirectly alter osteocytes by causing changes in the levels of hormones that regulate cell function, mainly 1,25-dihydroxyvitamin D3. Second, it impairs the function of osteocalcin by preventing the binding of osteocalcin with hydroxyapatite, and then, with its affinity for osteocalcin, even at submicromolar concentrations of free lead, it can inactivate osteocalcin. Thirdly, it can impair the cell's ability to produce or secrete components of bone matrix. Lastly, it has the ability to substitute for calcium in the calcium messenger system. Under normal circumstances, osteocalcin functions to maintain bone mineralization by binding with hydroxyapatite, and so, with the inactivation of osteocalcin, new bone formation may consequently be inhibited (Dowd et al. 1994). Lead inhibits glutathione synthesis and function as well as superoxide dismutase, both of which are important for protection against oxidative stress, thereby increasing the risk of atherosclerosis (Lamas et al. 2021). Mechanisms reported from studies suggest PM10 and PM2.5 have the ability to activate platelets and damage endothelium, the resulting effect being myocardial infarction and death. Heavy metal airborne particulates are thought to be more detrimental to the cardiovascular system. PM2.5 has been reported to be a source of inhaled metals present in particulate matter in urban and rural areas. From the study, it was found that of the annual number of deaths related to lead, more than half were cardiovascular-related, with 185,000 deaths being the resultant effect of ischemic heart diseases. Lead exposure, even in small quantities in the environment, acts as an important cardiovascular risk factor (Liu et al. 2017).

With the vast knowledge available regarding particulate matter and its detrimental effects on humans and the environment, as well as the quantities in which it is present in the environment; paired with the fact that those metals and chemicals can accumulate within certain tissues within the human body and remain in biological system for longer duration before they can be excreted; it should therefore be the ultimate goal to decrease the levels of these particulate matter by enforcing new laws into place to stringently control the use of them.

Cadmium

Cadmium (Cd) is a widely used toxic metal with strong evidence backing up the role of cadmium in the development of CVD. Cd is widely used industrially for the production of batteries, electroplating, solar panels, as a stabilizer in plastics, makeup, and in the production of pigments for paints and coatings is toxic to humans and animals, even in small quantities, and more so when inhaled (Zheng et al. 2020). Cadmium binds to proteins, mainly albumin, and by this mechanism is transported via the blood to soft tissues. Upon reaching the soft tissues, it begins to exert its effects by causing damage at a cellular level, promoting cell death, inflammation, and fibrosis. Cadmium also possesses the ability to cause genetic and epigenetic changes. Epidemiological studies found that the average concentration of cadmium in the air of cities had an impact on the rates of hypertension and atherosclerotic heart disease (Cakmak et al. 2014). Recent studies suggest that the effects of Cd on the cardiovascular systems of mice found that exposure to Cadmium led to histological damage of the cardiovascular tissue as well as cytotoxicity. Cd induces DNA damage leading to apoptosis and tissue damage; it can also lead to oxidative stress, endoplasmic reticulum stress, and mitochondrial damage. Cadmium affects cardiomyocytes, vascular endothelial cells, and smooth muscle cells. Histological findings were as follows: cardiac tissues from the groups exposed to Cd100 displayed early necrosis, disarrayed myofibrils, interstitial fibrosis, disarray and destruction of myofibrils, degenerated mitochondria with fragmented cristae, and elevated brain natriuretic peptide (BNP) levels. BNP normally rises when negative structural or functional changes of the left ventricle occur. The ejection fraction of the heart was minimally changed; however, BNP levels were reported to be considerably higher (Chou et al. 2023). The conclusion drawn was that Cd leads to myocardial damage, cardiomyocyte apoptosis, and ventricular dysfunction.

Arsenic

Arsenic (As) is a naturally occurring element located within the earth's crust, highly toxic in its inorganic form. A large number of countries have groundwater sources contaminated by arsenic. Arsenic can be found in many sources such as air, food from fertilizers and groundwater, water, piping, tobacco, etc., (Wang et al. 2009). It possesses the ability to contaminate both food and water sources and, via these mechanisms, finds a portal of entry into the human body, being absorbed either through inhalation or via the gastrointestinal tract (GIT). Arsenic exerts its effects on the cardiovascular system, causing CVDs such as hypertension, atherosclerosis, myocardial infarction (MI), arrhythmias, peripheral arterial disease, and increased carotid intima-media thickness; it does so by causing oxidative stress and epigenetic modifications and has been linked with a risk of CVD (Gump et al. 2023). The majority of people are affected by arsenic through exposure to groundwater. Arsenic is found in high concentrations in rice. People who live close to industrial areas, major roadways, and toxic waste sites are at a far greater risk of being exposed to these hazardous metals. Arsenic can also result in premature death as a result of increased CVD risk.

Arsenic uses the phosphate carrier system to enter the cell, glutathione then aids in the reduction of arsenate to arsenite, following which, it undergoes oxidative methylation with the final products of the methylation process being dimethylarsinic acid (DMAV) and dimethylarsinous acid (DMAIII). The inorganic arsenic and its methylated products are spread throughout the body and later excreted in the urine. As in comparison to Pb and Cd has a much shorter half-life. The susceptibility of an organ determines the level of toxicity that As can exert on it. The degree of toxicity of As can be determined by the urinary or blood serum concentration of As.

Arsenic affects the cardiovascular system causing oxidative stress in endothelium, smooth muscle, and epigenetic aberrations and also modifying the activity of enzymes involved in signalling pathways. The magnitude of the effect that Arsenic exerts is based on a combination of factors such as diet, lifestyle habits, as well as genetic factors. Arsenic causes injury to the arterial wall, resulting in endothelial dysfunction, laying the foundation for chronic vascular abnormalities by promoting the development of atherosclerosis, thus creating a favourable framework for CVD. Increased arterial stiffness and systolic blood pressure with decreased diastolic blood pressure have been associated with exposure to large quantities of arsenic, favoring the development of gangrene and PAD. It is important to note that the higher risk and increased CVD-related mortality in association with As is dose-dependent. As has also been found to cause changes in the shape of platelets and induce cytoskeletal changes, increasing thrombosis. Chronic exposure to Arsenic causes atherosclerosis and decreased arterial compliance, resulting in an increased workload for the heart and favouring the development of left ventricular hypertrophy. This increased arterial stiffness also promotes the development of hypertension. It can, therefore, be concluded that the effects of arsenic on the heart are quite significant.

Pesticides

Pesticides are substances used to control pests, either killing or repelling them. Types of pesticides include fungicides, herbicides, nematicides, insecticides, and rodenticides, among others. Exposure to these chemicals can result in the development of neurological disorders and developmental delays in children, endocrine disruption, an increases the risk of developing cancer, and can affect the reproductive system, leading to infertility (Bhatnagar 2017). The exact mechanism by which pesticides cause CVD is not fully understood. The results of numerous studies show very little evidence of an amplified risk of developing MI and mortality ensuing. However, studies of individual pesticides were shown to be highly associated with non-fatal MI. In cases of severe poisoning, cardiac abnormalities are not uncommon. Organochlorine pesticide exposure, especially in large quantities, possesses the ability to cause CVD. A Case control study conducted on individuals in which 829 people served as control 345 as MI group, 354 as ischemic stroke group, and 829 controls, determined a considerable association with increased odds of composite CVD. The conclusion is that the higher the exposure to organochlorides, the greater the likelihood of experiencing a cardiovascular event (Donat-Vargas et al. 2023). Observations have suggested association between environmental contamination and occupational exposure with CVD. It was put forth that occupational exposure to pesticides was related with a risk of 1.8–3.2 for the development of acute MI. On environmental exposure to tetrachlorodibenzo-p-dioxin, the CVD risk was 1.09–2.78, while the risk for CVD on exposure to organochlorine was 1.19–4.54 (Zago et al. 2020). Pesticides have been connected with a higher mortality, and have the capability to cause cardiovascular disease in individuals exposed to them, particularly in subjects working with little to no precautions. This was seen among factory workers dealing with the production of these pesticides, where they were found to have more cardiovascular diseases as well as endocrine.

A study was conducted on agrochemical particles and CVD among farmworkers. There was a high exposure of the farmworkers to agrochemicals due to the considerably high amounts of agrochemicals that were used for farming. Farm workers who had spent more time engaging in the farming processes of deweeding, harvesting, and planting had reported experiencing the highest number of cardiovascular symptoms, most likely due to exposure to pesticides. Edema, pallor, tachycardia, palpitations, and chest pain were among the most commonly experienced symptoms by those with a higher exposure to the agrochemicals. The results of several studies indicated that there was a positive association between the use of pesticides and the development of CVDs such as MI. Individuals with other cardiovascular confounders were at an even higher risk of developing exacerbations of their health complications on exposure to agrochemicals over time (Sekhotha et al. 2016).

A time-stratified case-crossover study in Hubei province (China) over a period of 5 years from 2013 to 2018 was conducted to investigate 151,608 MI death cases. The results were that people with exposure to particulate matter with aerodynamic diameters of ≤ 2.5 μm and ≤ 10 μm, and NO_2, or 1 day prior was greatly associated with higher odds of death due to myocardial infarction. In older adults ≥ 75 years of age, it was established that there was a greater association between exposure to NO_2 and MI-related mortality. The conclusion drawn was that short-term exposure to air pollution had been linked to an increase in MI mortality. Possible mechanisms of air pollution leading to increases MI mortality include oxidative stress, inflammation as it causes an increase in the circulatory proinflammatory biomarkers, as well as increasing platelet activation resulting in vascular dysfunction, atherosclerosis, and thrombosis eventually leading to MI, as well as abnormal regulation of the cardiac autonomic nervous system. The activation of platelets and coagulation factors as reported from animal studies which occur as a result of the exposure to particulate matter through inhalation has been thought to significantly reduce the total myocardial blood flow while increasing the resistance of coronary vessels. Increased coagulability of blood due to oxidative stress as a result of NO_2 exposure and the resultant nitrogen dioxide ion formation (Wang et al. 2015; Kim et al. 2020; Liu et al. 2021).

Agrochemicals also possess the ability to cause peripheral vascular disease as the fine particles may infiltrate the blood vessel, leading to blockage and obstructing the normal pathway of blood flow, laying the foundation for the progress of atherosclerosis, and is therefore associated with increased morbidity and mortality. Another mechanism by which the fine particulate matter may work is through being absorbed via the skin and moving directly to the bloodstream, where it may result in blockage of the vessel, resulting in the development of CVD. The mechanism by which particulate matter leads to the advancement of CVD is hypothesized to be through inflammation leading to increased plasma fibrinogen, this is done by causing endothelial dysfunction and vasoconstriction which results in increased blood pressure, which can result in changes with coagulability, oxidative stress which may trigger inflammation all of which may lead to, congestive heart failure (CHF), MI, stroke, arrhythmia and sudden death (Sekhotha et al. 2016).

Lifestyle Factors That Play a Role in the Development of CVD

Smoking

Smoking is a compulsive behavior and is associated with a wide range of health issues. One of the most significant effects of smoking is the development of cardiovascular diseases. Despite awareness of the harmful effects of smoking, a great number of individuals continue to smoke. It is recognized as the world's leading cause of avoidable premature death. It acts as one of the main risk factors for the development of cardiovascular disease, with nicotine and carbon monoxide being the two most dangerous substances found in cigarette smoke (Fig. 2). It increases the risk of development of cardiovascular disease through various mechanisms (Parmar et al. 2023). According to research, second-hand smokers are at an almost equally high risk of developing coronary heart disease. It is therefore imperative to enforce laws of no-smoking zones in public areas for the protection of the health of non-smokers (Barnoya and Glantz 2005; Bhatnagar 2006).

Cigarettes are made up of gases, tar-like liquid, as well as other particles, which are known to be poisonous and have injurious impacts on health. Among the contents of cigarettes are nicotine, tar, carbon monoxide, ammonia, acrolein, polyaromatic hydrocarbons (PAHs), cadmium, and so forth. Some of which are known environmental toxicants. Smoking can affect various systems of the body such as the respiratory system, resulting in respiratory diseases such as COPD, asthma, or even lung cancer, the cardiovascular system where it can cause coronary artery disease, stroke, or an aneurysm, and the reproductive system leading to

Fig. 2 Illustration of smoking and its contribution to the advancement of cardiac disease

difficulties falling pregnant, maintaining the pregnancy or even infertility in females, while in males it puts them at a two times higher risk of developing impotence. The mechanisms by which smoking increases the risk of development of coronary heart disease are as follows:

1. Increasing platelet activation, increasing fibrinogen levels, and increasing thromboxane, all of which act to increase the risk of thrombus formation.
2. Endothelial dysfunction by impairing the endothelium-dependent vasodilation, direct damage to the endothelium, and oxidant damage to the endothelium.
3. Increasing arterial stiffness.
4. Accelerating the development of atherosclerosis by decreasing high-density lipoprotein (HDL) and increasing low-density lipoprotein (LDL) levels.
5. Inflammation and infection in individuals exposed to smoke due to increased levels of inflammatory markers, and in the presence of chronic infection paired with exposure to tobacco smoke, people are at a higher risk of developing atherosclerosis.
6. Increase in the intima media thickness, which can accelerate the rate of development of atherosclerosis, as well as playing a role in the instability of atherosclerotic plaque, which can result in occlusion of the vessel, resulting in an acute vascular event.
7. Oxidative stress.
8. Mitochondrial damage.
9. Reduced heart rate variability.
10. Insulin resistance.

These substances, when burnt, are released into the air and affect not just active, but also passive smokers. Some of the main findings of studies are, that despite the dose of tobacco smoke delivered to active smokers being 100 times more than that delivered to passive smokers, the relative risk of developing coronary heart disease among second-hand smokers is 1.38, while that of smokers is 1.78; putting first and second-hand smokers at an almost equally high risk of developing CVD. Many vital responses in the cardiovascular system are quite sensitive to the toxins produced by secondhand smoking. The chemicals contained in cigarettes when burned can be released into the air and, once inhaled or ingested, begin to exert their effect at the cellular level by causing alterations to DNA, leading to genetic mutations that may ensue in the form of cancer (Barnoya and Glantz 2005).

Smoking has been found not only to increase the risk of developing CVD, but also to affect recovery post a major adverse cardiovascular event (MACE) as seen in the study conducted where out of a group of 4540 patients who had experienced an arterial ischemic stroke (AIS), 22% comprised of smokers who were younger and composed mainly of males who had several comorbidities which fall under the umbrella of CVD, along with the important finding of greater baseline neurological impairment in comparison to individuals who did not smoke. Individuals who smoke were also found to be at a 2 times increased risk of developing AIS. The conclusion was that smoking predicted a poor prognosis in relation to functional recovery in patients who had thrombolysed AIS, as well as an increased likelihood of recurrent vascular events. These patients were likely to have a final diagnosis of greater magnitude, such as large vessel occlusion or cardioembolism; this tribute may be paid to an acceleration of atherosclerosis and the increased risk of thrombus formation due to smoking. Possible mechanisms explaining these poor outcomes in patients who had thrombolysed AIS and smoke are that it may cause abnormal cardiopulmonary function as findings from a Taiwanese study showed that smokers had a twofold greater risk of mortality in smokers without a previous history of a stroke, as well as prolonged disability after stroke accounting for a period of approximately 11 years post the event (Arora et al. 2022; Sun et al. 2021).

Alcohol Consumption

Alcohol is one of the main risk factors in the development of hypertension, which is one of the major risk factors in the development of cardiovascular disease. Alcohol has both beneficial and harmful effects on the cardiovascular system, depending on the quantity and frequency of alcohol consumption. Low to moderate daily alcohol consumption is associated with a reduced risk of cardiovascular diseases and mortality, whereas large amounts of alcohol consumption along with a pattern of binge drinking have been associated with an increased risk (Zakhari 1997). Short-term alcohol consumption has minimal effect on blood pressure; however, binge drinking was found to have a substantial impact on blood pressure, causing a rise in systolic blood pressure ranging between 4 and 7 mmHg and diastolic blood pressure ranging between 4 and 6 mmHg. Lower amounts of alcohol have been found to exert a positive effect on the endothelium, stimulating the release of nitric oxide whereas a higher consumption of alcohol over time may alter the reactiveness of the arteriolar system, which may lead to a higher risk of developing hypertension. Alcohol consumption causes vascular wall oxidative stress, resulting in hypertension. It also possesses the ability to increase the levels of co-enzymes or reducing equivalents such as nicotinamide adenine dinucleotide phosphate (NADPH), leading to an in reactive oxygen species, and a decrease in endogenous nitric oxide synthase (Jerumbo 2024).

According to the latest WHO report, alcohol has been linked to 2.6 million deaths throughout 2019 alone. A large number of people suffer from alcohol disorders with more than half being accounted for by individuals suffering from alcohol dependence. Alcohol harms health not just physically, but also mentally, resulting in a multitude of avoidable deaths (Arora et al. 2022). Alcohol has been found to increase the risk factors for the development of CVD. It increases the risk of developing hypertension, stroke, and conduction abnormalities such as atrial fibrillation and flutter, as well as inducing the development of cardiomyopathies (Piano 2017).

Alcohol when used in large quantities over long periods exerts its effect by weakening the heart muscle predisposing the development of cardiomyopathies, more specifically dilated cardiomyopathy, which can lay the foundation for the development of congestive heart failure. According to studies conducted on animals, alcohol was found to decrease the contractile function of the heart by changing the manner in which the sarcoplasmic reticulum responds to calcium, which is responsible for the normal contractile function of the heart. It acts on the cellular level decreasing the production of cardiac proteins. Alcohol can also act to promote apoptosis, and can result in abnormal cardiac rhythms when consumed in large quantities over a short span of time, which may lead to sudden cardiac death. Consumed in large quantities it was also associated with a twofold increase in the development of atrial fibrillation, flutter, and other types of rhythm abnormalities. Possible mechanisms by which alcohol causes arrhythmias is through thickening and fibrosing the heart muscles, as this could block the normal electrical impulses of the heart leading to arrhythmias. High alcohol consumption has also been linked to the development of ischemic strokes; females have also been found to be at a higher risk of developing a hemorrhagic stroke despite minimal alcohol consumption (Genchi et al. 2017).

Sedentary Lifestyle

A sedentary lifestyle is one of the leading modifiable risk factors in the development of cardiovascular diseases. A large number of individuals worldwide continue living a sedentary lifestyle, engaging in little to no physical activity. The effects of sedentary behavior on health are quite detrimental, as it greatly increases the risk of CVD. However, certain cases may differ, as seen in individuals who actively participate in physical activity of moderate intensity daily. The high mortality rates of sedentary behavior are of little importance in individuals engaging in moderate to high-intensity exercise for 60–75 min per day, as it was found that the physical activities of this intensity were related to low CVD mortality regardless of the time-spent sitting. There was also a positive correlation between the number of hours spent watching television and CVD mortality, this does not apply to individuals who regularly engage in moderate to high-intensity physical exercise, regardless of the number of hours spent sitting and watching television. Physical inactivity is linked not only to CVD but also to other non-communicable diseases, and results in a large number of preventable premature deaths.

Physical inactivity has also been linked to metabolic disorders, greatly increasing the risk of CVD (Lin et al. 2021; Wang et al. 2015). Physical inactivity leads to increased oxidative stress, resulting in endothelial dysfunction, as well as reduced vascular compliance, both of which are detrimental to the cardiovascular system. This can in turn increase the workload of the heart leading to left ventricular remodeling and possibly ending in heart failure. Exercise reduces the amount of circulating inflammatory markers, helps to maintain good vascular compliance, and increases the number of antioxidant proteins, thereby decreasing inflammation, reducing the workload on the heart, and protecting the heart from undergoing remodeling and becoming dysfunctional (Lamas et al. 2023).

Diet

Diet is an important and modifiable risk factor in the development of CVD. An unhealthy diet can predispose to the development of obesity, diabetes, and hypertension, all of which may be preventable by eating a balanced diet that is high in fiber, and legumes contains lots of vegetables, and fruits, and is low in saturated fats, low in salt intake, monounsaturated, and polyunsaturated fatty acids and so forth. Diets that are high in saturated fats and trans-fatty acids can result in atherosclerosis, increasing the risk of MI. Diets high in salt lead to increased blood pressure promoting atherosclerosis and increasing the risk of CVD. Diets high in processed food and sugar increase the risk of developing diabetes and promote the development of atherosclerotic plaques, which will in turn, narrow the lumen of blood vessels leading to ischemia, it also favors calcification of vessels and media thickening of the artery (Corliss 2019). Healthy and well-balanced diet rich in antioxidants is essential for good cardiovascular and overall bodily well-being. Diets that are high in processed foods, fatty foods, and refined sugars result in increased oxidative stress and increased circulation of pro-inflammatory cytokines. This inflammatory process affects the sub-endothelial wall damaging it and creating a favorable environment for the development of atherosclerotic plaques thus playing an essential role in the development of coronary artery disease.

A healthy and balanced diet can help to decrease the levels of inflammation by eating food rich in antioxidants while also maintaining a healthy body weight; it must also be remembered that a healthy diet alone is not sufficient for good health and should be paired with lifestyle modifications such as exercise for a good outcome with overall health in the long run (Pounds et al. 1991).

Cardiovascular Disease

The cardiovascular system is composed of the heart and blood vessels. It is one of the most vital systems in the human body. CVD holds the title of the "leading cause of death globally" (Wahab et al. 2016), accounting for the loss of approximately 17.9 million lives each year according to WHO. There is an ever-increasing burden of CVD worldwide affecting not just individuals, but also the healthcare system. The increasing burden of CVD can be attributed to lifestyle factors, genetics, and environmental exposure to toxic substances, to name a few. Hypertension, hyperlipidemia, diabetes, and obesity are some of the major contributing factors to the development of CVD. The majority of these non-communicable diseases can be prevented or controlled simply through lifestyle modifications, and in certain cases, pharmacologic treatment alongside lifestyle interventions (Berg et al. 2019; States et al. 2008). The umbrella term CVD includes heart failure, coronary artery disease, cardiomyopathies, deep vein thrombosis, pericarditis, arrhythmia, peripheral vascular disease, rheumatic heart disease, stroke, heart valve disease, aortic aneurysm, and cerebrovascular disease. CVD is one of the biggest contributors to premature death among non-communicable diseases (Moon et al. 2012; Al-Forkan et al. 2021).

Conclusion

There is a link between cardiovascular disease and lifestyle as well as environmental toxicants, where the main findings have been that exposure to environmental toxicants and unhealthy lifestyle habits greatly increase the risk of developing CVD by contributing toward the risk factors such as hypertension and atherosclerosis that predispose to the development CVD. It is therefore important to combat these issues either by providing incentives that promote better lifestyle habits with the example of using financial incentives to promote physical activity, implementing new laws to stringently control alcohol consumption within countries, banning irresponsible disposal of toxic substances near residential areas or areas of the environment that should be protected such as freshwater sources or forests, lowering the limits of industrial air pollutants released, etc., which would collectively act toward promoting the well-being of individuals, as well as decreasing the burden on the health care systems.

References

Al-Forkan M, Wali FB, Khaleda L et al (2021) Association of arsenic-induced cardiovascular disease susceptibility with genetic polymorphisms. Sci Rep 11(1):6263

Arora M, ElSayed A, Beger B et al (2022) The impact of alcohol consumption on cardiovascular health: myths and measures. Glob Heart 17(1):45

Barnoya J, Glantz SA (2005) Cardiovascular effects of second hand smoke. Circulation 111(20): 2684–2698

Berg ZK, Rodriguez B, Davis J et al (2019) Association between occupational exposure to pesticides and cardiovascular disease incidence: the Kuakini Honolulu heart program. J Am Heart Assoc 8(19):e012569

Bhatnagar A (2006) Environmental cardiology. Circ Res 99(7):692–705

Bhatnagar A (2017) Environmental determinants of cardiovascular disease. Circ Res 121(2): 162–180

Bourdrel T, Bind MA, Béjot Y et al (2017) Cardiovascular effects of air pollution. Arch Cardiovasc Dis 110(11):634–642

Cakmak S, Dales R, Kauri L et al (2014) Metal composition of fine particulate air pollution and acute changes in cardiorespiratory physiology. Environ Pollut 189:208–214

Chou SH, Lin HC, Chen SW et al (2023) Cadmium exposure induces histological damage and cytotoxicity in the cardiovascular system of mice. Food Chem Toxicol 175:113740

Corliss J (2019) Lead exposure and heart disease. Available via https://www.health.harvard.edu/blog/lead-exposure-and-heart-disease-2019040516296. Accessed 12 Apr 2024

Donat-Vargas C, Schillemans T, Kiviranta H et al (2023) Blood levels of organochlorine contaminants mixtures and cardiovascular disease. JAMA Netw Open 6(9):e2333347

Dowd T, Rosen J, Gundberg C et al (1994) The displacement of calcium from osteocalcin at submicromolar concentrations of free lead. Biochim Biophys Acta 1226(2):131–137

Genchi G, Sinicropi M, Carocci A et al (2017) Mercury exposure and heart diseases. Int J Environ Res Public Health 14(1):74

Gump BB, Heffernan K, Brann LS et al (2023) Exposure to arsenic and subclinical cardiovascular disease in 9 to 11 year old children, Syracuse, New York. JAMA Netw Open 6(6):e2321379

Habeeb E, Aldosari S, Saghir SA et al (2022) Role of environmental toxicants in the development of hypertensive and cardiovascular diseases. Toxicol Rep 9:521–533

Jerumbo A (2024) Effects of alcohol on the cardiovascular system. Available via https://americanaddictioncenters.org/alcohol/risks-effects-dangers/cardiovascular. Accessed on 12 Mar 2024

Kim JB, Prunicki M, Haddad F et al (2020) Cumulative lifetime burden of cardiovascular disease from early exposure to air pollution. J Am Heart Assoc 9(6):e014944

Lamas GA, Ujueta F, Navas-Acien A (2021) Lead and cadmium as cardiovascular risk factors: the burden of proof has been met. J Am Heart Assoc 10(10):e018692

Lamas GA, Bhatnagar A, Jones MR et al (2023) Contaminant metals as cardiovascular risk factors: a scientific statement from the American Heart Association. J Am Heart Assoc 12(13):e029852

Lin HC, Hao WM, Chu PH (2021) Cadmium and cardiovascular disease: an overview of pathophysiology, epidemiology, therapy, and predictive value. Rev Port Cardiol (Engl Ed) 40(8): 611–617

Liu K, Shang Q, Wan C et al (2017) Characteristics and sources of heavy metals in PM2.5 during a typical haze episode in rural and urban areas in Taiyuan, China. Atmosfera 9(1):2

Liu Y, Pan J, Fan C et al (2021) Short-term exposure to ambient air pollution and mortality from myocardial infarction. J Am Coll Cardiol 77(3):271–281

Moon K, Guallar E, Navas-Acien A (2012) Arsenic exposure and cardiovascular disease: an updated systematic review. Curr Atheroscler Rep 14(6):542–555

Parmar MP, Kaur M, Bhavanam S et al (2023) A systematic review of the effects of smoking on the cardiovascular system and general health. Curēus 15(4):e38073

Pascual F (2020) Polychlorinated biphenyls as a cardiovascular health risk: a new threat from an old enemy? Environ Health Perspect 128(11):114003

Piano MR (2017) Alcohol's effects on the cardiovascular system. Alcohol Res 38(2):219–241

Pounds JG, Long GJ, Rosen JF (1991) Cellular and molecular toxicity of lead in bone. Environ Health Perspect 91:17–32

Sekhotha M, Monyeki K, Sibuyi M (2016) Exposure to agrochemicals and cardiovascular disease: a review. Int J Environ Res Public Health 13(2):229

States JC, Srivastava S, Chen Y et al (2008) Arsenic and cardiovascular disease. Toxicol Sci 107(2): 312–323

Sun L, Song L, Yang J et al (2021) Smoking influences outcome in patients who had thrombolysed ischaemic stroke: the ENCHANTED study. Stroke Vasc Neurol 6(3):e000493
Wahab A, Hod R, Ismail N et al (2016) The effect of pesticide exposure on cardiovascular system: a systematic review. Int J Commun Med Public Health 3:1–10
Wang C, Chen C, Hsiao CK et al (2009) Increased risk of QT prolongation associated with atherosclerotic diseases in arseniasis-endemic area in southwestern coast of Taiwan. Toxicol Appl Pharmacol 239(3):320–324
Wang C, Tu Y, Yu Z et al (2015) PM2.5 and cardiovascular diseases in the elderly: an overview. Int J Environ Res Public Health 12(7):8187–8197
Zago AM, Faria NMX, Fávero JL et al (2020) Pesticide exposure and risk of cardiovascular disease: a systematic review. Glob Public Health 17(12):3944–3966
Zakhari S (1997) Alcohol and the cardiovascular system: molecular mechanisms for beneficial and harmful action. Alcohol Health Res World 21(1):21–29
Zheng F, Gonçalves FM, Abiko Y et al (2020) Redox toxicology of environmental chemicals causing oxidative stress. Redox Biol 34:101475

Association of Environmental Toxicants and Lifestyle in the Advancement of Infertility

Haroon Habib Beigh

Abstract

Environmental toxicants and certain lifestyle factors have been reported to contribute to the decline in human fertility. It has been suggested that environmental toxicants mimic hormones, thus disrupting the endocrine system. Some known endocrine disruptors are pesticides, herbicides, industrial pollutants, heavy metals, and plastics having BPA (bisphenol A). On the other hand, lifestyle factors such as smoking, excessive alcohol consumption, diet, and stress; overweight are also contributing factors toward the advancement of infertility in the human population. Lifestyle factors and environmental toxicants contribute to DNA damage in sperms thus causing lower sperm count and reduced sperm motility in males. In females, environmental toxicants and lifestyle factors both pose a negative effect on reproductive capability, affect ovulation, and have been stated to cause miscarriages. Teratogenicity and congenital anomalies have also been associated with exposure to several environmental toxicants. However, the exposure from these components can be minimized by avoiding contaminated products and acquiring a healthy lifestyle along with medical advice. Environmental toxicants and routine environmental and dietary factors can considerably affect fertility and wakefulness. Appropriate measures can help alleviate these effects and improve homeostasis for longevity and positive health outcomes.

Keywords

Infertility · Environmental toxicants · Lifestyle · Teratogenicity · Endocrine description

H. H. Beigh (✉)
Faculty of Medicine, AlaToo International University, Bishkek, Kyrgyz Republic
e-mail: haroon.habibbeigh@alatoo.edu.kg

S. Raisuddin et al. (eds.), *Environmental Toxicants and Lifestyle Diseases*,
https://doi.org/10.1007/978-981-95-0575-3_16

Introduction

A substantial basis of progressive loss of fertility is due to cumulative body load of environmental toxins and life style factors in both men and women. While there is an appreciable increase in the number of couples worldwide but the rate of conception has decreased (Pizzorno 2018). There are several variable causes affecting male and female reproductive capability comprising mainly the use of pesticides, air pollution, exposure to excessive heat, harmful chemicals, and life style (Agarwal et al. 2021; Checa Vizcaíno et al. 2016). In males, sperm are regarded as basics of fertility while as in females egg quality and uterus health is accounted, when males come in contact with environmental toxins sperm quality, sperm shape, sperm morphology, and male hormones can be affected, while as in case of females tubal factors, endometriosis, ovulatory disorders, hormonal deregulation can be observed. Life style on the other hand can affect general health as well as reproductive capability of an individual (Rehman et al. 2019). Figure 1 illustrates some factors associated with infertility.

Ecological Toxicants and Their Contribution to Infertility

There are a number of factors that can contribute to male and female infertility, including the use of harmful chemicals, air pollution, exposure to pesticides, and extreme heat (Fig. 2).

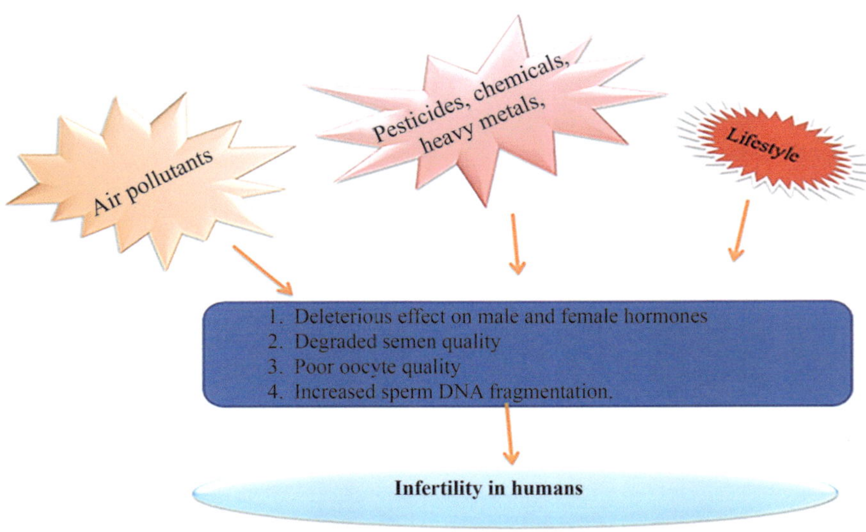

Fig. 1 Representation of environmental and lifestyle factors on human reproductive capability

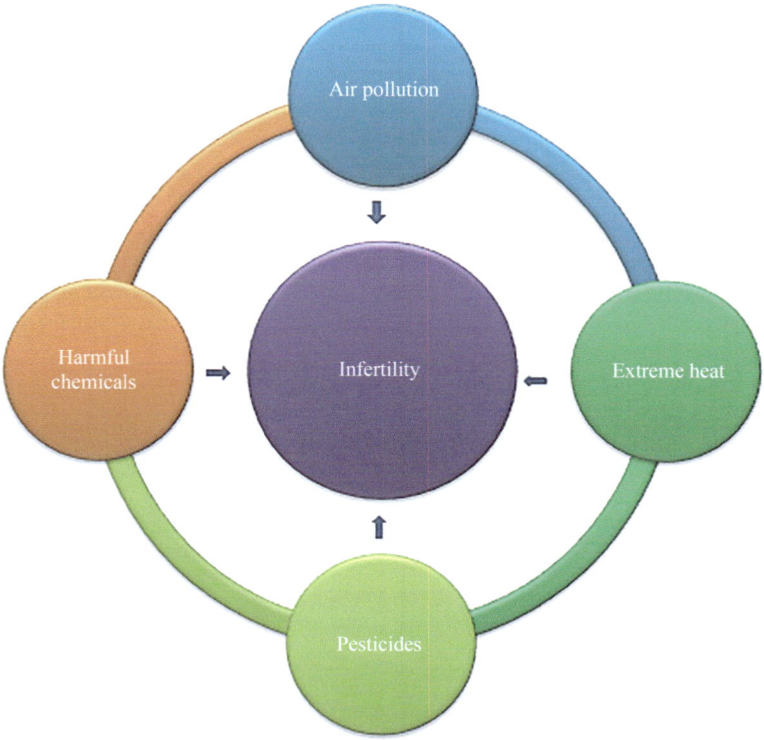

Fig. 2 Representation of common environmental toxicants leading to infertility

Harmful Chemicals and Its Association with Infertility

In daily life, humans are exposed to a number of chemicals. Several chemicals pose deleterious effect on human health. It has been found that chemicals can pose a direct threat to reproductive capability of humans (Deng et al. 2016). Exposure to chemicals such as dioxins, phthalates, plastic additives, heavy metals, especially during pregnancy can lead to endocrine disruption; it has been also suggested that male babies born after the exposure to EDC (endocrine disruptor chemicals) have shown the symptoms of low sperm count and deformed sperm morphology (Istvan et al. 2021).

Dioxins

These are the by-products of several industrial processes mainly produced during synthesis of herbicides, pesticides, burning of plastics, smelting and chlorination (Foster et al. 2010).

They are under great consideration because they are of their higher toxicity and ability to get deposited in fats; by this nature, they can remain in the biological system for a longer period. They have been linked to significant issues with cancer, development, and reproduction. In order to mediate their hazardous effects, dioxins, which are endocrine disruptors, bind to the aryl hydrocarbon receptor (AHR)/aryl hydrocarbon receptor nuclear translocator (ARNT) receptor complex that is present over human testicular cells. It remains unclear how precisely it influences human reproductive activities (Barrett 2011). According to a recent study, male mice exposed to dioxin showed a considerable decrease in sperm motility and count. In addition, testicular histology revealed decreased epithelial thickness and necrotic degeneration in dioxin-exposed animals relative to controls. Additionally, proving that male infertility and poor-quality semen are the results of dioxin exposure's significant effects on sperm functions in males (Mohammadi et al. 2019).

Plastic Additives (Bisphenol A)

Although plastic is now an essential part of our daily lives, its non-biodegradable nature has led to significant global concern. A significant component of plastic called bisphenol A (BPA) is released into the environment during the manufacture, use, or disposal of plastics as well as during the breakdown of wastes connected to the manufacturing of plastics in industry. In a recent study conducted in the United States, it was suggested that bisphenol F, bisphenol S, and bisphenol A are nearly universal alternatives for bisphenols. Due to its widespread distribution, extended environmental durability, and endocrine disrupting properties, it is currently regarded as harmful to human health. Numerous health issues, such as cancer, metabolic disorders, heart disease, and infertility, have been connected to it (Fig. 3). The hypothalamic–pituitary–gonadal axis can be disrupted by BPA due to its estrogenic, antiandrogen, and antithyroid properties (Cariati et al. 2019). Prolonged exposure to BPA causes damage to sperm DNA, mitochondrial malfunction and degeneration, a reduction in sperm motility and count, and a higher chance of sperm aneuploidies. Several rodent studies have demonstrated that male mouse

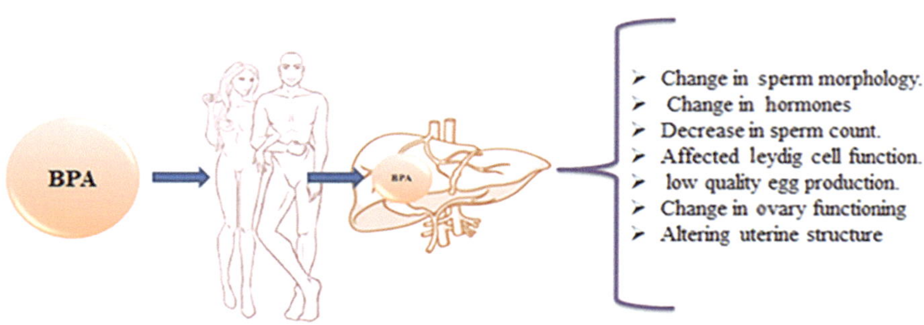

Fig. 3 Illustration of the exposure to BPA and possible effects on human reproductive capability

exposure to BPA causes a marked reduction in sperm motility, count, and normal morphology as well as increased sperm DNA damage and detrimental effects on the spermatogenesis process, ultimately leading to male infertility. According to a different recent study, sperm motility is hampered by high BPA exposure due to decreased sperm adenosine triphosphate levels and an early acrosome reaction, which impairs fertilization and causes issues with embryonic development. Additional research has also discovered a strong correlation between elevated BPA exposure and low semen quality, as well as other factors like motility and sperm quality (Caporossi et al. 2020).

Research is still ongoing to determine the precise mechanism by which BPA affects the quality of human sperm. However, it has been discovered that BPA is an endocrine disruptor that inhibits anti-apoptotic pathways like Bcl-2 and activates pro-apoptotic signaling pathways like mitogen-activated protein kinase (MAPK), Fas/FasL, Caspase 3 and 9, Bax, and reactive oxygen species-mediated damage. These actions result in decreased proliferation, increased damage caused by reactive oxygen species, and enhanced apoptosis of male gametes. It functions as an antagonist of the androgen receptor (AR), which reduces AR translocation and enhances AR transcriptional corepressors, which in turn suppresses Sertoli cell growth. Synthesis of testosterone is inhibited by BPA, according to a rodent study. Reduced expression of steroidogenic enzymes, cholesterol carrier protein in Leydig cells, and plasma levels of luteinizing hormone (LH) are caused by the drop in testosterone levels in plasma. Decreased Leydig cell counts in the testis are another effect of BPA. All of which result in dysfunctional male reproduction. Therefore, long-term exposure to high BPA concentrations may have an impact on male fertility (Bretveld et al. 2007). It has been found that in case of females BPA exposure leads to unsuccessful in vitro fertilization (IVF) also has an impact on the development of the ovary, the quality of the gametes (Chianese et al. 2018).

Phthalates

A class of synthetic, multipurpose chemicals, known as phthalic acid diesters or (phthalates), is utilized in a vast range of consumer and industrial goods. Examples of these products include plasticizers in polyvinyl chloride plastics, excipients in certain pharmaceuticals, and smell retainers in certain personal care items. Human exposure occurs everywhere in the lifespan, including in pregnancy due to the mother's exposures. For the general public, additional routes of exposure include ingestion, inhalation, and skin contact (Radke et al. 2018). Phthalate diesters are quickly converted to monoester metabolites following exposure, which are then eliminated in the urine. There are several different types of compounds in the phthalate group, each with unique structures, characteristics, and applications. Known are as following: butyl benzyl phthalate (BBP), dibutyl phthalate (DBP), diisobutyl phthalate (DIBP), di(2-ethylhexyl) phthalate (DEHP), and diethyl phthalate (DEP). Human biomonitoring data indicate that certain phthalates (e.g., DBP and DiBP) are moderately linked and share structural similarities with one another. Other phthalates, such as long, short chains, and DEP and DEHP, differ significantly from one another; relationships between these phthalates are generally weak (Howdeshell et al. 2008a, b). Phthalates have demonstrated well-documented anti-androgenic action, leading to decreased levels of circulating testosterone and abnormalities of the male reproductive system as per several animal studies (Howdeshell et al. 2008a, b).

Heavy Metals

Heavy metals are resistant to natural breakdown and can accumulate in soil, water, and the food chain. They pose a major threat to human health. Not every metal is inherently harmful. Certain elements, like zinc, copper, chromium, and manganese, are hazardous at greater quantities but necessary at extremely low ones; other elements, such as lead, mercury, and cadmium, have no metabolic functions and are toxic at all levels. The main cause of detectable human exposure to heavy metals is the ingestion of tainted drinking water, which frequently includes a combination of arsenic, cadmium, nickel, mercury, chromium, zinc, and lead (Chen et al. 2020).

Lead (Pb)

The toxicity of lead and its consequences on the ability to affect reproduction have been known since ancient times. Nonetheless, the first information indicating harmful effects on pregnancy was released in France in 1860. Blood lead levels are categorized into four groups: mild (5–9 μg/dL), moderate (10–14 μg/dL), high (15–20 μg/dL), and normal (<4 μg/dL). Severe amounts are defined as those that surpass these thresholds. Pb-dependent decreased the production of circulating gonadotropins and E2, resulting in irregular menstruation, spontaneous abortion, and fetal anomalies, which was caused by continuous lifetime exposure to Pb or in an adolescent monkey with approximately 35 μg/dL in the blood; Pb-treated mice also showed gametotoxicity and interference with the normally occurring steroid hormones during implantation. On the other hand, it is unclear how lead poisoning affects pregnancy. Some studies have found no significant correlation between pregnant women who were exposed to lead intoxication and those who spontaneously aborted, despite reports that females who experienced lead intoxication as children had a significantly higher rate of spontaneous abortion with a striking dose–response relation with blood lead levels. In contrast, women who had lead poisoning as children had a noticeably greater percentage of spontaneous miscarriages or stillbirths. In these situations, however, the individuals had no history of lead exposure throughout their childhood (Gulson et al. 2003). Lead can pass from mother to child during nursing and from the fetus to the mother during pregnancy. Research has shown that fetus can collect a significant amount of lead in their skeletons (RÍsovÁ 2019).

The primary cause of fetal poisoning during pregnancy can be attributed to the elevated demand for calcium during pregnancy, which can result in bone demineralization and the co-mobilization of Calcium and stored lead. The fact that women who smoked during their pregnancies had less fertility after giving birth indicates how important the time for fetal ovarian development is to future fertility. Therefore, it is not impossible to rule out the likelihood of reproductive abnormalities following childhood lead exposure, even though a direct link between lead exposure and miscarriage is not as clear in humans (Weinberg et al. 1989).

Cadmium (Cd)

Years of exposure cause the body to progressively absorb more cadmium. It is quantifiable in blood, urine, hair, nails, and saliva; for humans, the acceptable standard limit in saliva is 0.55 μg/L (Akesson et al. 2002). Batteries, particularly rechargeable nickel–cadmium batteries, and high-fiber meals (vegetables, cereals, potatoes, and spinach) are two sources of cadmium. Nonetheless, breathing in dust, fumes, and passive or active cigarette smoking are the main ways that people are exposed to lead. In fact, the median Cadmium contents in the smoking versus non-smoking females are higher (1.3 versus 0.32 μg Cd/L). While cadmium retention can impact people of any gender, it typically affects women more than males. This difference results from the increased gastrointestinal absorption of this metal iron reserves are being depleted, which is a common occurrence in women of reproductive age, particularly during pregnancy (Berglund et al. 1994).

It has been demonstrated that in vitro, Cd at as low as 5 μM can obstruct the biosynthesis of hormones by reducing the expression of key steroidogenesis-related enzymes in placental cells, such as 3β-hydroxy steroid dehydrogenase (3β-HSD) and P450 cholesterol side-chain cleavage (P450scc). Furthermore, it has been demonstrated that Cd binds to the hormone-binding domain of the estrogen receptor to activate it. These findings provide an explanation for the ensuing hormonal imbalances, early births, low birth weights, monthly irregularities, loss of pregnancy, and delayed puberty/menarche. Potential ways by which cadmium influences steroidogenesis include acting as an endocrine disruptor that can mimic or inhibit natural E2 activity, as well as interfering with DNA-binding Zn-finger motifs by exchanging Cadmium for Zinc (Henson and Chedrese 2004).

Cobalt, Nickel, and Zinc

Nutrient deficits in the mother's diet might affect the growth of her offspring, and elevated zinc concentrations can potentially have teratogenic and deadly effects due to faulty embryogenesis. Zinc deficiency is characterized by blood zinc levels below 70 μg/dL, which is between 70 and 120 μg/dL. Maternal zinc shortage may have detrimental effects on kids, which may manifest acutely during pregnancy or over the course of their lives by making them more susceptible to diseases as adults. A possible acute impact of zinc deficiency is increased congenital abnormalities, low birthweights, and premature deliveries, all of which have been linked to suboptimal zinc consumption in humans. Additionally, meiotic arrest and cumulus expansion are two crucial zinc-dependent ovarian processes that a zinc shortage can adversely impact oocyte maturation, cumulus expansion, and ovulation (Beaver et al. 2017). Nickel has been shown to be carcinogenic and embryo toxic; its effects on the female reproductive systems are still unknown, and there is a dearth of data from human or experimental investigations. Furthermore, it has been noted that the injection of 40 mg/kg can interfere with menstrual periods, reduce embryo implantation, and increase embryo resorption (Forgacs et al. 2012). One of the oligo elements that make up vitamin B12 is cobalt. In humans, it is poisonous in large quantities yet necessary at low levels for the production of new red blood cells. The range of daily consumption is 1.7–100 μg. Cobalt tends to collect mostly in the liver and kidneys, with an approximate 1.5 mg total level in the body. Because cobalt stabilizes hypoxia-inducible factor, which activates many responsive target genes involved in angiogenesis and the regulation of apoptosis/cell proliferation, cobalt can imitate hypoxic circumstances and accelerate the growth of cancer. Cobalt affects humans in a number of ways, including irregular menstruation, changed sexual behavior, infertility, delayed puberty, longer pregnancy times, issues with nursing, and altered menopause (Forgacs et al. 2012). Overdose of heavy metals can lead to the formation of Reactive oxygen species and this can affect humans at cellular level (Fig. 4).

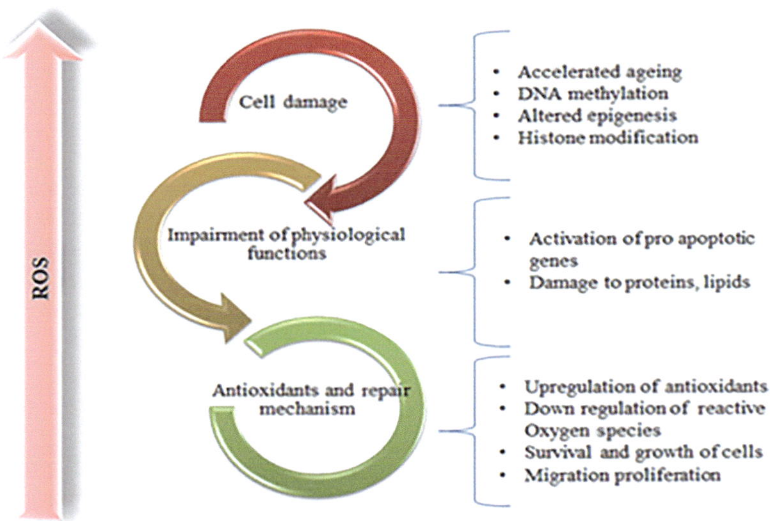

Fig. 4 Representation of different quantities of reactive oxygen species affecting cells. Low levels promote cell survival and recovery, whereas large levels change epigenetic systems and cause apoptosis or degenerative processes

Air Pollution

People in industrialized nations breathe and consume a variety of airborne particles and chemicals every day; many of these can also contaminate soil and water, which can lead to their entry into the food chain. Particulate matter (PM) with diameters of 10, 2.5–10, and 2.5 μm, ground-level ozone, benzo(a)pyrene, sulphur dioxide, carbon monoxide, nitrogen dioxide, polychlorinated biphenyls, organic compounds (such as dioxins and organic solvents), and heavy metals are among the many pollutants produced by transportation and industry. How to regulate exposure of toxins during post and pre conception that can result in anomalies in oocytes, embryos, and/or fetuses and can impair a baby's safe birth as well as general health and mental activity is a crucial question that remains unsolved. The primary air pollutants seem to affect gametogenesis in both humans and animals, which lowers reproductive success. Their ability to lead to Endocrine disruption, also to raise Reactive oxygen species and inflammation, and to activate particular sites that can trigger incorrect MAPK signaling are the main ways in which they impact the ovary (Iorio et al. 2014). According to a number of reports, women in vastly commercial areas have lower fertility rates, a higher rate of implantation failure, and fewer fertilizable oocytes overall than women in less industrialized areas. It has been suggested that particulate matter less than or equal to 2.5 $\mu g/m^3$ can impair sperm and oocyte quality, resulting in a 2% reduction in fertility for every 10 $\mu g/m^3$ increase in these tiny particles (Xue and Zhang 2018). A broad dysregulation of the endocrine system may be the cause of the dysfunction of gonadal function and, consequently, the entire process of gametogenesis.

The presence of a connection between lower fertility and increased air pollution by measuring antimullerian hormone (AMH) levels in the blood. The amount of fertile eggs that a woman's ovaries can produce is known as her ovarian reserve, and this hormone, which is generated by ovarian somatic cells, is used to measure it. Serum AMH levels were examined in commercial areas after daily values of PM2.5–10 and NO_2. The results showed a significant drop in AMH in women residing in the most polluted locations. These findings provide credence to the theory that environmental influences might disrupt ovarian physiology, even though the association between AMH levels and the likelihood of conceiving spontaneously is still up for debate (Gaskins et al. 2019).

The placenta undergoes morphological changes throughout pregnancy in response to the growing fetus, including increased uterine and umbilical blood flows and significant angiogenesis in the utero placental and feto-placental vasculatures. These modifications are necessary for the proper growth of the fetus. Therefore, fetal growth, development, and survival will be impacted by factors that affect vascular development and function. The functional morphology of the mouse placenta has been demonstrated to be impacted by air pollution, and it was postulated that changes may at least partially account for the lower fetal weights linked to air pollution exposure. It was recently suggested that gynecologists and reproductive endocrinologists encourage healthy pregnancies by teaching fertile women to have secure lifestyles prior to conception (Segal and Giudice 2019).

Extreme Heat

Overheating at work or as a result of climate change is another significant issue that may lead to male infertility. Maintaining a healthy temperature in the testes is essential for optimal spermatogenesis. The difference between the scrotal and core body temperatures is 2-4 °C. Male unproductiveness is the result of any influence that raises the scrotal heat because it distresses the spermatogenesis. A 1–1.5 °C rise in scrotal temperature has been reported to cause irregular sperm morphology and reduced sperm production (oligozoospermia, teratozoospermia, and azoospermia). Stresses from the environment, comprising a rise in temperature cause the heat shock protein (HSP) to activate, HSP70s, the key proteins stimulated by rising temperatures. They are in charge of assembling, folding, and disassembling other proteins. Any element that interferes with their regular expression and regulation has a negative effect on male fertility (Fig. 5). According to a study involving 13 fertile men (controls) and 37 infertile men (cases), the infertile group's HSP 70 levels were significantly higher than those of the fertile male group. This led researchers to the conclusion that the increased expression of HSP 70 in the spermatozoa of infertile men serves as a defence mechanism against apoptosis. Those who work for long time in laundries, kitchens, dry cleaning shops, or furnaces, bakeries, or other industrial settings where radiant heat is present, as well as drivers, may experience a loss of the scrotum's thermoregulatory function, which may impact one or more aspects of a man's semen quality (Erata et al. 2008; Mínguez-Alarcón et al. 2018).

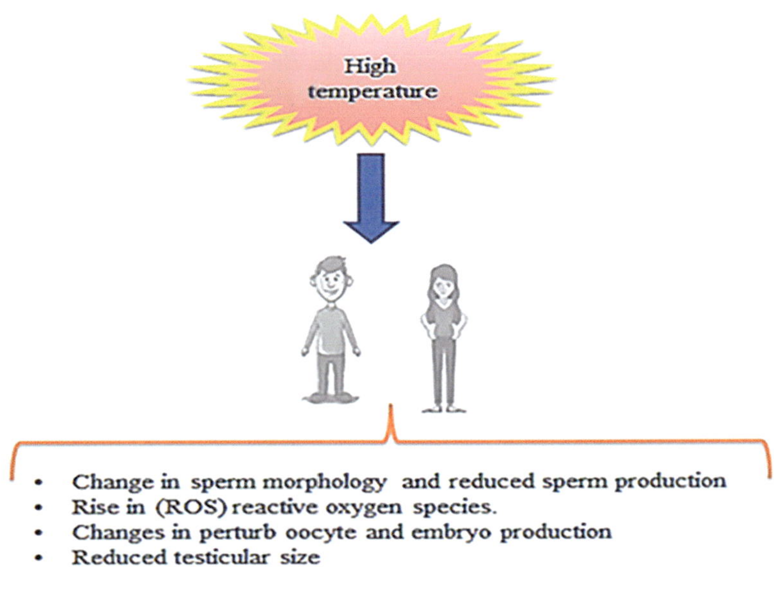

Fig. 5 Representation of possible effects of extreme temperature on humans leading to infertility

Higher scrotal temperatures cause less oxygen concentration in tissues and oxidative stress because they increase testicular metabolism without causing a spike in blood flow. Due to the high concentration of (PUFAs) polyunsaturated fatty acids in their plasma membrane, human spermatozoa are particularly vulnerable to oxidative stress-induced lipid peroxidation. This leads to a rise in reactive oxygen species formation, which further exacerbates sperm DNA breakage and male infertility. Additionally, prolonged exposure to heat decreases sperm motility by decreasing ATP levels and downregulating mitochondrial function (Vernet et al. 2004). A brief increase in scrotal temperature causes a reversible decrease in proteins necessary for gamete contact, sperm motility, and the process of spermatogenesis. A recent study on male rats found that heatstroke from exertion can result in azoospermia and infertility by causing erectile dysfunction, disrupting sperm quality, testicular temperature, poorly differentiated seminiferous tubules, reduced sperm quantity, and loss of interstitial Leydig cells and Sertoli cells. According to a related study on bovine sperm, heat stress in bulls causes seminal plasma oxidative stress, which in turn affects the sperm's motility, DNA fragmentation, mitochondrial activity, and plasma membrane integrity, finally resulting in infertility. In case of females, high temperatures can lead significant changes in embryo production and oocyte formation (Mahfouz et al. 2010; Lin et al. 2021).

An increase in testicular temperature has also been linked to reduced testicular size, lower sperm production, an increase in aberrant sperm forms, and impaired motility, all of which can contribute to male infertility, according to numerous animal studies. Therefore, exposure to high temperatures—whether as a result of one's job or the environment—has a negative effect on semen quality overall and can lead to infertility in men (Hassun 2007; Mao et al. 2017).

Life Style Factors

Numerous lifestyle factors can affect fertility in the present and in the future, including stress, nutrition, caffeine intake, body mass index (BMI), smoking, radiation and cellular phones, alcohol, and some medicines (Fig. 6). Stakeholder interests and questions have been raised about the role of lifestyle variables in fertility issues.

Alcohol Consumption

Alcohol intake is prevalent worldwide. Consistent alcohol consumption is widespread among couples planning to conceive or between women who are presently pregnant since a substantial proportion of the intake population is at maturity. Consumption of alcohol is linked to numerous reproductive vulnerabilities, such as an increased risk of abortions, the probability of a live delivery, and the delivery of a child with fetal alcohol spectrum disorders (FASD). It is generally recognized that consumption of alcohol while expecting might have a teratogenic effect. Alcohol

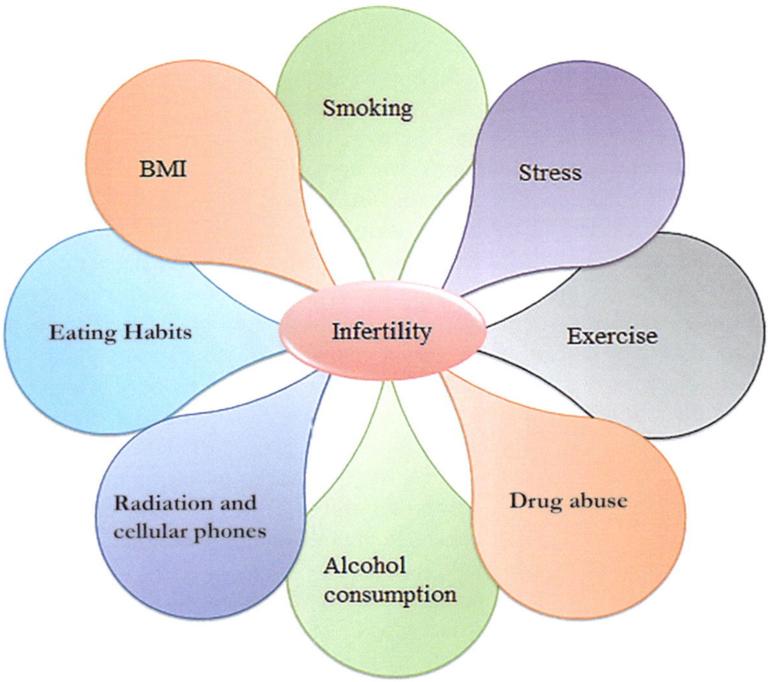

Fig. 6 An illustration of the lifestyle choices that lead to infertility in both men and women

effortlessly passes to the fetus through placenta and amniotic fluid. Alcohol and its metabolites get stored in the amniotic fluid, exposing the fetus to greater quantities of alcohol than the carrier parent and, in contrast, a lesser level of fetal metabolic enzyme movement. Some hypothesized teratogenicity pathways include decreased antioxidant capacity, elevated reactive oxygen species and free radicals, which lead to increased apoptosis in embryonic brain and cranial tissue (Gupta et al. 2016). FASD (fetal alcohol spectrum disorders) comprise alcohol-related birth anomalies, fetal alcohol syndrome (FAS), partial fetal alcohol syndrome (PFAS), and alcohol-related neurodevelopmental disorder (ARND). These conditions are brought on by alcohol consumption during gestation. Craniofacial deformities, development retardation, and behavioral and mental deficiencies are the traits of the FASD scale of illnesses. According to evaluations, the total U.S. inhabitants have a 2–5% frequency of FASD, and nearby 0.2–7 cases of FAS have been recorded for every 1000 children.

While alcohol consumption in the third trimester affected birth length, it is also linked to low birth weight and length in the second trimester. Additional research has verified that alcohol exposure throughout any trimester. It is frequently challenging result in growth insufficiency, neurobehavioral problems, and microcephaly;

nonetheless, the distinctive facial traits are probably the result of exposure during the first trimester. It is commonly challenging to determine in many investigations whether alcohol intake occurred in a single trimester or the entire gestation. As an outcome, at this phase, it is not reasonable to conclude the effects of alcohol on babies in cases where females refrain from using it during the first trimester and or second trimester and then ingest it during the third (del Campo and Jones 2017).

Male alcohol intake may also lead to infertility issues. Testosterone production, Testicular shrinkage, decreased gonadotropin release, and decreased sperm have all been detected in investigations on great, long-term alcohol drinking. Researches on men who drink disproportionately have regularly shown subordinate testosterone levels, beside an increase in estradiol and gonadotropins, irrespective of liver disease. Alcohol abuse is related to liver failure as well, and the incapability of the liver to produce estrogen can lead to hormonal inequalities. Also, with sporadic azoospermia, substantial alcohol consumers have also been frequently revealed to have a decline in the quality of sperm parameters. Additionally, there is ample evidence linking acute intoxication and alcohol addiction to sexual dysfunction, which encompasses problems with arousal, execration, and successful sex (Pendharkar et al. 2016).

Smoking

Many facets of reproduction are harmed by smoking. In the United States, an almost six million females conceive annually, and over 11,000 give birth every day. According to studies, women who smoke have a higher chance of experiencing both primary and secondary infertility as well as a delay in getting pregnant. Additionally, studies have indicated that pregnant women who smoke run the risk of difficulties, early delivery, low birth weight (LBW) babies, stillbirth, and infant death. One of the main causes of infant fatalities is LBW. Merely 18–25% of women cease smoking after becoming pregnant, despite growing awareness of the harmful health consequences of smoking during pregnancy. Fertility, smoking's negative effects on pregnancy, and the health of smokers' offspring (Ventura et al. 2000).

The definition of exposure for reproductive effects complicates the physiological basis of smoking's impacts on reproduction, and a methodologic framework may be the most useful for discussing this relationship. Two key methodological considerations when examining the effects of smoking on reproductive capability are determining the amount of smoking exposure and accounting for any confounding. Certain harmful reproductive consequences that may be causally linked to active smoking have brief acute revelation periods during gestation. For example, information about exposure to tobacco smoke in the early stages of gestation or during organogenesis is essential for analyzing the link between smoking and congenital abnormalities (Vine 1996).

Eating Habits

Nutritional issues in industrialized countries are often associated with eating disorders and obesity, whereas in underdeveloped countries, the main issues are deprivation, undernutrition, and malnutrition. Globalization has led to a rise in obesity rates in underdeveloped nations, nevertheless. The fertility is highly susceptible to environmental stimuli. This is due to the fact that it requires energy expenditures, and it seems sense that, as in living things, the physiological controls over the reproductive axis would be intimately correlated with nutritional condition. The rising trend in weight gain and obesity is attributed to a number of variables, including hereditary, environmental, and behavioral factors (Giahi et al. 2016).

Drug Abuse

Women are more prone than men to long-term drug dependence because of differences in weight, physiology, hormone levels, and other factors that can impact how these substances are metabolized in the body. Amenorrhea is known to be caused by methadone and heroin. Intravenous drug usage exposes a person to HIV and AIDS. It also allows a person to engage in unsafe sexual behavior that increases their chance of contracting STIs, such as the Human Papilloma Virus (HPV), when under the influence of such elements. This is associated with a higher incidence of cervical cancer in females. Substances like heroin, cocaine, methamphetamine, marijuana, and heroin can readily cross the placenta and enter the fetus, raising the risk of low birth weight, birth abnormalities, premature delivery, SIDS, and physical changes in the fetus (Acharya and Gowda 2017). This dangerous practice in guys results in effects comparable to STIs and infertility, which typically manifests as issues with sexual dysfunction. Methamphetamine and cocaine use can cause delayed orgasm and erectile problems in men who may have benefited initially from increased arousal. Long-term marijuana usage reduces spermatogenesis, motility, and the release of testosterone from Leydig cells. Additionally, marijuana can attach to receptors in reproductive organs like the uterus and tubes (Forray 2016).

BMI (Body Mass Index)

The prevalence of obesity (BMI $\geq$30 kg/m^2) in women who are fertile has dramatically increased during the past few decades. This is relevant since factors related to obesity decrease conception rates, lengthen the time it takes to conceive, and make patients less receptive to reproductive therapies. Obesity raises the risk of pregnancy problems such as gestational diabetes and hypertension, according to several studies. It also causes labor to last longer, raises the possibility of obstructed labor and caesarean birth, and has significant consequences on the developing fetus. Lean and fat animals are generally less fertile, but this is not always the case in people, according to recent animal tests. The nonlinear link between body mass index and infertility has not been thoroughly studied in many high-caliber studies. The association between body mass and fertility can be bimodal (Fontana and Torre 2016).

Caffeine Intake

A portion of foodstuffs and drinks, including tea, cola coffee, energy drinks, and chocolate, contain caffeine. Caffeine's effects on wellbeing are compound and complex. It may be defensive to liver disease, Parkinson's disease and diabetes. Cardiovascular disorders such, coronary heart disease, heart failure, and arrhythmia. Caffeine-based drinks indeed contain caffeine, an intoxicant. Additionally, products like sodas and energy drinks add it. It achieves this by activating the central nervous system, which keeps humans awake and stops us from falling asleep. Caffeine use in males has been associated to increased levels of the sex hormone binding globulin and testosterone. Coffee modifies the oxidative properties and Sertoli cells' glycolytic, which may lead to male infertility. However, the exact way via which caffeine may have an effect on the body is unfamiliar. Caffeine consumption during gestation and maturity may affect the hormones.

There is a major deal of misrepresentation out there about the effects of caffeine usage on infertility overall. Thorough research suggests that coffee consumption may have an influence on male reproductive capability. There are no results that will suggest that coffee use and reproductive health are related. Making the connection among coffee consumption and infertility is critical (Ricci et al. 2017).

Stress

Being diagnosed infertile can be extremely frightening for a couple, despite the fact that it is not a fatal illness. For both parties, infertility can lead to financial hardship, emotional strain, and psychological anguish. Emotions like melancholy, anxiety, depression, wrath, guilt, and loss of self-worth and confidence can all affect a couple. Additionally, the stress is exponentially amplified by the economic burden of handling infertility. In the social order, having offspring is seen as essential; thus, this societal pressure could aggravate already great levels of stress. In accumulation, it could be unfavorable to a person's relationships with their partner, friends and family. Disinterest from the family and deterioration in social engagement could be outcome from this. Another example is that emotional stress which could be a medical risk factor that depresses male fertility. It has been effectively established by several researchers that there is a contrary association between semen and psychological stress parameters. Additionally, stress could negatively impact how well a treatment works Effect of psychological stress on the L-arginine-nitric oxide pathway and semen quality (Eskiocak et al. 2006).

Radiation and Cellular Phones

There are rising evidences that cell phones have an undesirable influence on fertility. The extent of individuals using mobile phones is rising daily, and countless of them carry their phones in their trouser pockets. It has been projected that radioactivity from mobile phones can damage reproductive organs. There have been reports of advanced DNA disintegration index levels between cell phone operators. Cell phones send and receive radiofrequency electromagnetic waves, which harmfully impact the motility, quantity, and shape of sperm. Notable number of mobile phone operators stated greater levels of testosterone and lesser levels of LH incomparable to control people who have restricted access to cell phones (Ilacqua et al. 2018).

Exercise

Regular exercise has a positive impact on general physical health and may offer some resistance against disorders including type 2 diabetes, psychiatric stress, obesity, heart disease, and other conditions. Exercise improves ovarian function, raises insulin sensitivity, and may increase the chances of gestation. Exercise has been associated to a decline in energy balance, which may result to irregular ovulation and amenorrhea. This proposes a conceivable link among exercise and infertility. Obese infertile females benefit from exercise. The outcomes of an interventional trial revealed that improved fertility in obese females was linked to weight loss and physical activity. "Sedentary lifestyle behavior" is the term that defines the extent of period spent inactive with slight to no movement. There is natural evidence involving physical workout to infertility, it is a noteworthy risk factor for reproductive incapability. Due to implantation defects and anovulation, women who exercise vigorously may not be able to conceive (Wise et al. 2012).

Conclusion

There are two things that are certain: male fertility is dropping, and we are constantly exposed to environmental pollutants in our daily lives. Although there is some data that suggests a connection between male fertility and environmental exposures, conclusive proof connecting commonplace poisons to the recognized reduction in male fertility is hard to come by. This can be attributed to the complicated longitudinal investigations needed to prove causality as well as the ethical issues with performing prospective clinical trials involving human exposures, rather than a lack of proof that these substances cause reproductive harm. Furthermore, even though these toxins might affect the characteristics of semen, it is unclear if this would lead to a decrease in the likelihood of pregnancy and live births. Exposure to these substances should be reduced, and avoidance should be exercised where practical, until further information about the impact of our surroundings on male fertility is available. Future generations' ability to reproduce will probably depend on a number of things, such as the ability of governmental entities to set concentration limits, outlaw the use of specific drugs, and strictly regulate the use of substances that may be harmful to reproduction. Rather than adopting the present regulatory paradigm of waiting until harm manifests, which is reactive, this calls for a proactive and preventative approach. Emerging technology will allow consumers to measure their exposures, raising the prospect of greater public awareness of the risk and fostering advocacy for broader systemic reform.

References

Acharya S, Gowda CR (2017) Lifestyle factors associated with infertility in a rural area: a cross-sectional study. Int J Med Sci Public Health 6:502–550

Agarwal A, Baskaran S, Parekh N et al (2021) Male infertility. Lancet 397:319–333

Akesson A, Berglund M, Schutz A et al (2002) Cadmium exposure in pregnancy and lactation in relation to iron status. Am J Public Health 92:284–287

Barrett JR (2011) Window for dioxin damage: sperm quality in men born after the Seveso disaster. Environ Health Perspect 119:A219

Beaver LM, Truong L, Barton CL et al (2017) Combinatorial effects of zinc deficiency and arsenic exposure on zebrafish (*Danio rerio*) development. PLoS One 12:e0183831

Berglund M, Akesson A, Nermell B et al (1994) Intestinal absorption of dietary cadmium in women depends on body iron stores and fiber intake. Environ Health Perspect 102:1058–1066

Bretveld R, Brouwers M, Ebisch I et al (2007) Influence of pesticides on male fertility. Scand J Work Environ Health 33:13–28

del Campo M, Jones K (2017) A review of the physical features of the fetal alcohol spectrum disorders. Eur J Med Genet 60:55–64

Caporossi L, Alteri A, Campo G (2020) Cross sectional study on exposure to BPA and phthalates and semen parameters in men attending a fertility center. Int J Environ Res Public Health 17:489

Cariati F, D'Uonno N, Borrillo F, Iervolino S, Galdiero G, Rb T (2019) Bisphenol A: an emerging threat to male fertility. Reprod Biol Endocrinol 17:6

Checa Vizcaíno MA, González-Comadran M, Jacquemin B (2016) Outdoor air pollution and human infertility: a systematic review. Fertil Steril 106:897–904

Chen C, Xun P, Kaufman JD et al (2020) Erythrocyte omega-3 index, ambient fine particle exposure and brain aging. Neurology 95:e995–e1007

Chianese R, Troisi J, Richards S et al (2018) Bisphenol A in reproduction: epigenetic effects. Curr Med Chem 25:748–770

Deng Z, Chen F, Zhang M (2016) Association between air pollution and sperm quality: a systematic review and meta-analysis. Environ Pollut 208(Pt B):663–669

Erata GO, Koçak Toker N, Durlanik O et al (2008) The role of heat shock protein 70 (Hsp 70) in male infertility: is it a line of defense against sperm DNA fragmentation? Fertil Steril 90:322–327

Eskiocak S, Gozen AS, Taskiran A et al (2006) Effect of psychological stress on the L-arginine-nitric oxide pathway and semen quality. J Med Biol Res 39:581–588

Fontana R, Torre SD (2016) The deep correlation between energy metabolism and reproduction: a view on the effects of nutrition for women fertility. Nutrients 8:87

Forgacs Z, Massanyi P, Lukac N et al (2012) Reproductive toxicology of nickel-review. J Environ Sci Health A Tox Hazard Subst Environ Eng 47:1249–1260

Forray A (2016) Substance use during pregnancy. F1000Res 5:F1000 Faculty Rev-887

Foster WG, Maharaj-Briceño S, Cyr DG (2010) Dioxin-induced changes in epididymal sperm count and spermatogenesis. Environ Health Perspect 118(4):458–464

Gaskins AJ, Mínguez-Alarcón L, Fong KC et al (2019) Exposure to fine particulate matter and ovarian reserve among women from a fertility clinic. Epidemiology 30:486–491
Giahi L, Mohammadmoradi S, Javidan A et al (2016) Nutritional modifications in male infertility: a systematic review covering 2 decades. Nutr Rev 74:118–130
Gulson BL, Mizon KJ, Korsch MJ et al (2003) Mobilization of lead from human bone tissue during pregnancy and lactation-a summary of long-term research. Sci Total Environ 303:79–104
Gupta KK, Gupta VK, Shirasaka T (2016) An update on fetal alcohol syndrome-pathogenesis, risks, and treatment. Alcohol Clin Exp Res 40:1594–1602
Hassun PAF (2007) Re: Wet heat exposure: a potentially reversible cause of low semen quality in infertile men. Int Braz J Urol 33(2):269–270
Henson MC, Chedrese PJ (2004) Endocrine disruption by cadmium, a common environmental toxicant with paradoxical effects on reproduction. Exp Biol Med 229:383–392
Howdeshell KL, Rider CV, Wilson VS et al (2008a) Mechanisms of action of phthalate esters, individually and in combination, to induce abnormal reproductive development in male laboratory rats. Environ Res 108:168–176
Howdeshell KL, Wilson VS, Furr J et al (2008b) A mixture of five phthalate esters inhibits fetal testicular testosterone production in the Sprague Dawley rat in a cumulative, dose-additive manner. Toxicol Sci 105:153–165
Ilacqua A, Izzo G, Emerenziani GP et al (2018) Lifestyle and fertility: the influence of stress and quality of life on male fertility. Reprod Biol Endocrinol 16(1):115
Iorio R, Castellucci A, Ventriglia G et al (2014) Ovarian toxicity: from environmental exposure to chemotherapy. Curr Pharm Des 20:5388–5397
Istvan M, Rahban R, Dananche B et al (2021) Maternal occupational exposure to endocrine-disrupting chemicals during pregnancy and semen parameters in adulthood: results of a nation-wide cross-sectional study among Swiss conscripts. Hum Reprod 36:1948–1958
Lin PH, Huang KH, Tian YF et al (2021) Exertional heat stroke on fertility, erectile function, and testicular morphology in male rats. Sci Rep 11:3539
Mahfouz R, Sharma R, Thiyagarajan A et al (2010) Semen characteristics and sperm DNA fragmentation in infertile men with low and high levels of seminal reactive oxygen species. Fertil Steril 94:2141–2146
Mao H, Feng L, Yang WX (2017) Environmental factors contributed to circannual rhythm of semen quality. Chronobiol Int 34:411–425
Mínguez-Alarcón L, Gaskins AJ, Chiu YH et al (2018) Type of underwear worn and markers of testicular function among men attending a fertility center. Hum Reprod 33(9):1749–1756
Mohammadi S, Rahmani F, Hasanian SM (2019) Effects of dioxin on testicular histopathology, sperm parameters, and CatSper2 gene and protein expression in Naval Medical Research Institute male mice. Andrologia 51(11):e13411
Pendharkar S, Mattoo SK, Grover S (2016) Sexual dysfunctions in alcohol-dependent men: a study from North India. Indian J Med Res 144:393–399
Pizzorno J (2018) Environmental toxins and infertility. Integr Med (Encinitas) 17(2):8–11
Radke EG, Braun JM, Meeker JD et al (2018) Phthalate exposure and male reproductive outcomes: a systematic review of the human epidemiological evidence. Environ Int 121:764–793
Rehman H, Ullah I, David M et al (2019) Neonatal exposure to furan alters the development of reproductive systems in adult male Sprague Dawley rats. Food Chem Toxicol 130:231–241
Ricci E, Viganò P, Cipriani S et al (2017) Coffee and caffeine intake and male infertility: a systematic review. Nutr J 16:37
RÍsovÁ V (2019) The pathway of lead through the mother's body to the child. Interdiscip Toxicol 12(1):1–6
Segal TR, Giudice LC (2019) Before the beginning: environmental exposures and reproductive and obstetrical outcomes. Fertil Steril 112:613–621
Ventura SJ, Mosher WD, Curtin SC et al (2000) Trends in pregnancies and pregnancy rates by outcome: estimates for the United States, 1976–96. Vital Health Stat 21 Data Natal Marriage Divorce 56:1–47

Vernet P, Aitken RJ, Drevet JR (2004) Antioxidant strategies in the epididymis. Mol Cell Endocrinol 216:31–39

Vine MF (1996) Smoking and male reproduction: a review. Int J Androl 19:323–337

Weinberg CR, Wilcox AJ, Baird DD (1989) Reduced fecundability in women with prenatal exposure to cigarette smoking. Am J Epidemiol 129:1072–1078

Wise LA, Rothman KJ, Mikkelsen EM et al (2012) A prospective cohort study of physical activity and time to pregnancy. Fertil Steril 97:1136–1142

Xue T, Zhang Q (2018) Associating ambient exposure to fine particles and human fertility rates in China. Environ Pollut 235:497–504

Elucidating the Molecular Pathophysiology and Mechanistic Insights of Chronic Obstructive Pulmonary Disease

Saba Beigh

Abstract

Chronic obstructive pulmonary disease (COPD) is a major global health issue ranked the third leading cause of mortality worldwide. The disorder is characterized by progressive and irreversible airflow limitation resulting from chronic inflammation of the airways, lung parenchyma, and pulmonary vasculature. This chapter explores the intricate pathophysiological mechanisms of COPD, focusing on chronic inflammation driven by neutrophils, macrophages, and $CD8^+$ T-lymphocytes, which release pro-inflammatory cytokines and proteases, disrupting the balance between proteases and antiproteases and leading to tissue degradation and emphysema. Oxidative stress, exacerbated by reactive oxygen species from tobacco smoke and inflammatory cells, further contributes to cellular damage and disease progression. Structural lung changes, including airway remodeling and emphysema, are detailed, highlighting the consequences of goblet cell hyperplasia, subepithelial fibrosis, and loss of elastic recoil. The epidemiology of COPD is influenced by both genetic factors, such as alpha-1 antitrypsin deficiency, and environmental exposures, with smoking being the primary risk factor in developed countries and biomass fuel exposure in developing countries. Diagnostic approaches rely on symptomatic evaluation, spirometry, and imaging, with spirometry being the gold standard. Management involves a combination of pharmacological treatments, including bronchodilators and inhaled corticosteroids, and non-pharmacological strategies, such as smoking cessation and pulmonary rehabilitation. Advanced cases may require surgical interventions. We emphasize the complexity of COPD and the critical need for

S. Beigh (✉)
Department of Public Health, Faculty of Applied Medical Sciences, Al-Baha University, Al-Baha, Kingdom of Saudi Arabia
e-mail: sbeigh@bu.edu.sa

S. Raisuddin et al. (eds.), *Environmental Toxicants and Lifestyle Diseases*,
https://doi.org/10.1007/978-981-95-0575-3_17

early diagnosis and targeted therapies to improve patient outcomes and quality of life. Ongoing research into the genetic, environmental, and molecular factors driving COPD is essential for future advancements in treatment have also been highlighted in this chapter.

Keywords

Chronic obstructive pulmonary disease · Reactive oxygen species · Pathophysiology · Inflammation · Biomarkers

Introduction

Chronic obstructive pulmonary disease (COPD) stands as a prominent global health challenge, profoundly impacting both morbidity and mortality worldwide (Hattab et al. 2016). COPD encompasses a spectrum of progressive respiratory disorders characterized by persistent and irreversible airflow limitation, primarily including chronic bronchitis and emphysema. Both conditions contribute to a gradual and often debilitating decline in lung function (Ferrera et al. 2021). The prevalence of COPD in clinical settings underscores its severe implications for the quality of life of affected individuals, healthcare systems, and society at large (Ferrera et al. 2021). The pathogenesis of COPD begins with complex biochemical and cellular changes in the small airways and alveoli. Initial structural damage leads to a loss of elastic recoil, causing the lungs to expand and resulting in increased forced vital capacity (FVC) (Higham et al. 2019). These early physiological changes, detectable through spirometry, precede the onset of clinically apparent symptoms, which often emerge only when the disease has progressed to a moderate or advanced stage. COPD is primarily driven by chronic inflammatory responses to inhaled noxious particles or gases, with cigarette smoke being the predominant risk factor (Parris et al. 2019). The disease's pathophysiology is intricate, involving a synergy of genetic, environmental, and inflammatory factors. Chronic exposure to harmful substances initiates an inflammatory cascade that results in the activation of immune cells and the release of inflammatory mediators. This chronic inflammation induces structural damage to lung tissue, leading to alveolar destruction (emphysema) and airway wall remodeling (Hikichi et al. 2019). In the early stages of COPD, subtle yet significant molecular and cellular alterations occur, including elevated levels of inflammatory cytokines, oxidative stress, and an imbalance between proteases and antiproteases. These early changes offer a critical opportunity for potential early intervention, which may delay or prevent progression to more severe stages (Hikichi et al. 2019). As the disease advances, irreversible structural damage ensues. In emphysema, alveolar wall destruction impairs gas exchange, leading to symptoms of breathlessness and reduced exercise capacity. Conversely, chronic bronchitis is marked by persistent inflammation and hypertrophy of mucus-secreting glands, resulting in excessive mucus production, chronic cough, and airway obstruction. Epidemiologically, COPD represents a significant global burden. The World Health

Organization (WHO) reports COPD as one of the leading causes of mortality, ranking as the third leading cause of death worldwide, with over three million deaths annually (Safiri et al. 2022). Prevalence varies across regions and populations, influenced by smoking habits, air pollution, and socioeconomic conditions. In high-income countries, smoking remains the primary risk factor, whereas in low- and middle-income countries, indoor air pollution from biomass fuels and environmental exposures play a substantial role (Safiri et al. 2022). COPD's impact extends beyond mortality, significantly affecting quality of life and imposing a substantial burden on healthcare systems. Exacerbations-acute worsening of symptoms-are frequent and lead to increased healthcare utilization, including hospitalizations and emergency visits (Russell et al. 2018). These exacerbations exacerbate the decline in lung function and functional status, diminishing patients' ability to perform daily activities and affecting their social and occupational roles (Russell et al. 2018). Diagnosis of COPD primarily involves clinical history, symptom assessment, and spirometry. Spirometry, which measures lung function by assessing forced expiratory volume in 1 s (FEV1) and FVC, is crucial for confirming the diagnosis and determining disease severity. The Global Initiative for Chronic Obstructive Lung Disease (GOLD) classification system categorizes COPD into four stages based on spirometry measurements. Stage 1 (mild COPD) features an FEV1/FVC ratio <0.70 with an FEV1 $>80\%$ of the predicted normal value. Stage 2 (moderate COPD) is characterized by an FEV1/FVC ratio <0.70 with an FEV1 between 50% and 79% of predicted normal value. Stage 3 (severe COPD) is marked by an FEV1/FVC ratio <0.70 with an FEV1 between 30% and 49% of predicted normal value. Stage 4 (very severe COPD) includes an FEV1/FVC ratio <0.70 with an FEV1 $<30\%$ of predicted normal value or respiratory failure. Recent guidelines also propose a Stage 0 classification for individuals with chronic cough, sputum production, and breathlessness but without measurable airflow obstruction, aiming to identify at-risk individuals for early intervention (Ferrera et al. 2021). Research suggests that age-adjusted reference equations or the lower limit of normal (LLN) might improve diagnostic accuracy (Ferrera et al. 2021). Ongoing research aims to refine diagnostic criteria and treatment strategies by integrating factors such as ethnicity, sex, height, and age into reference equations (Ferrera et al. 2021). Chronic obstructive pulmonary disease remains a major public health concern with extensive implications for individuals, healthcare systems, and societies globally. Understanding its complex pathophysiology, diverse clinical presentations, and substantial impact on quality of life underscores the urgent need for continued research and improved management strategies. Advancements in COPD research and the refinement of diagnostic and therapeutic approaches are crucial for mitigating the disease's prevalence and impact, ultimately improving patient outcomes on a global scale. Figure 1 shows the COPD risk factors, pathophysiology, stages, impact, and its diagnosis.

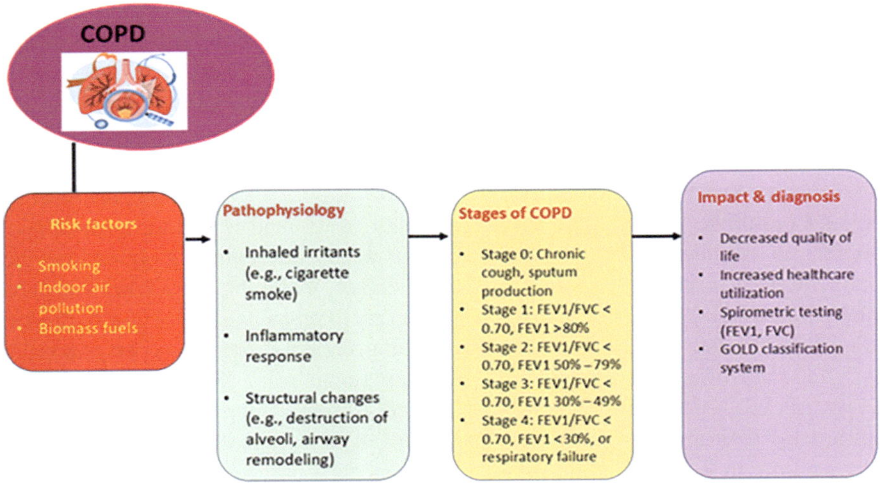

Fig. 1 Diagrammatic representation of COPD: risk factors, pathophysiology, stages, impact, and diagnosis

Epidemiology and Risk Factors

Prevalence and Global Burden

COPD is a major cause of chronic morbidity and mortality worldwide, ranking as the third leading cause of death globally (Alqahtani 2022). The prevalence of COPD varies significantly across different regions, influenced by factors such as smoking rates, air quality, and healthcare access (Alqahtani 2022). In developed countries, cigarette smoking is the predominant risk factor for COPD, while in developing regions, exposure to biomass fuels plays a critical role (Table 1). In clinical settings, the diagnosis of COPD typically begins with a comprehensive patient history, followed by a physical examination and functional testing. Airflow obstruction is commonly defined using post-bronchodilator spirometric measurements, with an FEV1/FVC ratio of 0.70 or less being the standard threshold for irreversible obstruction. However, this approach can result in false positives, as pre-bronchodilator spirometry may inflate COPD prevalence estimates by up to 30% (Bailey 2012). One study reported that fixed FEV1/FVC ratios can overestimate airflow obstruction, particularly in older populations, with false-positive rates increasing with age (Bailey 2012). Overdiagnosis of COPD is a well-recognized issue, with some studies indicating rates as high as 60%. The Burden of Obstructive Lung Disease (BOLD) study, an international initiative, reveals that fixed FEV1/FVC ratios often lead to higher prevalence estimates compared to alternative spirometry-based definitions. The study advocates for using the LLN for FEV1/FVC ratios and an FEV1 below 80% of the predicted value or below the LLN to reduce false positives (Ho et al.

Table 1 Global prevalence of COPD and its distribution across different regions

Region	Prevalence (%)	Number of cases (Millions)	Mortality rate (%)	Disability-adjusted life years (DALYs) per 100,000	Causes
Global	11.7	251	3.2	1800	Reflects the overall impact, including both smoking and environmental factors.
North America	10.6	16.4	2.7	1500	High prevalence linked to historical smoking trends.
Europe	10.1	36.4	3.3	1600	Aging population and high smoking rates are major contributors.
Sub-Saharan Africa	6.9	18.9	4.1	2200	Higher mortality due to limited access to healthcare resources.
South Asia	11.8	60.3	4.5	2300	Significant impact from biomass fuel use and air pollution.
East Asia & Pacific	13.0	85.3	2.9	1700	Rapid urbanization and air pollution are key factors.
Latin America & Caribbean	8.2	14.7	2.6	1400	Mixed factors including environmental exposure and smoking habits.
Middle East & North Africa	7.4	12.8	3.4	1900	Dust and pollution from arid regions increase prevalence.
Australia & New Zealand	7.1	1.9	2.5	1200	Effective healthcare systems contribute to lower prevalence and mortality.
India	12.1	78.6	4.0	2500	High prevalence due to both smoking and significant air pollution.

2019). The American Thoracic Society (ATS) and European Respiratory Society (ERS) now support this approach, emphasizing the need for periodic updates to the equations used for these calculations.

A meta-analysis of studies from 1990 to 2004 revealed geographical and methodological variations in COPD prevalence, estimated at 7.6% (95% CI 6–9.2%), with chronic bronchitis at 6.4% (95% CI 5.3–7.7%) and emphysema at 1.8% (95% CI 1.3–2.6%). These studies predominantly involved individuals over 40, showing a notable increase in prevalence with age, particularly among those over 65 years. Smoking status significantly influences COPD prevalence, with even nonsmokers exhibiting a prevalence of 4%, suggesting other risk factors such as passive smoking or occupational exposures (Jarhyan et al. 2022).

Geographically, Southeast Asia reports a higher COPD prevalence of 12.5%, while data from Africa and the Mediterranean are limited. In Africa and Asia, diagnosing COPD is challenging due to the high prevalence of respiratory infections. Modeling suggests a moderate-to-severe COPD prevalence of 6.3% in Southeast Asia, with China and Vietnam showing the highest rates. In South America, the PLATINO study estimates COPD prevalence between 7.8% and 20% among individuals over 40. In France, a 2002 study found chronic bronchitis prevalence at 4%, with underdiagnosis evident as only 24% of symptomatic individuals had been formally diagnosed. The ECRHS study highlights risk factors such as chronic cough and sputum, and the BOLD project provides global prevalence estimates, with stage II or higher COPD affecting 10.1 ± 4.8% of participants. This underscores the importance of considering additional exposures and potential genetic predispositions, particularly among females and non-smokers (Menezes et al. 2004). The complex epidemiology of COPD underscores the urgent need for continued research and improved management strategies. Advances in understanding COPD, along with the refinement of diagnostic and therapeutic approaches, are crucial for mitigating the disease's prevalence and impact, ultimately enhancing patient outcomes on a global scale.

Genetic Susceptibility

COPD is influenced by a complex interplay of genetic, environmental, and lifestyle factors. Among the genetic risk factors, alpha-1 antitrypsin (AAT) deficiency is the most well-documented and significantly impacts the development of COPD. Severe AAT deficiency, associated with early-onset emphysema, is the only genetic subtype of COPD with a specific treatment available (Silverman 2020). AAT, encoded by the SERPINA1 gene, functions as a primary plasma protease inhibitor of leukocyte elastase, an enzyme that contributes to tissue damage when not properly regulated. The most common genetic cause of severe AAT deficiency is the presence of two copies of the SERPINA1*Z allele, resulting from a single base pair mutation in the SERPINA1 gene (Silverman 2020). This mutation leads to a single amino acid substitution that causes AAT to polymerize abnormally in hepatocytes, where AAT is predominantly produced. The resultant accumulation of AAT polymers in the liver significantly reduces circulating levels of functional AAT, leading to an increased risk of COPD and liver diseases, including hepatitis, cirrhosis, and hepatocellular carcinoma. Approximately 1 in 3000 people in the United States inherit severe AAT deficiency, with significant variability in lung disease progression observed among affected individuals (Torres-Duran et al. 2018).

In Sweden, a newborn screening study from 1972 to 1974 identified 129 individuals with the PI Z genotype. A recent follow-up revealed that while some exhibited signs of lung hyperinflation and reduced diffusing capacity, variability in COPD risk among these individuals suggests the influence of yet-to-be-identified genetic modifiers (Torres-Duran et al. 2018). In the United States, augmentation therapy for AAT deficiency, administered as weekly intravenous infusions, has been shown to slow the progression of emphysema (Hersh 2017).

The impact of the SERPINA1 Z allele in individuals who are heterozygous for this allele, known as PI MZ, is also of significant interest. Studies have indicated that PI MZ individuals may have an increased risk of COPD, particularly when exposed to environmental risk factors such as smoking. A study conducted revealed significantly lower FEV1/FVC ratios in PI MZ individuals from both the GenKOLS study in Norway and families identified through COPD patients in the International COPD Genetics Network, although only the GenKOLS study found a significant association between PI MZ genotype and quantitative CT evidence of emphysema (Ragland et al. 2019). Similarly, studies involving families in Ireland and analyses from the COPD Gene and SPIROMICS studies demonstrate that PI MZ smokers have reduced lung function and increased emphysema compared to PI MM smokers (Molloy et al. 2014). However, the increased risk for COPD among PI MZ nonsmokers remains unclear.

Environmental and Occupational Exposures

COPD is predominantly linked to smoking, but non-smokers can also develop the disease due to various environmental and occupational exposures. Long-term exposure to air pollutants, such as particulate matter, sulphur dioxide, and nitrogen dioxide, has been consistently associated with an increased risk of COPD. Similarly, occupational exposures to dust, chemicals, and fumes are significant contributors to the disease (Duan et al. 2020). Despite the established connection between occupational exposure and COPD, the disease remains widely underdiagnosed. Recent studies indicate that nearly 70% of cases with persistent airflow obstruction go undiagnosed, a situation exacerbated by the underutilization of spirometry and a lack of awareness, particularly among younger individuals, current or non-smokers, and those with mild symptoms (Lin et al. 2023).

The underdiagnosis of occupational COPD has significant implications, affecting workers' compensation and potentially leading to disease progression and a decline in lung function. COPD, characterized by airflow limitation, respiratory symptoms, and comorbidities, results from exposure to harmful particles and gases. Smoking is the principal risk factor; occupational exposure to vapours, gases, dust, and fumes (VGDF) also plays a crucial role in the development of occupational COPD. Epidemiological studies support this association, particularly concerning inorganic dust, with population-based studies suggesting that workplace exposure to VGDF accounts for a significant population-attributable fraction (PAF) of 14%. The biological plausibility of this connection has been demonstrated through various experimental studies (Toren et al. 2017). However, further research is needed, particularly epidemiological cohort studies, to investigate additional risk factors that may increase COPD risk but lack a robust evidence base, such as certain irritant substances, types of organic dust, and specific metals (Grahn et al. 2021). Early diagnosis is crucial to prevent further lung function decline due to ongoing exposure and to mitigate the health and socioeconomic burden of COPD. Therefore, clinicians should remain vigilant regarding workplace exposures in COPD patients, even among smokers.

Pathophysiological Mechanisms

Chronic obstructive pulmonary disease is a progressive respiratory condition characterized by persistent airflow obstruction that is poorly reversible, accompanied by an abnormal inflammatory response in the lungs. This inflammatory response reflects both innate and adaptive immune responses to prolonged exposure to harmful particles and gases, most notably from cigarette smoke (Nurwidya et al. 2016). While all cigarette smokers experience some degree of lung inflammation, individuals who develop COPD exhibit an exaggerated or abnormal response to these toxic agents. This heightened inflammatory response can lead to a series of pathological changes, including mucous hypersecretion, which characterizes chronic bronchitis; tissue destruction, leading to emphysema; and disruption of normal repair and defence mechanisms, resulting in small airway inflammation and fibrosis, or bronchiolitis (Gupta et al. 2016). These changes cause increased resistance to airflow in the small airways, greater lung compliance, air trapping, and progressive airflow obstruction which are indeed hallmark features of COPD.

The cellular and molecular mechanisms driving these pathological changes in COPD are now well understood. COPD is marked by elevated numbers of neutrophils, macrophages, and T lymphocytes (with a predominance of CD8 over CD4 T cells) in the lungs. Generally, the severity of this inflammation correlates with the degree of airflow obstruction observed in patients. Inflammatory cells release a range of cytokines and mediators that contribute to disease progression. This inflammatory pattern is distinctly different from that observed in asthma, reflecting the unique pathophysiology of COPD (Saetta et al. 2001). The chronic inflammation in COPD is characterized by an imbalance between proteases and antiproteases, with an increase in proteolytic enzymes that contribute to tissue destruction and impaired repair processes. This imbalance, coupled with the recruitment and activation of inflammatory cells, drives the progressive nature of the disease and leads to the clinical manifestations of COPD, including persistent cough, sputum production, and progressive dyspnea.

Inflammatory Mediators in COPD

COPD is characterized by persistent airflow obstruction and an abnormal inflammatory response in the lungs, driven by various inflammatory mediators and mechanisms. One of the key mediators in COPD is leukotriene B4, a potent chemoattractant produced by macrophages, neutrophils, and epithelial cells. This mediator is crucial in directing immune cells, particularly neutrophils and T cells, to inflamed areas in the lungs, significantly contributing to the inflammatory response observed in COPD (Rovina et al. 2013). Several chemotactic factors, including CXC chemokines such as interleukin-8 (IL-8) and growth-related oncogene-alpha (GRO-α), play integral roles in COPD. These factors are produced by macrophages and epithelial cells and function to attract additional immune cells from the bloodstream to the lungs. This accumulation of immune cells amplifies the inflammatory response and exacerbates COPD symptoms.

Pro-inflammatory cytokines, such as tumor necrosis factor-alpha (TNF-α) and interleukins 1 and 6 (IL-1 and IL-6), are significantly elevated in COPD. These cytokines drive inflammation by activating and recruiting immune cells, thereby perpetuating the inflammatory cycle within lung tissue (1). Their increased levels contribute to ongoing tissue damage and dysfunction observed in COPD. Transforming growth factor-beta (TGF-β) is another crucial factor involved in fibrosis development within the disease. TGF-β can directly induce fibrosis by affecting lung cells or indirectly by stimulating the release of cytokines like connective tissue growth factor (CTGF). This fibrosis leads to airway remodeling and obstruction, further impairing lung function and worsening the disease. The interaction of these inflammatory mediators-leukotriene B4, chemotactic factors, pro-inflammatory cytokines, and growth factors-creates a complex network that drives inflammation, immune cell recruitment, and tissue remodeling in COPD. Understanding these mechanisms is essential for developing targeted therapies to manage and mitigate the impact of this debilitating disease. An imbalance between proteases and antiproteases is another critical aspect of COPD pathology. This imbalance, influenced by cigarette smoke and the inflammatory process, results from increased protease production or activity coupled with decreased antiprotease effectiveness (Pandey et al. 2017). Oxidative stress, driven by cigarette smoke and inflammatory cells, exacerbates this imbalance by promoting protease release and inhibiting antiproteases. Key proteases involved include neutrophil-derived serine proteases such as elastase, cathepsin G, and protease 3, as well as macrophage-released cysteine proteases and matrix metalloproteinases (MMPs) like MMP-8, MMP-9, and MMP-12. These proteases contribute to tissue damage and remodeling in COPD. Conversely, antiproteases such as alpha-1 antitrypsin, secretory leucoprotease inhibitor, and tissue inhibitors of metalloproteinases (TIMPs) protect tissue integrity, and reduced antiprotease activity further exacerbates tissue damage and disease progression. Oxidative stress is a significant component of COPD pathogenesis, with sources including cigarette smoke and reactive oxygen and nitrogen species from inflammatory cells. This stress disrupts the balance between oxidants and antioxidants, leading to increased oxidative damage, further inactivation of antiproteases, and enhanced inflammation through transcription factor activation, such as nuclear factor-kappa B (NF-κB) (Barnes 2020). These mechanisms culminate in various physiological abnormalities in COPD, including mucus hypersecretion, ciliary dysfunction, airflow obstruction, and hyperinflation. Mucous hypersecretion, a hallmark of chronic bronchitis, results from squamous metaplasia and increased mucus production. Ciliary dysfunction impairs the mucociliary escalator, leading to mucus accumulation and worsening symptoms. Airflow obstruction primarily affects small conducting airways, leading to hyperinflation and reduced lung capacity. Gas exchange abnormalities, such as hypoxemia and hypercapnia, result from impaired alveolar structure and function. Pulmonary hypertension, which develops in advanced COPD stages, strains the right side of the heart and can lead to right-sided heart failure. COPD also has systemic effects, including muscle wasting, weight loss, and increased cardiovascular risk. Exacerbations, often triggered by infections, air pollution, or temperature changes, are marked by heightened inflammation and worsening gas exchange, potentially leading to severe respiratory failure and complications like peripheral edema. Understanding the intricate interplay of inflammatory mediators, protease-antiprotease imbalance, and oxidative stress is crucial for developing effective interventions to manage and treat COPD. Figure 2 illustrates the complex interactions and pathological outcomes in COPD.

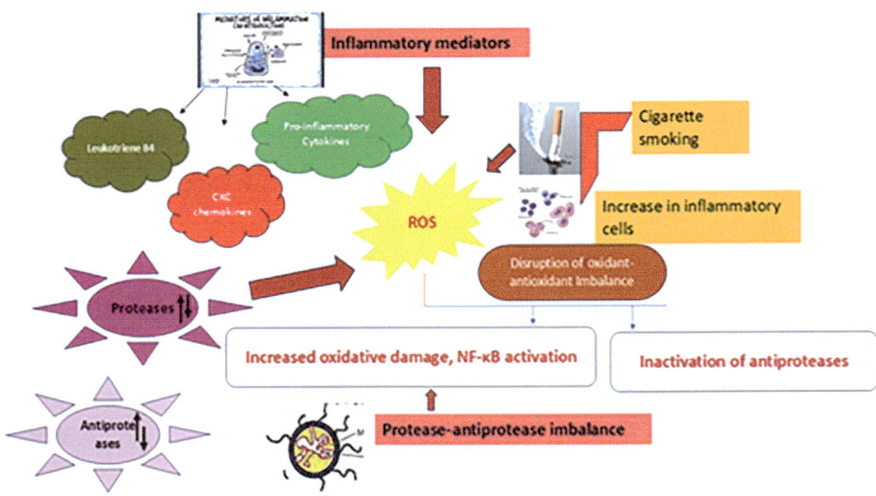

Fig. 2 Comprehensive diagram of inflammatory mediators, protease-antiprotease imbalance, and oxidative stress in COPD

Clinical Implications and Management

COPD is primarily diagnosed through clinical evaluations, spirometric measurements, and imaging studies. Spirometry, which measures the volume and flow of air during inhalation and exhalation, remains the gold standard for diagnosing COPD and assessing the degree of airflow obstruction. A definitive diagnosis of COPD is established when a post-bronchodilator FEV1/FVC (forced expiratory volume in 1 s/forced vital capacity) ratio of less than 0.70 is recorded, indicating persistent airflow limitation. Table 2 shows the clinical management of COPD. Standard therapeutic interventions for COPD include bronchodilator therapy, vaccination to prevent respiratory infections, smoking cessation, and pulmonary rehabilitation. These approaches are generally effective in controlling the disease in patients with less severe forms. However, some patients may continue to experience significant symptoms and frequent exacerbations despite adherence to these therapies.

Pathophysiological Mechanisms and Outcomes

The term "refractory COPD" describes cases where patients present with severe and progressive symptoms that persist despite adherence to standard medical therapies. Such individuals frequently encounter ongoing respiratory distress and recurrent exacerbations that are not adequately managed by conventional treatment methods. For these patients, a comprehensive, multidisciplinary management strategy is essential, which may involve advanced medical treatments and surgical interventions aimed at symptom relief and enhancing overall quality of life. Confirming the diagnosis of refractory COPD involves revisiting previous

Table 2 Clinical implications, management strategies, and approaches for managing COPD

Aspect	Clinical implications	Management strategies
Diagnosis	Spirometry, clinical evaluation and imaging studies	Post-bronchodilator FEV1/FVC <0.70 for COPD diagnosis
Refractory COPD	Severe, progressive symptoms despite standard therapy	Comprehensive, multidisciplinary approach; consider advanced treatments and surgical interventions (Murray et al. 2023)
Inhaler therapy	Poor adherence and improper use are common	Patient education, simplifying regimens, and ensuring proper usage
Exacerbations	Frequent, severe exacerbations impact quality of life.	Document history, manage promptly, and adjust therapy as needed.
Comorbidities	Contribute to symptom burden and reduced capacity.	Evaluate and manage cardiovascular disease, sleep-disordered breathing, and other comorbidities
Advanced testing	Worsening symptoms or disease progression.	Repeat PFTs, chest imaging, ABG analysis, and consider LTOT or nocturnal ventilation

pulmonary function test (PFT) results and analysing trends over time. Measuring the severity of dyspnea and history of exacerbations according to the Global Initiative for Chronic Obstructive Lung Disease (GOLD) guidelines is crucial. Tools such as the modified Medical Research Council (mMRC) dyspnoea scale or the COPD Assessment Test (CAT) can be used to evaluate dyspnoea severity. Documenting the frequency and severity of exacerbations, including treatments involving antibiotics and corticosteroids, and the need for hospitalization or intensive care, is also important. Reviewing inhaler therapy is necessary to ensure proper use of medications such as long-acting beta-agonists (LABA), long-acting antimuscarinics (LAMA), and inhaled corticosteroids (ICS). Poor adherence to inhalers, often due to complex regimens or improper usage, is common and associated with increased hospitalizations and economic burden. Addressing adherence challenges through patient education and follow-up supervision is critical. For patients with refractory COPD, repeat PFTs should be conducted to reassess airflow, lung volumes, and gas exchange capabilities. Evidence of worsening obstruction may indicate a progression of the disease, and such patients could be considered for lung volume reduction surgery. If new restrictive patterns or diffusion impairments are detected, chest imaging and a transthoracic echocardiogram (TTE) should be performed to rule out interstitial lung disease (ILD) or conditions like pulmonary hypertension. Arterial blood gas (ABG) analysis is necessary to identify hypercapnia or hypoxemia, which may require nocturnal non-invasive ventilation or long-term oxygen therapy (LTOT). Exercise pulse oximetry can further guide LTOT needs. Repeat CT imaging might be warranted to exclude potential comorbidities such as ILD, bronchiectasis, bronchiolitis, central airway obstruction, pulmonary edema, pulmonary hypertension, and lung cancer. COPD patients are often burdened with comorbid conditions that can significantly contribute to dyspnea and reduced exercise capacity. Common comorbidities include cardiovascular conditions such as coronary artery disease (CAD), congestive heart failure (CHF), cardiac arrhythmias, and pulmonary hypertension. Other frequent comorbidities include sleep-disordered breathing, obesity/metabolic syndrome, diabetes, anxiety, and depression. A comprehensive medical history and physical examination are essential for evaluating these conditions. A low threshold for further testing, such as transthoracic echocardiogram, cardiac stress tests, polysomnography, and chest CT, should be maintained to accurately diagnose and manage these common comorbidities.

Table 3 shows some pharmacological and non-pharmacological approaches towards COPD therapy. These are as follows:

Non-pharmacological Approaches for Managing COPD

Nutrition

Despite ongoing treatments, a significant number of patients with refractory COPD-up to 60%-experience severe weight loss or a decline in lean body mass due to malnutrition. This muscle wasting and cachexia contribute to adverse outcomes such

Table 3 Some pharmacological and non-pharmacological approaches for COPD

Clinical implications and management	*Diagnosis:* COPD is primarily diagnosed through clinical evaluations, spirometry (gold standard), and imaging studies. Post-bronchodilator FEV1/FVC <0.70 confirms COPD.
	Standard therapies: Include bronchodilators, vaccination, smoking cessation, and pulmonary rehabilitation.
	Refractory COPD: Patients with severe symptoms despite standard treatment. Requires comprehensive management, including advanced treatments and surgery.
	Co-morbidities: Often include cardiovascular issues, mental health disorders, diabetes, and liver disease. Managing these is crucial for better outcomes.
	Inhaler therapy: Challenges include poor adherence and incorrect use, particularly in patients with cognitive or physical impairments. Solutions involve simplifying regimens and patient education.
	Repeat testing: Essential to reassess COPD severity, especially in refractory cases. Includes pulmonary function tests, ABG analysis, and imaging to identify progression or comorbidities.
Non-pharmacological approaches	*Nutrition:* Malnutrition is common in COPD, leading to muscle wasting and poor outcomes. Oral supplements can help, but benefits are modest. Further research is needed on effective nutritional interventions.
	Pulmonary rehabilitation: Includes exercise training, education, and behavior modification. Shown to improve exercise capacity and quality of life. Programs typically last 8–12 weeks, with benefits plateauing after 12 weeks.
	Non-invasive positive pressure ventilation (NIPPV): Used in patients with hypercapnia or hypoxemia. Helps reduce mortality, but evidence on improving dyspnea and quality of life is mixed. Initiation typically occurs in a sleep lab or monitored setting.
Pharmacological approaches for refractory COPD	*Long-term oxygen therapy (LTOT):* Indicated for severe hypoxemia; improves survival but remains controversial for moderate hypoxemia.
	Macrolides: Reduce exacerbations by improving alveolar macrophage function. Risks include cardiac arrhythmias and antibiotic resistance.
	Phosphodiesterase-4 (PDE-4) inhibitors: Roflumilast reduces inflammation and exacerbations but has side effects like gastrointestinal and psychiatric issues.
	Systemic corticosteroids: Used to reduce exacerbation relapse in certain patients. Chronic use is not recommended due to adverse effects.
	Biologics: Target type 2 inflammation in eosinophilic COPD. Include anti-IL-5 therapies (mepolizumab, benralizumab) and anti-IL-4/IL-13 (dupilumab), which improve outcomes in selected patients.
	Lung volume reduction surgery (LVRS) and bronchoscopic lung volume reduction (BLVR): Surgical and less invasive options for severe COPD. LVRS improves lung function but carries risks. BLVR (e.g., endobronchial valves) is a less invasive alternative.

as accelerated deterioration in lung function, weakened immune response, reduced exercise capacity, and an elevated risk of mortality (De Brandt et al. 2022). Oral nutritional supplements have shown some effectiveness, with high-calorie supplements or enteral formulations contributing to modest weight gain and the preservation of fat-free mass. However, most studies in this area have been limited by small sample sizes or insufficient controls. Additionally, more comprehensive benefits, such as increased strength, improved health-related quality of life, or reduced healthcare utilization, are generally not observed. Two prospective studies combining rigorous nutritional supplementation with resistance training, anabolic steroids, and acute exacerbation therapy have demonstrated significant improvements in outcomes like exercise capacity and survival (Yu et al. 2014). Nonetheless, further research is necessary to determine the most effective nutritional supplements for the hypermetabolic state of refractory COPD. Certain nutrients, including probiotics, antioxidants, and omega-3 fatty acids, may play a role in maintaining gut integrity, reducing infection risk, and mitigating systemic inflammation. However, current evidence does not support the routine use of dietary therapies to improve long-term outcomes without addressing the underlying COPD disease activity, even though short-term stabilization may be achievable through these means. Future research priorities include randomized controlled trials comparing oral supplements and enteral feeding techniques within integrated, patient-centred care models.

Pulmonary Rehabilitation

Pulmonary rehabilitation is defined by the American Thoracic Society and European Respiratory Society as a comprehensive intervention that starts with a thorough patient assessment and continues with personalized therapies, including exercise training, education, and behavior modification, aimed at improving the physical and psychological well-being of individuals with chronic respiratory diseases. This program typically includes exercise training (such as endurance exercises and upper and lower extremity workouts), promotion of healthy behaviors (like smoking cessation, regular physical activity, healthy eating, adherence to prescribed medications, and effective disease management), and psychological support to enhance self-efficacy and coping strategies for chronic illness. The global initiative for chronic obstructive lung disease recommends pulmonary rehabilitation as part of an integrated approach for managing symptomatic patients with functional limitations and those who experience frequent exacerbations (Groups B and E) (Agusti et al. 2023). Pre-program assessments include evaluating respiratory impairment, exercise tolerance, and the presence of comorbidities, particularly cardiac, neurological, and musculoskeletal conditions. While the optimal duration of pulmonary rehabilitation is not clearly defined, the typical program lasts 8–12 weeks, with benefits tending to plateau after 12 weeks. Pulmonary rehabilitation has been shown to improve exercise capacity and quality of life. A meta-analysis of 65 randomized controlled trials revealed significant improvements in the 6-min walk distance compared to standard care and noted enhanced quality of life in domains such as dyspnea, fatigue, emotional function, and mastery. Although there is some evidence suggesting a potential mortality benefit, the quality of this evidence is low and inconsistent across studies (Tonga and Oliver 2023). Unfortunately, the benefits of pulmonary rehabilitation tend to diminish over time, and the timing and role of retraining are not well defined. However, an observational study suggests that maintenance with a self-monitored program can help sustain improvements in exercise tolerance and health status.

Non-invasive Positive Pressure Ventilation

Patients with refractory COPD who experience daytime hypercapnia or nocturnal hypoxemia unresponsive to nocturnal oxygen therapy may be candidates for non-invasive positive pressure ventilation (NIPPV). It is important to assess these patients for obstructive sleep apnoea (OSA) using polysomnography or through referral to a sleep medicine specialist, as OSA requires a different treatment approach. For patients with low clinical suspicion of OSA or those with a normal polysomnogram but ongoing nocturnal hypoxemia, further evaluation for NIPPV candidacy is warranted. These patients should undergo repeat pulmonary function tests and arterial blood gas (ABG) analysis. In advanced COPD, to counteract hypercapnia caused by ventilation-perfusion mismatch, the respiratory muscles must work harder to increase minute ventilation. This increased ventilatory load, coupled with changes in diaphragm configuration and insufficient nutritional support, can lead to respiratory muscle fatigue. NIPPV is hypothesized to improve respiratory function by providing rest to these overworked muscles, though the supporting evidence is still emerging. Early case series reported sustained improvement in daytime hypercapnia among patients using nocturnal NIPPV. More recent retrospective studies have shown improvements in spirometry indices, lung hyperinflation, and daytime hypercapnia over the course of a year. Consistent evidence suggests a reduction in mortality with NIPPV use, with a meta-analysis of over 800 patients showing a 14% reduction in mortality compared to usual care alone. However, the data on improvements in dyspnea, quality of life, and pulmonary function tests (PFTs) remain mixed (Rahi et al. 2024). Practical challenges associated with NIPPV vary by region and clinical scenario. For patients hospitalized with COPD exacerbation, bi-level positive airway pressure (BPAP) is typically used, allowing for close monitoring and determination of optimal settings. In non-hospitalized advanced COPD patients, initiating nocturnal NIPPV can be challenging. Efforts should be made to start NIPPV in a sleep laboratory or hospital setting where the patient can be closely monitored, which also helps exclude OSA and central sleep apnoea (CSA) (Menga et al. 2023).

Pharmacological Approaches for Refractory COPD Management

For patients with refractory COPD, particularly those experiencing frequent exacerbations despite standard therapy, various pharmacological options are available that target different aspects of the disease pathology to provide symptomatic relief, reduce exacerbations, and improve overall patient outcomes. Long-term oxygen therapy (LTOT) is indicated for patients with severe hypoxemia, defined as an oxygen saturation of ≤88% or an arterial partial pressure of oxygen (PaO_2) ≤55 mmHg. LTOT has been shown to improve survival in patients with severe hypoxemia, reduce pulmonary vascular resistance, and alleviate symptoms of hypoxemia-related complications (Elshof and Duiverman 2020). Chronic bacterial colonization in COPD patients leads to persistent inflammation and recurrent exacerbations. Macrolides, known for their immunomodulatory effects, are effective in reducing exacerbation frequency by improving alveolar macrophage function, decreasing airway inflammation, and reducing bacterial colonization (Rabe 2011). While macrolides have shown significant benefits, they also carry risks such as cardiac arrhythmias and antibiotic resistance, necessitating careful consideration, particularly in patients with co-existing risk factors. Phosphodiesterase-4 (PDE-4) inhibitors, such as roflumilast, reduce inflammation by increasing cyclic AMP levels, leading to bronchodilation and reduced exacerbations (Rabe 2011).

Roflumilast is particularly beneficial for patients with severe COPD who experience frequent exacerbations. However, its use is associated with gastrointestinal and psychiatric side effects, requiring a cautious approach to dosing and monitoring. Systemic corticosteroids can reduce the relapse of COPD exacerbations in certain patients, particularly those with elevated blood eosinophil levels. However, chronic use of systemic corticosteroids in refractory COPD is not recommended due to the high risk of adverse effects, including steroid myopathy, osteoporosis, and increased susceptibility to infections (Falk et al. 2008). Biologics targeting specific cytokines involved in type 2 inflammation, such as interleukins-4, -5, and -13, offer a novel therapeutic option for a subset of COPD patients with eosinophilic inflammation. Anti-IL-5 therapies like mepolizumab and benralizumab have shown efficacy in reducing exacerbations in patients with elevated eosinophil counts, while anti-IL-4/IL-13 therapy with dupilumab has demonstrated more consistent benefits, including reduced exacerbations, improved lung function, and enhanced quality of life. Patients with advanced or refractory COPD who are motivated may benefit from evaluating invasive management strategies, including lung volume reduction surgery (LVRS) and bronchoscopic lung volume reduction (BLVR) using endobronchial valves (EBVs). LVRS involves removing the most emphysematous parts of the lung to reduce hyperinflation and improve the function of the diaphragm and intercostal muscles. This approach can improve ventilation-perfusion matching and cardiac output and may have anti-inflammatory effects by reducing cytokines such as tumor necrosis factor-alpha and interleukins.

The national emphysema treatment trial (NETT) showed that while short-term mortality was higher in the surgical group, LVRS led to improved exercise capacity and lung function, particularly in patients with upper-lobe predominant emphysema and poor exercise tolerance. However, patients with very severe COPD (e.g., FEV1 ≤20% and low DLCO) were at higher risk and thus are contraindicated for LVRS (Criner and Sternberg 2008). Despite the risks, LVRS can significantly improve dyspnea, quality of life, and oxygenation in selected patients. BLVR is a less invasive alternative to LVRS, achieved through techniques such as the placement of EBVs, coils, sealants, or thermal airway ablation. EBVs are the most widely approved and used, designed to collapse emphysematous lung tissue and reduce hyperinflation. The success of EBV therapy largely depends on the integrity of lung fissures to prevent collateral ventilation, with complete fissures being essential for a good outcome. EBV therapy has been shown to improve lung function, exercise tolerance, and quality of life, with modest long-term survival benefits. However, it carries risks such as pneumothorax, hypoxemia, and exacerbations of COPD. Long-term complications can include granulation tissue formation, valve malfunction, and bacterial colonization (Garner and Desai 2021). For some patients with refractory COPD, lung transplantation may be the most suitable option. COPD is the leading indication for lung transplants, with the aim of extending life and improving quality of life.

Candidates for lung transplantation typically have severe, progressive disease despite maximal medical management and are often assessed using the BODE index, which predicts mortality based on body mass index, airflow obstruction, dyspnea, and exercise capacity. Lung transplantation is a complex decision requiring thorough evaluation and a strong support system. Contraindications include advanced age, obesity, active smoking, severe cardiovascular disease, and overall frailty. Despite the challenges, lung transplantation offers significant functional benefits and improved survival for suitable candidates. For patients with refractory COPD who are not candidates for invasive procedures, palliative care options are available to manage symptoms such as dyspnea and chronic cough. Opioids like morphine and dihydrocodeine have shown some effectiveness in relieving breathlessness, although evidence is mixed. Other approaches include exercise training, breathing techniques, cognitive-behavioral therapy, and medication for managing chronic cough and coexisting anxiety or depression. Historically, the management of COPD has revolved around inhaled bronchodilators for maintenance and corticosteroid use during exacerbations. For years, the therapeutic landscape saw little significant advancement. However, recent developments are beginning to shift the trajectory of COPD treatment, paving the way for more personalized and precise approaches. With the growing recognition of distinct COPD phenotypes, the future of COPD management is likely to see a significant move toward personalized medicine. This approach aims to tailor treatments to the specific characteristics of individual patients, potentially improving outcomes and minimizing the trial-and-error nature of current therapies. Biologics are increasingly being studied as potential therapeutic options in COPD. One promising area of research is the role of IL-33. High levels of IL-33 expression have been observed in patients with COPD, though its role in disease pathogenesis remains unclear. Animal models have shown that targeting IL-33 can reduce lung inflammation, providing a rationale for clinical trials.

Currently, three major phase III clinical trials-OBERON (NCT05166889), TITANIA (NCT05158387), and PROSPERO (NCT05742802)-are underway, investigating the efficacy of the anti-IL-33 agent tozorakimab in COPD patients. These trials could usher in a new class of biologics for COPD treatment, particularly for those with specific inflammatory profiles. Stem cell therapy represents a cutting-edge approach in the management of COPD, focusing on the potential to repair damaged lung tissue. Although early studies have shown promise, this therapy is not yet ready for widespread clinical application. Challenges remain in identifying the most effective types of stem cells, ensuring safe and targeted delivery, and achieving consistent therapeutic outcomes across diverse patient populations. Continued research and clinical trials will be crucial to advancing this promising area of COPD management. Technology has already begun to reshape COPD management, with telemedicine playing a key role in reducing hospital readmissions. Looking ahead, artificial intelligence (AI) is expected to make a significant impact on COPD management. AI has the potential to enhance our understanding of COPD heterogeneity, analyze the complex interactions between the heart and lungs, and develop personalized management algorithms. These innovations could lead to more precise treatment plans, better patient outcomes, and more efficient healthcare delivery.

Conclusion

Chronic obstructive pulmonary disease remains a formidable challenge to global public health, distinguished by its progressive nature and substantial impact on quality of life. Our research has explored the multifaceted pathophysiology of COPD, shedding light on the complex interplay between chronic inflammation, oxidative stress, and structural lung damage that characterizes the disease. As the third leading cause of mortality worldwide, COPD's burden is underscored by its debilitating effects, which arise from the cumulative damage to the airways, lung parenchyma, and pulmonary vasculature. Central to the development of COPD is the persistent inflammatory response that drives disease progression. Chronic inflammation in COPD is marked by the infiltration and activation of various inflammatory cells, such as neutrophils, macrophages, and CD8+ T-lymphocytes. These cells release a cascade of pro-inflammatory cytokines, chemokines, and proteases, which disrupt the balance between proteases and antiproteases. This imbalance contributes to the degradation of lung tissue and the development of emphysema. Concurrently, oxidative stress exacerbates the inflammatory process, with reactive oxygen species (ROS) generated from both external sources, such as tobacco smoke, and internal sources, including activated inflammatory cells. The resultant cellular and molecular damage perpetuates a vicious cycle of inflammation and tissue destruction. The structural alterations in the lungs associated with COPD are profound. Airway remodeling, characterized by goblet cell hyperplasia, subepithelial fibrosis, and smooth muscle hypertrophy, leads to airway narrowing and increased resistance. Emphysema, marked by the destruction of alveolar walls and loss of elastic recoil, results in the formation of bullae and impaired gas exchange. These structural changes significantly contribute to the clinical manifestations of COPD, including persistent cough, sputum production, and progressive breathlessness.

The epidemiology of COPD is influenced by a complex interplay of genetic and environmental factors. While smoking remains the primary risk factor in developed countries, exposure to biomass fuel and environmental pollutants is increasingly recognized as a significant contributor in developing regions. Additionally, genetic predispositions, such as alpha-1 antitrypsin deficiency, have been identified as crucial determinants in the early onset and progression of emphysema. From a clinical perspective, the diagnosis of COPD relies on a combination of symptomatic presentation, spirometric evaluation, and imaging studies. Spirometry, with a post-bronchodilator FEV1/FVC ratio of less than 0.70, remains the gold standard for assessing the severity of airflow limitation. The management of COPD necessitates a comprehensive approach, incorporating both pharmacological and non-pharmacological strategies. Pharmacological treatments, including bronchodilators, inhaled corticosteroids, and phosphodiesterase-4 inhibitors, are essential for symptom relief and reduction of exacerbations.

Non-pharmacological interventions, such as smoking cessation, pulmonary rehabilitation, and long-term oxygen therapy, play a critical role in improving patient outcomes and slowing disease progression. In advanced cases, surgical options like lung volume reduction surgery and lung transplantation may be considered. This comprehensive exploration underscores the complexity of COPD and highlights the urgent need for continued research. Unraveling the intricate genetic, environmental, and molecular interactions that drive COPD will be pivotal in developing more targeted and effective therapeutic strategies. By advancing our understanding of the disease, we can enhance treatment approaches, improve patient outcomes, and ultimately reduce the global burden of COPD. The ongoing commitment to research and innovation is crucial for achieving these goals and for improving the quality of life for those affected by this chronic and debilitating condition.

References

Agustí A, Celli BR, Criner GJ et al (2023) Global initiative for chronic obstructive lung disease 2023 report: GOLD executive summary. Am J Respir Crit Care Med 207(7):819–837

Alqahtani JS (2022) Prevalence, incidence, morbidity and mortality rates of COPD in Saudi Arabia: trends in burden of COPD from 1990 to 2019. PLoS One 17(5):e0268772

Bailey KL (2012) The importance of the assessment of pulmonary function in COPD. Med Clin North Am 96(4):745–752

Barnes PJ (2020) Oxidative stress-based therapeutics in COPD. Redox Biol 33:101544

Criner GJ, Sternberg AL (2008) National Emphysema Treatment Trial: the major outcomes of lung volume reduction surgery in severe emphysema. Proc Am Thorac Soc 5(4):393–405

De Brandt J, Beijers RJHCG, Chiles J et al (2022) Update on the etiology, assessment, and management of COPD cachexia: considerations for the clinician. Int J Chron Obstruct Pulmon Dis 17:2957–2976
Duan RR, Hao K, Yang T (2020) Air pollution and chronic obstructive pulmonary disease. Chronic Dis Transl Med 6(4):260–269
Elshof J, Duiverman ML (2020) Clinical evidence of nasal high-flow therapy in chronic obstructive pulmonary disease patients. Respiration 99(2):140–153
Falk JA, Minai OA, Mosenifar Z (2008) Inhaled and systemic corticosteroids in chronic obstructive pulmonary disease. Proc Am Thorac Soc 5(4):506–512
Ferrera MC, Labaki WW, Han MK (2021) Advances in chronic obstructive pulmonary disease. Annu Rev Med 72:119–134
Garner JL, Desai SR (2021) Lung fissural integrity: it's written in the genes. Am J Respir Crit Care Med 204(7):750–752
Grahn K, Gustavsson P, Andersson T (2021) Occupational exposure to particles and increased risk of developing chronic obstructive pulmonary disease (COPD): a population-based cohort study in Stockholm. Sweden Environ Res 200:111739
Gupta I, Ganguly S, Rozanas CR et al (2016) Ascorbate attenuates pulmonary emphysema by inhibiting tobacco smoke and Rtp801-triggered lung protein modification and proteolysis. Proc Natl Acad Sci USA 113:E4208–E4217
Hattab Y, Alhassan S, Balaan M et al (2016) Chronic obstructive pulmonary disease. Crit Care Nurs Q 39:124–130
Hersh CP (2017) Diagnosing alpha-1 antitrypsin deficiency: the first step in precision medicine. F1000Res 6:2049
Higham A, Quinn AM, Cançado JED et al (2019) The pathology of small airways disease in COPD: historical aspects and future directions. Respir Res 20:49
Hikichi M, Mizumura K, Maruoka S et al (2019) Pathogenesis of chronic obstructive pulmonary disease (COPD) induced by cigarette smoke. J Thorac Dis 11(Suppl 17):S2129–S2140
Ho T, Cusack RP, Chaudhary N et al (2019) Under- and over-diagnosis of COPD: a global perspective. Breathe (Sheff) 15(1):24–35
Jarhyan P, Hutchinson A, Khaw D et al (2022) Prevalence of chronic obstructive pulmonary disease and chronic bronchitis in eight countries: a systematic review and meta-analysis. Bull World Health Organ 100(3):216–230
Kany S, Vollrath JT, Relja B (2019) Cytokines in inflammatory disease. Int J Mol Sci 20(23):6008
Lin CH, Cheng SL, Chen CZ et al (2023) Current progress of COPD early detection: key points and novel strategies. Int J Chron Obstruct Pulmon Dis 18:1511–1524
Menezes AM, Victora CG, Perez-Padilla R et al (2004) The Platino project: methodology of a multicenter prevalence survey of chronic obstructive pulmonary disease in major Latin American cities. BMC Med Res Methodol 4:15
Menga LS, Delle Cese L, Rosa T et al (2023) Respective effects of helmet pressure support, continuous positive airway pressure, and nasal high-flow in hypoxemic respiratory failure: a randomized crossover clinical trial. Am J Respir Crit Care Med 207:1310–1323
Molloy K, Hersh CP, Morris VB et al (2014) Clarification of the risk of chronic obstructive pulmonary disease in α1-antitrypsin deficiency PiMZ heterozygotes. Am J Respir Crit Care Med 189(4):419–427
Murray MA, Mulryan K, Ní Chleirigh M et al (2023) Caring for patients with advanced COPD: beyond the inhaler. Breathe (Sheff) 19(1):220229
Nurwidya F, Damayanti T, Yunus F (2016) The role of innate and adaptive immune cells in the immunopathogenesis of chronic obstructive pulmonary disease. Tuberc Respir Dis (Seoul) 79(1):5–13
Pandey KC, De S, Mishra PK (2017) Role of proteases in chronic obstructive pulmonary disease. Front Pharmacol 8:512
Parris BA, O'Farrell HE, Fong KM et al (2019) Chronic obstructive pulmonary disease (COPD) and lung cancer: common pathways for pathogenesis. J Thorac Dis 11(Suppl 17):S2155–S2172

Rabe KF (2011) Update on roflumilast, a phosphodiesterase 4 inhibitor for the treatment of chronic obstructive pulmonary disease. Br J Pharmacol 163(1):53–67

Ragland MF, Benway CJ, Lutz SM et al (2019) Genetic advances in chronic obstructive pulmonary disease. Insights from COPD gene. Am J Respir Crit Care Med 6:677–690

Rahi MS, Mudgal M, Asokar BK et al (2024) Management of refractory chronic obstructive pulmonary disease: a review. Life (Basel) 14(5):542

Rovina N, Koutsoukou A, Koulouris NG (2013) Inflammation and immune response in COPD: where do we stand? Mediat Inflamm 2013:413735

Russell S, Ogunbayo OJ, Newham JJ et al (2018) Qualitative systematic review of barriers and facilitators to self-management of chronic obstructive pulmonary disease: views of patients and healthcare professionals. NPJ Prim Care Respir Med 28(1):2

Saetta M, Turato G, Maestrelli P et al (2001) Cellular and structural bases of chronic obstructive pulmonary disease. Am J Respir Crit Care Med 163(6):1304–1309

Safiri S, Carson-Chahhoud K, Noori M et al (2022) Burden of chronic obstructive pulmonary disease and its attributable risk factors in 204 countries and territories, 1990-2019: results from the Global Burden of Disease Study 2019. BMJ 378:e069679

Silverman EK (2020) Genetics of COPD. Annu Rev Physiol 82:413–431

Tonga KO, Oliver BG (2023) Effectiveness of pulmonary rehabilitation for chronic obstructive pulmonary disease therapy: focusing on traditional medical practices. J Clin Med 12(14):4815

Toren K, Vikgren J, Olin AC et al (2017) Occupational exposure to vapor, gas, dust, or fumes and chronic airflow limitation, COPD, and emphysema: the Swedish Cardio Pulmonary BioImage Study (SCAPIS pilot). Int J Chron Obstruct Pulmon Dis 12:3407–3413

Torres-Duran M, Lopez-Campos JL, Barrecheguren M et al (2018) Alpha-1 antitrypsin deficiency: outstanding questions and future directions. Orphanet J Rare Dis 13(1):114

Yu JG, Bonnerud P, Eriksson A et al (2014) Effects of long-term supplementation of anabolic androgen steroids on human skeletal muscle. PLoS One 9(9):e105330

Smoking as a Confounding Factor in Environmentally Induced Lifestyle Diseases

Syed Amaan Ali, Neha Agrawal, Mohammad Atif, Mohd Mustafa, and Safia Habib

Abstract

One important complicating element in the development and course of lifestyle illnesses is smoking, which greatly affects public health outcomes. The complex link between smoking and lifestyle disorders like cancer, diabetes, and cardiovascular disease is examined in this chapter. It delves into the biological mechanisms through which tobacco smoke exacerbates these conditions, including oxidative stress, inflammation, and endothelial dysfunction. This chapter also emphasizes how difficult it is to separate the negative consequences of smoking from other lifestyle factors, like nutrition and inactivity, which frequently overlap and make determining one's risk for health problems more difficult. Epidemiological studies consistently demonstrate that smoking exacerbates the severity and prevalence of lifestyle diseases, complicating efforts to manage and treat these conditions effectively. Effective public health strategies and interventions targeting smoking cessation are crucial in mitigating its role as a confounding factor. To address the interconnectedness of lifestyle disorders and smoking, this chapter ends by promoting comprehensive, multifactorial approaches in research

S. A. Ali · N. Agrawal
Department of Periodontics and Community Dentistry, Dr. Ziauddin Ahmad Dental College, Faculty of Medicine, Aligarh Muslim University, Aligarh, Uttar Pradesh, India
e-mail: saali.pu@amu.ac.in

M. Atif
Department of Paediatric and Preventive Dentistry, Dr. Ziauddin Ahmad Dental College, Aligarh Muslim University, Aligarh, Uttar Pradesh, India

M. Mustafa · S. Habib (✉)
Department of Biochemistry, Faculty of Medicine, Jawaharlal Nehru Medical College, Aligarh Muslim University, Aligarh, Uttar Pradesh, India
e-mail: safia.bh@amu.ac.in

S. Raisuddin et al. (eds.), *Environmental Toxicants and Lifestyle Diseases*,
https://doi.org/10.1007/978-981-95-0575-3_18

and healthcare practices. This will improve overall health outcomes and lessen the burden of chronic diseases worldwide.

Keywords

Smoking · Tobacco smoke · Lifestyle diseases · Non-communicable diseases

Introduction

Non-communicable diseases (NCDs) are another name for lifestyle diseases. They are ailments that mainly result from the choices and conduct of a person's way of living. On the contrary, communicable ailments are due to pathogens, which can be passed from one person to another. Lifestyle diseases arise from a complicated mixture of environment, genes, and behavior. These are cardiovascular conditions, cancers, diabetes, and chronic respiratory infections (Modi et al. 2023). Heart-related conditions, such as hypertension, coronary artery disease, stroke, and heart failure, are included in the category of conditions known as cardiovascular diseases (CVDs) (Roth et al. 2020). CVDs have four main risk factors: obesity, smoking, physical inactivity, and unhealthy eating habits. These behaviors result in high blood pressure, glucose intolerance, dyslipidemia, and overweight or excessive fat accumulation that significantly contribute to the progression of CVDs. Globally, according to WHO analysis, it was deduced that CVD accounts for 17.9 million deaths yearly, which is approximately 31% of all global deaths (Çomo et al. 2022).

Unusual cell growth and dissemination characterize the group of diseases known as cancer. Smoking, binge drinking, eating poorly, not exercising, and being around toxins are lifestyle choices that greatly increase the risk of developing cancer. Tobacco use stands out as the most significant risk factor, accounting for about 22% of cancer-related deaths. Additionally, dietary habits, like insufficient intake of fruits and vegetables and high consumption of processed or red meats, are linked to various types of cancer. The International Agency for Research on Cancer (IARC) suggests that adjusting or evading crucial risk factors and applying proven prevention methods could prevent 30–50% of cancer cases (Budrevičiūtė et al. 2020). The airways and other lung structures are impacted by chronic respiratory disorders (CRDs), which include asthma and chronic obstructive pulmonary disease (COPD). The primary risk factors for CRDs include tobacco smoke (including second-hand smoke), indoor and outdoor air pollution, occupational dust and chemicals, and frequent lower respiratory infections during childhood. According to WHO estimates, about three million people die from COPD annually, accounting for 6% of all deaths globally. Effective management of CRDs involves reducing exposure to risk factors, improving air quality, and enhancing healthcare systems to provide better diagnosis and treatment (Parums 2023). A metabolic disorder called diabetes is characterized by consistently elevated blood sugar levels brought on by problems with insulin secretion, action, or both. Type 2 diabetes, the most common form, is largely linked to being overweight and not being physically active (Fig. 1).

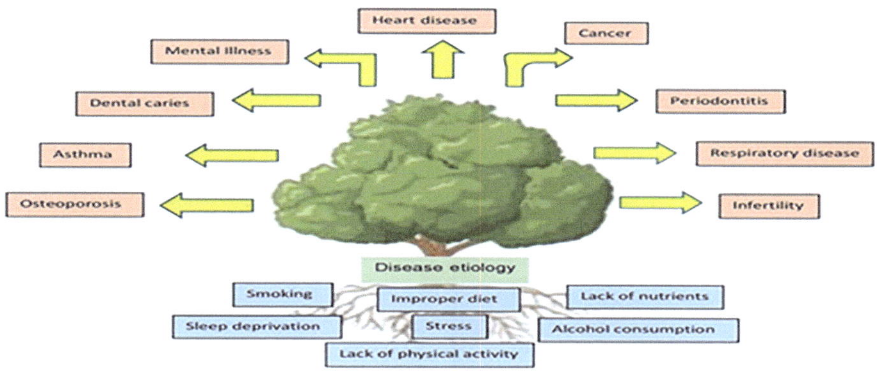

Fig. 1 Factors contributing to lifestyle disease

Globally, the incidence of diabetes is rising, with approximately 422 million adults affected in 2014, up from 108 million in 1980. Complications of diabetes include cardiovascular problems, kidney disease, vision loss, and amputation of lower limbs, significantly reducing quality of life and raising healthcare expenses (Cosentino et al. 2019).

Environmental Factors in Lifestyle Diseases

Environmental factors play a crucial role in the development and progression of lifestyle diseases. Unlike genetic and behavioral factors, which are often personal and individual, External conditions are environmental factors that affect groups of people and can often be mitigated through public policy and community efforts. Key environmental factors influencing lifestyle diseases include air pollution, exposure to toxins, urbanization, climate change, and the built environment (Reddy et al. 2022).

One of the biggest environmental factors that contributes to lifestyle disorders is air pollution. The presence of fine particulate matter (PM2.5) and other pollutants like sulfur dioxide (SO_2) and nitrogen dioxide (NO_2) has been related to cancer, respiratory disorders, and heart problems. Prolonged exposure to air pollution can cause oxidative stress and inflammation, which furthers the pathophysiology of many illnesses. According to the World Health Organization (WHO), ambient air pollution contributed to 4.2 million premature deaths globally in 2016. The bulk of these fatalities were attributed to ischemic heart disease and stroke (World Health Organization 2018). Exposure to environmental toxins, such as heavy metals (arsenic, lead, and mercury), pesticides, and industrial chemicals, is another critical factor in lifestyle diseases. These substances can contaminate air, water, and soil, leading to various health problems. For example, lead exposure has been correlated to hypertension and cardiovascular disease, while arsenic exposure increases the risk of skin, bladder, and lung cancers. Pesticides and industrial chemicals can interfere with endocrine function, leading to an increased risk of diabetes and obesity. The lasting impact of these toxins underscores the importance of strict environmental regulations and ongoing monitoring (Liu et al. 2022). The growth of cities and the migration of populations from rural to urban areas significantly impact lifestyle diseases. Urban environments often promote sedentary lifestyles due to limited space for physical activities and reliance on motorized transport. Moreover, the availability of unhealthy food options, such as fast food, and increased stress levels due to crowded living conditions and fast-paced lifestyles contribute to the rise in cardiovascular diseases, obesity, and diabetes. Studies have shown that urban residents are more likely to experience higher levels of air pollution and reduced access to green spaces, which further exacerbate these health issues (Balwan and Kour 2021).

Climate change is increasingly acknowledged as a significant factor affecting health, shaping the prevalence and intensity of lifestyle diseases. Higher temperatures and extreme weather conditions can worsen cardiovascular and respiratory issues. For example, heat waves heighten the risk of heatstroke and cardiovascular incidents, while shifting climate patterns can alter the spread of allergens and irritants, aggravating asthma and other respiratory ailments (Balwan and Kour 2021). The built environment, encompassing structures, parks, and transportation systems, plays a crucial role in shaping lifestyle diseases. Access to safe and walkable neighborhoods, availability of recreational facilities, and proximity to healthy food options are crucial for promoting physical activity and healthy eating habits. Poorly designed cities lacking sufficient green spaces and recreational facilities discourage physical activity, promoting sedentary lifestyles (Reddy et al. 2022).

Smoking remains a significant controllable risk factor for numerous lifestyle-related diseases, such as chronic respiratory conditions, cardiovascular diseases, cancers, and diabetes. Despite extensive public health campaigns and regulations designed to lower smoking rates, tobacco use persists as a major global health challenge, significantly contributing to illness and death worldwide. Recognizing the complex role of smoking as a confounding factor in lifestyle diseases is essential for creating effective prevention and intervention strategies (Dai et al. 2022). Smoking has a major impact on the development of CVDs, which are the leading cause of mortality globally. Tobacco smoke contains harmful chemicals like nicotine, carbon monoxide, and carcinogens that negatively impact the cardiovascular system. Smoking triggers atherosclerosis, the accumulation of fatty deposits in arteries, which can result in conditions such as coronary artery disease (CAD), heart attacks, and strokes. The toxic substances in cigarette smoke contribute to endothelial dysfunction, inflammation, and platelet aggregation, all of which are critical processes in the pathogenesis of atherosclerosis (Libby 2021). Studies reveal that the risk of coronary artery disease is two to four times higher in smokers than in non-smokers. Additionally, smoking is linked to a higher risk of hypertension and atherosclerosis, both of which are significant contributors to heart disease. Evidence suggests that smoking raises the chances of both ischemic and hemorrhagic strokes due to its impact on blood pressure and clotting mechanisms (Dai et al. 2022).

Tobacco smoke is the foremost contributor to multiple forms of cancer, making smoking a primary preventable factor for cancer globally. Over 7000 chemicals, many of which cause cancer, are found in cigarette smoke. These substances can harm cell DNA, causing mutations that elevate cancer risks. Tobacco smoke is categorized as a Group 1 carcinogen by the IARC, affirming strong evidence linking smoking to various cancers such as lung, oral, throat, oesophageal, pancreatic, bladder, kidney, and cervical cancers (Siegel et al. 2023). Lung cancer is predominantly linked to smoking, accounting for around 85% of cases (American Cancer Society 2019). The amount of cigarettes smoked each day and the length of time one smokes increase the risk of lung cancer. Moreover, smoking heightens the risk of other respiratory cancers like those affecting the larynx and trachea. Lifestyle factors such as diet and physical activity also interact with smoking, further influencing cancer risk (American Cancer Society 2019).

Chronic respiratory diseases (CRDs) like chronic COPD stem from prolonged exposure to tobacco smoke. The harmful components in cigarette smoke provoke chronic inflammation and harm to the airways and lung tissue. This damage reduces lung capacity, leading to breathing difficulties and raising the likelihood of respiratory infections and flare-ups (Lareau et al. 2019). Studies indicate that smoking is responsible for approximately 80–90% of COPD cases. The impact of smoking on chronic respiratory diseases highlights the necessity of smoking cessation as a crucial element in managing and preventing these conditions (Leung et al. 2020).

The connection between smoking and diabetes is intricate, as smoking not only increases the incidence of type 2 diabetes but also worsens its complications. Smoking impairs insulin sensitivity and encourages insulin resistance, which further contributes to the onset of type 2 diabetes. Research indicates that smokers have a 30–40% higher risk of developing diabetes compared to non-smokers (Lee and Hamling 2014). Additionally, smoking can complicate the management of diabetes by increasing the risk of cardiovascular diseases, neuropathy, and nephropathy, thereby worsening the overall prognosis for diabetic patients.

Confounding Effects and Public Health Implications: Smoking not only directly contributes to the onset of lifestyle diseases but also serves as a confounding factor when analyzing other risk factors. For instance, the link between physical inactivity and cardiovascular diseases might be skewed by smoking, as smokers are more prone to being physically inactive. Likewise, the connection between a poor diet and cancer risk may be affected by smoking, since smokers often have different dietary patterns than non-smokers. Hence, it is crucial to account for smoking in epidemiological studies to accurately evaluate the independent effects of other risk factors (Santos et al. 2022). Efforts in public health aimed at lowering smoking rates have been successful, incorporating strategies like tobacco taxation, smoking bans in public areas, impactful health warnings on cigarette packaging, and cessation programs. These initiatives have resulted in notable reductions in smoking rates across several nations. Yet, challenges persist, particularly in low- and middle-income countries, where smoking prevalence remains high and tobacco control policies may be inadequate. Strengthening worldwide tobacco control endeavors and ensuring their thorough and consistent implementation is essential to lessen the impact of smoking-related illnesses (Masud et al. 2023).

Understanding Confounding Factors

Confounding factors, or confounders, are variables that impact both the dependent variable (outcome) and the independent variable (exposure) in a study, creating a misleading connection between the two. In epidemiological research, confounders can obscure or amplify the actual relationship between the variables being examined. For example, when investigating the association between environmental exposures (such as air pollution) and lifestyle diseases, confounding factors like smoking can alter the observed relationships. This distortion happens because smoking independently influences the risk of developing lifestyle diseases and is also related to the degree of environmental exposure (Binette et al. 2023).

Confounding factors can be broadly categorized into three types: Known confounders, unknown confounders, and time-related confounders. Each type plays a critical role in the interpretation of epidemiological studies (McGowan 2022).

1. *Known confounders:* Known confounders are the variables that have been identified and measured in a study, allowing researchers to relate their effects through statistical adjustments. In the context of smoking as a confounding factor in environmentally induced lifestyle diseases, known confounders could include, age, sex, and socioeconomic factors. Age is a well-recognized confounder in studies of lifestyle diseases because the potential risk of diseases like cancer, cardiovascular disease, and respiratory conditions increases with age. Older individuals may also have longer exposure to both smoking and environmental pollutants (McGowan 2022). Differences in disease prevalence and smoking habits between men and women make sex an important confounder. For instance, men historically have higher smoking rates, which could affect studies linking environmental exposures to lifestyle diseases (Reimer et al. 2020). Socioeconomic status influences both smoking behavior and environmental exposures. Lower SES is often associated with high smoking rates and greater exposure to environmental hazards, complicating the assessment of their independent effects on health (Reimer et al. 2020).
2. *Unknown confounders:* Unknown confounders are the variables that researchers have not identified or measured, making it difficult to account for their effects. These confounders can lead to residual confounding, where the association between exposure and outcome remains biased even after adjusting for known confounders (Cao et al. 2023). For example, genetic predispositions and behavioral factors. Genetic factors that make individuals more susceptible to both smoking and lifestyle diseases may not always be fully understood or accounted for in studies. These genetic predispositions can affect the apparent relationships between environmental exposures and health outcomes. Lifestyle behaviors,

such as diet and physical activity, might act as confounders if they are linked with both smoking and exposure to environmental pollutants. If not measured, these behaviors can lead to biased results.

3. *Time-related confounders:* Time-related confounders involve changes in due course that affect both the exposure and the outcome. These confounders are particularly relevant in longitudinal studies where both smoking habits and environmental exposures can vary over time (Cao et al. 2023). Examples include changes in smoking patterns and environmental exposures. Over a person's life, smoking habits may change, potentially influencing the timing and severity of disease onset. For instance, individuals who quit smoking might experience different health outcomes compared to those who continue smoking, complicating the analysis of environmental exposures. Environmental regulations and changes in industrial practices can alter levels of pollutants over time. Individuals exposed to higher pollution levels in the past might experience different health outcomes compared to those exposed to lower levels more recently.

Addressing smoking as a confounding in research: Smoking is a particularly challenging confounder in studies of environmentally induced lifestyle diseases due to its strong association with multiple health outcomes and its interaction with various environmental exposures. Here is how smoking functions as a confounding factor in these studies (US Department of Health and Human Services 2014). Smoking directly increases the threat of numerous lifestyle diseases, including cancers, cardiovascular diseases, and chronic respiratory diseases (Table 1). These direct effects can overshadow or obscure the impacts of environmental exposures (US Department of Health and Human Services 2014). Smoking is often associated with other risk behaviors, such as poor diet, physical inactivity, and alcohol consumption. These correlated behaviors can confound the relationship between environmental exposures and lifestyle diseases if not properly adjusted for in the analysis (US Department of Health and Human Services 2014). Researchers use several strategies to mitigate the effects of confounding factors like smoking (Cao et al. 2023):

1. *Statistical adjustment:* Techniques such as multivariable regression models can adjust for the effects of known confounders. Including smoking status as a covariate helps isolate the impact of environmental exposures on health outcomes.
2. *Stratification:* Dividing the study population into subgroups based on smoking status (e.g., current smokers, former smokers, never smokers) allows researchers to assess the association between environmental exposures and health outcomes within each subgroup, reducing confounding.
3. *Matching:* In case-control studies, matching cases and controls based on smoking status ensures that the comparison groups are similar for smoking, thereby reducing confounding.
4. *Sensitivity analysis:* Conducting sensitivity analyses to assess how robust the study results are to potential confounding by smoking can provide insights into the reliability of the findings.

Table 1 The confounding effect of smoking in lifestyle diseases (Li and Hecht 2022)

S. no.	Lifestyle diseases	Pathophysiology	Environmental triggers	Role of smoking
1.	Cardiovascular diseases	The process starts with endothelial injury, triggering inflammation Plaques rupture, causing thrombosis, leading to myocardial infarction	Air pollution (PM2.5 and NO_2) increases CVD risk by causing systemic inflammation and oxidative stress. Compromised endothelial function	Nicotine and carbon monoxide cause hypertension and reduce oxygen delivery to tissues
2.	Respiratory diseases	Chronic obstructive pulmonary disease (and lung cancer) are driven by inflammatory processes and genetic mutations	Environmental pollutants, particularly PM2.5, ozone, and industrial chemicals, are major contributors Long-term exposure to these pollutants causes chronic inflammation and oxidative damage Occupational exposures, such as asbestos and silica, are strongly linked to lung cancer and respiratory conditions	Inhaling tobacco smoke causes chronic inflammation and impaired immune response Smoking accelerates alveoli destruction. Smoking cessation slows COPD progression
3.	Cancers	Cancer cells evade apoptosis, sustain angiogenesis, and develop the ability to metastasize	Air pollution, particularly fine particulate matter and polycyclic aromatic hydrocarbons, has been linked to lung cancer and other malignancies Lifestyle factors, such as diet, alcohol, and smoking, modulate cancer risk	Smoking is a leading cause of various cancers Tobacco smoke contains over 60 carcinogens, which cause DNA damage and mutations Smoking cessation reduces the risk, but former smokers have a higher risk
4.	Diabetes	Diabetes mellitus is a group of metabolic disorders characterized by hyperglycemia The progression of diabetes involves complex interactions	Air pollution has been linked to an increased risk of diabetes Lifestyle factors are critical determinants of diabetes risk Urbanization and	Smoking is a significant modifiable risk factor for type 2 diabetes Chemicals in tobacco smoke impair insulin

(continued)

Table 1 (continued)

S. no.	Lifestyle diseases	Pathophysiology	Environmental triggers	Role of smoking
		between genetic factors, lifestyle, and environmental exposures	changes in dietary patterns have contributed to the global rise in diabetes prevalence	sensitivity Smokers have abdominal obesity, a risk factor for insulin resistance

Mechanisms of Smoking-Related Health Effects

Chemical Composition of Tobacco Smoke: Tobacco smoke is an intricate blend of thousands of different chemicals, many of which have significant health impacts. These chemicals can be broadly categorized into several groups, including gases, particulate matter, and various organic compounds. Understanding the chemical composition of tobacco smoke is crucial for grasping its health risks and the mechanisms by which it contributes to diseases (Li and Hecht 2022). Major chemical components like nicotine give it its addictive qualities. It is a potent alkaloid that stimulates the release of neurotransmitters such as dopamine. Nicotine contributes to cardiovascular diseases by increasing heart rate, and blood pressure, and promoting atherosclerosis. It also affects brain development and function, particularly in adolescents and pregnant women (Li and Hecht 2022). Another component is tar, which is sticky and brown, contributing to the staining of teeth and lungs. It damages the cilia in the lungs, impairing the body's ability to remove harmful substances. Carbon monoxide reduces the blood's oxygen-carrying capacity, leading to hypoxia. This contributes to cardiovascular diseases, such as myocardial infarctions and strokes, and can be particularly harmful to individuals with pre-existing heart conditions (Li and Hecht 2022). Formaldehyde is a known carcinogen, causing cancers of the nasopharynx and leukemia. It also irritates the respiratory system, eyes, and skin. Benzene exposure can lead to bone marrow suppression and immune system dysfunction and toxic heavy metals like cadmium is linked to kidney disease and bone damage, lead to neurological disorders, and arsenic to skin, lung, and bladder cancers (Li and Hecht 2022).

Smoking's Impact on Studying Environmental Diseases: Smoking has profound effects on the body's biological systems, making it difficult to study diseases caused by environmental factors and lifestyle choices. The widespread impact of smoking on various physiological functions complicates efforts to identify and understand how environmental factors alone contribute to these diseases. Smoking affects numerous aspects of health, from the cardiovascular system to the respiratory system and beyond. This makes it challenging for researchers to isolate the role of environmental factors without accounting for the significant influence of smoking. Therefore, understanding the specific biological mechanisms through which smoking affects health is essential. This knowledge is crucial for researchers to effectively separate the confounding effects of smoking from other environmental influences in epidemiological studies (US Department of Health and Human Services 2014).

Pathophysiology and the Role of Smoking in Lifestyle Diseases

Smoking introduces numerous toxins into the bloodstream, including nicotine, carbon monoxide, and oxidizing chemicals. Nicotine stimulates the release of catecholamines, which increases heart rate and blood pressure, leading to chronic hypertension. Carbon monoxide binds with hemoglobin more effectively when compared with oxygen, reducing the oxygen-carrying capacity of the blood and causing hypoxia. The oxidizing chemicals induce endothelial dysfunction and promote atherosclerosis by damaging the inner lining of blood vessels, facilitating the formation of plaques. These processes elevate the risk of heart attacks and strokes, which are also influenced by environmental pollutants like particulate matter (PM2.5). This overlap makes it challenging to distinguish the cardiovascular impacts of smoking from those caused by air pollution (US Department of Health and Human Services 2014). The inhalation of tobacco smoke directly damages the respiratory tract. The particulate matter and carcinogens in smoke induce chronic inflammation and oxidative stress in the lungs, leading to COPD and lung malignancies. Smoking impairs the function of cilia in the respiratory epithelium, reducing the clearance of inhaled pollutants and increasing susceptibility to respiratory infections and diseases. Environmental exposures, such as air pollution, similarly exacerbate respiratory conditions, creating a complex interplay between smoking and environmental factors. This interaction complicates the attribution of respiratory diseases solely to environmental exposures, as smoking intensifies the effects of pollutants (Li and Hecht 2022).

Smoking induces systemic inflammation, characterized by elevated levels of inflammatory cytokines. This chronic inflammatory state is a common pathway in the progression of various lifestyle diseases, including cardiovascular diseases and diabetes. It is challenging to separate the effect of smoking in investigations of environmentally generated diseases since environmental factors, such as exposure to chemicals and pollutants, also cause immunological responses and inflammation (Li and Hecht 2022). Table 1 shows the role of smoking in lifestyle diseases.

Smoking causes cancer through several harmful mechanisms. It exposes the body to carcinogens like PAHs and nitrosamines, which damage DNA and lead to mutations. These mutations can affect critical genes, turning normal cells into cancerous ones. Smoking also induces chronic inflammation, which further damages DNA and promotes cancer. Additionally, it suppresses the immune system, allowing mutated cells to survive and proliferate. Smoking stimulates angiogenesis, aiding tumor growth, and causes epigenetic changes that can activate oncogenes and inactivate tumor suppressor genes. Together, these mechanisms significantly increase the risk of cancer. These mechanisms contribute to cancers, such as lung, mouth, throat, esophagus, bladder, pancreas, kidney, liver, stomach, cervix, and certain leukemias (Kobets et al. 2022).

Smoking affects metabolic processes by inducing insulin resistance, which is a precursor to type 2 diabetes. Nicotine and other chemicals in tobacco smoke alter glucose metabolism and lipid profiles, leading to dyslipidemia and increased abdominal fat. These metabolic disturbances are similarly influenced by environmental factors like endocrine-disrupting chemicals, which can also hinder glucose and lipid metabolism (Li and Hecht 2022).

Impact of Smoking Cessation

Smoking cessation is one of the highest significant steps individuals can take to improve their overall health and longevity. The benefits of quitting smoking are substantial and begin almost immediately, continuing to accrue over time. This section explores the profound positive impacts on various aspects of health following smoking cessation (Sadarang 2021). The cardiovascular system is one of the first areas to experience improvements after quitting smoking. Within just 20 min of cessation, heart rate and blood pressure start to normalize. The risk of a heart attack starts to decline after a day. Within a year, the risk of coronary heart disease is halved compared to that of a smoker. Long-term cessation significantly suppresses the risk of stroke, peripheral artery disease, and overall cardiovascular mortality (Heber et al. 2019). Along with the cardiovascular system, the respiratory system is also heavily impacted by smoking, showing remarkable recovery post-cessation. In as little as 2 weeks to 3 months, lung function improves, and respiratory symptoms such as coughing and shortness of breath diminish. After 1–9 months, ciliary function in the lungs is restored, enhancing the lungs' ability to handle mucus, clean the airways, and reduce the risk of infection (Fig. 2). Over several years, the risk of COPD and lung cancer decreases substantially (Bowerman et al. 2022).

Lung, mouth, throat, esophageal, bladder, and pancreatic cancers are among the many malignancies for which smoking is a major risk factor. The risk of these diseases is considerably lower when one stops smoking. For instance, the chance of developing lung cancer is approximately half that of an ongoing smoker after 10 years of quitting. Risks for mouth, throat, and oesophagus cancers also decrease significantly within 5 years of quitting (US Department of Health and Human Services 2014). Quitting smoking results in improved immune function, enhanced reproductive health, improved mental health status, and overall quality of life (US Department of Health and Human Services 2014).

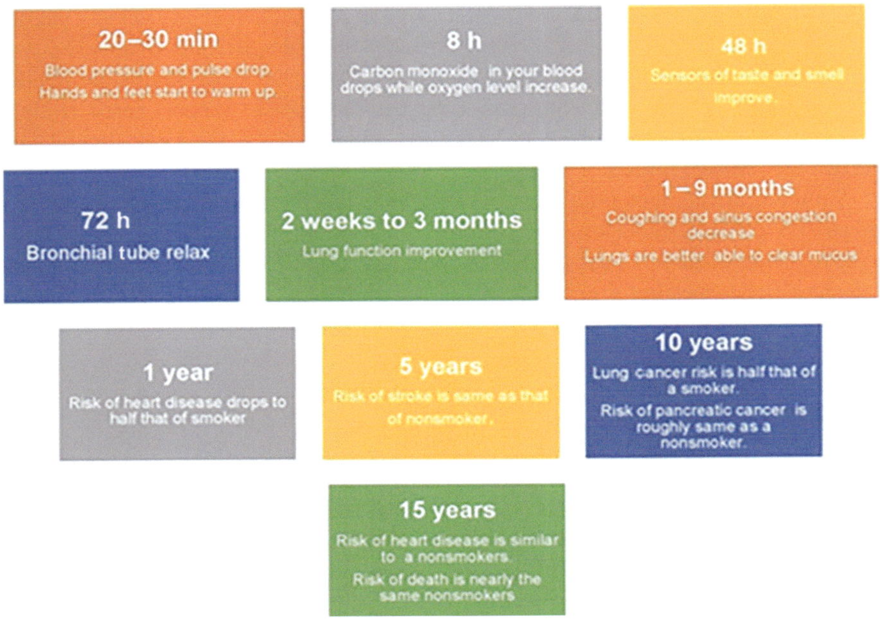

Fig. 2 Timelines to indicate how the body changes with respect to different diseases after quitting smoking

Public Health Implications

Strategies for Smoking Prevention and Cessation: Effective strategies for smoking prevention and cessation are critical to reducing the prevalence of smoking-related diseases and improving public health. These strategies encompass a wide range of approaches, from public policies and community programs to individual interventions and support systems. This section explores various evidence-based strategies designed to prevent smoking initiation and aid individuals in quitting smoking. Governments and public health organizations play a critical role in smoking prevention and cessation through the implementation of comprehensive policies (Pichón Riviere et al. 2023). Key policies include:

1. *Tobacco taxes:* Taxing tobacco goods more heavily is one of the best strategies to lower the number of people who smoke. Increased costs entice existing smokers to give up and deter young people from starting.
2. *Smoking bans:* Enforcing smoking bans in public places, such as workplaces, restaurants, and public transport, reduces exposure to second-hand smoke and de-normalizes smoking behavior.
3. *Advertising restrictions:* Banning tobacco advertising, promotion, and sponsorship helps reduce the appeal of smoking, particularly among young people. Plain packaging laws also diminish the attractiveness of tobacco products.

4. *Education campaigns:* Public awareness campaigns that emphasize the hazards of smoking and the benefits of quitting can influence attitudes and behaviors. These campaigns often use mass media, social media, and community events to disseminate information.
5. *School programs:* Integrating smoking prevention education into school curricula helps deter smoking initiation among youth. Programs that educate life skills and provide peer support are particularly effective.
6. *Support groups:* Community support groups provide a forum for people trying to give up smoking to exchange stories, get support, and obtain information. Peer counseling and scheduled meetings are two common ways that these organizations offer continuous support.

Advancements in technology have introduced innovative tools for smoking cessation, including (Sawyer et al. 2022). Smoking cessation apps provide resources, tracking tools, and support directly to users' smartphones. These apps often comprise features such as goal setting, progress tracking, and access to virtual support communities. Text messaging interventions send regular motivational messages, tips, and reminders to individuals trying to quit. Research reflects that these programs can significantly improve quit rates by providing continuous support. Virtual support groups and forums offer a platform for individuals to join with others who are also attempting to quit. These groups provide a sense of community and shared experience, which can be instrumental in maintaining motivation and commitment. Workplace-based smoking cessation programs can attract a large number of smokers and provide convenient access to resources (Poole et al. 2023). Employers can organize workshops that offer education on the health risks of smoking, the advantages of quitting, and available cessation resources. Employee Assistance Programs often include smoking cessation support, such as counseling services and access to cessation medications. These programs can be tailored to fulfill the requirements of employees. Creating a smoke-free workplace environment and offering incentives for employees who quit smoking can foster a culture of health and wellness.

Environmental factors significantly contribute to the prevalence and severity of lifestyle diseases. Effective policies to reduce environmental exposures are essential components of comprehensive approaches to lifestyle disease prevention. These policies target various sources of pollution, hazardous substances, and unhealthy environments to protect public health and promote healthier communities (Poole et al. 2023). Air pollution is a main environmental risk factor correlated to respiratory and cardiovascular diseases (Lelieveld et al. 2015). To limit the release of dangerous pollutants such as PM, SO_2, NO_x, and volatile organic compounds (VOCs), tight emission standards should be set for businesses, power plants, and automobiles. Establishing air quality monitoring networks to provide real-time data and certify compliance with air quality standards. Public access to air quality information helps communities stay informed and take protective measures. Also, motivating the use of renewable energy sources, such as wind, solar, and hydroelectric power, to decrease reliance on fossil fuels will contribute towards air quality improvement. Policies that incentivize clean energy technologies and improve energy efficiency contribute to long-term air quality improvements.

Preventing waterborne illnesses requires having access to clean, safe water, and promoting overall health (Lelieveld et al. 2015). Safe Drinking Water Standards: Implementing and enforcing regulations that set the highest contaminant levels for pollutants in drinking water, ensuring water is safe for consumption. Regular testing and reporting are critical to maintaining water quality. To achieve appropriate water quality, proper treatment of industrial and municipal wastewater should be mandated before its discharge into water bodies. Upgrading aging infrastructure and adopting advanced treatment technologies help reduce pollution and protect aquatic ecosystems. Policies should be framed and implemented to gain best management practices in order to reduce runoff from agricultural conduct, which can carry pesticides, fertilizers, and other pollutants into water sources. Policies may include buffer zones, cover crops, and controlled use of agrochemicals. Exposure to hazardous chemicals in the environment, workplace, and consumer products can lead to chronic diseases, including cancers and endocrine disorders. Policies to enhance chemical safety include (Sharifi Rad et al. 2020), requiring manufacturers to register chemicals and assess their safety before market entry. Comprehensive risk assessments and toxicological studies help identify and mitigate potential health risks. Safety data sheets and proper labelling provide critical information for the safe handling and use of chemicals. Banning or restricting the use of particularly dangerous substances, such as certain pesticides, flame-retardants, and plasticizers. Policies promoting safer alternatives and green chemistry innovations help reduce chemical exposures (Semeraro et al. 2021; Mata et al. 2022).

Future Research Directions and Conclusion

Approximately 1.3 billion people worldwide smoke, with the highest prevalence in Europe and Southeast Asia (Flor et al. 2021). Smokers are 2–4 times more expected to develop coronary heart disease and stroke. Smoking causes about 1.9 million deaths per annum from heart disease globally. The risk of dying from COPD is 12–13 times higher in smokers (Spencer et al. 2022). COPD caused approximately 3.23 million mortality in 2019, with smoking being the primary risk factor. Smoking is responsible for 85% of global lung cancer deaths (Safiri et al. 2022). The risk of type 2 diabetes is increased among smokers by 30–40% (Zinman et al. 2015; Bhat et al. 2023).

Despite substantial progress in understanding the links between smoking and lifestyle diseases, significant gaps in knowledge remain. These gaps impede the development of comprehensive prevention and intervention strategies. Addressing these gaps is crucial for advancing public health efforts and improving health outcomes related to smoking and lifestyle diseases such as chronic respiratory disease, cardiovascular disease, diabetes, and cancer. Smoking is one of the major risk factors for numerous lifestyle diseases, including chronic respiratory diseases, cardiovascular diseases, cancer, and diabetes. The role of smoking as a confounding factor complicates the assessment and management of these diseases due to its widespread impacts on health. Smoking significantly confounds the risk and management of lifestyle diseases. Comprehensive and multifaceted approaches are necessary to mitigate its impact. Continued research, policy implementation, and innovative intervention strategies are essential to reduce smoking prevalence and mark improvement in public health outcomes related to lifestyle diseases.

References

American Cancer Society (2019) Cancer facts & figures 2019. Available via https://www.cancer.org/research/cancer-facts-statistics/all-cancer-facts-figures/cancer-facts-figures-2019.html. Accessed 10 Mar 2024

Balwan WK, Kour S (2021) Lifestyle diseases: the link between modern lifestyle and threat to public health. Saudi J Med Pharm Sci 7(4):179–184

Bhat TA, Kalathil S, Leigh NJ et al (2023) Do alternative tobacco products induce less adverse respiratory risk than cigarettes? Respir Res 24:261

Binette AP, Janelidze S, Cullen N et al (2023) Confounding factors of Alzheimer's disease plasma biomarkers and their impact on clinical performance. Alzheimers Dement 19:1403–1414

Bowerman C, Bhatka NR, Brazzale D et al (2022) A race-neutral approach to the interpretation of lung function measurements. Am J Respir Crit Care Med 207(6):768–774

Budrevičiūtė A, Damiati S, Sabir D et al (2020) Management and prevention strategies for non-communicable diseases (ncds) and their risk factors. Front Public Health 8:574111

Cao D, Enouen J, Wang Y et al (2023) Estimating treatment effects from irregular time series observations with hidden confounders. In: Williams B, Chen Y, Neville J (eds) Proceedings of the AAAI conference on artificial intelligence. AAAI Press, Washington, D.C.

Çomo N, Tikkinen K, Medina MT (2022) Remdesivir and three other drugs for hospitalised patients with COVID-19: final results of the WHO solidarity randomised trial and updated meta-analyses. Lancet 399:1941–1953

Cosentino F, Grant P, Aboyans V et al (2019) ESC guidelines on diabetes, pre-diabetes, and cardiovascular diseases developed in collaboration with the EASD. Eur Heart J 41(2):255–323

Dai X, Gakidou E, Lopez AD (2022) Evolution of the global smoking epidemic over the past half century: strengthening the evidence base for policy action. Tob Control 31:129–137

Flor LS, Reitsma M, Gupta V (2021) The effects of tobacco control policies on global smoking prevalence. Nat Med 27:239–243

Heber S, Sallaberger LM, Hausharter M et al (2019) Exercise-based cardiac rehabilitation is associated with a normalization of the heart rate performance curve deflection. Scand J Med Sci Sports 29:1364–1374

Kobets T, Smith BPC, Williams G (2022) Food-borne chemical carcinogens and the evidence for human cancer risk. Foods 11(18):2828

Lareau S, Fahy BF, Meek P et al (2019) Chronic obstructive pulmonary disease (COPD). Am J Respir Crit Care Med 199(1):P1–P2

Lee PN, Hamling J (2014) Systematic review of the relation between smokeless tobacco and cancer in Europe and North America. BMC Med 7:36

Lelieveld J, Evans J, Fnais M et al (2015) The contribution of outdoor air pollution sources to premature mortality on a global scale. Nature 525:367–371

Leung J, Yang C, Tam A et al (2020) ACE-2 expression in the small airway epithelia of smokers and COPD patients: implications for COVID-19. Eur Respir J 55:2000688

Li Y, Hecht S (2022) Carcinogenic components of tobacco and tobacco smoke: a 2022 update. Food Chem Toxicol 165:113179

Libby P (2021) The changing landscape of atherosclerosis. Nature 592:524–533

Liu Z, Xu Z, Xu L et al (2022) Modified biochar: synthesis and mechanism for removal of environmental heavy metals. Carbon Res 1:8

Masud H, Sekalala S, Gill P et al (2023) Critical analysis of tobacco taxation policies in Pakistan after two decades of FCTC: policy gaps and lessons for low- and middle-income countries. Tob Induc Dis 21:156

Mata TM, Martins A, Calheiros C et al (2022) Indoor air quality: a review of cleaning technologies. Environments 9(9):118

McGowan LD (2022) Sensitivity analyses for unmeasured confounders. Curr Epidemiol Rep 9: 361–375

Modi K, Singh I, Kumar Y (2023) A comprehensive analysis of artificial intelligence techniques for the prediction and prognosis of lifestyle diseases. Arch Comput Methods Eng 30:4733–4756

Parums D (2023) Editorial: global initiative for chronic obstructive lung disease (GOLD) 2023 guidelines for COPD, including COVID-19, climate change, and air pollution. Med Sci Monit 29:e942672

Pichón Riviere A, Bardach AE, Cairoli FR et al (2023) Health, economic and social burden of tobacco in Latin America and the expected gains of fully implementing taxes, plain packaging, advertising bans and smoke-free environments control measures: a modelling study. Tob Control 4:tc-2022-057618

Poole NL, Nagelhout G, Magnée T et al (2023) A qualitative study assessing how reach and participation can be improved in workplace smoking cessation programs. Tob Prev Cessat 9:07

Reddy M, Dattaprakash B, Kammath S et al (2022) Application of MLOps in prediction of lifestyle diseases. ECS Trans 1:01

Reimer P, Austin W, Bard E et al (2020) The IntCal20 northern hemisphere radiocarbon age calibration curve (0–55 cal kBP). Radiocarbon 62:725–757

Roth GA, Mensah GA, Johnson CO et al (2020) Global burden of cardiovascular diseases and risk factors, 1990–2019: update from the GBD 2019 study. J Am Coll Cardiol 76:2982–3021

Sadarang R (2021) Factors associated with quitting smoking in Indonesia. J Prev Med Public Health 54:137–144

Safiri S, Nejadghaderi SA, Abdollahi M et al (2022) Global, regional, and national burden of cancers attributable to tobacco smoking in 204 countries and territories, 1990–2019. Cancer Med 13:2662–2678

Santos A, Willumsen J, Meheus F et al (2022) The cost of inaction on physical inactivity to public health-care systems: a population-attributable fraction analysis. The Lancet Glob Health 11(1): e32–e39

Sawyer C, Hassan L, Guinart D et al (2022) Smoking cessation apps for people with schizophrenia: how feasible are m-health approaches? Behav Sci 12(8):265

Semeraro T, Scarano A, Buccolieri R et al (2021) Planning of urban green spaces: an ecological perspective on human benefits. Land 10(2):105

Sharifi Rad M, Anil Kumar NV, Zucca P et al (2020) Lifestyle, oxidative stress, and antioxidants: back and forth in the pathophysiology of chronic diseases. Front Physiol 11:694

Siegel R, Miller K, Wagle NS et al (2023) Cancer statistics, 2023. CA Cancer J Clin 73:17–48

Spencer M, Miniño A, Warner M (2022) Drug overdose deaths in the United States, 2001–2021. NCHS Data Brief 457:1–8

US Department of Health and Human Services (2014) The health consequences of smoking-50 years of progress: a report of the Surgeon General. National Center for Chronic Disease Prevention and Health Promotion (US). Available via https://www.ncbi.nlm.nih.gov/books/NBK179276/. Accessed 12 Mar 2024

World Health Organization (2018) Ambient air pollution: A global assessment of exposure and burden of disease. Available via https://iris.who.int/bitstream/handle/10665/250141/9789241511353-eng.pdf. Accessed 6 Mar 2024

Zinman B, Wanner C, Lachin J et al (2015) Empagliflozin, cardiovascular outcomes, and mortality in type 2 diabetes. N Engl J Med 373(22):2117–2128

Washed Covert Hair Dye Components are a Threat to Environmental Sustainability and a Potential Risk Factor for Increasing Cases of Breast Cancer

Hamda Khan, Safia Habib, Syed Amaan Ali, and Moinuddin

Abstract

Hair dye residues, particularly hair colour pigments, couplers and precursors that are carelessly dumped into the sewage system, have been detected in water resources. These pollutants have the potential to endanger living things by damaging DNA and inducing mutation, resulting in the development of cancer through mutagens or tumour promoters. The risk of breast cancer is also influenced by these chemicals that lead to mammary tumours. Numerous environmental chemicals either raise mammary gland susceptibility to carcinogenesis or activate key hormonal pathways, making them mammary gland carcinogens. The complexity, high concentration and bright colour of the dye make it difficult for biodegradable substances to break down. Because of this, the wastewater treatment plants are a shield to local ecosystems and humans from harmful substances present in wastewater. Many nations use a lot of untreated wastewater for agricultural purposes. Nowadays, it has become a very significant and persuasive global environmental and public health issue. In this chapter, we discussed the harmful effects of hair dye pollution on health. Keeping our attention on the rise of breast cancer, we further discussed how hair dye pollution may be a contributing factor in this life style disease.

H. Khan · S. Habib (✉) · Moinuddin
Department of Biochemistry, Jawaharlal Nehru Medical College, Aligarh Muslim University, Aligarh, Uttar Pradesh, India
e-mail: safia.bh@amu.ac.in

S. A. Ali
Department of Periodontics and Community Dentistry, Dr. Ziauddin Ahmad Dental College, Faculty of Medicine, Aligarh Muslim University, Aligarh, Uttar Pradesh, India

S. Raisuddin et al. (eds.), *Environmental Toxicants and Lifestyle Diseases*,
https://doi.org/10.1007/978-981-95-0575-3_19

Keywords

Hair dye contamination · Breast cancer · Aquatic pollution · Wastewater treatment

Introduction to Hair Dye

Hair colour change by dye application is a common procedure among women. In the last two decades, the interest in hair colour dye, its usage and hair dye contamination has increased immensely, particularly since 2010. A significant number of people, primarily women, utilize cosmetics, such as hair dye products, to change the colour of their hair. Many cosmetics, including creams for skin care and hair care products, are designed to be applied to the complex surfaces of the skin and hair. Colour resistance divides hair dyes into four categories: semi-permanent, permanent, demi-permanent and temporary (Fig. 1). The first two rely on pre-coloured molecules. Semi-permanent dyes can slightly enter the cortex, allowing the colour to withstand up to six washings, while temporary dyes work by depositing dye on cuticles. The semi-permanent molecules with small aromatic amine-containing molecules enable the shaft to inflate, lift the cuticle and let the amines penetrate the cortex, thereby colouring the cortex. Temporary dyes enable the largest molecules to coat the shaft rather than penetrating it. Demi-permanent has the same hair colour influence as semi-permanent, but it stays a little bit more than semi-permanent dye. Demi-permanent and permanent dyes are based on colour precursors known as oxidation dyes, and the final shade is generated by their interactions with an oxidizing agent (França et al. 2015). Despite believing that natural colours were harmless, the majority employed semi-permanent dyes. Numerous investigations determined the dye molecule's diffusion path to the inner hair fibre (França et al. 2015; Lewis et al.

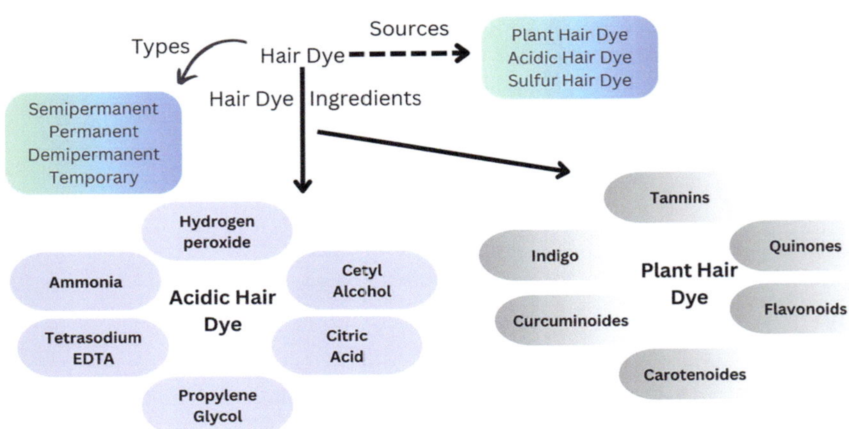

Fig. 1 Hair dye ingredients

2013; Palaniappan et al. 2023). It involves the molecules entering the intercuticular spaces and traveling through intracellular and non-keratinized areas of the endocuticle. It finally reaches the macro fibrils after migrating to keratinized areas in later stages before becoming a part of the matrix (França et al. 2015). The product must penetrate the cuticle to deposit or remove colour in the cortex to provide permanent colour.

In the past, people utilized natural hair dyes, but as the cosmetics business becomes advanced, new varieties have been developed. In oxidative dyes, the tiniest molecular structure makes them persistent for hair colouring (Palaniappan et al. 2023). The process includes the hydrogen peroxide and dye reach the cortex of the hair because the aromatic amines cause the cuticle to swell, allowing the colour to be removed and the hair to become lighter until it falls out (Lewis et al. 2013). Longer colour resistance in oxidation systems is facilitated by an intense chemical diffusion into the cortex. There are variations in dyes and colour precursors in terms of chromophore groups, an affinity for hair fibre, water solubility and photostability (França et al. 2015). The safety of several dye ingredients in hair colour has long been a source of worry due to their reputation as animal carcinogens. Several reports suggested the association between dye component exposure and cancer (Seydi et al. 2019; Yu et al. 2021). The majority of the research on exposure routes focuses on inhalation and skin contact (Watt et al. 2004; Fowles et al. 2013; Diana et al. 2018; Palaniappan et al. 2023). Thus, applying hair color causes clinical damage to the hair shaft. It is interesting to know if this damage can be fixed and how long it will take, as this could help calculate the ideal gap between applications (Palaniappan et al. 2023).

Hair Dye Ingredients

The chemicals used in hair dye formulation are diverse in nature, and the toxicities of these chemicals are critical. Certain ingredients are consistently present, and these are (i) ammonia, which is the chemical that opens the hair cuticle to allow colour to soak in deeply. Ammonia can, however, be harsh and harmful to hair, depleting it of its natural oils. Because of this, ammonia-free hair dye has been developed that works just as ammonia-based dye, (ii) alcohol is frequently used as a solvent for other components in hair colouring. It can dry out the hair and scalp, causing frizz and irritation, even if it aids in the colour's equal distribution and (iii) pigment, the colorants that give hair colouring its vivid hues are called pigments. Pigments come in a variety of forms, including both natural and manufactured varieties (Fig. 1). While natural pigments made from plants can be a wonderful option if an individual chooses a more delicate and environment friendly approach to hair colouring, synthetic pigments tend to be more long-lasting and offer a wider spectrum of colours. Darker dyes are the most potentially concerning because certain chemicals that can cause cancer are present in higher concentrations in them. According to several studies, it is determined that hair dyes do not raise the risk of cancer in consumers, clients, or professional hairdressers. But the majority of the studies does prove the possible role of hair dye ingredients in the incidence of cancer (Seydi et al. 2019; Yu et al. 2021). Hair dyes can trigger allergic responses in certain sensitive persons, just like other goods like specific foods or pharmaceuticals (Sankar et al. 2017). Temporary hair colours, like Victoria Blue B, Basic Red 2 and Fuchsin Basic, are particularly well-liked due to their affordability and ease of application. There has been a lot of research on the association between permanent hair dyes and cancer but not much on the possibility that temporary hair dyes used at home increase the risk of cancer (Zhang et al. 2020). Large dinuclear, trinuclear, or polynuclear complexes that become trapped inside the hair shaft are formed when colouring couplers react with quinonediimine intermediates, which are generated when H_2O_2 oxidizes the primary colouring intermediates (Kim et al. 2016). Nonetheless, the majority of evidence points to hair dyes and their constituents having moderate-to-low acute toxicity, such as being the source of allergic contact dermatitis (Diana et al. 2018; Lloyd et al. 1977). While certain ingredients in hair dye have been linked to cancer in animals, besides there is not enough data to extrapolate comparable findings from human trials (Seo et al. 2012). Therefore, more investigation is needed to thoroughly examine the importance of this matter, particularly in light of the safety of the hair colour components.

Hair Dye-Induced Adverse Health Effects

Hair dye products currently constitute one of the personal care and beauty businesses that is expanding and capturing the market at the fastest pace. The greying of hair is a part of natural human aging process. Many people who have grey hair coating their hair with dye very frequently. Hair colouring products are used extensively and may have negative consequences. The user's awareness and understanding of hair colouring and its associated adverse effects are therefore crucial. Dermatologists often treat hair dye contact dermatitis as a common acute problem. It is supposed to be delayed hypersensitivity reaction that typically affects the neck, hairline and scalp. The most prevalent allergy directly linked to hair dye contact dermatitis is para-phenylenediamine (PPD), which is a synthetic aromatic amine. The gold standard test for diagnosing contact dermatitis caused by hair colouring is epicutaneous patch testing. Nevertheless, there is a chance that PPD will cause co-sensitization or cross-sensitivity to allergens. Thus, a great deal of work has been done into trying to figure out what adverse effects, including cancer risk, could result from this kind of exposure (Table 1) (Diana et al. 2018; Lloyd et al. 1977; Palaniappan et al. 2023). Increased risk of BC, ovarian carcinoma and basal cell carcinoma was seen in substantial prospective cohort research involving 1,17,000 personal hair dye users. Individual or pooled relative risks were higher in cancers such as lung carcinoma, prostate carcinoma and salivary gland carcinoma, though the data are still insufficient (Zhang et al. 2020). Permanent hair dye use is also linked to a higher prevalence of non-Hodgkin's lymphoma, especially follicular lymphoma (Palaniappan et al. 2023). While certain ingredients in hair dye have been linked to cancer in animals, although there is not enough data to make comparable

Table 1 Adverse health effect of hair dye usage

Hair dye ingredient	Possible adverse effects	Action	Study model	References
PPD	Associated with cancer and mutagenicity	ROS generation and Lipid peroxidation		Seydi et al. (2019)
N-monoacetyl-p-phenylenediamine	Ovarian Cancer	Downregulates the expression of FSH[a] receptor and LH[a] receptor	Mice	Yu et al. (2021)
N-phenyl-p-phenylenediamine	Cardiotoxicity	Calcium signal pathway	zebrafish larvae	Kim et al. (2016)
N,N-bis (hydroxyethyl)-p-phenylenediamine	Reduced body weight	Decreased serum iron	Rat or Mice	Fang et al. (2023)
4-methoxy-m-phenylenediamine	Irritant effect on skin	Oral dosing	Rabbit	Lloyd et al. (1977)
p-aminophenol				
m-aminophenol, *p*-phenylenediamine				
toluene-2,5-diamine				
Hydrogen peroxide	Irritation of the gastrointestinal tract with nausea, vomiting	Corrosive damage oxygen gas formation and lipid peroxidation	General review analysis	Watt et al. (2004)
Propylene glycol	Adverse respiratory effects	Intermediate inhalation exposure	In vivo study	Fowles et al. (2013)
Ammonia	Irritations to the skin, eyes and respiratory tract	Exposure to human	Human	Diana et al. (2018)

[a]*FSH* follicle-stimulating hormone, *LH* luteinizing hormone

conclusions from human research. Therefore, more investigation is needed to thoroughly to examine the importance of this matter (He et al. 2022). The majority of evidence indicates the hair dyes and their constituents having moderate-to-low acute toxicity involved in allergic contact dermatitis (Diana et al. 2018; Fowles et al. 2013; Lloyd et al. 1977; Palaniappan et al. 2023). In this chapter, we summarize the effects of hair colouring on human health in order to offer a critical evaluation of this matter. This initiative will contribute to a better understanding of conflicting or contentious myths around the use of hair colour and its potential link to cancer (He et al. 2022). Hence, the adverse effects of the hair dye are allergy, hair loss, respiratory sensitization, hair dye poisoning, reproductive toxicity, disruption of thyroid hormone synthesis, bladder cancer, BC, hematopoietic cancer and maternal hair dyeing-induced childhood tumours (Kim et al. 2016).

Hair Dye Use and Cancer

One well-designed study with a thorough assessment found associations between the use of personal hair colour and non-Hodgkin's lymphoma, multiple myeloma, acute leukaemia and bladder cancer. Previous findings from research on bladder cancer indicate that certain populations may have genetic susceptibilities to hair dye exposures; however, these conclusions are based on small subgroups from a single, well-planned case-control study. To ascertain, if the suggested correlations are genuine or fictitious, these findings must be replicated (Fang et al. 2023). It has been shown that there is rising concern about the potential role of hair dyes in the aetiology of breast cancer (BC). Numerous hair dye products contain residues of illegal aromatic amines, also hair dye users' breast epithelial cells have been shown to contain higher concentrations of these amines' DNA-adducts. While, another study came to the conclusion that there is insufficient evidence to support the carcinogenicity of using personal hair colour and little evidence to support the carcinogenicity of hair colorants in experimental animals (Rollison et al. 2006). Personal use of hair colouring might not have been linked to an increased risk of cancer or of common malignancies including lung, breast or colorectal cancer, despite extensive research over the past 40 years (Heikkinen et al. 2015). Case-control studies carried out in Caucasian populations have accounted for the majority of research that has indicated favourable relationships. Higher risks have been observed in these investigations, mostly associated with long-term usage of dark hair dye and hair colour applied before 1980 (Correa et al. 2000). A potential link between hair colours and haematological malignancies has been suggested by retrospective research (Grodstein et al. 1994). Using permanent hair colour has been associated with a higher risk of certain cancers, yet epidemiologic data have not consistently supported this idea.

A case-control study including patients who visited a New York City screening centre was carried out to determine whether using hair colour raises the risk of BC. 398 cases of BC that were discovered at the screening facility between 1977 and 1981 were included in the investigation along with 790 randomly chosen controls who were also screened during that time. The majority of participants (77%) had used hair colour at least once, and 38 percent had done so at least 100 times (Koenig et al. 1991). Hence, the study suggested strong association with hair dye usage and cancer. Numerous hair products have carcinogens and endocrine disruptors that may be linked to BC. Previously, researcher investigated and successfully associated the relationship between hair colour and chemical relaxer/straightener use and BC risk by ethnicity in a nationwide prospective cohort research (Kogevinas et al. 2006). It is also reported that hair dye users have a higher risk of bladder cancer. In a Spanish hospital-based case-control research (166 female controls, 152 female incident cases), the study assessed this connection among females (Kelsh et al. 2008). Both toxicologists and epidemiologists have disputed whether permanent hair dyes pose a carcinogenic danger (Hennekens et al. 1979).

Hair Dye Contamination

Cosmetic residues have been found in water resources, particularly minute amounts of hair dye pigments, couplers and precursors that are carelessly dumped into the sewage system. These pollutants have the potential to endanger living things because they are bioaccumulative, persistent and bioactive. The removal of leftover hair colour from the environment is a growing concern as a consequence of the recent increase in hair colour usage. A variety of chemicals are included in hair colour, including the genotoxic substance PPD. Synthetic dyes, which are present in water at extremely low concentrations (less than 1 part per million), are highly visible and can impact the aesthetic appeal, transparency and gas solubility of water bodies. Colours are typically the first contaminant detected in wastewater. They impede photosynthesis and interfere with the growth of aquatic animals by reflecting and absorbing sunlight that enters the water (Pereira and Alves 2012). A previous study explored the effect of hair dye contamination on bacteria and plant system. It summarizes that the cytotoxic, clastogenic and genotoxic contaminants will poison the entire aquatic ecosystem. Therefore, brief evaluation must concentrate on the colorants and their pollutants in light of the toxicological effect of hair dye in order to ascertain the ecological effect and its involvement in human health.

Hair Dye Usage Versus Hair Dye Contamination

The possible health effects of hair dye use for personal use have received significant attention from both the scientific community and general population throughout the last three decades. Epidemiological studies have been carried out to look into the connection between the usage of hair colour and the risk of cancer in people. However, the link varies depending on the type of cancer, and the reported results have been inconsistent (Hüttl et al. 2016). While on the other hand, hair dye contamination is something different with hair dye usage. Hair dye contamination does not only affect one person like the hair dye usage but the whole environment, and it is linked to environmental pollution and can be toxic to aquatic life when they enter water systems (Fig. 2).

Hair Dye Contamination Is more Lethal than Hair Dye Usage

Hair dye usage and the wastewater generated should be managed and assessed through some toxicological assessment because the contamination produced through the waste of hair dye or hair dye washes is more lethal. Since personal usage can only affect a person responsible but the environmental pollution through hair dye contamination can harm the human and the animal population. Nowadays, due to high beauty standards, it is difficult to limit the use of hair dye cosmetics, but we can manage the associated contamination by limiting, continuously monitoring and

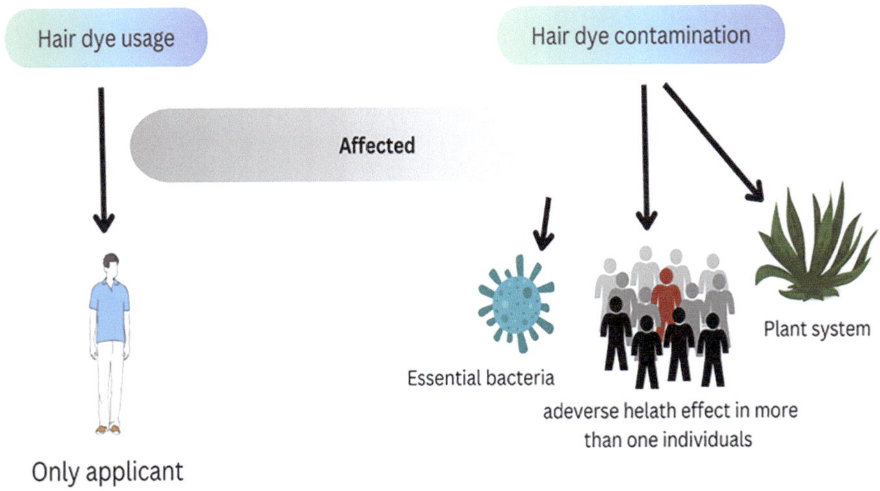

Fig. 2 Hair dye usage versus hair dye contamination

treating the washed covert hair dye components, which are a threat to environmental sustainability and responsible for aquatic pollution.

Impact of Hair Dye Contamination on the Environment

The consumers nowadays are becoming more aware of the negative consequences of synthetic hair dyes worldwide; therefore, there has been a persistent demand in recent years for safer and "greener" hair colouring products. The scientific community's attention has been drawn to natural colorants due to their health benefits and sustainability (Banat et al. 1996). In addition to harming individuals, synthetic colours also devastate the ecosystem, which in turn affects the population. Apart from hair dyeing, synthetic colours are also excessively used in the textile industry. The hurdle with handling them is that producing and applying synthetic colours to fabric require a lot of water. Water often runs out in the vicinity of dye plants, particularly in China. Simultaneously, the factories' chemical waste is frequently just thrown into the rivers. As they use the water for farming, the soil in which people grow their food becomes poisonous. Because fish and plants cannot tolerate the water when the rivers are dead. The majority of those factories are situated in nations with a lack of environmental protection legislation, like China and India. Therefore, they must ensure that their labour and environmental rules are not negligent as most of these businesses are not directly managed by Indian or Chinese companies. American and European businesses locate their factories in these nations in order to manufacture their products as cheaply as possible (and to maximize profits) (Bessegato et al. 2018). Since the majority of the clothing we purchase is made in those nations, we may enjoy affordable clothing without worrying about the harm that these low-quality goods do to people and the environment.

Studies conducted on animals have confirmed the direct involvement of substances such as PPD and 2,4-diaminoanisole sulfate in the growth of tumours. Similarly, the active component in hair dyes, 4-aminobipheny, is likewise to blame for BC. H_2O_2, PPD, para-aminophenol, meta-aminophenol, resorcinol, pyrimidine, pyridine, ammonia, 1,4-diaminobenzene or 2,5-diaminotoluene, are the approximate chemical components of synthetic hair dyes (Umair et al. 2018). A number of health effects, including pharyngeal and laryngeal irritation, asthma, skin and eye irritation, lightheadedness, sleepiness, etc., are caused by all of these chemicals. More than 5000 chemical components, including carcinogenic and endocrine-disrupting substances such as aromatic amines, are found in hair dyes (He et al. 2022). Certain findings seem less encouraging, despite the fact that almost all toxicological and epidemiological investigations have shown little or very little damage to human health (Zhang et al. 2012).

Assessments of cosmetic safety were formerly mostly carried out using animal models. Previous study based on Organization for economic cooperation and development (OECD)-validated in vitro techniques, such as the bovine corneal opacity and permeability (BCOP) assay in conjunction with histopathological analysis suggested that the Hen's egg test on chick chorioallantoic membrane (HET-CAM), and a test on reconstructed human epidermis models, researcher assessed the potential for eye and skin irritation associated with four plant hair dyes: henna, Chinese gallnuts, sappanwood and walnut husks (Umair et al. 2018). Similarly, the sensitization potential of henna-based hair dye products containing PPD was reported to be evaluated using a combination of in vitro techniques, including the micro-direct peptide reactivity assay (mDPRA), HaCaT keratinocytes-associated IL-18 assay, U937 cell line activation test (USENS)/IL-8 levels, blood monocyte-derived dendritic cell test and genomic allergen rapid detection (GARD skin) (Bhagavathula et al. 2019). The development of hair dyes with ecologically friendly properties and excellent washing and UV light durability is crucial to overcoming the market's limits with traditional permanent hair colours (Sampathkumar and Yesudas 2009). The toxicological effects of synthetic hair dyes are a major source of concern (Pereira and Alves 2012).

Industries and Sectors Contributing to Hair Dye Contamination

Recently, more focus has been placed on the environmental impact of chemical substances in aquatic environments. Several studies have demonstrated the presence of chemicals, including as pesticides, steroid hormones and antibiotics, in the aquatic environment (Ardila-Leal et al. 2021). The textile, leather and paper industries discharge a lot of coloured effluent into water bodies. Service providers like hospitals, colleges and the food industry employ artificial colours. The maximum discharge limits and toxicological effects on the environment and human health are unknown for the majority of colours. National laws generally set a colour restriction without mentioning dye restrictions, in a few cases, they just give colour measurement and omit coloured water limits (Tkaczyk et al. 2020). Colour removal techniques such as chemical, physical, biological and hybrid have all shown to be highly effective. Conversely, the disadvantage of chemical and physical treatments is that they result in costly, hard-to-handle sludge that requires large treatment spaces. On the other hand, when it comes to aerobic treatments, the anaerobic approach results in less sludge, is less costly, and poses less environmental risk (Bhatia et al. 2017). Contamination of the aquatic environment is becoming an increasingly important problem.

Currently regarded as micropollutants due to their low concentration in aquatic settings, synthetic organic dyes are undervalued as environmental toxins. They are widely used and produced on a massive scale, which makes them ubiquitous (Pereira and Alves 2012). Recent years have seen the publication of several research on the subject of artificial organic dyes in sediment, water and wild fish. It proves that synthetic organic dyes are detrimental to aquatic ecosystems. Some microorganisms that have the ability to completely mineralize or change colour include yeasts, fungi, bacteria and algae (Yang et al. 2015). White rot fungi are the best at destroying artificial colours because they provide the enzymes needed to catalyse the dye removal and degradation processes. How thoroughly colour is removed depends on a number of factors, including pH, nutrient load, treatment duration, aeration, C/N ratio, biomass shape, inoculum concentration, co-substrate addition and the production of hazardous byproducts (Choi 2021). Because some phenolic chemicals hinder fungal development during bioremediation procedures, enzymes are utilized as an option to minimize the pollution and environmental impact of industrial effluents they generate (Adrio and Demain 2014).

The textile industry releases pollutants that are tainted with dyes, plastic, polyester, fibres, yarns and other hazardous materials. These wastes can be either liquid or solid. These polymeric compounds have contaminated neighbouring waste ecosystems and agricultural areas, especially in underdeveloped countries. Plant growth is inhibited by this soil pollution by oxidative stress induction, reduction of protein content, photosynthesis and CO_2 absorption rates (Vikrant et al. 2018). The dyeing processes used by the textile industry require a lot of water, which makes handling the enormous volumes of wastewater discharge challenging (Ahmad et al. 2019). Consequently, several countries have instituted measures to guarantee that the effluent satisfies a threshold prior to its release into the environment or repurposing. Physical, chemical and biological methods have been employed to protect the environment and public health from discharge problems. These strategies could be applied singly or in combination to achieve positive results (Dai et al. 2020). It is impossible to overestimate the importance of indirect sources of synthetic dye content in aquatic settings as a factor in the ecosystem's overall dye cycling. Fish farming wastes used as fertilizer and dye-containing sewage sludge from sewage treatment plants (STPs) can directly contaminate farmlands. Additionally, aquatic life and plants that grow on the bottom of water bodies may be at risk from STP effluents (Kant 2011).

The wastewater that runs down the drains causes the sewerage pipes to crust and corrode. If allowed to run in rivers and drains, it degrades drinking water from hand pumps and makes it unsafe for human consumption. It also results in drain leaks, which drive up maintenance expenses. Such tainted water can act as a refuge for bacteria and viruses. Water pollution affects the textile manufacturing process in a number of ways. An epidemiological study conducted recently suggests that the aromatic amines in hair colours may increase the incidence of bladder cancer. Additionally, the use of hair colour was linked to high levels of 4-aminobiphenyl (4-ABP), which involves the formation of DNA adducts in a preliminary study conducted on human epithelial breast cells. The brown dye did not contain ABP, but the black, red and blonde hair dyes did. PPD is a necessary ingredient for colour development in many permanent hair colours. According to the test results, 4-ABP is present in several hair hues. Research on the dermal absorption and bioavailability of 4-ABP from hair dyes is required to determine whether this aromatic amine is responsible for the higher risk of bladder cancer seen in regular users (Turesky et al. 2003).

Hair Dye Contamination-Induced Adverse Health Effects

An increasing number of studies in underdeveloped countries have connected hair dye toxicity to purposeful self-harm. Hair dyes include PPD along with several other compounds that can cause rhabdomyolysis, acute metabolic acidosis, laryngeal oedema and rapid renal failure. A few examples of the numerous diverse ecosystems that make up the incredibly complex aquatic environment are rivers, lakes, estuaries, seas and oceans (Gonçalves et al. 2023). However, this classification is too simplistic, as these ecosystems are dynamic structures with unique characteristics. Aquatic ecosystems are now less resilient due to anthropogenic activities brought about by changes in social context, agriculture, industry and lifestyle. These environments experience direct or indirect alteration, and the ensuing eco-toxicological damage is a complex and unanticipated event. Furthermore, it was emphasized that (Coronado-Apodaca et al. 2023; Hüttl et al. 2016) in order to preserve environmental quality, multidisciplinary research that addresses the complexity of the environmental problem from multiple angles is required. Consequently, the toxicity of hair colour is becoming acknowledged as a major etiological problem. Dye pollution has a major negative influence on human health as well as the environment. Water pollution in the environment, induced by dyes and dye-containing effluents, affects both terrestrial and aquatic ecosystems (Banat et al. 1996). By decreasing sunlight penetration, dissolved oxygen levels and the aesthetic value of water bodies, these contaminants have an adverse effect on primary food production, food chains and food webs. Dyes accumulate and biomagnified in the food chain as well, endangering aquatic life and ultimately human health (Umair et al. 2018). Moreover, synthetic dyes damage the environment since they are utilized in a wide range of industries and are very resistant to deterioration (Bhagavathula et al. 2019). Human health has been linked to mutagenic and carcinogenic effects of colour pollution exposure (Sampathkumar and Yesudas 2009).

Hair Dye Contamination and Cancer

Human hair dye is regarded as possible biowaste, and inappropriate hair dye disposal can result in several environmental problems. However, the biggest challenge with the dyes is how to dispose of them. A significant amount of dye is contained in salon waste when it finds its way into nearby water bodies, which can have negative effects on aquatic life, soil fertility, aquatic assets and human health through food chains. Concerns about dye disposal have been highlighted by these environmental difficulties (Gatti et al. 2023). Pollutants in the environment could play a significant role in the development of this illness. Long-term exposure to environmental pollutants, especially those with estrogenic effects, may have a major role in the development of cancer, especially hormone-related malignancies like BC (Sharma et al. 2023). Not only does the release of these colours have an unfavourable visual impact, but their chemical composition is also harmful. In the past few decades, decolourization of textile dye has garnered a lot of interest. The removal of dye from the environment has become more of a concern in recent times due to the increase in dye usage. It is well known that microorganisms can thrive in environments with lots of resistant material. Microbes may offer useful resources for dye remediation that are also reasonably priced. There are findings on the microbial breakdown of several contaminants (Gatti et al. 2023). Numerous studies have demonstrated a connection between certain environmental pollutants and BC. Understanding the different underlying risk factors for BC may help to prevent the illness. Serious issues arise from aqueous waste effluent that contains colouring chemicals. As a result, the production and application of synthetic dyes for fabric dyeing have grown significantly in the modern era. Consequently, during their synthesis and later during the dyeing of fibres, they have turned into frequent industrial environmental contaminants. This is the primary prerequisite for both the research into suitable and ecologically friendly treatment technologies as well as for advancements in associated processes (Gürses et al. 2016).

Breast Cancer and Its Global Incidence

Globally, BC is the most prevalent cancer among women. Planning preventive strategies requires knowledge of the incidence and mortality of BC. With 192,000 cases and 40,000 fatalities from cancer in 2001, it is the second most prevalent cause of cancer-related deaths overall and the most common incident cancer (apart from superficial skin tumours). Older age, obesity or high body mass index, tobacco use, physical inactivity, high-fat diet, early menarche, late age at first full-term pregnancy, shorter breastfeeding durations, use of oral contraceptives or hormonal menopausal therapy, breast density and family history of BC are the main risk factors for BC (IARC 2023; Zhang et al. 2020). Of the various nations, China accounted for the greatest proportion of BC cases worldwide, making up about 18.4% of all instances. The United States came in second, with 11.8% of cases worldwide. While developing nations (Iran, China, Mexico, Cameroon, and Costa Rica) had lower incidence rates than the world age-standardized incidence rates (47.8 per 100,000 population), highly developed nations (Belgium, Denmark, Australia, USA, United Kingdom and Italy) had substantially higher incidence rates. The age-specific incidence rates of BC were comparatively low for females under 25 years old and sharply increased beyond this age. Surprisingly, different parts of the world have different BC peak ages. At 40 years old, South Korea and Cameroon had the youngest onset peak ages. China, Japan, Iran, Fiji and Morocco had the highest percentage of female 55–60 year olds. The United States, Belgium, Australia and the United Kingdom had the oldest peak beginning ages of BC, at 70 years old (Lei et al. 2021). Although the prevalence of BC is higher in advanced nations, the proportion of deaths from the disease is highest in less developed countries.

In less developed nations, it is particularly critical to raise awareness of risk factors and promote early detection (Ghoncheh et al. 2016). The World Health Organization (WHO) introduced the Global Breast Cancer Initiative (GBCI) Framework in 2023. It places a strong emphasis on evaluating, reinforcing and expanding services for the early diagnosis and treatment of BC (Ong et al. 2024). In 2020, 10% of cancer-related deaths and 13.5% of new cancer cases in India were due to female BC, which is the primary cause of cancer incidence and mortality in the country (Kulothungan et al. 2024). China has a high disease burden from BC, which is considered as the most frequent malignant tumour worldwide. The general public is becoming more and more likely to get their health information from the internet, particularly from video social media sites. To reduce the prevalence of BC and improve the prognosis, it is crucial to educate the public about accurate facts. Fertility presents special hurdles for young women with BC who carry hereditary pathogenic mutations in either BRCA1 or BRCA2. There was little information available on BRCA carriers in earlier research that showed the safety and viability of pregnancy in BC survivors (Liu et al. 2024).

Hair Dye Contamination and Breast Cancer

Pollutants in the environment could play a significant role in the development of this BC. Long-term exposure to environmental pollutants, especially those with estrogenic effects, may have a major role in the development of cancer, especially hormone-related malignancies such as BC (Sharma et al. 2023). Recent examples from the literature on BC serve as illustrations of the basic categories of epidemiological inquiry (Skegg 1995). Previous analysis examines the potential link between some developing environmental pollutants, including triclosan, alkylphenols (APs), bisphenol A (BPA), parabens, phthalates, perfluoroalkyl substances (PFASs), synthetic musks and phthalates, and its role in the development of BC. It is suggested that the newly discovered environmental pollutants may contribute to the onset and/or progression of BC, according to both in vitro and in vivo data. The risks of BC and other cancers related to these exposures remain unclear, despite the fact that exposures have been evaluated in various human populations (Siddique et al. 2016). The assessment of these environmental exposures in human populations, as well as their interactions with other genetic and behavioural risk factors, should be the main focus of efforts. A study suggested that the environmental pollutants have been demonstrated in laboratory to either create mammary gland tumours in animals or to alter the mammary gland's susceptibility to carcinogenesis. Additionally, several of the pollutants are hormonally active, notably imitating the risk factor for BC like oestrogen (Siddique et al. 2016). Future research and exposure reduction efforts focused at prevention can be guided by an evaluation of epidemiologic research on these pollutants found in toxicological studies. The evidence till date, is based on a relatively small number of studies, generally supports a link between certain genetic polymorphisms involved in carcinogen activation and steroid hormone metabolism and BC. These polymorphisms include polycyclic aromatic hydrocarbons (PAHs) and polychlorinated biphenyls (PCBs). There is minimal and methodologically restricted evidence about dioxins and organic solvents, although it does point to a possible correlation. However, existing information suggests that PCBs have little potential for genotoxicity and do not cause cancers in investigations with adult rodents. Apart from initiating the disease, PCBs may also cause BC growth (Onat et al. 2024). Many of the substances linked to BC in toxicological study have not been studied in people. To close this gap, more effective exposure assessment techniques must be created. We are now in a position to better understand how environmental exposures, including air and water contaminants, nutrition, personal and consumer care goods and social variables, may contribute to the development of aggressive breast tumours thanks to new developments in single-cell approaches (Aguilar and Colacino 2024; Bhardwaj et al. 2024).

Possible Mechanism of Breast Cancer Incidence Through Aquatic Pollution Via Hair Dye

Around 50% of female cases of BC lack particular identified risk factors, even though environmental factors are considered to be more significant than hereditary factors. The association of environmental pollutants with BC has received special interest due to the geographic diversity seen in worldwide clusters of BC cases. The quality of drinking water is a crucial mediator between environmental exposure and human health, as it influences both the overall incidence and mortality risk of cancer. Meanwhile, persistent doubts over the association of water contamination and BC have hindered efforts to prevent the disease (Liu et al. 2022). Drinking water tainted by effluent may expose one to substances that alter hormone balance, including natural and synthetic hormones, and carcinogens that can cause BC. It is hypothesized that these pollutants raise the risk of BC (Brody et al. 2006a). In addition to increasing the risk in women, long-term exposure to estradiol raises the risk of BC in many animal species.

The exact mechanisms causing this effect are still up for debate. According to the widely accepted idea, oestrogens speed up DNA replication by promoting transcription mediated by oestrogen receptors, which in turn increases the amount of oversights that occur during DNA replication (Yue et al. 2003). There is an ongoing debate on the exact processes by which oestrogens induce BC, and these mechanisms have not been proven beyond a reasonable doubt. There is a theory that states that oestrogens increase the transcription of genes involved in cell proliferation by binding to ERα or ERβ (Santen 2002). Numerous reproductive and hormonal traits have been linked to BC risk factors (Newcomb 1997). It makes sense to think about how early life experiences like menarche, parity and nursing can impact the development of ER^- versus ER^+ BC in the future. A plausible mechanism could involve effects on the mammary gland progenitor cells. Strong evidence suggests that luminal progenitor cells, whether basal-like (ER^-) or luminal (ER^+), are the primary source of most breast malignancies (Gross et al. 2016). The following mechanisms might have involved in the rise in BC cases.

Drinking Water Contamination by Wastewater

The main causes of ailment from contaminated water through waterborne pathogens are disease-causing bacteria and viruses from human and animal waste. When hazardous chemicals or threatening microbes contaminate water body system such as ocean, lake and river, the water quality is reduced and the water becomes poisonous to people or the aquatic environment. Wastewater is water that has been used in many different purposes. It originates from commercial, industrial and agricultural activity (metals, solvents and hazardous sludge) as well as from our sinks, showers and toilets (Currie et al. 2013). It also refers to storm water runoff, which happens when precipitation enters our waterways from impermeable surfaces and transports debris, oil, grease and road salts. This is how the chemicals in hair colour, which are released from our homes, businesses and salons, can contaminate drinking water. A possible source of exposure to endocrine-disrupting substances from commercial items, excreted natural and pharmaceutical hormones and mammary cancers is drinking water contaminated by wastewater (Brody et al. 2006b). A prior study found a correlation between exposure to drinking water pigmented by wastewater effluent and incidence of BC (Gallagher et al. 2010). It is thought that groundwater contamination by anions, such as sulfate ($SO4^{2-}$), nitrate (NO^{3-}), fluorides (F^-) and heavy metals, such as arsenic (As), lead (Pb), uranium (U), mercury (Hg), cadmium (Cd), nickel (Ni) and chromium (Cr), poses a major risk to human health (Kaur et al. 2021). Consequently, there is strong evidence linking numerous environmental factors to BC. Variations in the distribution of socio-demographic variables, such as parity, race/ethnicity and maybe nutritional status, could account for some of the discrepancies observed in different research (Rudel et al. 2007).

Environmental Chemicals Are Mammary Gland Carcinogens

Currently, the primary approach for predicting the impact of chemicals on cancer in humans is to identify carcinogens in animal research. Because environmental and occupational epidemiology research on BC is lacking, animal studies assessing possible chemical carcinogenicity are essential (Rudel et al. 2007). Not all substances that cause mammary gland cancers in animals provide a serious danger to human health or are equally carcinogenic. The previous evidence linking the few chemicals to human carcinogenesis varies, and each needs to be assessed in terms of potency, dose-response, target locations, tumour incidence (Fig. 3) multiplicity, latency and human exposure routes and levels (Rudel et al. 2007). Geographic heterogeneity in BC condition implies that a person's environment can have a significant etiological role in the development of cancer. Exposure to wastewater or sewage, which contains a complex mixture of infections, mutagens and carcinogens, can be one of the environmental risk factors (du Plessis et al. 2023). According to a recent study, chemicals that cause mammary tumours in rodents or trigger the signalling of oestrogen or progesterone are likely to raise the risk of BC. When chemicals are linked to these behaviours, action can be taken to safeguard public health (Kay et al. 2024).

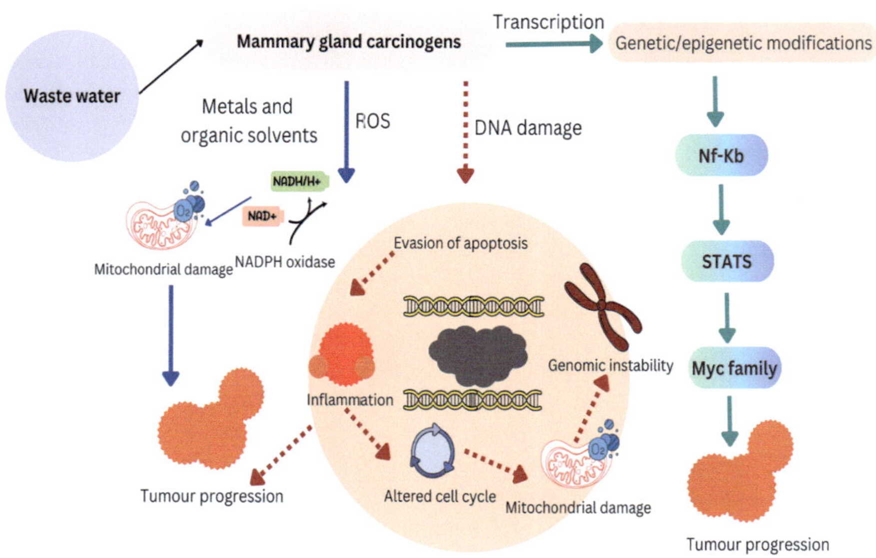

Fig. 3 Possible pathways involved in the progression of breast tumour through environmental pollutants

Breast Cancer Risk Near Polluted River

The potential risk factor of BC has been recognized by the American Cancer Society, out of which strong genetic variables account for 5–10% of hereditary BC cases. Other risk factors include ageing, long-term hormone replacement therapy use, obesity, inactivity and alcohol (Brody et al. 2007). Several studies also indicate the link between environmental pollution and BC. It was suggested that living near a river was linked to the increasing risk of BC (Gallagher et al. 2010). The prevalence of BC has been observed to be on the rise, with reasonably elevated risk assessments at less than five kilometres from the river. The research cannot totally rule out the possibility that living close to a river may have slightly increased the risk of cancer, particularly for farmers (Verkasalo et al. 2004). Research has also connected carcinogens in contaminated rivers to BC and gallbladder cancer (Sreen et al. 2021).

Treatment of Coloured Wastewater

The ecology and public health suffer greatly when effluent from several companies is continuously released into the environment without being treated. A vast number of commercial permanent dye formulations contain the precursor p-aminophenol (PAP), which is very toxic and common effluent from the industries in sewage system. Based on toxicological testing, this dye class can induce lung irritation, bronchial asthma, methemoglobinemia, severe allergies, dermatitis, rash, tremors, migraines, seizures, loss of consciousness and harm to the liver, blood, kidneys and brain (de Souza et al. 2020). Several other dyes are involved in the aquatic pollution (Table 1) leading the uneasiness in body. In order to lessen the detrimental impacts on our ecosystem through contaminated water, dye effluent from industry needs to be treated. The technologies used in the treatment of coloured water are coagulation, membrane treatment, biological therapy and enhanced oxidation process (Gonawala and Mehta 2014).

Coagulation

Coagulation is a simple technique, which is used for decolorization in wastewater treatment through chemicals. In a simple way, the usage of chemical agents in wastewater in order to alter the physical condition of suspended and dissolved solids and encourage their settling is known as the coagulation process.

The selection of a coagulant plays a crucial part in the removal of contaminants during the coagulation/flocculation process. Coagulants exist into a variety of forms, both organic and inorganic (Dotto et al. 2019). Inorganic coagulants precipitate iron or aluminium when introduced to water. By absorbing contaminants in the water as they were mixed, these aid in its purification. The "sweep-floc" mechanism is the term for this process. For the formation of sludge and the separation of solids and liquids, organic coagulants are typically utilized. The most popular organic

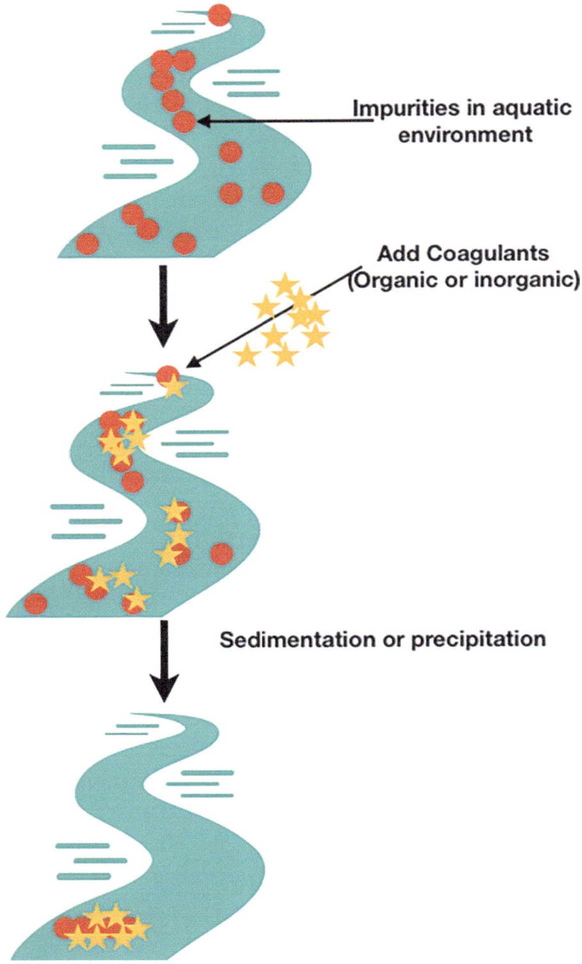

Fig. 4 Mechanism of coagulation treatment

coagulants are PolyAMINEs and PolyDADMACs, which are cationic by nature and react only by neutralizing charges (Dotto et al. 2019). In conclusion, the mechanism involved the addition of coagulant in the aquatic environment, which further sediments the impurities by trapping them (Fig. 4).

Membrane Treatment of Aquatic Environment

Membrane-based filtering technology has been acknowledged as a possibly successful treatment strategy for treating water and wastewater, capable of handling various organic micropollutants including hair dye contamination or pesticides. Membrane-based filtering technology includes microfiltration, nanofiltration and reverse

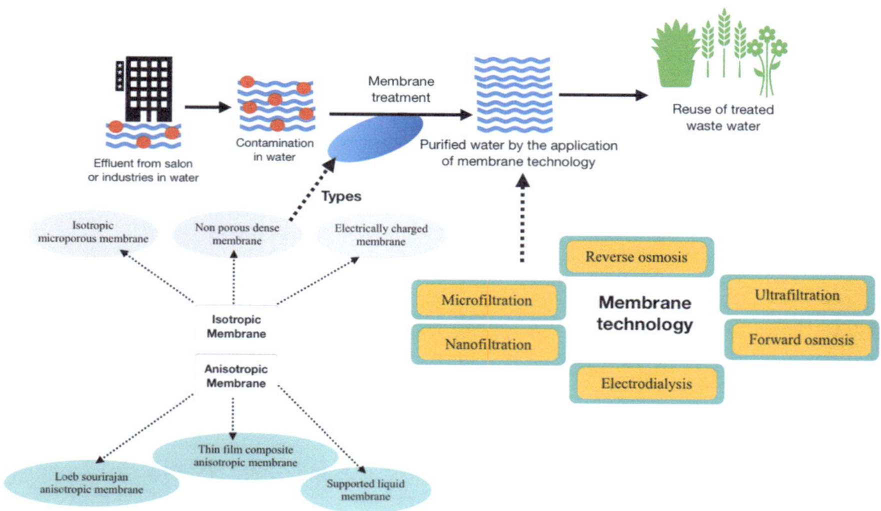

Fig. 5 Overview of membrane technology

osmosis. Because of their high treatment efficiency and adaptability, nanofiltration (NF), reverse osmosis (RO) and forward osmosis (FO) have been investigated more and more for the removal of pesticides from aquatic environments (Goh et al. 2022). A membrane is a phase that separates two neighbouring phases and controls the flow of materials between them by acting as a barrier. This special separation principle makes the membrane's transport selectivity relate to the primary benefits of membrane technology when compared to other chemical engineering unit operations (Ulbricht 2006).

The membrane materials that are used for wastewater treatment are quite diverse, with a greater variation in chemical content than bulk morphology (Maheshwari et al. 2021) (Fig. 5). Sensitivity and toxicity of membranes are the two limitations of this technology. Due to this, researchers have developed a number of research methods to get around membrane technology. Thus far, research has shown that membrane technology is a highly promising technique for treating wastewater and improving the quality of the water in the process (Fig. 5).

Biological Therapy

Biological therapy for wastewater treatment includes anaerobic biological treatment, aerobic and aerobic facultative biological treatment. Aerobic biological treatments are the most widely utilized biological techniques for wastewater treatment. While anaerobic and aerobic degradation have different applications in the treatment of dye wastewater (Maheshwari et al. 2021) (Fig. 6). It is well recognized that microbes are essential to the mineralization of xenobiotics and biopolymers. When there is

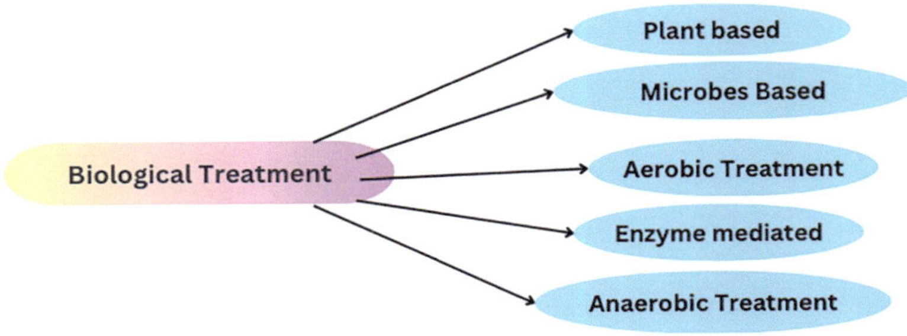

Fig. 6 Biological therapy

enough dissolved oxygen in the wastewater, anaerobic processes use microorganisms to remove contaminants. It uses bacteria that don't have oxygen to remove pollutants from wastewater. The choice of microbes and the activity of the enzymes govern how well the biodegradation process proceeds. Different kinds of microorganisms, including fungi, bacteria and algae, extract different kinds of colour from textile effluent water (Shindhal et al. 2021). Since they contain enzymes, fungi act as a good option to remove colour from water system. Colour degradation is aided by enzymes such as laccase, manganese peroxidase (MnP) and lignin peroxidase (LiP). The pigments or colours from textile wastewater have been treated by *Phanerochaete chrysosporium* and the white-rot fungus. Fungal cultures that cause white rot break down azo colours. Fungi culture can be used to get rid of Solvent Red 24's hue.

Traditionally, several bacterial species remediate textile effluent biologically in anaerobic, facultative anaerobic and aerobic environments. Pure cultures of microorganisms or mixtures of microbial populations can be employed in the therapy procedure. Due to their combined metabolic activity, mixed microbial cultures can be more successful during the therapy process than pure cultures. When treating coloured wastewater, mixed microbial cultures can be more effective than pure cultures due to their synergistic metabolic activities (Alsukaibi 2022).

Enhanced Oxidation Process

This process has its great efficiency in removing refractory organic contaminants. Advanced oxidation processes, or AOPs, have gained increasing attention in light of the ever-growing challenges surrounding water pollution. The majority of currently available technologies (such as Fenton, plasma, electron beam, electrolysis, ionizing radiation, ultrasound irradiation and wet air oxidation) are based on the in situ generation of highly reactive oxygen species (ROS, such as hydroxyl ion ($^{\cdot}OH$), $SO4^{\cdot -}$, O_2 and $O_2^{\cdot -}$), which show excellent potential in the efficient decomposition and satisfactory mineralization of toxic and biorefractory pollutants from the contaminated aqueous system (Li et al. 2021).

Conclusion

Overall, hair dye contamination produces aquatic pollution, which is involved in the BC. It induces contamination in sewage through complex mixture of pathogens and toxins, including potential mutagens and carcinogens. Considering the increase in the incidence of BC, it is must to evaluate and investigate all the possible mechanism involved in its acceleration. Hair dye contamination is one of them. In this chapter, we mentioned the evidence evaluating environmental chemical exposures and BC rise. Although, the previous researchers and this chapter are unable to provide a clear insight into the hair dye contamination risk factors that increase a woman's chance of developing BC. Hence, the extended study is required in this field. Further, we concluded the treatment of wastewater system facilities to protect the aquatic pollution by removing toxins in wastewater. Wastewater treatment might limit the environmental pollution in the future.

References

Adrio JL, Demain AL (2014) Microbial enzymes: tools for biotechnological processes. Biomolecules 4(1):117–139

Aguilar D, Colacino JA (2024) Single cell approaches to understand environmental impacts on aggressive breast cancers. Curr Opin Toxicol 10:100459

Ahmad M, Yousaf M, Nasir A et al (2019) Porous eleocharis@ MnPE layered hybrid for synergistic adsorption and catalytic biodegradation of toxic Azo dyes from industrial wastewater. Environ Sci Technol 53(4):2161–2170

Alsukaibi AK (2022) Various approaches for the detoxification of toxic dyes in wastewater. PRO 10(10):1968

Ardila-Leal LD, Poutou-Piñales RA, Pedroza-Rodríguez AM et al (2021) A brief history of colour, the environmental impact of synthetic dyes and removal by using laccases. Molecules 26(13): 3813

Banat IM, Nigam P, Singh D et al (1996) Microbial decolorization of textile-dyecontaining effluents: a review. Bioresour Technol 58(3):217–227

Bessegato GG, De Souza JC, Cardoso JC et al (2018) Assessment of several advanced oxidation processes applied in the treatment of environmental concern constituents from a real hair dye wastewater. J Environ Chem Eng 6(2):2794–2802

Bhagavathula AS, Bandari DK, Khan M et al (2019) A systematic review and meta-analysis of the prevalence and complications of paraphenylenediamine-containing hair dye poisoning in developing countries. Indian J Pharmacol 51(5):302–315

Bhardwaj PV, Dulala R, Rajappa S et al (2024) Breast cancer in India: screening, detection, and management. Hematol Oncol Clin North Am 38(1):123–135

Bhatia D, Sharma NR, Singh J et al (2017) Biological methods for textile dye removal from wastewater: a review. Crit Rev Environ Sci Technol 47(19):1836–1876

Brody JG, Aschengrau A, McKelvey W et al (2006a) Breast cancer risk and drinking water contaminated by wastewater: a case control study. Environ Health 5:1–1

Brody JG, Aschengrau A, McKelvey W et al (2006b) Breast cancer risk and drinking water contaminated by wastewater: a case control study. J Environ Health 5:1–1

Brody JG, Rudel RA, Michels KB et al (2007) Environmental pollutants, diet, physical activity, body size, and breast cancer: where do we stand in research to identify opportunities for prevention? Cancer 109:2627–2634

Choi KY (2021) Discoloration of indigo dyes by eco-friendly biocatalysts. Dyes Pigments 184:108749

Coronado-Apodaca KG, González-Meza GM, Aguayo-Acosta A et al (2023) Immobilized enzyme-based novel biosensing system for recognition of toxic elements in the aqueous environment. Top Catal 66(9):606–624

Correa A, Jackson L, Mohan A et al (2000) Use of hair dyes, hematopoietic neoplasms, and lymphomas: a literature review. II. Lymphomas and Multiple Myeloma: Environmental Carcinogenesis. Cancer Investig 18(5):467–479

Currie J, Graff Zivin J, Meckel K et al (2013) Something in the water: contaminated drinking water and infant health. Can J Econ 46(3):791–810

Da França SA, Dario MF, Esteves VB et al (2015) Types of hair dye and their mechanisms of action. Cosmetics 2(2):110–126

Dai Q, Zhang S, Liu H et al (2020) Sulfide-mediated azo dye degradation and microbial community analysis in a single-chamber air cathode microbial fuel cell. Bioelectrochemistry 131:107349

de Souza JC, da Silva BF, Morales DA et al (2020) Assessment of p-aminophenol oxidation by simulating the process of hair dyeing and occurrence in hair salon wastewater and drinking water from treatment plant. J Hazard Mater 387:122000

Diana MP, Roekmijati WS, Suyud WU et al (2018) Why it is often underestimated: historical study of ammonia gas exposure impacts towards human health. E3S Web Conf 73:06003

Dotto J, Fagundes-Klen MR, Veit MT et al (2019) Performance of different coagulants in the coagulation/flocculation process of textile wastewater. J Clean Prod 208:656–665

du Plessis M, Fourie C, Stone W et al (2023) The impact of endocrine disrupting compounds and carcinogens in wastewater: implications for breast cancer. Biochimie 209:103–115

Fang C, Fang L, Di S et al (2023) Characterization of N-(1, 3-dimethylbutyl)-N′-phenyl-p-phenylenediamine (6PPD)-induced cardiotoxicity in larval zebrafish (*Danio rerio*). Sci Total Environ 882:163595

Fowles JR, Banton MI, Pottenger LH (2013) A toxicological review of the propylene glycols. Crit Rev Toxicol 43(4):363–390

Gallagher LG, Webster TF, Aschengrau A et al (2010) Using residential history and groundwater modeling to examine drinking water exposure and breast cancer. Environ Health Perspect 118(6):749–755

Gatti RC, Di Paola A, Monaco A et al (2023) The spatial association between environmental pollution and long-term cancer mortality in Italy. Sci Total Environ 855:158439

Ghoncheh M, Pournamdar Z, Salehiniya H (2016) Incidence and mortality and epidemiology of breast cancer in the world. Asia Pac J Cancer Prev 17(S3):43–46

Goh PS, Ahmad NA, Wong TW et al (2022) Membrane technology for pesticide removal from aquatic environment: status quo and way forward. Chemosphere 307:136018

Gonawala KH, Mehta MJ (2014) Removal of color from different dye wastewater by using ferric oxide as an adsorbent. Int J Eng Res Appl 4(5):102–109

Gonçalves LC, Roberto MM, Peixoto PV et al (2023) Toxicity of beauty salon effluents contaminated with hair dye on aquatic organisms. Toxics 11(11):911

Grodstein F, Hennekens CH, Colditz GA et al (1994) A prospective study of permanent hair dye use and hematopoietic cancer. J Natl Cancer Inst 86(19):1466–1470

Gross K, Wronski A, Skibinski A et al (2016) Cell fate decisions during breast cancer development. J Dev Biol 4(1):4

Gürses A, Açıkyıldız M, Güneş K et al (2016) Colorants in health and environmental aspects. In: Dyes and pigments, pp 69–83

He L, Michailidou F, Gahlon HL et al (2022) Hair dye ingredients and potential health risks from exposure to hair dyeing. Chem Res Toxicol 35(6):901–915

Heikkinen S, Pitkäniemi J, Sarkeala T et al (2015) Does hair dye use increase the risk of breast cancer? A population-based case-control study of Finnish women. PLoS One 10(8):e0135190

Hennekens C, Speizer F, Rosner B (1979) Use of permanent hair dyes and cancer among registered nurses. Lancet 313(8131):1390–1393

Hüttl RF, Bens O, Bismuth C et al (2016) Introduction: a critical appraisal of major water engineering projects and the need for interdisciplinary approaches. In: Hüttl RF, Bens O, Bismuth C, Hoechstetter S (eds) Society-water-technology: a critical appraisal of major water engineering projects. Springer Nature, Cham, pp 3–9

IARC (International Agency for Research on Cancer) (2023) Agents classified by the IARC monographs on the evaluation of carcinogenic risks to humans. http://monographs.iarc.fr/ENG/Classification/ Accessed on 4 Mar 2024

Kant R (2011) Textile dyeing industry an environmental hazard. Sci Res J 4:17027

Kaur G, Kumar R, Mittal S et al (2021) Ground/drinking water contaminants and cancer incidence: a case study of rural areas of South West Punjab, India. Hum Ecol Risk Assess 27(1):205–226

Kay JE, Brody JG, Schwarzman M et al (2024) Application of the key characteristics framework to identify potential breast carcinogens using publicly available in vivo, in vitro, and in silico data. Environ Health Perspect 132(1):017002

Kelsh MA, Alexander DD, Kalmes RM et al (2008) Personal use of hair dyes and risk of bladder cancer: a meta-analysis of epidemiologic data. Cancer Causes Control 19:549–558

Kim KH, Kabir E, Jahan SA (2016) The use of personal hair dye and its implications for human health. Environ Int 89:222–227

Koenig KL, Pasternack BS, Shore RE et al (1991) Hair dye use and breast cancer: a case-control study among screening participants. Am J Epidemiol 133(10):985–995

Kogevinas M, Fernandez F, Garcia-Closas M et al (2006) Hair dye use is not associated with risk for bladder cancer: evidence from a case-control study in Spain. Eur J Cancer 42(10):1448–1454

Kulothungan V, Ramamoorthy T, Sathishkumar K et al (2024) Burden of female breast cancer in India: estimates of YLDs, YLLs, and DALYs at national and subnational levels based on the national cancer registry programme. Breast Cancer Res Treat 205(2):323–332

Lei S, Zheng R, Zhang S (2021) Global patterns of breast cancer incidence and mortality: a population-based cancer registry data analysis from 2000 to 2020. Cancer Commun 41(11): 1183–1194

Lewis D, Mama J, Hawkes J (2013) A review of aspects of oxidative hair dye chemistry with special reference to N-nitrosamine formation. Materials 6(2):517–534

Li Y, Dong H, Li L et al (2021) Recent advances in waste water treatment through transition metal sulfides-based advanced oxidation processes. Water Res 192:116850

Liu M, Graham N, Wang W et al (2022) Spatial assessment of tap-water safety in China. Nat Sustain 5(8):689–698

Liu H, Peng J, Li L et al (2024) Assessment of the reliability and quality of breast cancer related videos on TikTok and Bilibili: cross-sectional study in China. Front Public Health 11:1296386

Lloyd GK, Ligget MP, Kynoch SR et al (1977) Assessment of the acute toxicity and potential irritancy of hair dye constituents. Food Chem Toxicol 15(6):607–610

Maheshwari K, Agrawal M, Gupta AB (2021) Dye pollution in water and wastewater. Novel materials for dye-containing wastewater treatment. In: Khadir A, Muthu SS (eds) Novel materials for dye-containing wastewater treatment. Springer Nature, Singapore, pp 1–25

Newcomb PA (1997) Lactation and breast cancer risk. J Mammary Gland Biol Neoplasia 2:311–318

Onat İ, Kocagöz R, Öz MD et al (2024) Blood and tissue levels of persistent organic pollutants and genetic susceptibility in patients with breast cancer. Environ Toxicol Pharmacol 108:104433

Ong SK, Haruyama R, Yip CH et al (2024) Feasibility of monitoring global breast cancer initiative framework key performance indicators in 21 Asian National Cancer Centers Alliance member countries. EClinicalMedicine 67:102365

Palaniappan V, Karthikeyan K, Anusuya S (2023) Dermatological adverse effects of hair dye use: a narrative review. Indian J Dermatol Venereol Leprol 27:1–7

Pereira L, Alves MD (2012) Dyes-environmental impact and remediation. Environmental protection strategies for sustainable development. In: Malik A, Grohmann E (eds) Environmental protection strategies for sustainable development. Springer, Dordrecht, pp 111–162

Rollison DE, Helzlsouer KJ, Pinney SM (2006) Personal hair dye use and cancer: a systematic literature review and evaluation of exposure assessment in studies published since 1992. J Toxicol Environ Health B Crit Rev 9(5):413–439

Rudel RA, Attfield KR, Schifano JN et al (2007) Chemicals causing mammary gland tumors in animals signal new directions for epidemiology, chemicals testing, and risk assessment for breast cancer prevention. Cancer 109:2635–2666

Sampathkumar K, Yesudas S (2009) Hair dye poisoning and the developing world. J Emerg Trauma Shock 2(2):129–131

Sankar J, Sawarkar S, Malankar J et al (2017) Mechanism of hair dying and their safety aspects. Res J Topic Cosmet Sci 8(2):72–77

Santen RJ (2002) To block estrogen's synthesis or action: that is the question. J Clin Endocrinol Metab 87(7):3007–3012

Seo JA, Bae IH, Jang WH et al (2012) Hydrogen peroxide and monoethanolamine are the key causative ingredients for hair dye-induced dermatitis and hair loss. J Dermatol Sci 66(1):12–19

Seydi E, Fatahi M, Naserzadeh P et al (2019) The effects of para-phenylenediamine (PPD) on the skin fibroblast cells. Xenobiotica 49:1143–1148

Sharma A, Hooda N, Sharma R et al (2023) A review of environmental pollutants as breast cancer risk factor. AIP Conf Proc 2558(1):020075

Shindhal T, Rakholiya P, Varjani S et al (2021) A critical review on advances in the practices and perspectives for the treatment of dye industry wastewater. Bioengineered 12(1):70–87

Siddique S, Kubwabo C, Harris SA (2016) A review of the role of emerging environmental contaminants in the development of breast cancer in women. Emerg Contam 2(4):204–219

Skegg DC (1995) Risk assessment issues in breast cancer. Mutat Res 333(1–2):51–58

Sreen A, Singh HP, Guleria V et al (2021) Verma N. Relation of gallbladder cancer, gallbladder stones and breast cancer with polluted rivers–a case control study in the Indo-Gangetic plains. J Evid Based Med Health 8:657–661

Tkaczyk A, Mitrowska K, Posyniak A (2020) Synthetic organic dyes as contaminants of the aquatic environment and their implications for ecosystems: a review. Sci Total Environ 717:137222

Turesky RJ, Freeman JP, Holland RD et al (2003) Identification of aminobiphenyl derivatives in commercial hair dyes. Chem Res Toxicol 16(9):1162–1173

Ulbricht M (2006) Advanced functional polymer membranes. Polymer 47(7):2217–2262

Umair SF, Amin I, Urrehman A (2018) Hair dye poisoning:"an early intervention". Pak J Med Sci 34(1):230

Verkasalo PK, Kokki E, Pukkala E et al (2004) Cancer risk near a polluted river in Finland. Environ Health Perspect 112(9):1026–1031

Vikrant K, Giri BS, Raza N et al (2018) Recent advancements in bioremediation of dye: current status and challenges. Bioresour Technol 253:355–367

Watt BE, Proudfoot AT, Vale JA et al (2004) Hydrogen peroxide poisoning. Toxicol Rev 23:51–57

Yang J, Yang X, Lin Y et al (2015) Laccase-catalyzed decolorization of malachite green: performance optimization and degradation mechanism. PLoS One 10(5):e0127714

Yu L, Zhai J, Wang Y et al (2021) Exposure to N-monoacetyl-p-phenylenediamine impaired ovarian function in mice. J Appl Toxicol 41(12):2031–2041

Yue W, Santen RJ, Wang JP et al (2003) Genotoxic metabolites of estradiol in breast: potential mechanism of estradiol induced carcinogenesis. J Steroid Biochem Mol Biol 86(3–5):477–486

Zhang Y, Kim C, Zheng T (2012) Hair dye use and risk of human cancer. Front Biosci (Elite edition) 4:516–528

Zhang Y, Birmann BM, Han J et al (2020) Personal use of permanent hair dyes and cancer risk and mortality in US women: prospective cohort study. BMJ 2:370

Interactions Between Genetic and Environmental Risk Factors in Lifestyle Diseases

Faizyana Ali, M. Ashraf Ganie, Sami Jan, and Fouzia Rashid

Abstract

Lifestyle diseases also called as non-communicable diseases are caused by the unhealthy lifestyle choices and behaviours of an individual. The complexity of lifestyle diseases, highlighting their causes, effects on public health and socio-economic burden are prime areas to focus on. The development of these chronic diseases can be attributed to the factors such as unhealthy lifestyle, poor diet, physical inactivity, stress and poor mental health, toxicant exposure, excessive consumption of alcohol and smoking. Lifestyle diseases thus primarily are caused by interplay of unhealthy behaviours, genetic predisposition and environmental factors. A fundamental concept of the onset and progression of lifestyle diseases is the interaction between hereditary and environmental risk factors. These disorders include cardiovascular disease, diabetes, obesity, polycystic ovary syndrome (PCOS) and certain cancers. While genetic predisposition can have a major impact on an individual's susceptibility to certain diseases, environmental factors such as nutrition, exercise and exposure to toxins are crucial in regulating the risk of acquiring lifestyle diseases. The complex connection that exists between exposure to the environment and genetic vulnerabilities illustrates the synergy between these elements and the development of diseases. Prevention of

F. Ali · F. Rashid (✉)
Department of Clinical Biochemistry, University of Kashmir, Srinagar, Jammu and Kashmir, India

M. A. Ganie
Department of Endocrinology, Sher I Kashmir Institute of Medical Sciences, Soura, Srinagar Jammu and Kashmir, India

S. Jan
Department of Gynaecology & Obstetrics, Sher I Kashmir Institute of Medical Sciences Medical College Hospital, JVC, Bemina, Srinagar, Jammu and Kashmir, India

S. Raisuddin et al. (eds.), *Environmental Toxicants and Lifestyle Diseases*,
https://doi.org/10.1007/978-981-95-0575-3_20

lifestyle disease involves a comprehensive approach that involves genetic and environmental risk management and public health campaigns.

Keywords

Lifestyle diseases · Genetic and environmental factors · Diabetes and polycystic ovary syndrome

Introduction

Lifestyle disorders refer to the health issues that are triggered by the unhealthy lifestyle choices and habits of an individual. They are also called as non-communicable disorders as they are not transmitted from one person to another. These disorders are chronic in nature and are triggered by factors such as unhealthy lifestyle, poor diet and physical inactivity, sedentary lifestyle, stress, excessive consumption of alcohol, smoking and use of excessive social media. Continuous exposure to three modifiable lifestyle behaviours; smoking, unhealthy diet and physical inactivity share similar risk factors with lifestyle diseases, which results in the development of chronic diseases such as hypertension, stroke, diabetes, obesity, metabolic syndrome, chronic obstructive pulmonary disease and many types of cancers. These chronic diseases result in inability to make decisions, debilitation or even death of an individual and inflict a substantial economics strain on healthcare services. The causes of NCDs are multifaceted, involving various several risk factors and therapy require a protracted course of care. Conditions such as cardiovascular diseases, stroke and obesity impact our way of life. The incidence of lifestyle diseases, which include obesity, cardiovascular diseases and stroke, is on rise globally and is starting to pose a serious threat to public health (Mathur et al. 2023). According to global estimates, diabetes affects 415 million people worldwide, while depression impacts approximately 350 million people; hence both conditions are expensive for national health care budgets as per International Diabetes Federation and World Health organization (WHO) reports. As countries become more industrialized and people live longer, certain diseases seem to become more common. These conditions include a range of serious health issues such as osteoporosis, polycystic ovary syndrome (PCOS), stroke, depression, Alzheimer's disease, arthritis, atherosclerosis, asthma, cancer, chronic obstructive pulmonary disease, type 2 diabetes, cardiovascular disease, hypertension, metabolic disorders, chronic kidney failure and bone disease. The development of these diseases is influenced by a complex array of factors, encompassing genetics, physiological processes, environmental influences and behavioural patterns. It is well recognized that many NCDs can manifest as a cluster of illnesses leading to more complicated problems like metabolic syndrome (Vallgarda 2011). The leading factors contributing to lifestyle diseases include poor dietary habits, lack of regular physical activity, disturbed circadian rhythm, poor body posture, smoking, excessive alcohol consumption, melancholy and anxiety resulting from heavy workloads (Hussain et al. 2021).

According to the WHO, adopting a balanced diet, exercising on a regular basis and abandoning tobacco use could prevent at least 80% of coronary heart disease, strokes, type 2 diabetes and 40% of malignancies.

Causes of Lifestyle Diseases

Lifestyle diseases can be caused by multiple risk factors, which include behavioural risk factors like excessive consumption of alcohol, poor dietary habits, chewing and smoking tobacco, low physical activity, bad postures, disturbed biological clock and environmental parameters. In addition to this age, race, genetics and metabolic risk factors like hypertension, obesity, hyperglycaemia and hyperlipidaemia increase the possibility of these diseases. High blood pressure is considered to be the leading metabolic risk factor worldwide responsible for 19% of global deaths, followed by obesity and hyperglycaemia (Tabish 2017).

Genetic and Environmental Risk Factors Influencing Main Lifestyle Disorders

Obesity

Obesity is a complex and multifaceted health condition, characterized by an excessive amount of body fat. Obesity exacerbates numerous health problems, both independently and in correlation with other diseases. Body mass index (BMI) is used to estimate the accumulation of fat in an individual. According to clinical sciences, the body fat of individual is calculated based on their height and body mass, using BMI. A graded classification of excessive body weight and obesity on the basis of BMI values provides valuable information about increasing body fat (Kopelman 2000). Obesity is an epidemic disease that is linked with a number of comorbid conditions including Type 2 diabetes mellitus (T2DM), hypertension, hyperlipidaemia, cardiovascular disorders, acid reflux disease and obstructive sleep apnoea. The prevalence of certain cancers such as breast or prostrate tumours is also increasing the epidemic of obesity (Kumar et al. 2023). It can disrupt the insulin secretion and affect a number of physiological processes, including respiratory and cardiac functions (Hussain 2021). Genetic, environmental and physiological factors interact to influence body weight through the physiological mediators of energy intake and expenditure. The idea that obesity is a neuroendocrine and metabolic illness that arises from the interaction of "obesogenic" environmental variables and a polygenic predisposition has gained support in recent years. To put it in a simplistic way, the genetic background loads the gun, but environmental factors pulls the trigger (Hebebrand and Hinney 2009). The metabolic and behavioural aspects are assumed to be influenced by the genetic predisposition. Thus, a number of hereditary and non-genetic risk factors contribute to obesity, making it a genuinely complicated condition. While genetic differences are of undoubtedly play an important role, the significant rise in the occurrence of obesity is best explained by behavioural and environmental changes that have resulted from technological advances.

Genetic Factors Influencing Obesity

Although obesity is a hereditary condition, non-genetic factors may mitigate or amplify the effects of genotype's influence on the aetiology of obesity. In addition to rare obesity-associated syndromes, the genetic factors appear to work through susceptibility genes. These genes increase the risk of developing a characteristic but are not necessary for its expression. A multifaceted approach is needed to find the genes responsible for obesity. This includes investigating potential candidate genes obtained from animal models, examination of human obesity syndromes and genome-wide search utilizing microsatellites that span the entire human genome. While choosing candidate genes for obesity, their potential effects on food intake, body fat proportion, fat distribution and utilization of energy may be taken into consideration. Multiple genes that are linked with human obesity or its related metabolic disorders like receptors are involved in hunger regulation and those crucial to thermogenesis mechanisms (such as the β3-adrenergic receptor gene and the family of uncoupling proteins). Obesity is consistently found in numerous single-gene disorders of humans. Prader–Willi syndrome (PWS) is one such example, which is characterized by upper body obesity, short stature, mental impairment and hypogonadism. This condition is prevalent in approximately 1 in 25,000 births, which may occur occasionally, but commonly it is linked with familial inheritance, which is caused by the deletion of paternal segment of chromosome 15. Bardet–Biedl syndrome is the second less common disorder. Despite similarities with PWS, research among affected families has identified multiple chromosomal loci (Chromosome16,11,3,15) responsible for the syndrome, indicating its heterogeneity. Obesity-related genes encode proteins that regulate body weight. The hormone Leptin plays a significant role in this system. Leptin, produced by the fat tissue, conveys nutritional information to important regulatory centres in the brain called hypothalamus. Decreased body fat leads to decreased levels of this hormone, which in turn stimulates intake of food. Furthermore, low leptin levels trigger hormonal response associated with starvation. Increased body fat is related to higher levels of leptin, which acts to suppress food intake.

Environmental Factors Influencing Obesity

The susceptible gene hypothesis proposes that environmental factors have a significant role in revealing hidden genetic tendencies that contribute to obesity. However, predicting how genes interact with the environment is challenging due to factors like:

- Delayed exposure to environments that promote obesity (obesogenic environment)
- Lifestyle modifications resulting from changes in living conditions
- Uncertainty surrounding the precise timing and duration of environmental exposure that may lead to weight gain.

Energy Expenditure Physical activity is the most variable factor in energy expenditure, accounting for 20–50% of total energy expenditure. Subject groups with BMIs of less than 20, 20–25 and 25–35 showed similar levels of habitual exercise after studying their physical activity levels. Studies examining the correlation between physical activity and BMI across cultural boundaries suggest that people with physical activity level ratio (Total energy expenditure/RMR) of less than 1.8 had a seven-fold greater chance of being overweight (BMI $>$ 25). There is a correlation between obesity and physical inactivity.

Increased Energy Intake Inaccurate food consumption reports complicate the understanding of function of energy consumption in the aetiology of obesity. Research has repeatedly found that individuals with obesity tend to underreport their food intake with studies comparing energy consumption and expenditure revealing a significant discrepancy of 30% between reported food intake and actual energy needs. The data strongly suggest that the effects of three macronutrients—proteins, fats and carbohydrates have different effects on eating behaviours, mainly due to their impact on satiety. Fat has a weak satiating capacity as compared to proteins, and subjects in experimental conditions overeat when offered with high-fat food. It appears plausible that environmental factors work by either raising or lowering energy use.

Culture Studies conducted on migrants and westernization of diet and lifestyle in developing nations provide evidence that environmental influences play a critical role in the onset of obesity in these countries. A considerable rise in the age-standardized prevalence of obesity (over 60% in both men and women) among the Polynesians in Western Samoa and Nauruans in Micronesia is closely correlated with dietary and lifestyle changes. A considerable change is BMI is witnessed in migrant studies, where populations sharing a common ancestry live under new and different environmental conditions. It is observed that Pima Indians living in the United States are on average 25 kg heavier than Pima Indians living in Mexico. African American who resides in the United States shows a similar pattern. The average BMI for men and women living in Nigeria is 21.7 and 22.6, respectively, but the average BMI for Nigerian men and women in the United States is 27.1 and 30.8, respectively (Jackson et al. 2007). This rising rate of obesity is linked to a higher risk of severe health problems. Overweight/obesity increases with age, accelerating between 20 and40 years and peaking around 50–60 years, in both sexes. As low-income groups become more affluent, their obesity rates increase rapidly.

Decreased Physical Activity Sedentary lifestyle and lack of physical activity are closely related to an increased risk of obesity, as they contribute to an energy imbalance and reduced calorie expenditure resulting into fat accumulation and weight gain. Studies using cross sectional data have frequently examined the relationship between excessive TV watching and obesity. According to Gortmaker and colleagues, there was a dose-dependent relationship between the amount of time spent watching TV each day and the prevalence of obesity. Children and adolescents who watched TV for more than 5 hours a day had a 5-fold higher rate of obesity than those who watched for less than 2 hours a day. In a more recent long-term study, 980 children were tracked up biannually from age 3 to 15 years as well as at ages 21 to 26 years, found that excessive TV viewing between ages 5 and 15 contributed to a significant 17% of cases of overweight at age 26, even after accounting for other potential factors (Hancox et al. 2004). Children and adolescents of industrialized countries are severely inactive. A representative survey from 2002 found that 61.5% of American children between the ages 9 and 13 years did not take part in any structured physical activity during non-school hours (Reilly et al. 2004). Obesity and overweight are linked with decreased body gross motor development and endurance conduct (Duke et al. 2003). A study comparing the motor skills of German Children aged 10 years in 1976 and 1996, using standardized tests such as 6-minute run, 50-meter sprint and 30 second push-up challenge, found a drastic decline in physical performance among latter age groups (Graf et al. 2004).

Diabetes Mellitus

In antiquity, diabetes was among the first diseases to be documented and described. Throughout the fascinating history of diabetes mellitus (DM), numerous brilliant scientists have contributed to its identification as a disease and much more advancements in its treatment. Globally, type 2 diabetes is the most common type and is caused by controllable behavioural risk factors. It is leading cause of mortality worldwide and has emerged as the dreadful disease of twenty-first century (Lefebvre and Pierson 2004). Obesity is also one of the main causes of T2DM they result from poor dietary habits or improper nutritional intake. Studies have shown that people in many Asian countries are affected by diabetes because of rapid economic growth, resulting in greater occurrence of this disorder. These communities are affected by this disorder at younger ages and lower BMI levels as compared to Caucasians (Hu 2011). The incidence has risen to almost 20% in southern India (Ramachandran et al. 2008). There are numerous reports that report that Indian people are genetically predisposed to diabetes, which is made worse by unfavourable environmental factors. About 75% of patients in the country have family history of diabetes indicating a strong genetic predisposition (Ramachandran and Snehalatha 2009). Adopting healthier lifestyle changes such as eating healthy food, increased physical activity and avoiding the use of harmful substances such as artificial sweeteners, alcohol, cigarettes and so on can help to prevent individuals from this disorder (Shrivastava et al. 2013).

Gene–Environment Interplay in DM

Type 1 and Type 2 diabetes mellitus are intricate and complex diseases that arise from the dynamic interplay of genetic predisposition, environmental influences and epigenetic factors. Because of this, the effects of T1D and T2D vary among populations based on factors such as age, race, ethnicity and socioeconomic level (Gonzalez et al. 2016). The pathophysiology of type 1 diabetes is predicted on the view that environmental stimuli, either separately or in combination trigger an autoimmune process in the islets of pancreas, leading to death of beta cells that secrete insulin in genetically vulnerable individuals.

Genetic Factors

People with type 1 diabetes typically do not have a family history of this disease. Only 10% to 15% of patients have a first- or second-degree relative with such condition. The rate of developing T1DM is higher in relatives of patients; with 6% of children, 5% of siblings and 50% of monozygotic twins have T1DM, as compared to 0.4% prevalence of disease in general population (Bryan et al. 2012). Genome-wide association studies and meta-analyses have found over than 50 genetic risk loci for type 1 diabetes (T1DM). The primary genes that predispose to type 1 diabetes are found on chromosome 6 in the major histocompatibility complex (MHC) region, also referred to as HLA (Human leucocyte antigen). Polymorphic alleles of HLA complex account for 40% to 50% of the hereditary risk of developing T1DM. About 15% of the genetic propensity is attributed to polymorphism in cytotoxic T lymphocyte-associated antigen-4 (CTLA-4) gene located on chromosome 2 and the insulin gene (Ins-VNTR, T1DM) polymorphism on chromosome 11. It has been found that many other different genetic loci, with alone or in conjugation with other autoimmune diseases, contribute to the genetic susceptibility to T1DM to a lesser extent (Storling and Pociot 2017). Some of them are briefly described below.

Major Histocompatibility Complex The MHC is a cluster of genes that code for proteins expressed on the surface of immune cells and other cells, playing a critical role in regulating the immune system's ability to recognize and respond to specific antigens by regulating T-cell specificity. In T1DM, specific HLA class II alleles or haplotypes are associated with the disease, as is the prevalence in autoimmune disorders. The ability of Class II molecules to present antigens is determined by the amino acid sequence of their alpha and beta chains. About 90% of T1DM patients have either HLA-DR3, DQB1*0201 (also referred as DR3-DQ2) or HLA-DR4, DQB1*0302 (also referred as DR4-DQ8) haplotypes, in contrast to 40% healthy individuals. Moreover, about 30% of the patients have both haplotypes (i.e. DR3/4 heterozygotes).

Insulin-VNTR (Variable Number Tandem Repeat) Variations in the insulin gene promoter region contribute to around 10% of the genetic risk for T1DM. This region is located close to the proinsulin region on short arm of chromosome 11 and contains a variable number of tandem repeats (VNTRs) and exists in two main classes: a small class (26–63 repeats, type I) and a large class (140–243 repeats, type III). Although this genetic loci does not itself encode for any protein, yet it is essential for the initiation of transcription of insulin. The large class (type III), present in approximately 30% of the general population, offers protection against T1DM regardless of HLA haplotypes. This protection is thought to occur because the large class allele enhances insulin transcription and expression in the foetal thymus, leading to the elimination of insulin-reactive T-cells and preventing autoimmune beta-cell destruction (Barratt et al. 2004).

Other Genetic Associations

PTPN22 (Protein Tyrosine Phosphatase Non-receptor Type 22) Research has identified a link between the C1858T (Arg620Trp) polymorphism in the PTPN22 gene and an increased risk of Type 1 Diabetes (T1DM) and other autoimmune diseases. The PTPN22 gene produces the lymphoid tyrosyl phosphatase (LYP), a potent inhibitor of T-cell activation. Normally, LYP prevents excessive T-cell activation by dephosphorylating and inactivating the CSK kinase. However, a missense mutation in PTPN22 reduces LYP's binding affinity to CSK, leading to uncontrolled T-cell activation and autoimmune responses (Smyth et al. 2004; Zheng and She 2005).

STAT3 (Signal Transducer and Activator of Transcription 3) CD4+ T cells have two major subpopulations Th1 or Th2 cells. These cells derive from the same precursor T lymphocytes depending on the specific cytokines they receive. Th1 cells are formed when IL-12, a protein made by immune cells like macrophages and dendritic cells activates transcription factors STAT1, STAT3 and STAT4. Problems with one of these genes called STAT3 can lead to autoimmune diseases like Type 1 Diabetes and thyroid problems. Sometimes changes in the STAT3 gene can cause these diseases to start early in life (Flanagan et al. 2014).

Environmental Factors

The rising incidence level of TDM in migrant populations and the occurrence of 'epidemics' of T1DM highlight the significant impact of environmental factors on the development of the disease. Identifying the environmental risk factors has proved vital, as it could lead to the implementation of preventive measures and potentially decrease the rising trend of T1DM cases. The key environmental contributors that are pivotal for the development of disease are viral infections, some dietary factors, toxins and stressful life events.

Viral Infections

The role of viruses in initiating T1DM was initially uncovered through epidemiological investigations. Children exposed to rubella virus during foetal life have higher risk of developing T1DM, and other autoimmune disorders like autoimmune thyroiditis. Enterovirus has been found to play a pivotal role in the early stages of development of T1DM by the activation of innate immune response (Hober and Sauter 2010). There is a connection between T1DM and the interferon induced helicase 1 gene, which in turn encodes for MDA5 Protein. When a virus infects the cell, MDA5 works as a cytoplasmic sensor especially for coxsackieviruses B and triggers the activation of innate immune response (Hober and Alidjinou 2013). Molecular mimicry is another mechanism by which viruses induce autoimmune response. Based on molecular mimicry hypothesis, cells are destroyed when the immune system attacks auto antigens that imitate the antigens of the virus. In Type T1DM, a well-studied example of molecular mimicry is the similarity between the Coxsackie B4 virus's P2-C protein and the glutamic acid decarboxylase (GAD65) enzyme in pancreatic β cells. They share a similar amino acid sequence, which may lead to the immune response mistakenly attacking the β cells (Dotta et al. 2007). Studies have shown that certain T-cells can recognize and respond to both GAD65 and P2C peptides, without cross-reacting, using different molecular mechanisms (Schloot et al. 2001). However, intriguingly, some studies suggest that enteroviruses like Coxsackie may also have a protective effect against T1DM under specific circumstances (Jaidane et al. 2010).

Dietary Factors

The relationship between diet and the onset of T1DM is still in debate. However, various studies indicate that introducing cow's milk to an infant's diet in the early age may increase their risk of developing T1DM. Research has identified a specific fragment of albumin protein present in cow's milk, ABBOS (peptide 17, positions 152–167), as a potential self-reactive epitope due to its similarity to the p69 protein on pancreatic β cells. In addition, other nutrition-related factors have been associated to an increased risk of T1DM, these include early introduction of cereals in diet, nitrate contamination in water, low intake of omega-3 fatty acid and Vitamin D deficiency in body.

Type 2 Diabetes Mellitus

Type 2 diabetes mellitus (T2DM) is a multifaceted condition that impacts over 90% of diabetes cases, resulting from a dynamic interaction between genetic susceptibility and environmental factors (Park 2011). Few environmental factors such as excessive food consumption, sedentary lifestyle and obesity have led to increased diabetes epidemic. However, genetic factors also play a pivotal role in determining an individual's predisposition to developing T2DM. Two important findings indicate the significance of genetic risk factors responsible for T2DM. First, the development of T2DM varies by ethnicity. In comparison to native Europeans, Asians or Pima Indians living in Western nations have at least a twofold higher incidence of type 2 diabetes. Second, a significant family history of type 2 diabetes exists. Children of parents having T2DM face a significantly higher risk, with a 40% likelihood of developing this condition, which is six times greater than the general population's risk of 7% (Doria et al. 2008). Three primary strategies have been utilized to discover genetic risk factors: (1) concentrating on family study linkage peaks, (2) focusing on candidate genes based on their biological properties and (3) conducting genome-wide association analysis. The genome-wide association studies (GWAS) have revolutionized our understanding of the genetic factors critical for T2DM. About 40 genetic variants have been discovered that influence an individual's susceptibility for developing T2DM.

Family-Based Linkage Analysis

In a family pedigree, genetic markers are used in family-based linkage analysis to find which chromosomal regions are linked to Type 2 diabetes. This method works well in cases where the condition being studied has a monogenic inheritance pattern. Using this method, it was possible to identify causative genetic changes of various monogenic forms of diabetes, such as maturity-onset diabetes of the young, neonatal diabetes, maternally inherited diabetes and deafness. This approach failed to identify the genes responsible for the typical form of T2DM. Although some findings pinpointed that linkage peaks are located close to CAPN10 and ACRP30 (Horikawa et al. 2000; Vionnet et al. 2000).

The Candidate Gene Approach

The candidate gene approach involves conducting case-control association studies that target specific genes or genetic regions selected due to their known biological relevance. This approach has been used to investigate over a hundred candidate genes. But only a few genes like PPAR and KCNJ11 have been identified to be linked with T2DM (Gloyn et al. 2003). PPARG encodes for a protein called Peroxisome proliferator-activated receptor Y, which plays a vital role in the processes of adipogenesis and insulin sensitivity, by regulating transcriptional activity of various genes. A mere change with proline at 12 amino acid is linked with modest but significant increase in the development of T2DM (Altshuler et al. 2000). The KCNJ11 gene (Potassium inwardly rectifying channel subfamily J member 11), situated on the short arm of chromosome 11, codes for pore-forming of the ATP-sensitive potassium channel Kir 6.2 of the pancreatic β-cells. Mutations that increase the function of KCNJ11 cause the potassium channel to remain open, inhibiting β-cell depolarization and leading to impaired insulin secretion. The important discovery being made by using candidate gene approach is the strong association of TCF7L2 (Transcription factor 7-like 2) (Grant et al. 2006). This gene has been found to be associated with an increased risk of developing T2DM, by attempting to identify the previously documented linkage peak (Reynisdottir et al. 2003). The role of TCF7L2 in the pathogenesis of T2DM is under study. It plays a pivotal role in Wnt signalling and is connected to insulin production and β-cell proliferation (Shu et al. 2008).

Genome-Wide Association Studies (GWAS)

The completion of GWAS in 2007 marked a significant milestone in the identification of genetic risk factors for type 2 diabetes. GWAS involve scanning hundreds of thousands of single-nucleotide polymorphisms across the genomes of hundreds to thousands of individuals to identify their potential associations with complex diseases like T2DM (Manolio et al. 2009). More than 10 additional genetic loci were found to alter the risk of T2DM in Europeans with genome-wide significance from the first six GWAS results published in 2007 (Sladek et al. 2007). These included differences in or close to SLC30A8, IGF2BP2, HHEX, FTO, CDKAL1, CDKN2A/2B and FTO. On the basis of first GWAS report by Sladek et al., mutations in SLC30A8 and near HHEX were identified to be strongly linked to T2DM. The genetic variation of SLC30A8, situated on chromosome 8q24, encodes ZnT8, a zinc transporter specifically expressed in islet cells that plays an important role in facilitating insulin production and secretion. Notably, the mutations in *SLC30A8*, rs13266634 lead to non-synonymous mutation of the protein (Sladek et al. 2007). One locus at 10q23-25 with a substantial linkage disequilibrium block that includes HHEX, IDE and KIF11 was also significantly associated with T2DM. Meta-analyses with large-scale replication analyses were conducted after the initial reports of GWAS (Zeggini et al. 2008). Interestingly, GWAS carried out in Asians identified novel genetic risk factors associated with T2DM that were not previously known in Europeans (Yamauchi et al. 2010). These include *KCNQ1*, *UBE2E2* and *C2CD4A/4B*. One subunit of a voltage-gated potassium channel is encoded by KCNQ1, which is specifically expressed in pancreatic β-cells. Mutations in KCNQ1 are believed to contribute to the risk of T2DM by impairing β-cell function (Unoki et al. 2008).

Cancer

Cancer is a disease in which abnormal cells in the body multiply and grow uncontrollably, invade and damage other surrounding tissue and potentially spread to distant organs in the body. It is a genetic disease in which growth-regulatory genes are altered genetically, causing tumour cells to differ from their normal progenitors. Malignant transformation is caused by a series of mutations in particular cellular genes that activate oncogenes and deactivate tumour suppressor genes. Despite extensive amount of research and great advancements in recent years, cancer remains a global killer (Futreal et al. 2004).

Genetic and Environmental Risk Factors for Cancer

Both internal (genetic mutations, immune system disorders and hormones) and external (dietary, radiation, tobacco use and infectious organisms) factors are responsible for causing cancer. Environmental and genetic factors account for 90–95% and 5–10%, respectively, for causing cancer. The correlation between diet and cancer is clear from the substantial variations in cancer rates across different countries and the changes in cancer incidence rates observed in populations that migrate to new regions, adopting different dietary habits. For instance, research has revealed that Asians have lower incidence rates of prostate cancer (25 times lower) and breast cancer (10 times lower) compared to Western populations. However, when Asians migrate to Western countries, the incidence rate for these cancers increases drastically. Lifestyle factors also play a crucial role in the development of cancer as was shown when a study was carried out on monozygotic twins. The genetic predispositions responsible for cancer cannot be altered; however, lifestyle factors can be modified. The fact that cancer is less influenced by genetics and more by environmental factors, which can be changed, shows that cancer can be prevented. The crucial lifestyle factors that have profound impact on cancer incidence and mortality rates include smoking, alcohol, diet, obesity, harmful pathogens, environmental pollutants and radiation.

Genetic Factors

Breast and Ovarian Cancer Mutations in breast cancer 1 (BRCA1) and BRCA2 genes greatly increase the likelihood of breast and ovarian cancer. Women with such mutations are more likely of developing cancer when compared to the general population. Other genetic factors and family history also play a role. Lynch syndrome is found to cause cancer.

Colorectal Cancer Lynch syndrome (hereditary non-polyposis colorectal cancer) and familial adenomatous polyposis can lead to colorectal cancer. Also, mutations in APC and MLH1 gene cause such cancers.

Prostate Cancer Mutation in BRCA2 gene and other genes related to androgen receptor pathway increases the risk of prostate cancer.

Environmental Factors

Lung Cancer Lung cancer is caused mainly by environmental factors, with smoking being the leading cause of 87% of deaths. Other environmental exposures that significantly contribute to lung cancer risk include: Radon gas, outdoor air pollution (e.g. carbon), indoor air pollution (caused by smoke, formaldehyde, benzene, 1, 3-butadiene), food pollution (e.g. food additives, nitrates, pesticides, dioxins, organ chlorines), carcinogenic metals and metalloids, certain pharmaceuticals and cosmetics.

Skin Cancer (Melanoma and Non-melanoma) It is caused by exposure to UV radiation during sun tanning.

Oesophageal Cancer Smoking, heavy alcohol consumption and chronic gastro-esophageal reflux disease (GERD) are crucial contributors responsible for developing oesophageal cancers.

Interaction Between Genetic and Environmental Factors

Cancer risk may be enhanced by the interaction of genetic predispositions and environmental exposures. Individuals with specific genetic mutations may be more vulnerable to cancer when subjected to particular environmental risk factor. A better understanding of genetic and environmental factors contributes to the correct assessment of an individual's cancer risk as well as modification of treatments and preventive measures. In conclusion, a person's chance of developing cancer is influenced by both hereditary and environmental variables. Genetic predispositions can set the stage for cancer, but environmental factors ultimately determine the expression of these risks.

Polycystic Ovary Syndrome (PCOS)

PCOS is a complex and diverse endocrine disorder affecting approximately 1 in 15 women globally, typically characterized by three key features: (1) Hyperandrogenism (elevated androgens), evident through clinical symptoms and/or biochemical markers. (2) Ovarian dysfunction, often accompanied by polycystic ovary morphology. (3) Associated metabolic disorders, present in a considerable proportion of affected women. Although the exact cause of PCOS remains unclear, a growing body of research suggests that a complex interplay of genetic and environmental factors contributes to the development, prevalence and severity of PCOS.

Genetic Basis of PCOS

In the aetiology of PCOS, there is compelling evidence of a significant genetic component, including the following:

A. *Familial aggregation:* The number of studies documenting the association between cases of hyperandrogenism, PCOS and related metabolic illnesses within the same families. As reported by Kahsar-Miller et al., about 35% of the mothers and 40% of the sisters of women with PCOS displayed characteristics of the syndrome themselves, although, these studies have not depicted the inheritance pattern of syndrome.
B. *Male phenotype:* Limited data on male relatives of PCOS patients suggest that the male counterpart of the syndrome shows signs of insulin resistance, glucose intolerance, pathologic responses to gonadotropin releasing hormone and adrenal corticotropic hormone, increased hairiness, elevated serum dehydroepiandrosterone sulfate levels and early onset of baldness.
C. *Twins' studies:* According to research on the syndrome's frequency in twins, PCOS is thought to be caused by a combination of hereditary and environmental factors. In a sizable Dutch twin research, monozygotic twins' occurrence of PCOS symptoms in monozygotic twins (70%) was double that of dizygotic twins. Since PCOS is a "oligogenic" condition, it is likely that several genes are involved in its genesis.

Genome-Wide Association Studies

Genome-wide association studies (GWAS) does not rely on prior knowledge of the underlying aetiology. This is especially important for conditions like PCOS, which have a complex and diverse phenotype. The GWAS data that have been released so far on individuals of European and Chinese ancestry have nevertheless shed light on the aetiology of PCOS. These are described as follows:

- Gonadotrophin-secreting genes may be important in the development of PCOS, according to the implication of these genes. Anovulation and hyperandrogenism, which are characteristics of PCOS, may be enhanced by these genes.
- The role of gonadotropin activity in the development of PCOS. Variants of the gene that alter gonadotrophin action at the FSH and LH receptors (encoded by FSHR and LHCGR, respectively) have been linked (Hayes et al. 2015).
- Potential genetic role of GATA4 variations in controlling steroidogenesis.

In order to fully understand the aetiology of PCOS, additional functional research is necessary to determine the cellular function of the detected genetic variations and their potential effects on physiological, reproductive and metabolic processes. There have been published functional studies using GWAS data from PCOS. DENND1A, LHCGR, INSR, RAB5B and other adapter proteins are a few of these functions.

Research has particularly focused on how genetic variations affect androgen production in theca cells. Based on GWAS data, additional functional investigations will be presented in the future, offering a deeper understanding of the aetiology of PCOS. Future reports of more functional investigations will offer a greater understanding of the aetiology of PCOS based on GWAS data.

Environmental Risk Factors

It is evident that PCOS is a hereditary disorder, with potential epigenetic influences on its development. However, it is also evident that genetics and epigenetics cannot fully explain the development of PCOS in women. Weight gain also contributes to its development, particularly in genetically predisposed women. Recent data have revealed a wide range of potential environmental factors that may lead to the development of PCOS. Several prominent theories and supporting data include the "foetal programming hypothesis", environmental contaminants, diet-related weight gain, gut microbiota, inactivity and lack of exercise and advanced glycation end products.

Foetal Programming Hypothesis

Research suggests that PCOS may be linked to early life factors, with studies showing that adolescent girls with PCOS tend to have lower birth weights compared to their peers without PCOS. Additionally, nutritional deficiencies during foetal development may also contribute to the future development of PCOS (Gur et al. 2015). The foetal programming hypothesis proposes that the intrauterine environment plays a crucial role in the development of PCOS, supported by both human observational and animal-based studies. This hypothesis suggests that two key mechanisms contribute to PCOS risk: (1) Intrauterine nutritional restriction, leading to changes in foetal insulin levels, insulin resistance and potential epigenetic modifications (gene methylation). (2) Foetal exposure to excess androgens, resulting in "programming" and altered tissue differentiation. These mechanisms are thought to predispose the foetus to develop PCOS later in life.

Environmental Toxins

Endocrine Disrupting Chemicals (EDCs) mostly found in plastics, pesticides and personal care products may play a role in the hormonal imbalances associated with PCOS. EDCs can enter the body through various routes, including Intravenous exposure (through medical treatments) and by oral ingestion (through food and water).

Triclocarban Triclocarban also known as 3,4,40-trichlorocarbanilide, a chemical found in daily use products such as plastics, toys, clothing and carpets may contribute to the development of PCOS. It enhances the effects of testosterone by increasing transcriptional activity on the androgen receptor mimicking hyperandrogenaemia and has been detected in plasma of umbilical cord suggesting prenatal exposure. It has ability to amplify androgen effects, and detection in umbilical cord blood contributes to prenatal androgenization of female foetuses potentially leading to the development of PCOS later in life.

Bisphenol A BPA is an estrogenic monomer found in polycarbonate plastics. Research has shown that women with PCOS have considerably higher serum levels of BPA compared to control women. It has been found to be associated with androgens such as testosterone in women with and without PCOS to demonstrate estrogenic receptor binding.

Phthalates Phthalates are used to increase the flexibility of plastics, which are widely present in various products including food packaging, personal care items (toiletries) and medical equipment (intravenous tubing). These have been detected in amniotic fluid, indicating they can cross the placenta and reach the foetus. Another EDC di (2-ethylhexyl) phthalate (DEHP) has been shown to increase ovarian weights in female mouse offspring following prenatal exposure. Research has shown that adult female rats when exposed to DEHP experience suppressed ovulation and development of polycystic ovarian morphology, which are typical features of PCOS.

Dietary-Related Weight Gain

Weight gain and obesity are the most significant contributors to the development of PCOS. Obesity plays a pivotal role in shaping the PCOS phenotype, and excess weight worsens many of its symptoms, particularly insulin resistance. In fact, weight gain can exacerbate nearly all PCOS symptoms, making it a critical factor in the severity and progression of the disorder. Obesity is not solely an environmental factor but rather a complex trait resulting from the interplay of genetic and environmental influences. Genetic variants that regulate fat mass likely contribute to the heritability of PCOS, and epidemiological data confirm a strong link between PCOS and obesity, with 38–88% of women with PCOS being overweight or obese. The significance of fat mass in triggering PCOS symptoms is further emphasized by the finding that a modest weight loss of just 5% in obese women with PCOS can lead to significant improvements in, reproductive features, hyperandrogenic symptoms, cardiometabolic risk factors. According to current research, PCOS typically develops in women who are genetically susceptible to the condition and experience weight gain, often during puberty. Obesity and PCOS are known to be linked with insulin resistance; obese women with PCOS commonly show higher insulin resistance as a result of these two conditions working together, creating a metabolic "double-whammy". This scenario explains why women with PCOS have a higher chance of developing type 2 diabetes (about 10%) and early-onset impaired glucose tolerance (around 30%–40%). Insulin resistance, (made worse by weight gain), is a main cause of PCOS. This means that managing PCOS effectively requires addressing insulin resistance, which can be achieved through weight loss and healthy lifestyle choices.

Lack of Exercise and Physical Activity

Promoting physical exercise is a pivotal component of healthy lifestyle advice for PCOS women. Exercise plays a vital role in understanding and managing the condition.

Advanced Glycation End Products

AGEs are produced endogenously, when lipids, nucleic acids and free amino groups of proteins react with carbohydrates through glycoxidation. AGEs are also found in certain foods including: Cooked and processed fast foods, raw animal products, high-protein and high-fat foods, foods cooked at high temperatures. Diet and tobacco are major external sources of AGEs. Recent research suggests that dietary AGEs may contribute to the development of PCOS, including its metabolic and reproductive problems. Research has shown that high levels of AGEs (due to dietary intake) in the blood serum are related to higher levels of inflammatory markers such as C-reactive protein (CRP) and HOMA-IR (marker of IR). It been suggested that consuming dietary AGEs may exacerbate the chronic inflammation seen in PCOS, potentially playing a role in the development and progression of this condition.

Interplay Between Genetic and Environmental Factors

Gene–Environment Interaction Environment may exert an influence on genetic susceptibility. For example, individuals with genetic tendency towards insulin resistance may be more vulnerable to the negative consequences of a poor diet or lack of exercise. Environmental factors may impact gene expression through epigenetic changes, which could potentially lead to the development of PCOS. For instance, dietary decisions or exposure to certain chemicals may change the expression of genes involved in metabolism and hormonal regulation. Overall, PCOS is a multifaceted disorder that arises from the interplay of genetic and environmental elements, requiring a comprehensive approach that incorporates lifestyle changes and personalized medical care to effectively manage its symptoms and severity.

Conclusion

A lifestyle disorder is the interplay between genetic predisposition and environmental factors. The findings suggest that genetic susceptibility (genetic factors) of these diseases is significantly influenced by the environmental factors such as diet, physical inactivity, chemical exposure and socioeconomic status. The results suggest that better understanding of genetic and environmental factors contributed to the correct assessment, prevention and management of these diseases. Ultimately, a comprehensive approach that involves genetic risk management, lifestyle changes and environmental modifications can offer the most effective solution for the prevention and management of these diseases.

References

Altshuler D, Hirschhorn JN, Klannemark M et al (2000) The common PPARγ Pro12Ala polymorphism is associated with decreased risk of type 2 diabetes. Nat Genet 26(1):76–80

Barratt BJ, Payne F, Lowe CE et al (2004) Remapping the insulin gene/IDDM2 locus in type 1 diabetes. Diabetes 53(7):1884–1889

Bryan H, Riese H, Hawa MI et al (2012) Glycotoxin and autoantibodies are additive environmentally determined predictors of type 1 diabetes: a twin and population study. Diabetes 61(5): 1192–1198

Doria A, Patti ME, Kahn CR (2008) The emerging genetic architecture of type 2 diabetes. Cell Metab 8(3):186–200

Dotta F, Censini S, van Halteren AG et al (2007) Coxsackie B4 virus infection of β cells and natural killer cell insulitis in recent-onset type 1 diabetic patients. Proc Natl Acad Sci USA 104(12): 5115–5120

Duke J, Huhman M, Heitzler C (2003) Physical activity levels among children aged 9-13 years-United States, 2002. MMWR Morb Mortal Wkly Rep 52(33):785–789

Flanagan SE, Haapaniemi E, Russell M et al (2014) Activating germline mutations in STAT3 cause early-onset multi-organ autoimmune disease. Nat Genet 46(8):812–814

Futreal PA, Coin L, Marshall M et al (2004) A census of human cancer genes. Nat Rev Cancer 4(3): 177–183

Gloyn AL, Weedon MN, Owen KR et al (2003) Large-scale association studies of variants in genes encoding the pancreatic β-cell KATP channel subunits Kir 6.2 (KCNJ11) and SUR1 (ABCC8) confirm that the KCNJ11 E23K variant is associated with type 2 diabetes. Diabetes 52(2): 568–572

Gonzalez-Zacarias AA, Mavarez-Martinez A, Arias-Morales CE et al (2016) Impact of demographic, socioeconomic, and psychological factors on glycemic self-management in adults with type 2 diabetes mellitus. Front Public Health 4:195

Graf C, Koch B, Kretschmann-Kandel E et al (2004) Correlation between BMI, leisure habits and motor abilities in childhood (CHILT-project). Int J Obes Relat Metab Disord 28(1):22–26

Grant SF, Thorleifsson G, Reynisdottir I et al (2006) Variant of transcription factor 7-like 2 (TCF7L2) gene confers risk of type 2 diabetes. Nat Genet 38(3):320–323

Gur EB, Karadeniz M, Turan GA (2015) Fetal programming of polycystic ovary syndrome. World J Diabetes 6(7):936

Hancox RJ, Milne BJ, Poulton R (2004) Association between child and adolescent television viewing and adult health: a longitudinal birth cohort study. Lancet 9430:257–262

Hayes MG, Urbanek M, Ehrmann DA et al (2015) Genome-wide association of polycystic ovary syndrome implicates alterations in gonadotropin secretion in European ancestry populations. Nat Commun 6(1):7502

Hebebrand J, Hinney A (2009) Environmental and genetic risk factors in obesity. Child Adolesc Psychiatr Clin N Am 18:83–94
Hober D, Alidjinou EK (2013) Enteroviral pathogenesis of type 1 diabetes: queries and answers. Curr Opin Infect Dis 26(3):263–269
Hober D, Sauter P (2010) Pathogenesis of type 1 diabetes mellitus: interplay between enterovirus and host. Nat Rev Endocrinol 6(5):279–289
Horikawa Y, Oda N, Cox NJ et al (2000) Genetic variation in the gene encoding calpain-10 is associated with type 2 diabetes mellitus. Nat Genet 26(2):163–175
Hu F (2011) Globalization of diabetes: the role of diet, lifestyle, and genes. Diabetes Care 34(6): 1249–1257
Hussain MD (2021) Obesity and higher risk for severe complications of COVID-19. Curr Trends Pharmacol Clin Trials 4:180029
Hussain S, Kaur G, Deb A (2021) Mini-review on personalized medicine: a revolution in health care. Precis Med Res 3(4):1–3
Jackson M, Walker S, Cruickshank JK et al (2007) Diet and overweight and obesity in populations of African origin: Cameroon, Jamaica and the UK. Public Health Nutr 2:122–130
Jaidane H, Sauter P, Sane F et al (2010) Enteroviruses and type 1 diabetes: towards a better understanding of the relationship. Rev Med Virol 20(5):265–280
Kopelman PG (2000) Obesity as a medical problem. Nature 6778:635–643
Kumar M, Hussain MS, Raj S et al (2023) An overview of treatment modalities and management aspects for obesity. Curr Nutr Food Sci 19:105–113
Lefèbvre P, Pierson (2004) The global challenge of diabetes. World Hosp Health Serv 40(3):37–39
Manolio TA, Collins FS, Cox NJ et al (2009) Finding the missing heritability of complex diseases. Nature 461(7265):747–753
Mathur K, Sharma S, Hussain MS (2023) Mini-review on the management of lifestyle disorders: attempting to keep Indians healthy for a bright future. Dis Diagn 12:144–150
Park KS (2011) The search for genetic risk factors of type 2 diabetes mellitus. Diabetes Metab J 35(1):12–22
Ramachandran A, Snehalatha C (2009) Current scenario of diabetes in India. J Diabetes 1(1):18–28
Ramachandran A, Mary S, Yamuna A et al (2008) High prevalence of diabetes and cardiovascular risk factors associated with urbanization in India. Diabetes Care 31(5):893–898
Reilly JJ, Jackson DM, Montgomery C et al (2004) Total energy expenditure and physical activity in young Scottish children: mixed longitudinal study. Lancet 9404:211–212
Reynisdottir I, Thorleifsson G, Benediktsson R (2003) Localization of a susceptibility gene for type 2 diabetes to chromosome 5q34–q35. 2. Am J Hum Genet 73(2):323–335
Schloot NC, Willemen SJ, Duinkerken G et al (2001) Molecular mimicry in type 1 diabetes mellitus revisited: T-cell clones to GAD65 peptides with sequence homology to Coxsackie or proinsulin peptides do not cross react with homologous counterpart. Hum Immunol 62(4):299–309
Shrivastava S, Shrivastava P, Ramasamy J (2013) The necessity of a balanced diet to prevent the emergence of lifestyle disorders. S Afr J Clin Nutr 26(3):156–157
Shu L, Sauter NS, Schulthess FT et al (2008) Transcription factor 7-like 2 regulates β-cell survival and function in human pancreatic islets. Diabetes 57(3):645–653
Sladek R, Rocheleau G, Rung J (2007) A genome-wide association study identifies novel risk loci for type 2 diabetes. Nature 445(7130):881–885
Smyth D, Cooper JD, Collins JE et al (2004) Replication of an association between the lymphoid tyrosine phosphatase locus (LYP/PTPN22) with type 1 diabetes, and evidence for its role as a general autoimmunity locus. Diabetes 53(11):3020–3023
Størling J, Pociot F (2017) Type 1 diabetes candidate genes linked to pancreatic islet cell inflammation and beta-cell apoptosis. Genes 8(2):72
Tabish SA (2017) Lifestyle diseases: consequences, characteristics, causes and control. J Cardiol Curr Res 9:00326
Unoki H, Takahashi A, Kawaguchi T et al (2008) SNPs in KCNQ1 are associated with susceptibility to type 2 diabetes in East Asian and European populations. Nat Genet 40(9):1098–1102

Vallgarda S (2011) Why the concept "lifestyle diseases" should be avoided. Scand J Public Health 39:773–775

Vionnet N, Dupont S, Gallina S et al (2000) Genome wide search for type 2 diabetes–susceptibility genes in French Whites: evidence for a novel susceptibility locus for early-onset diabetes on chromosome 3q27-qter and independent replication of a type 2–diabetes locus on chromosome 1q21–q24. Am J Hum Genet 67(6):1470–1480

Yamauchi T, Hara K, Maeda S et al (2010) A genome-wide association study in the Japanese population identifies susceptibility loci for type 2 diabetes at UBE2E2 and C2CD4A-C2CD4B. Nat Genet 42(10):864–868

Zeggini E, Scott LJ, Saxena R (2008) Meta-analysis of genome-wide association data and large-scale replication identifies additional susceptibility loci for type 2 diabetes. Nat Genet 40(5): 638–645

Zheng W, She JX (2005) Genetic association between a lymphoid tyrosine phosphatase (PTPN22) and type 1 diabetes. Diabetes 54(3):906–968

Effects of Dust Particles on Provoking the Risk of Lung Cancer

Prasad Sherekar, Sanvidhan G. Suke, Pooja Patel, and Basu Dev Banerjee

Abstract

Carcinogenic dust is a key contributor to the global lung cancer burden, as the dust particles (PM 2.5) may readily pass through the upper respiratory barriers and enter lung alveoli. Chromosomal instability, cell cycle arrest, mdig expression, apoptosis and necrosis are the subsequent events leading to lung cancer once the dust particles are engulfed by the alveolar macrophages. Small cell lung cancer (SCLC) and non-small cell lung cancer (NSCLC) are the two subtypes of lung cancer accounting for over 20% of all cancer deaths worldwide. The International Agency for Research on Cancer (IARC) has classified all of the dusts that contain crystalline silica, outdoor air pollutants, cadmium (Cd), vinyl chloride monomer (VCM), welding fumes, radon, asbestos, hardwood dust, polycyclic aromatic hydrocarbons (PAHs), diesel exhausts (DEs), chromium-VI (Cr-VI), and nickel (Ni) as 'Group 1' carcinogens. This chapter discusses the various dust particles, their exposure pattern, and carcinogenic effects leading to the development of lung cancer, along with the findings of recent investigations from various countries regarding dust-induced lung cancer incidences, mortality rate, and death toll. Additionally, the potential pathophysiological mechanism of lung malignancy caused by inhaling carcinogenic dusts has been revealed. Thus,

P. Sherekar · S. G. Suke (✉) · P. Patel
Department of Biotechnology, Priyadarshini College of Engineering, Nagpur, India

B. D. Banerjee
Department of Medical Elementology and Toxicology, School of Chemical and Life Sciences, Jamia Hamdard, New Delhi, India

Department of Biochemistry, Central Research Facility, Santosh Deemed to be University, Ghaziabad, Uttar Pradesh, India
e-mail: drbdbanerjeecrf@santosh.ac.in

S. Raisuddin et al. (eds.), *Environmental Toxicants and Lifestyle Diseases*,
https://doi.org/10.1007/978-981-95-0575-3_21

this chapter warns policymakers to reconsider the safety recommendations for workers in occupations that are assumed to involve dust.

Keywords

Environmental contaminants · Dust particles · Lung cancer

Introduction

'Dust' is a fraction of particulate matter (PM), which can be easily respirable and or breathable by humans and animals. The characteristic of dust contains particulate matter of various sizes with several kinds of harmful contaminants (Jenwitheesuk et al. 2020). The particles of size 30–100 microns are largest inhalable fraction, 10–30 microns are the thoracic fraction, whereas particles below 10 microns are the respirable fraction of the dust (Fig. 1). The airborne respirable dust has classified according to its pattern of toxicity. Fibrogenic dust, carcinogenic dust, toxic aerosols, radiological dust, explosive dust and nuisance dust are the main classifications of respirable dusts that contain numerous toxic components (Pingle et al. 2022). Long-term inhalation of these respirable dust fraction leads to the development of several fatal lung diseases including asthma, bronchitis, silicosis, asbestosis, chronic obstructive pulmonary diseases (COPD), coal workers pneumoconiosis (CWP) and lung cancer (Sherekar et al. 2022). Most of the world population exposed to polluted air with comprising various types of toxic dusts causes around 4.2 million of deaths worldwide every year (Jenwitheesuk et al. 2020).

Cancer is the leading cause of death, where cancer of respiratory tract includes trachea, bronchus and lung, which further stated as LBC accounting for over 20% of all cancer deaths globally. Small cell lung cancer (SCLC) and non-small cell lung cancer (NSCLC) are the two subtypes of lung cancer (Jiˇrík et al. 2021; Huang 2020). Although smoking is confirmed key risk factor, almost 25% of lung cancer patient are non-smoker, indicating that the aetiology of lung cancer also has an environmental and genetic origin (Minina et al. 2022). The major environmental risk factors (lung carcinogens) are respirable crystalline silica (RCS) (Gardner et al. 2022; Kim et al. 2021; Pavan et al. 2020; Rahimimoghadam et al. 2023), outdoor

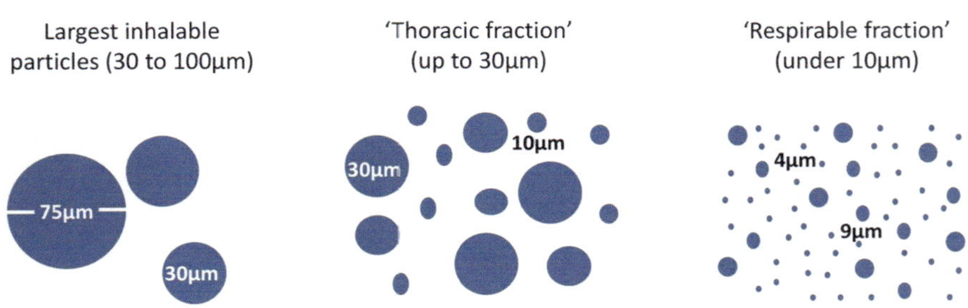

Fig. 1 Combined airborne dust particles make up the 'Inhalable fraction'

air pollutant (Jenwitheesuk et al. 2020; Ryu et al. 2023), cadmium (Cd) (Lombaert et al. 2023), vinyl chloride monomer (VCM) (Girardi et al. 2022), welding fumes (Dauter et al. 2022), radon, benzo[a]pyrene (Jiˇrík et al. 2021), hardwood dust (Matrat et al. 2019; Scarabelli et al. 2021; Collatuzzo et al. 2023), asbestos (chrysotile), (Shi et al. 2021; Schuz et al. 2024), polycyclic aromatic hydrocarbons (PAH) (Zhang et al. 2023), diesel engine exhaust, hexavalent chromium (Cr VI) and/or nickel (Ni) (Minina et al. 2022; Collatuzzo et al. 2023; Hanona et al. 2023). The vast evidence provoked the International Agency for Research on Cancer (IARC) to classify above environmental risk factors as a 'Group 1' carcinogen, indicating that there is adequate proof of carcinogenicity for lung cancer in humans. Recently, two different investigations in experimental mice model defined accelerated lung tumorigenesis followed by chronic exposure to dust from swine confinement facilities (Dominguez et al. 2022) and house dust mites (HDM) (Wang et al. 2023), suggesting both can be a strong carcinogenic agent leading to lung cancer in humans. Co-exposure to some of the above lung carcinogens has associated with higher risk of lung cancer development compared to individual exposures (Olsson et al. 2024).

The aim of this chapter is to shed light on the effects of occupational exposure to carcinogenic dusts on lung cancer risk. Thus, the relationship between occupational exposure of prominent lung carcinogens and exposure factors, evidence-based studies and the dust-induced lung cancer mechanisms is discussed in the following sections.

Carcinogens Provoking the Risk of Lung Cancer

Mineral mining such as coal, asbestos, marble and heavy metal was the key occupation for the workers in the twentieth century, whereas due to continuous globalization in twenty-first century, the industries like bricklayer, welding, painting, hardwood, construction and artificial stone are booming. The professionals associated with these industries experience heavy dust exposure for long active periods and at a high risk of development of respiratory tract cancer including larynx, tracheal and lung.

Respirable Crystalline Silica Dust

Silica is the highest mineral in the Earth's crust and more than 230 million workers globally engaged in mineral mining exposed to RCS dust yearly (Sherekar et al. 2024). From 1997, crystalline silica has been concluded as Group 1 carcinogen by IARC when inhaled in the form of cristobalite or quartz (Kleinschmidt et al. 2022). Several epidemiological, case control and experimental investigations conducted in different parts of the world demonstrated the severe fibrogenic and tumorigenic effects of RCS dust in occupational workers, animal and cell line models. Toxicity due to long-term exposure to RCS dust results in silicosis, still lung cancer has been a final outcome.

The interaction between biomolecules and exogenous inhaled silica particles is critical to predict and manage expected or adverse outcome of these foreign dust particles. Silica has crystalline and amorphous surface properties, where RCS is leading cause of lung cancer. Nearly free silanols (NFS) has found to be energetically favourable moiety at the surface of silica. NFS moiety is responsible to establish interactions with the components of pulmonary cell membrane and initiate toxicity via membranolysis and inflammation (Pavan et al. 2020).

Workers involved in the collection, recycling and safe management of municipal household waste (MHW) are at an increased risk of respiratory diseases due to the presence of RCS particles in the MHW dust. According to study conducted in South Korean cities, workers who collects coal briquette ash (CBA) exposed to highest RCS geometric mean (GM) concentration (7.3 μg m^{-3}), followed by blowing workers (5.1 μg m^{-3}), and MHW collectors (2.6 μg m^{-3}) who is more than the recommended exposure limit for the RCS (2.6 μg m^{-3}) by California's Office of Environmental Health Hazard Assessment (Kim et al. 2021).

Spodumene ($LiAlSi_2O_6$) is the most commercially exploited due to its high lithium content and respirable crystalline quartz present in α-spodumene concentrates in significant amounts (14%). Though presently no toxicological data report potential health effects of exposure to respirable α-spodumene cleavage fragments, the concentration of crystalline quartz can be a matter of concern (Gardner et al. 2022).

The interesting cross-sectional study conducted by Margan et al. (2022) estimates the intensity of exposure to RCS in various high-risk industries from Slovenia. Of the 2131 listed companies, author received the data of number of employees exposed to silica from only 277 (21.6%) companies. Among 18,064 workers who work in these 277 companies, 2072 (11.5%) workers were exposed to RCS according to self-reported estimates by company representatives, whereas, study authors based on literature data according to job description, 5007 (27.7%) workers estimated to expose RCS from same companies. Moreover, the data on measurement of total and silica dust are made available to this cross-sectional study by only 8.3% and 1.8% of companies, respectively, working with silica. Author concluded that such discrepancy in RCS exposure estimation by companies is most likely owed to a lack of knowledge and awareness about the health of RCS-exposed workers (Margan et al. 2022).

The cross-sectional study is conducted on 72 male workers of a concrete manufacturing plant in Nishapur city, Iran. The 4-year occupational exposure to silica dust at concrete plant was found to be 0.025 mg m^{-3} and the overall mortality risk due to lung cancer in various occupational groups was 10.11 per 1000 workers with a range of 3.75–30.66 (Moghadam et al. 2020). Another cross-sectional study conducted on 79 stone carvers of Nishapur city, Iran to measure occupational exposure of crystalline silica and evaluated the associated health risks among the workers. The mean exposure of total inhalable dust was 1.44 mg m^{-3} from which 0.5 mg m^{-3} was found to be RCS concentration, which was higher than recommended occupational standard (0.05 mg m^{-3}) (Table 1). Such high RCS exposure is associated with the risk of silicosis (3 to 12 per 1000) and lung cancer (150 per 1000) among stone carvers (Rahimimoghadam et al. 2023).

The seven-decade (1945–2004) cohort study of 2650 workers in open pit quarry mining (roofing granule) of rock and mill from the USA reported the lung cancer mortality according to cumulative RCS exposure (mg m^{-3}) per year. Total 116 deaths attributed to lung cancer with hazard ratio (HR) of 1.00 (referent), 1.20, 1.85 and 0.92, respectively, for 29 (<0.089), 28 (0.089–<0.224), 30 (0.224–<0.456) and 29 (≥ 0.456) deaths (mg/m^{-3}- years) (Kleinschmidt et al. 2022).

Coal is the important fossil fuel mined from several decades for energy needs. Along with carbon, crystalline silica is the major component of coal and one of the key risk factors associated with coal dust exposure to coal miners. A recent meta-analysis includes nine studies from the USA, seven from the United Kingdom, three from China, two from Czech Republic and one from the Netherlands, suggested strong association between coalmine dust exposure and increased mortality risk of lung cancer. The pooled data of 22 studies estimated the lung cancer relative risk (RR) of 1.16 (at 95% CI). The studies from China indicated a higher lung cancer risk for coalminers with RR of 4.94 (at 95% CL). Overall, this meta-analysis confers the potential carcinogenicity of RCS containing coalmine dust on the lung (Li et al. 2021).

Association of cigarette smoking with silica dust exposure increases the risk of mortality for lung cancer. This phenomenon was confirmed in a cohort of 44,708 workers belong to metal mines and pottery factories from central and southern China. The hazard ratio with mortality and proportions for additive interaction were reported 4.51 (95% CI, 3.23–6.29) and 21.63%, respectively, for lung cancer under combined exposure of silica dust and cigarette smoking (Wang et al. 2020) (Table 1).

Outdoor Air Pollutants

Ambient air pollution comprises of harmful inhalable particles; especially, a particulate matter of size 2.5 micron (PM 2.5). PM 2.5 owns the carcinogenic potential as they induce oxidative damage, lung inflammation and continuous proliferative signalling leading to lung cancer. The outdoor air pollution was also recognized as lung carcinogen by IARC in 2013 (Jenwitheesuk et al. 2020; Ryu et al. 2023). The mineral dust, organic and black carbon, sea-salt and SO_4 are major chemical types of aerosols; parts of ambient air pollution. A correlation between lung cancer incidences (Thai population cancer data) and the six-year accumulation of each aerosol was investigated by Jenwitheesuk et al. (2020). The results suggested that 21%, 26% and 61% excess risk of lung cancer with every 1 μg m^{-3} increase in organic carbon compound, average SO_4 and combined dust levels, respectively. The urban dust particles fraction of PM 2.5 found to delays the mitotic progression inhibits short and long-term proliferation and leads to subsequent chromosomal instabilities in A549 lung cancer cells (Ryu et al. 2023). Both studies validated the

Table 1 The number of deaths and or mortality risk and ΔQALY for lung cancer by cumulative exposure of carcinogenic agents (mg m^{-3}/years) estimated by different studies

Studies	Study period (Country)	Agents	Cumulative exposure (mg m^{-3}/years)	No. deaths	$QALY_{dea}$ /person	MR /1000
Wang et al. (2020)	1960–1974 (China)	RCS	3.64 ± 4.17	917	–	–
Moghadam et al. (2020)	2012–2015 (Iran)	Silica dust	9.12 ± 2.92	–	–	10.11
Kleinschmidt et al. (2022)	1945–2015 (USA)	RCS	<0.456	77	–	–
Rahimimoghadam et al. (2023)	2020 (Iran)	RCS	≥28.1	–	–	63.6
Liu et al. (2019)	2017 (China)	PM 2.5	242.26 ± 14.23	–	13.9	–
Raaschou-Nielsen et al. (2023)	1991–2015 (Denmark)	PM 2.5	1.29	67,756	–	–
Schuz et al. (2024)	1975–2015 (Russia)	Asbestos dust	≥150	564	–	–

Abbreviation: *RCS* respirable crystalline silica, $QALY_{dea}$ quality-adjusted life years lost due to premature death from lung cancer, *MR* mortality rate

capability of outdoor air pollutant to increase the risk of lung cancer through DNA damage and chromosomal disruption.

Cadmium, Vinyl Chloride Monomer and Welding Fumes

All these three agents were classified as lung carcinogens by IARC based on several experimental and epidemiological study results. Workers associated with industries including pigment manufacture and formulation plants, cadmium alloy production, and mechanical plating are exposed to Cd-contaminated air. The Cd at workplace has direct effect on lung respiratory tract through inhalation. Using the data from various epidemiological studies, the dose–response relationship for Cd exposure and lung cancer risk was 1:10,000 at exposure of 4 μg m^{-3} (respirable). Hence, 4 μg $m^{-3}$of observed effect level (OEL) was suggested for implementation to protect workers from lung cancer risk (Lombaert et al. 2023).

A cohort study evaluated the relationship between lung cancer mortality and VCM exposure to workers (PVC baggers) from 1973 to 2017 in Porto Marghera (Venice, Italy). After adjusting the smoking mortality rate ratios (MMR) of 1.99, (95% CI 1.04–3.81) for lung cancer was found to be associated with VCM exposure in PVC baggers (Girardi et al. 2022). The activation of alveolar macrophages, pulmonary inflammation and release of growth factors under prolonged exposure to VCM dust may promote the carcinogenesis in lung tissues.

The respirable particles of size 20–1000 nm and the mixture of iron, chromium, manganese and nickel (metals) are the components that regulate the carcinogenicity of welding fumes. A longitudinal investigation for 6 years determines the association among early cancer-related biomarkers and levels of exposure to mild steel welding fumes in 78 welders from southern Sweden (Dauter et al. 2022). The 0.7 mg $m^{-3}$of median respirable dust exposure was recorded at both time points (at the beginning and after 6 years). A significant decrease in DNA methylation of *B3GNTL1* CpG1 and CpG4 was found over time in welders compared to controls. Moreover, decrease in methylation of *F2RL3* CpG2 among all welders was found to be allied with cumulative respirable dust exposure. This indicates that early lung cancer-related changes like lower methylation of some CpG sites of cancer-related genes are associated with low-to-moderate exposure to mild steel welding fumes.

Radon and Benzo[a]pyrene

Epidemiological studies have provided compelling evidence of the impact of radon on the incidence of lung cancer; hence, Radon was a considered as a human carcinogen by IARC. The lifetime carcinogenic risks of radon, benzo[a]pyrene and fibrogenic dust were assessed in combination with a longitudinal cross-sectional population study (from 1994 to 2016) on the incidence of lung, bronchial and tracheal (LBC) cancer, in the Industrial and Non-Industrial Regions of the Czech Republic (Jiˇrík et al. 2021). The study reported that the absolute share of radon and benzo[a]pyrene risk in the total incidence of LBC in the participated population was overestimated. Significant share of smokers in the Czech Republic might be reflected on the occurrence of LBC. However, the study further suggested that the radon and benzo[a]pyrene risks can produce major impact on the overall incidence of LBC if the large portion of population has associated with high-risk occupations that involve exposure to fibrogenic dust.

Hardwood Dust

Although the IARC classified wood dust as carcinogenic to humans with strong evidence for a link between hardwood dust exposure and sinonasal cancer, there was inconsistent evidence for lung cancer (Matrat et al. 2019). However, a meta-analysis by Hancock et al. (2015) provided compelling evidence of a link between wood dust and lung cancer. Previous longitudinal health studies have recognized 5 mg $m^{-3}$of an occupational threshold for inhalable wood dust (Holm and Festa 2019).

Depending on these findings, a case-control population-based study including 5056 men population (2780 controls and 2276 cases) from France investigated the association among lung cancer risk and wood dust exposure. The results showed a clear increase in lung cancer risk in wood dust-exposed population having co-exposures of smoking with squamous cell carcinoma (35.3%) and adenocarcinoma (34.9%) as the two dominant histological subtypes. After adjusting for smoking, and exposure to other group 1 carcinogens, there was insufficient evidence to support the claim that wood dust exposure contributed to the incidence of lung cancer (Matrat et al. 2019). The role of wood dust in incidence of sinonasal and upper respiratory tract cancer was also mentioned in the meta-analysis. The attributable fraction of sinonasal cancer was 11.8% to wood dust exposure with prevalence of 2.3% among other occupational carcinogens (Collatuzz et al. 2023).

Another case study on 58-year-old male patient (never smoker) of follicular B-cell lymphoma showed the presence of foreign fibres in the pleural fluid suggesting the direct effects of inhalation of toxic air pollutant. As the patient was lived nearby the paper mill 10 years preceding the diagnosis, the wood dust in the surrounding environment might be responsible for carcinogenic pollutant (Scarabelli et al. 2021). These heterogeneous results are associated with the relation between lung cancer risk and exposure to wood dust. Therefore, strong longitudinal cohort studies are needed to validate such ambiguous findings.

Asbestos and Polycyclic Aromatic Hydrocarbons

Humans are susceptible to asbestos in all forms (amosite, anthophyllite, chrysotile, crocidolite, tremolite and actinolite)-related cancers including cancers of the lung, larynx, ovary and mesothelioma. Recent large-scale epidemiological study investigated the cancer-related mortality in workers (from 1975 to 2010 and follow-up to 2015) from world's largest asbestos (chrysotile) mine located in the town of Asbest, Russian Federation. Among 30,445 workers that were found eligible for Asbest Chrysotile Cohort risk analysis, 32% were women. The cumulative mean exposure to dust in mg m^{-3}/year and fibres in cm^{-3}/year in men were 61.2 and 29.2 and in women were 77.8 and 42.8, respectively. Mortality for all cancers irrespective of cumulative exposure to dust (0 to $\geq$150 mg m^{-3}/year) and fibre (0 to $\geq$80 cm^{-3}/year) in men and women was 1526 and 535, respective (Table 1). Among all cancer-related deaths, the share of lung cancer in men was 36.95% and 7.7% in women (Schuz et al. 2024). This cohort analysis confirmed that men who were exposed to high chrysotile dust had a higher mortality rate from lung cancer. On the other hand, there was no palpable rise in female lung cancer mortality.

Another carcinogenic pollutant, PAHs are associated with the augmented risk of lung cancer in never smokers. The particulate enrichment factor (EF_P) of 436 (54–835) for PAHs accumulated in pulmonary PM was found by the cohort study of 68 lung cancer patients from an air-polluted region of China. The study further revealed that the carcinogenic potential of PAHs increased with age since patients between the ages of 70 and 80 had a 17-fold greater rate of monoclonal tumorigenesis in their lungs than patients under the age of 50 (Zhang et al. 2023).

Diesel Exhausts (DEs), Chromium-VI and Nickel

All of the above components were classified as IARC group 1 carcinogen. The attributable fraction (AF) of lung cancer related to occupational exposure to DEEs, Cr-VI, and Ni in Italy was estimated by Collatuzzo et al. (2023). The second most prevalence of exposure was diesel exhaust (4.3%), followed by Cr (1.30) and Ni (0.81%), which has been reported. The AF of 3.8% for lung cancer reflected due to 1152 and 1390 attributable deaths in 2017 and cases in 2020, respectively. Moreover, the number of deaths in 2017 attributable to DEs, Cr-VI and Ni was 489, 191 and 35, respectively. The population AF of 1.4%, 0.5% and 0.6% was observed for lung cancer, respectively, due to occupational exposure of DEs, Cr-IV and Ni. The calculated values of population AF for lung cancer were supported with the data of attributable deaths in 2021 and cases in 2020 reported for individual occupational exposure to mentioned carcinogens. Furthermore, the carcinogenic effects of Ni have been highlighted in later case study of 40-year-old male patient of thoracic undifferentiated SMARCA4-deficient tumour having work history related to dust and melted nickel exposure (Hanona et al. 2023).

World Trade Center (WTC) Dust

Upon the collapse of the World Trade Center (WTC) towers on September 11, 2001, residents and workers in the area were exposed to dust for short and long-term periods, which has been associated with pulmonary disorders, especially lung cancer. In recent years several epidemiological studies shed light on the mutagenicity of WTC dust (DeMarini et al. 2022), associated lung cancer characteristics

(Durmus et al. 2021) and incidences (Sigel et al. 2022). Following this, a latest study focused on the key cancer screening information, cancer diagnostic and treatment best practices (Calverta et al. 2023).

The carcinogenic effect of large inorganic fraction of WTC dust has been studied widely; whereas, DeMarini et al. (2022) identified the carcinogens in the smaller organic fraction of WTC dust and determined the mutagenicity in different *Salmonella* strains. According to the study findings, only 0.74% of the mass of PM below 53 μm was extractable organic matter (EOM) that comprised 37 previously analysed PAHs with 5.4% of the EOM mass and 0.04% of the PM mass. This data confirm the PAH-associated elevated cancer risk among the population exposed to WTC dust. The characteristics of lung cancer in the WTC Environmental Health Center (WTC EHC) were discussed earlier by (Durmus et al. 2021). There were a total of 248 WTC EHC patients; women (57%) and men (43%) had been diagnosed for lung cancer up to the end of 2019, among which 37% were never-smoker (≤1 pack-years). The adenocarcinoma (72%) was found to be dominant in never-smoker irrespective of gender. This study provides primary depiction of lung cancers in non-smoking local community members that documented for WTC dust exposure. The lung cancer incidence of 39.5 (95% CI: 30.7–49.9) per 100,000 person-years among WTC rescue and recovery workers (WTC RRWV) was estimated by another study using the data from the WTC Health Program General Responder Cohort (WTCGRC) during 2002–2018 (Sigel et al. 2022).

Under the influence of suggestions for elevated cancer risk by various epidemiologic studies of 9/11 WTC collapse-exposed persons, US Preventive Services Task Force (USPSTF) recommends early diagnostic screening in asymptomatic persons for lung and other cancers. WTC Health Program covered the expense of cancer screening and diagnosis as well as medically necessary treatment for enrolled members, assuming WTC collapse exposures were linked to, contributed to or caused the cancer among enrolled WTC members (Calverta et al. 2023).

Pathophysiological Mechanisms of Lung Cancer Development

The impact of dust storm particles (Montmorillonite) on human lung epithelial cell (A549) was studied by (Ardon-Dryer et al. 2020). The study found that, even in reaction to the identical dust concentration, individual cells exhibit distinct cellular fates (survival or death). In comparison to surviving cells, dying cells engaged with dust particles for longer periods of time and ingested more of them. Higher dust concentrations had a dose-dependent reduction in viability, but they had a non-monotonic effect on cell death, with a larger fraction of dying cells occurring at intermediate dust concentrations than at lower and higher concentrations (Fig. 2). The results of the study, which showed that high dust concentrations prevent cell development, explained this non-monotonic relationship. Based on morphological characteristics of the cells and evidenced by immunoblots and pro-inflammatory cytokines, the study concluded that at low dust concentrations, apoptosis is the

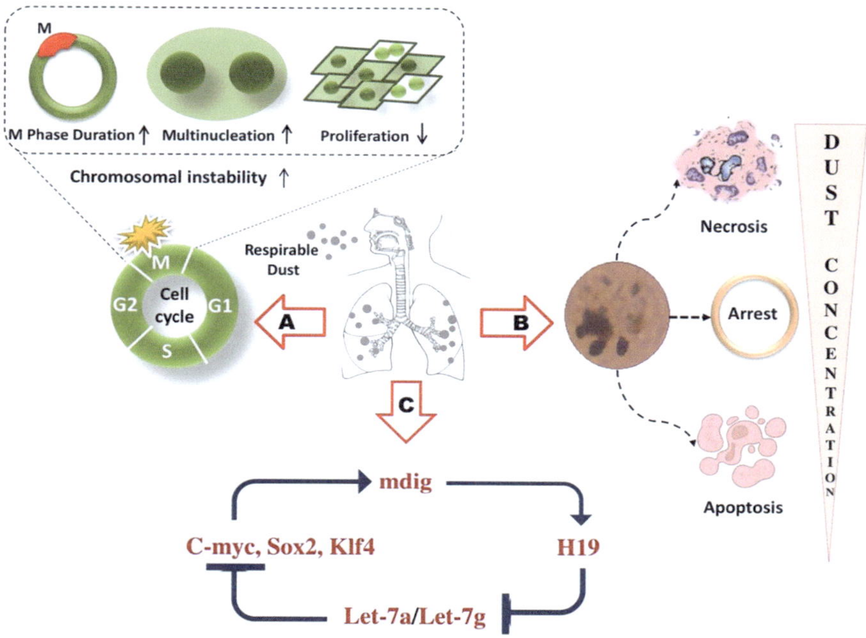

Fig. 2 Pathophysiological mechanisms of respirable dust-induced lung cancer development. (A) UDPs-mediated chromosomal instability leading to mitotic progression and inhibition of lung cell proliferation, (B) Dust particles-induced mechanism of lung cell death and (C) Mineral dust-associated forward feedback circuit of mineral dust induced gene (mdig) and oncogene (c-myc)

predominant mode of death, but greater dust concentrations trigger necrosis (Ardon-Dryer et al. 2020).

The mineral dust-induced gene (mdig), which is triggered by mineral dust, was initially discovered in the late 1990s in the alveolar macrophages of patients who had developed chronic lung disorders as a result of being exposed to mineral dust in mining operations. There have been reports of increased mdig expression in several human malignancies, including lung cancer. Despite lacking the structural identities of the demethylases, emerging data indicate that the carcinogenic activity of mdig is most likely done by its modulation of histone protein demethylation. C-myc, an oncogene, causes mdig transcription. Consequently, mdig de-represses H19, which can irritate Let-7a and Let-7 g. Tumour suppressive miRNAs that target c-myc, Sox2 and Klf4 are Let-7a and Let-7g (Fig. 2) (Zhang et al. 2019). High expression of mdig was also noted in different subtypes of lung cancer patient, which confirms the role of mdig in transforming non-cancerous lung cells to the full-grown cancer cells (Shi et al. 2021).

A recently published study by Ryu et al. (2023) reveals how chromosomal instability and impacts on mitotic progression triggered by urban dust particles (UDP) contribute to the development of lung cancer. Researchers found that UDPs in the PM2.5 range strongly inhibited both short-term and long-term cell proliferation in A549 lung cancer cells by delaying mitotic advancement and perturbing mitotic structures (Fig. 2). Multiple centrioles and hyper-stabilized microtubules were the outcomes of UDP treatment. Deficient Aurora B localization and activation at kinetochores, the spindle midzone, the equatorial cortex and the midbody cause SAC inactivation and hinder the ingression of cleavage furrows. At the end, chromosomal instability, including multinucleation, was caused by UDPs. All things considered, Aurora B's spatiotemporal activities are essential for halting the buildup of mitotic abnormalities and the chromosomal instability that follows, which are caused by UDPs (Ryu et al. 2023). These different molecular mechanisms of lung cancer development are part of recent experimental investigations. It gives strong insights to design new protective measures, early diagnostic criteria and novel therapeutic management for the respirable dust-induced lung cancer.

Conclusion

In summary, lung cancer has a major share in the global disease burden attributed to cancer. Numerous studies reviewed here support the claim that dust particles raise the risk of lung cancer. With the help of evidence-based research, the molecular mechanisms behind the development of lung cancer have been updated. These mechanisms include chromosomal instability, cell cycle arrest and mdig expression in relation to different environmental and occupational dusts, known as lung carcinogens. This chapter discusses each and every lung carcinogen exposure pattern and associated lung cancer incidence with mortality worldwide. It also illustrated the meaningful link between different mechanisms of lung cancer development and gives insight to reframe the safety recommendations for workers in occupations that are assumed to involve dust.

References

Ardon-Dryer K, Mock C, Reyes J et al (2020) The effect of dust storm particles on single human lung cancer cells. Environ Res 181:108891

Calverta GM, Lillya G, Cochran J (2023) The World Trade Center Health Program: cancer screening and cancer care best practices. Arch Environ Occup Health 78(4):222–228

Collatuzzo G, Turati F, Malvezzi M et al (2023) Attributable fraction of cancer related to occupational exposure in Italy. Cancers (Basel) 15(8):2234

Dauter UM, Alhamdow A, Cediel-Ulloa A et al (2022) Cancer-related changes and low-to-moderate exposure to welding fumes: a longitudinal study. Scand J Work Environ Health 48(1):21–30

DeMarini DM, Warren SH, Brooks LR (2022) Mutagenicity of the organic fraction of World Trade Center dust. Environ Mol Mutagen 64(1):16–25

Dominguez EC, Phandthong R, Nguyen M et al (2022) Aspirin-triggered resolvin D1 reduces chronic dust-induced lung pathology without altering susceptibility to dust-enhanced carcinogenesis. Cancers (Basel) 14(8):1900

Durmus N, Pehlivan S, Zhang Y et al (2021) Lung cancer characteristics in the world trade center environmental health center. Int J Environ Res Public Health 18(5):2689
Gardner M, Cross M, Reed S et al (2022) Pathogenic potential of respirable spodumene cleavage fragments following application of regulatory counting criteria for asbestiform fibres. Int J Environ Res Public Health 19(24):16649
Girardi P, Barbiero F, Baccini M et al (2022) Mortality for lung cancer among PVC baggers employed in the vinyl chloride industry. Int J Environ Res Public Health 19(10):6246
Hancock DG, Langley ME, Chia KL et al (2015) Wood dust exposure and lung cancer risk: a meta-analysis. Occup Environ Med 72(12):889–898
Hanona PF, Ezekwudo D, Fullmer J et al (2023) An aggressive case of a thoracic undifferentiated SMARCA4-deficient tumor with extensive pleural involvement. Thorac Cancer 15(10): 847–851
Holm SE, Festa JL (2019) A review of wood dust longitudinal health studies: implications for an occupational limit value. Dose Response 17(1):1559325819827464
Huang X (2020) Cotton dust exposure and risk of lung cancer: a meta-analysis of observational studies. Medicine (Baltimore) 99:14(e19565)
Jenwitheesuk K, Peansukwech U, Jenwitheesuk K (2020) Construction of polluted aerosol in accumulation that affects the incidence of lung cancer. Heliyon 6(2):e03337
Jiˇrík V, Tomášek L, Fojtíková I et al (2021) Lifetime carcinogenic risk proportions from inhalation exposures in industrial and non-industrial regions. Int J Environ Res Public Health 18(24): 13295
Kim B, Kim E, Cha W et al (2021) Occupational exposure to respirable crystalline silica in municipal household waste collection and road cleaning workers. Sci Rep 11(1):13370
Kleinschmidt SE, Andres KL, Holen BM et al (2022) Mortality among mine and mill workers exposed to respirable crystalline silica. PLoS One 17(10):e0274103
Li L, Jiang M, Li X et al (2021) Association between coalmine dust and mortality risk of lung cancer: a meta-analysis. Biomed Res Int 2021:6624799
Liu M, Zhou G, Saari RK et al (2019) Quantifying $PM_{2.5}$ mass concentration and particle radius using satellite data and an optical-mass conversion algorithm. ISPRS J Photogramm Remote Sens 158:90–98
Lombaert N, Gilles M, Verougstraete V (2023) Cadmium monitoring at the workplace: effectiveness of a combination of air- and biomonitoring. Toxics 11(4):354
Margan A, Verlak D, Roj G et al (2022) Occupational exposure to silica dust in Slovenia is grossly underestimated. Arh Hig Rada Toksikol 73(4):297–302
Matrat M, Radoï L, Févotte J et al (2019) Occupational exposure to wood dust and risk of lung cancer the ICARE study. Occup Environ Med 76(12):901–907
Minina V, Timofeeva A, Torgunakova A et al (2022) Polymorphisms in DNA repair and xenobiotic biotransformation enzyme genes and lung cancer risk in coal mine workers. Life (Basel) 12(2): 255
Moghadam SR, Khanjani N, Mohamadyan M et al (2020) Changes in spirometry indices and lung cancer mortality risk estimation in concrete workers exposed io crystalline silica. Asian Pac J Cancer Prev 21(9):2811–2817
Olsson A, Bouaoun L, Schüz J et al (2024) Lung cancer risks associated with occupational exposure to pairs of five lung carcinogens: results from a pooled analysis of case-control studies (SYNERGY). Environ Health Perspect 132(1):17005
Pavana C, Santalucia R, Leinardi R et al (2020) Nearly free surface silanols are the critical molecular moieties that initiate the toxicity of silica particles. Proc Natl Acad Sci USA 117(45):27836–27846
Pingle S, Sherekar P, Thakkar L et al (2022) Goblet, club and alveolar cells: front-line defenders of the airways in chronic obstructive pulmonary disease, a most common lung disease in miners. In: Randive K, Pingle S, Agnihotri A (eds) Medical geology in mining: health hazards due to metal toxicity. Springer Nature, Cham, pp 83–100

Raaschou-Nielsen O, Antonsen S, Agerbo E et al (2023) $PM_{2.5}$ air pollution components and mortality in Denmark. Environ Int 171:107685

Rahimimoghadam S, Ganjali A, Khanjani N et al (2023) Application of multiple occupational health risk assessment models for crystalline silica dust among stone carvers. Asian Pac J Cancer Prev 24(11):3999–4005

Ryu J, Lee SH, Kim S et al (2023) Urban dust particles disrupt mitotic progression by dysregulating Aurora kinase B-related functions. J Hazard Mater 459:132238

Scarabelli TM, Corsetti G, Chen-Scarabelli C et al (2021) Follicular B-cell lymphoma and particulate matter associated with environmental exposure to wood dust. Am J Case Rep 22:e929396

Schuz J, Kovalevskiy E, Olsson A et al (2024) Cancer mortality in chrysotile miners and millers, Russian Federation: main results (Asbest Chrysotile Cohort-study). J Natl Cancer Inst 116(6): 866–875

Sherekar P, Jain R, Pingle S et al (2022) Role of pyrite in aggravating coal worker's pneumoconiosis. In: Randive K, Pingle S, Agnihotri A (eds) Medical geology in mining: health hazards due to metal toxicity. Springer Nature, Cham, pp 15–47

Sherekar P, Suke SG, Dhok A et al (2024) Nano-enabled delivery of diosgenin and emodin ameliorates respirable silica dust-induced pulmonary fibrosis silicosis in rats. Ecotoxicol Environ Saf 279:116483

Shi J, Thakur C, Zhao Y et al (2021) Pathological and prognostic indications of the mdig gene in human lung cancer. Cell Physiol Biochem 55(S2):13–28

Sigel K, de la Hoz RE, Markowitz SB et al (2022) Lung cancer incidence among world trade center rescue and recovery workers. Cancer Med 11(16):3136–3144

Wang D, Yang M, Liu Y et al (2020) Association of silica dust exposure and cigarette smoking with mortality among mine and pottery workers in China. JAMA Netw Open 3(4):e202787

Wang D, Li W, Albasha N et al (2023) Long-term exposure to house dust mites accelerates lung cancer development in mice. J Exp Clin Cancer Res 42(1):26

Zhang Q, Thakur C, Shi J et al (2019) New discoveries of mdig in the epigenetic regulation of cancers. Semin Cancer Biol 57:27–35

Zhang X, Leng S, Qiu M et al (2023) Chemical fingerprints and implicated cancer risks of Polycyclic aromatic hydrocarbons (PAHs) from fine particulate matter deposited in human lungs. Environ Int 173:107845

Air Pollutants Aggravate Some Major Lifestyle Diseases and Beneficial Role of Medicinal Phytocompounds

Debosree Ghosh, Sourav Das, and Partha Sarathi Singha

Abstract

Lifestyle diseases are caused due to the habits or manners a person adopts in his/her life. These diseases are non-contagious, caused by unhealthy practices, unhealthy diet and physical inactivity. Their early symptoms need to be addressed carefully and seriously with clinical interventions if needed. Certain environmental toxicants and pollutants are known to aggravate these lifestyle diseases. Toxic air pollutants and particulate matter are known to initiate and aggravate some major lifestyle diseases like lung cancer, heart disease, asthma, stroke and chronic obstructive pulmonary disease (COPD). Therapies for mitigating these lifestyle disorders include modification in lifestyle, changing daily habits, good diet, exercise, and some conventional medicines. Bioactive phytocompounds are always a better alternative choice to combat health issues as they come with no cytotoxicity, side effects and also work efficiently in a unique manner. These phytochemicals are natural compounds having potent beneficial roles against air pollutant-aggravated lifestyle diseases. Their mechanism of action is multifaceted and also known to work to enhance the impact of conventional drugs. Most of these bioactive phytocompounds are antioxidants and thus have an overall health benefit when administered in a specific dose.

D. Ghosh (✉)
Department of Physiology, Government General Degree College, Kharagpur II, Paschim Medinipur, West Bengal, India

S. Das
Department of Zoology, Government General Degree College, Kharagpur II, Paschim Medinipur, West Bengal, India

P. S. Singha
Department of Chemistry, Government General Degree College, Kharagpur II, Paschim Medinipur, West Bengal, India

S. Raisuddin et al. (eds.), *Environmental Toxicants and Lifestyle Diseases*,
https://doi.org/10.1007/978-981-95-0575-3_22

Keywords

Lifestyle diseases · Environmental toxicants · Toxic air pollutants · Phytocompounds · Phytochemicals

Introduction

The way one lives and the daily routine one follows comprise one's lifestyle. Food habits and work patterns are also significant components of one's lifestyle. The World Health Organization (WHO) says that a big part of the health and quality of one's life is dependent on the lifestyle of the person (Ziglio et al. 2004). Millions of people around the globe follow unhealthy lifestyles due to unawareness and busy life. This leads to the onset and progression of different health issues including life-threatening diseases. Thus, lifestyle has a direct impact on the physical health and mental health of an individual. Lifestyle medicine is a branch of the medical science that is known to deal with prevention, treatment and management of lifestyle diseases. Certified healthcare providers also try to promote health as a part of lifestyle medicine (Phillips et al. 2020).

The lifestyle diseases are mostly chronic and are classified under non-communicable non-infectious diseases (Mathur and Mascarenhas 2019). There are certain diseases that begin at an initial stage of life and continue and intensify with increasing age, leading ultimately to fatal consequences at advanced age. Strokes, ischaemic heart disease, hypertension, cancer, chronic obstructive pulmonary disease (COPD), cataracts and aggravated asthma are some such diseases (Mathur and Mascarenhas 2019).

Studies reveal that several factors are involved in occurrence of such lifestyle diseases and impacts the pattern and speed of progression of them. Smoking, alcohol use, lack of physical activity, bad diet, untimely eating habits, overweight, underweight, stress, irregular and inadequate sleeping habits and environmental pollution (air, water, and soil) have been recognized as some of the primary contributing factors in developing and aggravating lifestyle diseases (Ng et al. 2020). Most of these factors and their determinants are basically human-behaviour dependent and thus can be easily modified for better with motivation at individual, family and social levels. Environmental pollutants are mostly toxic and adversely affect the health of living organisms including human (Singha et al. 2023). Studies show that exposure to air pollution may lead to certain diseases like asthma, COPD, lung and cancer, and most of these are lifestyle diseases also (Ghosh and Parida 2015a). Thus, it is evident that both unhealthy lifestyle and exposure to environmental toxic pollutants are common contributors in the etiology of certain such diseases. Studies also report that exposure to toxic air pollutants may enhance the risk for onset and progression of the COVID-19 disease (Ghosh et al. 2022). Emphysema, neurodegenerative diseases like the Alzheimer's and Parkinson's diseases, psychological complications, low birth weight, retinopathy, developmental disorders, diabetes, obesity, atherosclerosis, chronic liver disease or cirrhosis, non-alcoholic liver diseases and depression are some of the other major lifestyle diseases. Genetic predisposition is also a strong determinant for the susceptibility of lifestyle diseases and their aggravation with the influence of other factors like air pollutants (Said et al. 2019). Thus, it is now evident that lifestyle diseases may create a major critical impact on human health. Unfortunately, it is probable that lifestyle diseases will increase its ill effects in the upcoming years. According to United Nations Organization, lifestyle diseases are going to lead to an economic burden on the health services. Studies reveal that about 99% of the global population residing in the areas with the air pollution above the recent standards set by WHO for air quality do have a connection between air pollution and metabolic diseases of the heart. It is extremely necessary for adapting human health risk assessment techniques regularly for assessing the impacts of pollution on the health of people of a particular region (Singha et al. 2023). With proper assessment of the burden and impacts of the toxic air pollutants, adequate measures and proper clinical interventions need to be taken in order to prevent and treat those air pollutant aggravated lifestyle diseases (Singha and Ghosh 2023). One significant treatment and preventive measure is use of potent bioactive phytocompounds from medicinal plants. Phytochemicals are recommended for the prevention of diseases and for maintaining health (Zhang et al. 2015). Most of these phytocompounds are medicinally potent in various manners, and they have antioxidant potentials (Zhang et al. 2015). The most important factors to consider in phytocompounds are their ability to maintain good health without causing side effects and cytotoxicity (Zhang et al. 2015). In this chapter, the effect of air pollutants on some major lifestyle diseases and their possible prevention and treatment using phytocompounds has been briefed (Table 1).

Impact of Air Pollutants on Lung Cancer

Exposure to environmental toxicants, such as pesticides, heavy metals and volatile organic compounds (VOCs), has been reported to impart an enhanced risk of various types of cancer, including lung, breast and prostate cancer. Long-term impacts of exposure to PM2.5, PM10 and PM1 are associated with cardiovascular and pulmonary diseases (Guo et al. 2023). SPM can cause premature death of the patients who have chronic heart and lung diseases. SPM also creates decreased lung function and growth in children. In 2015, International Agency for Research on Cancer published a review which explained that SPM can induce pulmonary malignancy. The SPM, especially fine particulates and sulfur oxide, which are released from diesel engines, cause lung cancer and cardiopulmonary mortality. The points of entry of metal ions inside body are the most vulnerable parts where tumours are developed. So, the lung cancer will be developed if large quantities of metal ions are inhaled (Tokar et al. 2011). Lead, antimony, beryllium, cadmium, cobalt, zinc, copper and iron may be carcinogenic for human lung. Nickel, chromium, cadmium, arsenic and inorganic arsenic compounds induce risk of developing pulmonary malignancy. The zinc-containing enzymes are structurally and functionally changed after their encounter

Table 1 Some major lifestyle diseases and effects of air pollutants on them

Diseases	Effects of Air Pollutants							References
	Suspended particulate matters (SPM)	Metal ions	Carbon monoxide (CO)	Hydrocarbons	Ozone (O_3)	Oxides of nitrogen (NOx)	Oxides of sulphur (Sox)	
Lung cancer	Cause lung cancer	Adversely affect pulmonary malignancy	Increases the risk of lung cancer.	PAHs such as Benzo(a) pyrene & Benz (a) anthracene, cause lung cancer	Ozone is associated with lung cancer	NOx and NO_2 are known to be accountable for lung cancer	Exposure to SOx has correlation with the occurrence of lung cancer in humans	Guo et al. (2023), Tokar et al. (2011), Moorthy et al. (2015), Hamra et al. (2015), Ghosh and Parida (2017), Singha et al. (2022)
Cardiovascular Diseases (CVD)	Both systolic and diastolic blood pressures, LDL, HDL levels and cholesterol are affected	Heavy metals like lead, arsenic, copper, and cadmium are the main risk factors	Causes short-term effects on cardiovascular system	Positive correlation between PAH biomarker and CVD in general population	Mortality rate of the patients suffering from CVD increases with exposure to ozone	Short-term and long-term exposure of NOx can develop CVD, Arteriosclerosis early aortic and coronary calcification	Long-term risk for CVD is caused by SOx, mainly SO_2	Lee et al. (2014), Ghosh et al. (2013), Bandyopadhyay et al. (2014), Gaalema (2022), Mallah et al. (2021), Gao et al. (2023)
Asthma	Airway inflammation, swelling and tightening due to SPM2.5 exposure leads to breathing difficulty	Metals like lead, cadmium, mercury can act as environmental allergens and induce asthma	CO exposure increases asthma-related health problems and increasesrisk of COPD	Early onset of asthma is triggered by exposure to PAHs	Asthma symptoms like chest pain, chest tightness, coughing and wheezing are increased by ozone	Long-term NOx exposure increases the risk for asthma.	Existing symptoms worsen by SO2 exposure. Risk of COPD increases by 0.44% for every 10 $\mu g/m^3$ increase in SO_2 concentration.	Zhang et al. (2023), Wu et al. (2019), Karimi et al. (2015), Kowalska et al. (2020)
Chronic obstructive pulmonary	Short-term exposure to PM_{10} worsens	Aggravate COPD symptoms	Long-term exposure to	The risk of COPD increases	Risk of COPD and mortality increases with	Short-term and long-term exposure to	Emphysema is caused by SOx	Ling and Van (2009), Valkova et al. (2023),

disease (COPD)	the condition of COPD		CO may induce COPD	with exposure to PAHs	short-term expo -sure to high O_3 concentration	NOx increase the risk of COPD and worsen symptoms		Duan et al. (2020), Yang et al. (2017), Kim et al. (2020), Ghozikali et al. (2015)
Emphysema	$PM_{2.5}$ exposure causes inflammatory emphysematous lesion in both lungs	Metal ions, especially cadmium may cause emphysema through oxidative stress	Long-term exposure of CO cause emphysema	PAHs exposure causes irritation and leads to air way obstruction	Exposure to O_3 accelerates progression of emphysema	Emphysema and bronchitis are caused by NOx pollution	SO_2 exposure may cause central retinal artery occlusion in patients with hypertension	Yan et al. (2023), Mohamed et al. (2011), Nwaozuzu et al. (2021), Wiegman et al. (2020)
Retinopathy	Cause retinal arteriolar narrowing, venular widening and adverse retinal structural features	Toxic metals may damage the retinal tissue	CO causes an increase in the diameters of retinal blood vessels	No direct association between hydrocarbon pollution and retinopathy	No direct association has been found between O_3 pollution and retinopathy	Free radicals generated by NO may damage the biomolecules and lead to retinal damage	Increased risk of neuronal toxicity and neurodegeneration may occur with exposure to SO_2 and its derivatives	Jeong et al. (2023), Pamphlett et al. (2020), Resch et al. (2005), Lin et al. (2022), Chiou (2001)
Neurological Disorder	PM2.5 can also directly enter brain tissue via the olfactory nerve, leading to neural damage therein	Metals can cause neurodegenerative diseases and mitochondrial dysfunction in neurological disorder	CO exposure is known to cause neurotoxicity	Cause learning & memory problem, neuronal damage, impaired neurotransmitter regulation and parasympathetic dysregulation	Cause decreased cognitive response, neurochemical changes, reduced motor activity, headaches, sleep–wake cycle disturbances, neuronal dysfunctions, neurodegeneration	NO_2 exposure causes neuronal damage and leads to cognitive impairment and neurodegentaion	Neurotoxicity, damage to central nervous system	Fu et al. (2022), Bandyopadhyay (2016), Chen et al. (2016a, b), Levy (2015), Olasehinde and Olaniran (2022), Martínez-Lazcano et al. (2013), Roy and D'Angiulli (2024)

(continued)

Table 1 (continued)

Diseases	Effects of Air Pollutants							References
	Suspended particulate matters (SPM)	Metal ions	Carbon monoxide (CO)	Hydrocarbons	Ozone (O_3)	Oxides of nitrogen (NOx)	Oxides of sulphur (Sox)	
Skin Infection	PM exposure may worsen skin infection by inducing generation of free radicals	Acrodynia, hyperkeratosis, hyperpigmentation and skin cancer may occur due to toxic metal exposure	Causes eczema, dehydration, disruption of skin microbiome, damage of skin barrier, psoriasis and flexural eczema	Extrinsic skin aging, pigmentation, cancers and acneiform eruptions are caused by PAHs	Eczema, contact dermatitis, rashes, and urticaria are caused by exposure to ozone	Psoriasis, eczema, dermatitis, acne and vitiligo are aggravated by NOx	SO_2 can cause skin irritation and blisters	Dijkhoff et al. (2020), Torne et al. (1991), Puri et al. (2017), Chao et al. (2021)
Diabetes	One fifth of the Type-II Diabetes is caused by prolonged exposure of Suspended Particulate matter (PM2.5)	Metals induce free radical generation that may cause oxidative damage to pancreas and glucose receptors	Patients with CO poisoning have higher risk to develop diabetes mellitus	PAH may increase the risk of development of diabetes	Long-term exposure to ozone enhances the risk of diabetes	NOx pollution is associated with insulin resistance and development of diabetes mellitus	Adults are at increased risk of developing diabetes with exposure to SOx	He et al. (2017), Chen et al. (2016a, b), Zhang et al. (2022), Huang et al. (2017), Mallah et al. (2023), Li et al. (2021), Bai et al. (2018)

with the metal ions that lead to pulmonary cancer. Studies show that occupational exposure to nickel, chromium VI and cadmium enhances the risk for pulmonary cancer. Carbon monoxide is a highly poisonous gas and toxic air pollutant. CO is associated with cancers of various organs including lung cancer, bladder cancer, oesophageal cancer and larynx cancer. Pancreatic cancer and stomach cancer are also associated with CO poisoning. Polycyclic aromatic hydrocarbons (PAHs) such as Benzo(a)pyrene and Benz(a)anthracene that are released from smokers, car emission and incinerators are carcinogens (Moorthy et al. 2015). They can cause several cancers that include lung cancer when inhaled. PAHs upregulate the enzymes CYP1A1 and CYP1B1. These enzymes cause mutation in p53 gene which eventually results lung tumour (Moorthy et al. 2015). There are no direct evidences that explain the role of ozone (O_3) to cause lung cancer. However, according to some studies, O_3 is known to bear a positive association with lung cancer in females. O_3, along with nitrogen dioxide, is known to bear an association with pulmonary cancer. Some studies interestingly report that ozone may selectively inhibit the growth of human cancer cells in certain conditions. Studies show that risk of developing lung cancer is enhanced by oxides of nitrogen (NOX) exposure. An increase of almost 4% and 3% in rates of lung cancer has been reported after exposure to 10 $\mu g/m^3$ of NO_2 and Nox, respectively. Pollution from vehicles is reported to account for several diseases that including lung cancer. Nitrogen dioxide is also known to bear a significant correlation with lung cancer (Hamra et al. 2015). Studies reveal that the risk of developing lung cancer may increase by the exposure to SO_2. It is reported that death due to lung cancer increased in workers in pulp and paper industries who were exposed to sulphur dioxide (SO_2). Lung cancer is classified as non-small cell lung cancer and small cell lung cancer (Ghosh and Parida 2017). Studies conducted on professional drivers those who got exposed to air pollution show that the risk of developing lung cancer increases with exposure to toxic pollutant gases. There exists a relationship between exposure to toxic air pollutants and pulmonary cancer.

Exposure to toxic environment pollutant gasses like the nitrogen dioxide, carbon dioxide, sulphur dioxide, methane, ozone and particulate matter enhances the risk of respiratory diseases, namely COPD, asthma and lung cancer. Studies also report that exposure to toxic environmental pollutants increases the risk for respiratory infectious diseases like COVID-19 (Ghosh et al. 2022; Singha et al. 2022).

Impact of Air Pollutants on Cardiovascular Disorders (CVDs)

Cardiovascular ailments like coronary heart disease, heart attack, strokes and arrhythmias are reported to be caused and adversely affected by the toxic environmental pollutants like heavy metals, particulate matter, toxic gases and pesticides (Ghosh et al. 2022; Singha and Ghosh 2023).

Studies show that fine particulate matter that is 2.5 microns or less in diameter ($PM_{2.5}$) enhances the levels of cholesterol and low-density lipoprotein (LDL). They reduce HDL level. PM exposure can change the arrhythmia and cause blood clotting disorders. Cardiovascular disorders, premature death in people with cardiac and pulmonary diseases are very common problems associated with exposure to PM (Singha and Ghosh 2023). Particulate matters may contain toxic substances that may enter our body and cause adverse health impacts. The effect of PM on health depends largely on the size of the PM that is on the other hand dependent on the aerodynamic diameter (AD) of the particles. PM_{10} particles mostly have AD of 2.5–10 μm and are mostly deposited in the nasal cavities and upper respiratory tract, whereas $PM_{2.5}$ and $PM_{0.1}$ particles having ADs <2.5 and <0.1 μm may penetrate the alveoli of the lungs and reach other vascularized organs. Low-level exposure to heavy metals i.e. lead, arsenic, mercury and calcium metal ions causes cardiac ailments. CVDs including ischaemic heart disease (IHD) to heart attacks, strokes and premature death from cardiac causes occur due to exposure to toxic heavy metals. These heavy metals are known to enter the body in dust form through respiration (Ghosh et al. 2013). Industrial exposure is the prime risk factor for heavy metal-induced cardiovascular disorders in man (Bandyopadhyay et al. 2014). Metals are known to be responsible for hypertension. As per the evidence from the American Heart Association, environmental pollutants, such as toxic metals, are associated with CVD. The EPA (Environmental Protection Agency) recommends that barium concentrations not exceed 2 mg/litre of water.

Carbon monoxide is known to form irreversible compound called carboxyhaemoglobin with haemoglobin and adversely affect the circulatory and cardiovascular system. CO has more affinity to bind with haemoglobin than O_2, which leads to decreased O_2 supply throughout body that eventually results in severe cardiovascular problems (Gaalema 2022). People with no reported CVD have been found to have decreased peak O_2 uptake and exercise duration on exposure to CO (Gaalema 2022). Atherosclerosis may be developed due to long-term exposure to CO. CO is known to increase plasma cholesterol level thereby adding to the risk of occurrence of atherosclerosis and hypertension. In China, 1 mg/m^3 increase CO concentration per day would result almost 23,000 deaths from CVD. Study also reports that a 1 mg/m^3 increase in CO concentrations caused 4.4% increase in hospital admission. Ischaemic heart failure may result through oxidative stress pathway triggered by CO exposure.

Exposure to PAHs can add to the risk of cardiovascular disorder. Exposure to PAH metabolites is reported to cause atherosclerosis, hypertension and altered heart rates. There exists a positive relationship between CVD and exposure to PAH (Mallah et al. 2021). Exposure to PAH may cause an increase in the concentration of PAH in the body, leading to excretion of PAH metabolites in the urine. Studies show that Urinary PAH metabolites have a positive association with CVDs like IHD, stroke, angina and heart failure (Mallah et al. 2021). Aged people are known to be more susceptible to cardiovascular ailments that are induced and aggravated by exposure to ozone. Ozone exposure is reported to be associated with increased hospitalization due to CVD. Studies report that exposure to high temperature rather that to a warm climate may add to the CVD mortality due to ozone in population below the age of 65 years. Short-term exposure to environment NOx leads to onset of CVD. Preexisting heart disease may become worse by NOx exposure which may culminate to premature death. A recent study shows that the risk of CV mortality is increased 13% with annual NOx exposure of 10 μg/m^3 (Luo et al. 2016). SO_2 exposure leads to CVD morbidity and mortality is increased due to IHD, heart failure and arrhythmia that are induced by SOx exposure. Cardiopulmonary ailments are also reported due to SOx exposure (Lin et al. 2018). Sulphur dioxide is reported to cause cardiac and mitochondrial dysfunctions in murine experimental models.

Impact of Air Pollutants on Asthma

Air pollutants cause irritation in the airways, thereby causing inflammation, swelling leading to tightening of the airways and breathing problem called asthma. There exists a clear correlation between exposure to air pollutants and asthma. Lung exposure to PM2.5 causes inflammation and immune responses eventually leading to asthma and other cardiopulmonary diseases. Asthma attack is also caused by short-term exposure to PM2.5—infants, children and older adults are more susceptible to asthma-related health issue developed due to SPM (Zhang et al. 2023).

As per the reports of the agency for toxic substance and disease registry, certain toxic heavy metals like lead, cadmium and mercury can act as environmental allergens and induce asthma. Heavy metal exposure may cause childhood asthma (Wu et al. 2019). In the children of age 2–15 years, higher concentration of blood lead load can cause asthma. In air, fine particles can absorb heavy metals, enter our respiratory tract and cause inflammatory asthma Wu et al. (2019). Studies show that short-term and long-term exposures to high concentration of CO lead to mutation of some genes associated with asthma. CO exposure is known to cause tissue hypoxia. Early onset of asthma and increasing symptoms of asthma are found to be caused due to exposure to PAHs (Karimi et al. 2015). Childhood asthma is associated with PAH exposures. Asthma symptoms are aggravated with ozone exposure. The tissue in the respiratory tract is damaged by ozone and leads to inflammation and irritation. The muscles in the airways are constricted by ozone exposure, thereby trapping the air in the alveoli, which leads to shortness of breath. Long-term exposure to NOx may increase the risk of asthma. Children and elder person suffering from asthma are at greater risk after NOx exposure (Kowalska et al. 2020). Existing symptoms of asthma worsen with exposure to SOx. Asthma attack, coughing, wheezing, phlegm, nose, throat and lung irritation are caused by SO_2.

Impact of Air Pollutants on Chronic Obstructive Pulmonary Disease

The symptoms of COPD enhance with SPM in the air. They cause pulmonary inflammation and decreased lung activity. The COPD symptoms worsen by short-term exposure to PM10 (Ling and Van 2009). Chemokines and proinflammatory cytokines are released from lung by the influence of SPM, which leads to elastic degradation and alveolar destruction eventually leading to distressed breathing and causing COPD (Ling and Van 2009). Exposure to metal ions including copper, iron and zinc have been reported to have significant role in the development of COPD. Serum level of heavy metals like Cu, Cd, Zn and Pb have been reported to be associated with progression of COPD (Valkova et al. 2023). Cadmium and lead may exacerbate COPD symptoms (Valkova et al. 2023). Dysregulation of essential transition metal ions has connection to oxidative stress and inflammatory changes in lungs of COPD patient. Risk of COPD with co-morbidity is known to be increased with exposure to CO. Short-term exposure of low concentration of CO is reported to be associated with reduced risk of COPD. The risk of COPD increases with exposure to PAHs. Increased Benzo(a)pyrene, 9-Hydroxyfluorene and 3- Hydroxyfluorene concentration is known to be associated with the occurrence of COPD. Studies show that the risk of COPD enhances with exposure to high concentration of ozone. Mortality due to COPD also increases with exposure to ozone (Duan et al. 2020). Risk of COPD increases with short-term and long-term exposure to nitrogen oxide and nitrogen dioxide. Studies show that eventually the COPD symptoms become worsen when NO_2 penetrates the alveolar tissue. Exposure to increased SO_2 concentration increases the risk of COPD and is positively correlated with increased hospitalization (Ghozikali et al. 2015).

Impact of Air Pollutants on Emphysema

The risk of developing emphysema in lung alveoli is enhanced by the inhalation of SPM. The particulate matters corrode the alveolar wall and eventually alveolar wall loses its elasticity. Alveolar wall may rupture, and one large air space may form instead of many smaller air spaces. This condition reduces the surface area, which is available for gas exchange and cause emphysema. PM2.5 causes emphysematous lesions and inflammatory changes in the pulmonary tissue. PM2.5 deposition in each lung lobe is associated with the degree of emphysema. Exposure to metal ions like cadmium may cause emphysema. The heavy metals are known to deplete the indigenous antioxidant enzymes like superoxide dismutase and catalase, thus speeding up the deterioration of elastin and collagen proteins. This damages the cellular barrier in the lungs leading to emphysema (Yan et al. 2023). Emphysema development has been linked with long-term exposure to CO. The combined effects of several air pollutants along with CO are reported to be greater than the effect of individual CO (Tian et al. 2015). Studies show that a decreased transfer coefficient of the lung for CO is correlated to emphysema (Mohamed et al. 2011). Exposure to PAHs is associated with developing airway obstruction as in emphysema. These hydrocarbons act as airborne irritants and cause emphysema. The progression of emphysema accelerates by long-term effects of ozone exposure. The muscles in the airways are constricted by ozone and trapping the air in the alveoli, which leads to shortness of breath and cause emphysema. Non-smokers may develop emphysema by ozone pollution (Kim et al. 2020). Emphysema and bronchitis are caused by NOx pollution. PM2.5 pollution occurs by the compound produced by the reaction of NOx, VOC and ammonia. PM2.5 enters into the deep and sensitive parts of the lung, which eventually results in emphysema. Emphysema is caused by SOx. Several other respiratory problems associated with emphysema are also caused by SOx exposure.

Impact of Air Pollutants on Retinopathy

Retinal arteriolar narrowing, venular widening and adverse retinal structural features are caused by SPM. PM2.5 exposure can also damage the inner blood–retinal barrier (iBRB), which is a common pathological process of various retinal vascular diseases, such as diabetic retinopathy and retinal vein obstruction (Jeong et al. 2023). PM2.5 exposure can cause cell damage in human retinal microvascular endothelial cells (HRMECs) in a time- and dose-dependent manner. PM2.5 exposure can also cause prolonged ocular inflammation and endoplasmic reticulum stress in the retina. They also cause ferroptosis in retinal vascular endothelial cells. Studies show that exposure to PM2.5 may lead to suppression of cell proliferation and promotion of cell apoptosis, thus causing abnormal human retinal development. Toxic metals may have adverse effects on the retina due to oxidative damage (Pamphlett et al. 2020). The diameter of retinal arteries and veins are increased by CO inhalation. CO can penetrate the alveolar wall, enter the blood circulation and cross the blood–retinal barrier (BRB), causing pathological changes in retina (Resch et al. 2005). Retinopathy, as well as other eye diseases such as diabetic retinopathy, age-related macular degeneration and glaucoma is caused by NO. NO is converted into NO_2, nitrite, peroxynitrite and free radicals, which can damage the retina and optic nerve. The retina is sensitive to reactive oxygen species (ROS) and reactive nitrogen species (RNS), which can react with biomolecules like lipids, proteins and DNA, potentially leading to cell death (Lin et al. 2022, Chiou 2001). Elevated SO_2 levels for 1 day can significantly increase the risk of central retinal artery occlusion (CRAO) in patients with hypertension.

Impact of Air Pollutants on Neurological Disorders

Exposure to environmental toxicants, such as heavy metals and pesticides, has been linked to an increased risk of neurological disorders, including Parkinson's disease, Alzheimer's disease and attention deficit hyperactivity disorder. Neurodegenerative diseases like Alzheimer's disease, dementia, Parkinson's disease, autism spectrum disorder (ASD) and multiple sclerosis (MS) are caused by long-term exposure to PM2.5. PM2.5 can also directly enter brain tissue via the olfactory nerve and damage brain nerves (Bandyopadhyay 2016). PM2.5 can induce brain damage (Li et al. 2021). Metals can cause neurodegeneration. Alzheimer's disease, amyotrophic lateral sclerosis, autism spectrum disorders, Guillain–Barré disease, Gulf War syndrome, Huntington's disease, multiple sclerosis, Parkinson's disease and Wilson's disease are reported to be caused by metal ion pollution. Metals can cause neurotoxicity by increasing ROS and oxidative stress in the nervous system. This can lead to mitochondrial dysfunction, which can disrupt brain function and result in neuronal cell death. Neurotoxicity is caused by CO pollution.

CO can impair neurodevelopment in infants and children and disrupt critical processes within the developing brain (Levy 2015). The risk of neurological diseases is increased by long-term exposure to hydrocarbon pollution. PAHs are a class of environmental pollutants that can cause neurological deficits by crossing the blood–brain barrier and reaching the central nervous system. Some of the neuropathological mechanisms of PAHs are learning and memory problems, neuronal damage, impaired neurotransmitter regulation, neurodegeneration and parasympathetic dysregulation (Olasehinde and Olaniran 2022). Decreased cognitive response, neurochemical alterations, decreased motor activity, headaches, sleep–wake cycle disturbances, neuronal dysfunctions and neurodegeneration are caused by the exposure of O_3 (Martínez-Lazcano et al. 2013). Cognitive impairment, neurodegeneration and oxidative stress-induced inflammation in the brain are caused by NO_2 exposure that leads to neuronal damage (Roy and D'Angiulli 2024). Neurological diseases, neurotoxicity, central nervous system diseases and an increased risk of brain disorder are caused by the effects of abnormal levels of SO_2 and its derivatives.

Impact of Air Pollutants on Skin Disorders

The development or worsening of inflammatory skin infection by PM pollution has been reported. Especially atopic dermatitis, eczema, psoriasis and acne are caused by PM. PM stimulates the formation of ROS, which leads to oxidative stress and inflammation and eventually causes skin infection (Dijkhoff et al. 2020). Skin infections like acrodynia, hyperkeratosis, hyperpigmentation and skin cancer may be developed by exposure to toxic metal ion. Acrodynia is known to be caused due to mercury toxicity. Percutaneous absorption of metals occurs due to exposure to road dust. Eczema, dehydration, disruption of skin microbiome, damage of skin barrier, psoriasis and flexural eczema are caused by CO pollution. Skin lesion may occur due to carbon monoxide intoxication (Torne et al. 1991). Extrinsic skin aging, pigmentation, cancers and acneiform eruptions are caused by PAHs. Eczema, contact dermatitis, rashes, urticaria and infected skin are caused by exposure to ozone. Ozone oxidizes biological systems at the skin's surface. Skin diseases like psoriasis, eczema, dermatitis, acne and vitiligo are aggravated by NOx. Higher oxides of nitrogen are corrosive and skin irritants. A study found that short-term exposure to low concentrations of NO2 can affect the epidermal barrier function in people with atopic eczema, which can change the structure of the skin surface. NO_2 exposure can also disturb the cutaneous microbiome, which can induce psoriasis. Severe skin irritation, including stinging pain, redness and blisters, especially on mucous membranes are caused by SO_2. Compressed gas or liquid SO_2 can cause frostbite and irritation injury.

Impact of Air Pollutants on Diabetes

Studies show that one fifth of the Type-II diabetes is caused by long-term exposure to Suspended Particulate matter (PM2.5). Risk of diabetes increases with exposure to SPM. No direct association is reported between metal ion pollution and diabetes. But air pollution may reduce the insulin secretion and trigger inflammation of pancreas that leads to diabetes. Heavy metals may induce oxidative damage to pancreatic cells and glucose receptors on the cells, thereby complicating the situation of diabetes. Multimetal exposures like Cd, Fe, Zn, copper and Pb enhance the risk for diabetes (Zhang et al. 2022). Inflammatory and immunological reactions in brain and other organs are caused by CO that adds to the risk of development of diabetes mellitus. It is found that patients with CO poisoning have higher risk to develop diabetes mellitus than non-CO poisoning patients (Huang et al. 2017). PAH may increase the risk of development of diabetes (Mallah et al. 2023). The risk of development of type 2 diabetes increases by long-term exposure to ozone. Adults of age 30–50 years are at higher risk of developing type 2 diabetes with long-term exposure to ozone with $SO_{2.}$ These gases may create abnormal glucose tolerance, fat metabolism disorder and increased insulin resistance index which lead to development of diabetes. NOx pollution is associated with insulin resistance and development of diabetes mellitus.

Impact of Air Pollutants on Other Lifestyle Diseases

Exposure to environmental toxicants, such as endocrine-disrupting chemicals (EDCs), has been linked to an increased risk of metabolic disorders, including obesity and diabetes. The EDcs are also called environmental obesogens and play role in obesity as well as other metabolic disorders (Haverinen et al. 2021). Exposure to environmental toxicants, such as toxic pesticides and heavy metals, are linked to an enhanced risk of reproductive problems that includes infertility, miscarriage and birth defects (Pizzorno 2018). Environmental toxins may cause inflammation in reproductive organs that may lead to infertility (Ghosh and Ghosh 2020). Exposure to environmental toxins may cause risk of spontaneous abortion, birth defects and developmental deformities (Frazier 2007).

Pollutants may also cause rise in blood cholesterol and result in hypertension. Constant exposure to hazardous toxic air pollutants may also increase the risk for cancer in various tissues, primarily the lungs and the skin.

Phytocompounds Against Air Pollutants-Mediated Lifestyle Diseases

Air pollution is an increasing threat to human health and well-being around the world (Ghosh and Parida 2015a). Phytocompounds are compounds found in plants. They are primarily the secondary metabolites of plants and possess medicinal properties. The bioactive medicinal Phytocompounds are beneficial in treating various health ailments (Ghosh and Parida 2015b). Plants like *Murraya koenigii, Allium sativum, Ocimum sanctum* and *Coriandrum sativum* are rich sources of phytocompounds with potent antioxidant activities. They scavenge the free radicals generated by exposure to air pollutants like the SPM, transition metals and heavy metals. Most of the plants are rich sources of bioactive phytocompounds, and a major part of these phytocompounds have antioxidant potentials. These phytocompounds are also reported to act synergistically to enhance the antioxidant potential of other synthetic and natural antioxidant compounds (Ghosh and Parida 2015b). Thousands of medicinal plants have been in use in folk medicine and traditional medicine since time immemorial. These medicinal plants help to treat and heal various health conditions including the lifestyle diseases like neural disorders, memory impairment, infertility issues, inflammations, cardiovascular ailments and certain respiratory pathological conditions. Plants from under the ocean and certain mushrooms are even known to have pronounced medicinal properties (Ghosh and Parida 2014; Patra et al. 2021). All these plants impart their medicinal impacts by virtue of the bioactive phytocompounds they contain. Various health conditions including infections and inflammations associated with lifestyle get aggravated due to the exposures to air pollutants. These health conditions may be effectively improved with immunomodulatory and immune-boosting phytocompounds (Patra et al. 2021). Studies also report beneficial role of antioxidant phytocompounds in treating and dealing with oxidative stress-induced male infertility (Ghosh et al. 2023b). Besides, the phytocompounds are rich source of essential nutrients that help to fight back certain health conditions. Table 2 represents some of the phytocompounds, examples of their source plants, their use and mechanism of action against some common lifestyle diseases. Thus, these medicinal phytocompounds may not only help to combat the onset of lifestyle diseases but may also help to mitigate air pollutant-induced aggravation of the lifestyle diseases.

Conclusion

In a nutshell, the way we live our daily life is called our lifestyle and that highly impacts our health. A good way of living, a balanced diet and a healthy lifestyle helps us to maintain good health, whereas, bad lifestyle, bad food habit and unhealthy habits adversely affect our health and well-being. Thus, the bad effects of our lifestyle on our health cause various types of diseases and those are termed as lifestyle diseases. As discussed in this chapter, exposure to various toxic air pollutants like SPM, ozone, metal dust, SOx and NOx not only initiates various lifestyle diseases but also aggravates them. These pollutants enter the human body primarily through respiratory pathway, enter the blood stream, reach different parts of the body and get stored in the highly vascularized soft organs of the body. They not only impart toxic affects at the site of storage, rather they generate free radicals, and these highly reactive free radicals react with the biomolecules they get in touch with. They cause peroxidation of lipids that build the cell membranes and

Table 2 Some of the phytocompounds, examples of their source plants, medicinal properties and their mode of action against certain lifestyle disease

Beneficial phytocompounds	Plants containing compounds	Scientific names	Medicinal properties	Mode of action	References
Catechins	Tea	*Camellia sinensis*	Prevents upper respiratory tract infections and COPD	Catechins alleviate conditions associated with vascular dysfunctions, including vascular inflammation and smooth muscle cell proliferation, blood platelet aggregation, lipoprotein oxidation, altered lipid profile, and vascular reactivity. They also have antioxidant properties	Babu and Liu (2008)
	Cocoa	*Theobroma cacao*			
Caffeine	Coffee	*Coffea arabica* & *Coffea canephora*	Protect from pulmonary disease	Has anti-inflammatory activity	Paiva et al. (2019)
Curcumin	Turmeric	*Curcuma longa*	Prevent skin disease and development of tumours in skin	By quenching free radicals and reducing inflammation through nuclear factor-KB inhibition	Thangapazham et al. (2017)
Gallic acid	Pomegranate	*Punica granatum,*	Prevent inflammatory disorders in gastro intestinal tract, cardiovascular system and neurological system	Ant-inflammatory, antioxidant and antineoplastic properties	Kahkeshani et al. (2019)
	Tea	*Camellia sinensis*			
Quercetin	Oak	*Quercus leucotrichophora*	May have inhibitory activity against COPD biomarkers and improve emphysema	By reducing the release of inflammatory cytokines, chemokines, adhesion molecule and cell infiltration	David et al. (2016)
	Grapes	*Vitis vinifera.*			
	Onions	*Allium cepa*			
	Apricot vine	*Passiflora incarnata*			
Theobromine	Cocoa	*Theobroma cacao*	Prevent CVD and maintains skin health	Reduces blood pressure and beneficial in treating skin ailments	Latif (2013)

(continued)

Table 2 (continued)

Beneficial phytocompounds	Plants containing compounds	Scientific names	Medicinal properties	Mode of action	References
Anthocyanins	Grapes	*Vitis vinifera.*	They prevent cancer, diabetes, CVD, etc.	They have antioxidant potentials	Khoo et al. (2017)
	Eggplants	*Solanum melongena*			
Monacolin	Red yeast	*Monascus purpureus*	Prevent CVD and hypertension	By reducing blood cholesterol level	Minamizuka et al. (2021)
Flavonoids	Blackberry	*Rubus fruticosus*	Prevent CVD	Prevent low-density lipoprotein oxidation, which promotes vasodilatation, prevents atherosclerosis	Ciumărnean et al. (2020)
	Cherry	*Prunus avium*			
Plant sterol	Almond	*Prunus dulcis*	Prevent CVD, hypertension, etc.	By decreasing cholesterol absorption in the blood	Ciumărnean et al. (2020)
	Indian Olive	*Elaeocarpus floribundus*			
Sulphur phytocompounds	Chick pea	*Cicer arietinum*	Prevent CVD	By activation of nuclear factor-erythroid factor 2-related factor 2 (Nrf2) and inhibition of cholesterol synthesis	Bachheti et al. (2022)
Quinone	Kiwi	*Actinidia deliciosa*	Decreases the risk of CVD and reduces risk of cancer and microbial infections	By increasing ATP production in mitochondria. They are anticancer, antimicrobial and anti-inflammatory	Sarewicz and Osyczka (2015)
	Spinach	*Spinacia oleracea*			
Terpenoids	Broccoli	*Brassica oleracea*	Decreases the risk of CVD and prevents cancer, inflammation, infection, etc	By decreasing atherosclerotic lesion in the aortic valve. They are reported to have antibacterial, antiviral, anti-inflammatory, anticancer, anti-allergic, antihyprglycemic and immumomodulatory potentials	Thoppil and Bishayee (2011)

Polyphenols	Walnut	*Juglans regia*	Have immense beneficial effect on entire living system. They potentially improve pulmonary function. They are vastly used in preventive medicine	Acts as antioxidants to protect the body's tissues from oxidative stress and related pathologies	Pandey and Rizvi (2009)
	Pistachios	*Pistacia vera*			
Coumarin	Tonka Beans	*Dipteryx odorata*	They are used as potent anti-inflammatory agents	They are anti-inflammatory, antioxidant and anticonvulsants	Ghosh et al. (2023a)
Carbazoles	Curry Leaves	*Murraya Koenigii*	Prevent tumour growth and diabetes	Anti-inflammatory, anti-bacterial, anti-oxidant, anti-tumour, anti-convulsant, anti-psychotic and anti-diabetic	Thawabteh et al. (2019)
Withanolides	Aswagandha	*Withania somnifera*	Prevent asthma, diabetes, arthritis, impotence, amnesia, hypertension, anxiety, stress, cancer, CVD and neurodegeneration	They have immunomodulatory and neuroprotective potentials	Singha et al. (2024)
Aloin	Aloe vera	*Aloe vera*	Decreases the risk of diabetes mellitus and CVD	Prevent CVD by lipid regulation. Anti-inflammatory, anticancer and immunomodulatory potentials	Xiao et al. (2022)
Allicin	Garlic	*Allium sativun*	Prevents cancer and CVD. Lower blood pressure, glucose and cholesterol	Allicin can induce apoptosis in cancer cells and inhibit tumour growth. Allicin can lower the risk of cardiovascular disease by preventing hyperlipidaemia and cardiac hypertrophy, and by inducing vasorelaxation	Borlinghaus et al. (2014)
Alkaloids	Drumstick	*Moringa oleifera*	Significant beneficial effect on DM, metabolic syndrome, lipid disorders	Hypoglycemic, cardioprotective, Hepatoprotective, hypolipidemic;	Heinrich et al. (2021)
	Turmeric	*Curcuma longa*			
	Aloe	*Aloe vera,*			
	Tulsi	*Ocimum sanctum*			
	Curry Patta	*Murraya koenigii*			

(continued)

Table 2 (continued)

Beneficial phytocompounds	Plants containing compounds	Scientific names	Medicinal properties	Mode of action	References
Vinca alkaloids	Madagascar periwinkle	*Catharanthus roseus*	Used to treat non-small-cell lung cancer, diabetes, hypertension, CVD etc.	Anti-mitotic and anti-microtubule alkaloid agents. They block beta-tubulin polymerization in a dividing cell. They are antioxidants, antidiabetic and anti-inflammatory	Moudi et al. (2013)
Azadirachtin	Neem	*Azadirachtin indica*	Used to treat skin problems, dandruff and various health issues	Antibiotic, antiviral, antimicrobial, cardioprotective, antioxidant, etc.	Subapriya and Nagini (2005)
Vasicine	Vasaka	*Adhatoda vasica*	Used primarily to heal respiratory tract inflammation, cold, cough, asthma	Antiviral, antibacterial, anti-inflammatory, immunomodulatory, etc.	Singha et al. (2021)

membranes of the sub-cellular organelles. They also cause carbonylation of proteins and damage the structural and functional proteins of the human body. These free radicals may also cause genetic mutations and alter genes. Regular exposure to air pollutants due to habits, way of living or occupational reason causes continuous adverse impact of the toxicants on the body. Further, lack of healthy food habits and healthy life practices adds to the toxic impacts of these toxicants. Good diet rich in antioxidants, vitamins and vital nutrients may help to scavenge and remove the free radicals from the system thereby mitigating their adverse effects. The vital nutrients may help to repair the damages, wear and tear caused by oxidative damages incurred by the air pollutant-induced generation of the free radicals in the body. Regular exercise and yoga may improve the blood circulation and metabolism of the body thereby increasing perfusion of the vital organs helping in removing the accumulated metabolic wastes and toxins. Besides, the nutraceuticals in food help to improve various molecular signalling pathways thereby aiding in healing of the diseases. Bioactive phytocompounds from medicinal plants have established antioxidant activities, and some of them directly help in removing the toxins. Also, many of these phytocompounds have immunomodulatory, anti-inflammatory and antimicrobial potentials by virtue of which they act to heal poor health situations associated with lifestyle diseases. The phytocompounds may thus help to prevent and protect against the lifestyle diseases through their direct and indirect mode of action.

Acknowledgement The authors acknowledge their respective departments and the institutes under the Department of Higher Education, Government of West Bengal, India.

Conflict of Interests None.

References

Babu PV, Liu D (2008) Green tea catechins and cardiovascular health: an update. Curr Med Chem 15(18):1840–1850

Bachheti RK, Worku LA, Gonfa YH (2022) Prevention and treatment of cardiovascular diseases with plant phytochemicals: a review. Evid Based Complement Alternat Med 2022:5741198

Bai L, Chen H, Hatzopoulou M et al (2018) Exposure to ambient ultrafine particles and nitrogen dioxide and incident hypertension and diabetes. Epidemiology 9(3):323–332

Bandyopadhyay A (2016) Neurological disorders from ambient (urban) air pollution emphasizing UFPM and PM2.5. Curr Pollut Rep 2:203–211

Bandyopadhyay D, Ghosh D, Chattopadhyay A (2014) Lead induced oxidative stress mediated myocardial injury: a review. Int J Pharm Sci Rev Res 29(2):67–71

Borlinghaus J, Albrecht F, Gruhlke MC et al (2014) Allicin: chemistry and biological properties. Molecules 19(8):12591–12618

Chao L, Lu M, An Z et al (2021) Short-term effect of NO_2 on outpatient visits for dermatologic diseases in Xinxiang, China: a time-series study. Environ Geochem Health 43(9):1–11

Chen P, Miah MR, Aschner M (2016a) Metals and neurodegeneration. F1000Res 5:F1000 Faculty Rev-366

Chen Z, Salam MT, Toledo-Corral C (2016b) Ambient air pollutants have adverse effects on insulin and glucose homeostasis in Mexican Americans. Diabetes Care 39(4):547–554

Chiou GC (2001) Review: effects of nitric oxide on eye diseases and their treatment. J Ocul Pharmacol Ther 17(2):189–198
Ciumarnean L, Milaciu MV, Runcan O (2020) The effects of flavonoids in cardiovascular diseases. Molecules 25(18):4320
David AAV, Arulmoli R, Parasuraman S (2016) Overviews of biological importance of quercetin: a bioactive flavonoid. Pharmacogn Rev 10(20):84–89
Dijkhoff IM, Drasler B, Karakocak BB (2020) Impact of airborne particulate matter on skin: a systematic review from epidemiology to in vitro studies. Part Fibre Toxicol 17(1):35
Duan RR, Hao K, Yang T (2020) Air pollution and chronic obstructive pulmonary disease. Chronic Dis Transl Med 6(4):260–269
Frazier LM (2007) Reproductive disorders associated with pesticide exposure. J Agromedicine 12: 27–37
Fu C, Kuang D, Zhang H (2022) Different components of air pollutants and neurological disorders. Front Public Health 10:959921
Gaalema DE (2022) Carbon monoxide and its effects on those with cardiovascular disease. J Cardiopulm Rehabil Prev 42(5):E55–E56
Gao P, Wu Y, He L et al (2023) Adverse short-term effects of ozone on cardiovascular mortalities modified by season and temperature: a time-series study. Front Public Health 11:1182337
Ghosh S, Ghosh D (2020) Current perspectives of male infertility induced by immunomodulation due to reproductive tract infections. Chem Biol Lett 7(2):85–91
Ghosh D, Parida P (2014) Drugs from the ocean: a review. World JPharm Pharm Sci 3(12): 1437–1442
Ghosh D, Parida P (2015a) Air pollution and India: current scenario. Int J Curr Res 7(11): 22194–22196
Ghosh D, Parida P (2015b) Medicinal plants of Assam, India: a mini review. Int J Pharmacol Pharm Sci 2(6):5–10
Ghosh D, Parida P (2017) Interferon therapy in lung cancer: current perspectives. Cur Cancer Ther Rev 12(4):237–245
Ghosh D, Mitra E, Firdaus SB et al (2013) Melatonin protects against lead-induced cardio toxicity: involvement of antioxidant mechanism. Int J Pharm Pharm Sci 5(3):806–813
Ghosh D, Singha PS, Ghosh S (2022) Impact of air pollution on the pathophysiology of COVID 19 in Indian population: a brief account. In: Das S (ed) Environment in 21st century. Kripadrishti Publishers, India, pp 114–121
Ghosh R, Singha PS, Das LK et al (2023a) Anti-inflammatory activity of natural coumarin compounds from plants of the Indo-Gangetic plain. AIMS Mol Sci 10(2):79–98
Ghosh S, Singha PS, Acharyaa A et al (2023b) Beneficial impacts and limitations of antioxidant supplements on male infertility. J Integr Sci Technol 12(3):751
Ghozikali MG, Mosaferi M, Safari GH et al (2015) Effect of exposure to O_3, NO_2, and SO_2 on chronic obstructive pulmonary disease hospitalizations in Tabriz, Iran. Environ Sci Pollut Res Int 22(4):2817–2823
Guo J, Chai G, Song X (2023) Long-term exposure to particulate matter on cardiovascular and respiratory diseases in low- and middle-income countries: a systematic review and meta-analysis. Front Public Health 11:1134341
Hamra GB, Laden F, Cohen AJ (2015) Lung cancer and exposure to nitrogen dioxide and traffic: a systematic review and meta-analysis. Environ Health Perspect 123(11):1107–1112
Haverinen E, Fernandez MF, Mustieles V et al (2021) Metabolic syndrome and endocrine disrupting chemicals: An overview of exposure and health effects. Int J Environ Res Public Health 18(24):13047
He D, Wu S, Zhao H et al (2017) Association between particulate matter 2.5 and diabetes mellitus: a meta-analysis of cohort studies. J Diabetes Investig 8(5):687–696
Heinrich M, Mah J, Amirkia V (2021) Alkaloids used as medicines: structural phytochemistry meets biodiversity-an update and forward look. Molecules 26(7):1836

Huang CC, Ho CH, Chen YC et al (2017) Increased risk for diabetes mellitus in patients with carbon monoxide poisoning. Oncotarget 8(38):63680–63690
Jeong S, Shin EC, Lee JH et al (2023) Particulate matter elevates ocular inflammation and endoplasmic reticulum stress in human retinal pigmented epithelium cells. Int J Environ Res Public Health 20(6):4766
Kahkeshani N, Farzaei F, Fotouhi M et al (2019) Pharmacological effects of gallic acid in health and diseases: a mechanistic review. Iran J Basic Med Sci 22(3):225–237
Karimi P, Peters KO, Bidad K et al (2015) Polycyclic aromatic hydrocarbons and childhood asthma. Eur J Epidemiol 30(2):91–101
Khoo HE, Azlan A, Tang ST et al (2017) Anthocyanidins and anthocyanins: colored pigments as food, pharmaceutical ingredients, and the potential health benefits. Food Nutr Res 61(1): 1361779
Kim SY, Kim E, Kim WJ (2020) Health effects of ozone on respiratory diseases. Tuberc Respir Dis (Seoul) 83(Supple 1):S6–S11
Kowalska M, Skrzypek M, Kowalski M et al (2020) Effect of NOx and NO2 concentration increase in ambient air to daily bronchitis and asthma exacerbation, silesian voivodeship in Poland. Int J Environ Res Public Health 17(3):754
Latif R (2013) Health benefits of cocoa. Curr Opin Clin Nutr Metab Care 16(6):669–674
Lee BJ, Kim B, Lee K (2014) Air pollution exposure and cardiovascular disease. Toxicol Res 30(2): 71–75
Levy RJ (2015) Carbon monoxide pollution and neurodevelopment: a public health concern. Neurotoxicol Teratol 49:31–40
Li YL, Chuang TW, Chang PY et al (2021) Long-term exposure to ozone and sulfur dioxide increases the incidence of type 2 diabetes mellitus among aged 30 to 50 adult population. Environ Res 194:110624
Lin CK, Lin RT, Chen PC et al (2018) A global perspective on sulfur oxide controls in coal-fired power plants and cardiovascular disease. Sci Rep 8:2611
Lin CC, Chiu CC, Lee PY et al (2022) The adverse effects of air pollution on the eye: a review. Int J Environ Res Public Health 19(3):1186
Ling SH, Van Eeden SF (2009) Particulate matter air pollution exposure: role in the development and exacerbation of chronic obstructive pulmonary disease. Int J Chron Obstruct Pulmon Dis 4: 233–243
Luo K, Li R, Li W et al (2016) Acute effects of nitrogen dioxide on cardiovascular mortality in Beijing: An exploration of spatial heterogeneity and the district-specific predictors. Sci Rep 6: 38328
Mallah MA, Mallah MA, Liu Y et al (2021) Relationship between polycyclic aromatic hydrocarbons and cardiovascular diseases: a systematic review. Front Public Health 9:763706
Mallah MA, Basnet TB, Ali M et al (2023) Association between urinary polycyclic aromatic hydrocarbon metabolites and diabetes mellitus among the US population: a cross-sectional study. Int Health 15(2):161–170
Martínez-Lazcano JC, González-Guevara E, del Carmen RM (2013) The effects of ozone exposure and associated injury mechanisms on the central nervous system. Rev Neurosci 24(3):337–352
Mathur P, Mascarenhas L (2019) Life style diseases: keeping fit for a better tomorrow. Indian J Med Res 149(Suppl):S129–S135
Minamizuka T, Koshizaka M, Shoji M (2021) Low dose red yeast rice with monacolin K lowers LDL cholesterol and blood pressure in Japanese with mild dyslipidemia: a multicenter, randomized trial. Asia Pac J Clin Nutr 30(3):424–435
Mohamed Hoesein FAA, Zanen P, Van Ginneken B et al (2011) Association of the transfer coefficient of the lung for carbon monoxide with emphysema progression in male smokers. Eur Respir J 38:1012–1018
Moorthy B, Chu C, Carlin DJ (2015) Polycyclic aromatic hydrocarbons: from metabolism to lung cancer. Toxicol Sci 145(1):5–15
Moudi M, Go R, Yien CY et al (2013) Vinca alkaloids. Int J. Prev Med 4(11):1231–1235

Ng R, Sutradhar R, Yao Z et al (2020) Smoking, drinking, diet and physical activity-modifiable lifestyle risk factors and their associations with age to first chronic disease. Int J Epidemiol 49(1):113–130

Nwaozuzu CC, Partick-Iwuanyanwu KC, Abah SO (2021) Systematic review of exposure to polycyclic aromatic hydrocarbons and obstructive lung disease. J Health Pollut 11(31):210903

Olasehinde TA, Olaniran AO (2022) Neurotoxicity of polycyclic aromatic hydrocarbons: a systematic mapping and review of neuropathological mechanisms. Toxics 10(8):417

Paiva C, Beserra B, Reis C et al (2019) Consumption of coffee or caffeine and serum concentration of inflammatory markers: a systematic review. Crit Rev Food Sci Nutr 59(4):652–663

Pamphlett R, Cherepanoff S, Too LK (2020) The distribution of toxic metals in the human retina and optic nerve head: implications for age-related macular degeneration. PLoS One 15(10): e0241054

Pandey KB, Rizvi SI (2009) Plant polyphenols as dietary antioxidants in human health and disease. Oxidative Med Cell Longev 2(5):270–278

Patra S, Maity P, Chakraborty I et al (2021) Structural studies of immunomodulatory (1 → 3)-, (1 → 4)-α glucan from an edible mushroom *Polyporus grammocephalus*. Int J Biol Macromol 168: 649–655

Phillips EM, Frates EP, Park DJ (2020) Lifestyle medicine. Phys Med Rehabil Clin N Am 31(4): 515–526

Pizzorno J (2018) Environmental toxins and infertility. Integr Med (Encinitas) 17(2):8–11

Puri P, Nandar SK, Kathuria S et al (2017) Effects of air pollution on the skin: a review. Indian J Dermatol Venereol Leprol 83:415–423

Resch H, Zawinka C, Weigert G et al (2005) Inhaled carbon monoxide increases retinal and choroidal blood flow in healthy humans. Invest Ophthalmol Vis Sci 46(11):4275–4280

Roy R, D'Angiulli A (2024) Air pollution and neurological diseases, current state highlights. Front Neurosci 18:1351721

Said MA, Van de Vegte YJ, Zafar MM et al (2019) Contributions of interactions between lifestyle and genetics on coronary artery disease risk. Curr Cardiol Rep 21(9):89

Sarewicz M, Osyczka A (2015) Electronic connection between the quinone and cytochrome C redox pools and its role in regulation of mitochondrial electron transport and redox signaling. Physiol Rev 95(1):219–243

Singha PS, Jana AK, Ghosh D, Firdaus SB (2021) Antiviral phytochemicals from Adhatoda vasica, atraditional medicinal plant of India. In: Sinha D, De RP, editors. Plants: A Valuable Resource for SustainableAgriculture, Food, and Medicine. Chapter 15. Delhi: ABS Books 2022:219-31

Singha PS, Ghosh D (2023) Human health risk assessment (HHRA) for environmental exposure: a brief account. In: Ahmad MI, Mahamood M, Javed M, Alhewairini SS (eds) Toxicology and human health. Springer, Singapore, pp 251–262

Singha PS, Ghosh D, Ghosh S et al (2022) Air pollution & its health impact in the urban population of India: current scenario in three major metropolitan cities. In: Das D (ed) Biodiversity in our mother earth. Bharti Publications, India, pp 57–74

Singha PS, Ghosh D, Ghosh S (2023) Impacts of environment pollutants on pregnancy & preventive measures. In: Ravichandran S (ed) Sustainable solutions for a changing world. Kripadrishti Publishers, India, pp 72–81

Singha PS, Ghosh R, Ghosh D (2024) Immunomodulatory phytocompounds from *Withania somnifera* (L.) Dunal (Ashwagandha) against COVID-19. Trends Immunother 8(1):2703

Subapriya R, Nagini S (2005) Medicinal properties of neem leaves: a review. Curr Med Chem Anticancer Agents 5(2):149–146

Thangapazham RL, Sharma A, Maheshwari RK (2017) Beneficial role of curcumin in skin diseases. Adv Exp Med Biol 595:343–357

Thawabteh A, Juma S, Bader M et al (2019) The biological activity of natural alkaloids against herbivores, cancerous cells and pathogens. Toxins (Basel) 11(11):656

Thoppil RJ, Bishayee A (2011) Terpenoids as potential chemopreventive and therapeutic agents in liver cancer. World J Hepatol 3(9):228–249

Tian L, Qiu H, Pun VC, Ho KF, Chan CS, Yu IT (2015) Carbon monoxide and stroke: A time series study of ambient air pollution and emergency hospitalizations. International journal of cardiology 201:4–9.https://doi.org/10.1016/j.ijcard.2015.07.099

Tokar EJ, Benbrahim-Tallaa L et al (2011) Metal ions in human cancer development. Met Ions Life Sci 8:375–401

Torne R, Soyer HP, Leb G et al (1991) Skin lesions in carbon monoxide intoxication. Dermatologica 183(3):212–215

Valkova E, Atanasov V, Vlaykova T et al (2023) The serum levels of the heavy metals cu, zn, cd, and pb and progression of COPD-A preliminary study. Int J Environ Res Public Health 20(2): 1427

Wiegman CH, Li F, Ryffel B et al (2020) Oxidative stress in ozone-induced chronic lung inflammation and emphysema: a facet of chronic obstructive pulmonary disease. Front Immunol 11:1957

Wu KG, Chang CY, Yen CY et al (2019) Associations between environmental heavy metal exposure and childhood asthma: a population-based study. J Microbiol Immunol Infect 52(2): 352–362

Xiao J, Chen S, Chen Y et al (2022) The potential health benefits of aloin from genus Aloe. Phytother Res 36(2):873–890

Yan Z, Xu Y, Li K et al (2023) Heavy metal levels and flavonoid intakes are associated with chronic obstructive pulmonary disease: an NHANES analysis (2007-2010 to 2017-2018). BMC Public Health 23:2335

Yang L, Wang WC, Lung SC et al (2017) Polycyclic aromatic hydrocarbons are associated with increased risk of chronic obstructive pulmonary disease during haze events in China. Sci Total Environ 574:1649–1658

Zhang YJ, Gan RY, Li S (2015) Antioxidant phytochemicals for the prevention and treatment of chronic diseases. Molecules 20(12):21138–21156

Zhang J, Yin H, Zhu X et al (2022) Effects of multi-metal exposure on the risk of diabetes mellitus among people aged 40-75 years in rural areas in southwest China. J Diabetes Investig 13(8): 1412–1425

Zhang Y, Yin X, Zheng X (2023) The relationship between PM2.5 and the onset and exacerbation of childhood asthma: a short communication. Front Pediatr 11:1191852

Ziglio E, Currie C, Rasmussen VB (2004) The WHO cross-national study of health behavior in school aged children from 35 countries: findings from 2001–2002. J Sch Health 74(6):204–206

Management of Lifestyle Diseases Caused by Exposure to Toxicants

Lifestyle Modification Strategies as the First Line of Chronic Disease Management

Saika Manzoor and Fouzia Rashid

Abstract

Non-communicable diseases (NCDs) are considered as lifestyle disorders that eventually turn chronic and potentially fatal. Significant changes in lifestyles and food preferences lead to an imbalance between energy expenditure and calorie intake, which eventually results in obesity. NCDs are caused by a variety of risk factors, including biological and behavioral ones. It is significant to note that majority of deaths from NCDs are thought to be avoidable by eliminating four behavioral risk factors, i.e., unhealthy diets, alcohol abuse, tobacco use and physical inactivity, which in turn contribute to the biological risk factors. NCDs are threat to global economy and healthcare industry as the process of diagnosis, prevention, and treatment involves a significant wastage of infrastructure, money, time, and human resources. Implementing NCD prevention measures is believed to be the most cost-effective, sustainable, and affordable approach to reduce the incidence of chronic diseases worldwide. Adopting healthy lifestyle, providing societal awareness, deciding on national health policies, overseeing the global health strategy, and putting specific initiatives and innovations into practice are some of the ways to prevent NCDs. Further, the "invisible epidemic" of NCDs can be effectively controlled by addressing modifiable risk factors, early detection programs and adopting healthy lifestyles (screening high-risk populations).

Keywords

Non-communicable disease · Metabolic syndrome · Cardiovascular disease · Keto-diet

S. Manzoor · F. Rashid (✉)
Department of Clinical Biochemistry, University of Kashmir, Srinagar, Jammu and Kashmir, India

S. Raisuddin et al. (eds.), *Environmental Toxicants and Lifestyle Diseases*,
https://doi.org/10.1007/978-981-95-0575-3_23

Introduction

Non-communicable diseases (NCDs) are considered chronic diseases because they last for a long term and have a slow progression. As per the report of the WHO, NCDs account for 71% of all fatalities annually, which is found to be the major cause of death. The four leading NCDs that cause the highest number of deaths are cardiovascular diseases (17.9 million), cancer (9 million), respiratory illness (3.9 million), and diabetes (1.6 million). Nevertheless, NCDs have been expanded to incorporate a wide range of diseases, namely mental disorders, trauma, hepatic, renal, and gastroenterological diseases; endocrine, hematological, dermatological, genetic, and neurological disorders; and other disabilities (blindness and deafness). Genetic, physiological, behavioral, and environmental variables are among the many causes of the majority of NCDs (Budreviciute et al. 2020).

Metabolic risk factors are intimately linked to the leading causes of NCDs (Sakharoff 2019). The four major metabolic risk factors, which are considered as the primary cause of metabolic syndrome (MetS) and have a significant causative involvement in NCDs, are as follows:

1. Obesity/overweight
2. Hypertension
3. Hyperglycemia
4. Hyperlipidemia

Metabolic syndrome is also known by several other names including cardio-metabolic syndrome, Reaven's syndrome, deadly quartet, syndrome X, and insulin resistance syndrome (Esmailnasab et al. 2012). The etiology of metabolic syndrome was mostly attributed to insulin resistance when it was first identified as a clinical disorder. As a result, IR was previously necessary for the diagnosis of metabolic syndrome. More composite criteria were introduced in later formulations (Chandrasekaran and Sagili 2018). Increased circumference of the waist, high levels of triglycerides (TG), low HDL, increased levels of blood pressure (BP), and poor fasting glucose levels are the five risk variables that, in accordance with the NCEP-ATPIII criteria, must be present in order to diagnose metabolic syndrome (McCullough 2011; Alexander et al. 2006).

Non-communicable Disease Risk Factors

Environmental Risk Factors

Numerous environmental risk factors cause NCDs. Air pollution is one of them, which constitutes 6.7 million fatalities, out of which 5.7 million are related to NCDs like heart attack, lung cancer, and chronic obstructive pulmonary disease (Bloom et al. 2011). Air pollution also affects unborn children in addition to maternal habits including drug and alcohol use, stress, nutrition, and tobacco use during pregnancy. According to Barker's Developmental Origin of Health and Disease (DOHaD) hypothesis, environmental changes during pregnancy and the first few months of postnatal life can alter gene expression patterns permanently, influencing a person's health and vulnerability to disease in later stages of development. As per this theory, environmental changes during critical phases of tissue and organ development, namely, gestation cause long-lasting structural and physiological disruptions that may make diseases more likely to emerge later in life (Barker and Osmond 1986). It has been stated that developmental programming may be a pathogenic component of polycystic ovary syndrome (PCOS). The pathophysiology of PCOS has also been associated with environmental variables. This endocrine condition may have an etiology that includes all environmental elements that can change insulin function, cause overweight or obesity, or both. According to research, dietary and lifestyle modifications that increase insulin sensitivity are able to improve anovulation and irregular menstrual cycles, both indicators of reproductive problems in obese women with PCOS. This implies that environmental factors play a part in the pathophysiology of PCOS (Sanchez-Garrido and Tena-Sempere 2020).

Social and Demographic Elements

Age Despite the fact that NCDs are usually associated with the elderly, anyone can have them at any age of life, including before birth. The life course viewpoint provides evidence for the uterine origins of adult NCDs. Barker has shown that a major contributing factor to adult diseases is maternal nutrition. He discovered that human fetuses permanently altered their shape and metabolism in order to adapt to a limited supply of nutrition. These planned alterations, therefore, may have contributed to various illnesses later in life including diabetes, heart attack and hypertension (Barker 1998). At kindergarten and schools, children have easy access to unhealthy food and drinks, which can cause them to develop new NCDs risks early in life. As a result, a large proportion of youngsters are overweight or obese. Adolescents can develop new and dangerous habits like smoking and consuming alcohol after that point in their lives, these unhealthy practices may persist into adulthood, which can elevate their chance of developing NCDs. Many workplace issues that adults encounter as they age, including low social engagement, joblessness, financial stress, and unsatisfactory occupations, can impact the progression of NCDS development (Mikkelsen et al. 2019). The risk of age-related chronic diseases (NCDs) in older adults has been shown to be significantly increased by poor nutrition, consumption of tobacco and alcohol, social exclusion, inactivity, and financial stress (Budreviciute et al. 2020).

Poverty/Low Income NCDs and poverty are strongly associated. Initiatives to reduce poverty in low-income nations are expected to be hampered by the rapid increase in NCDs, especially because huge health care costs would increase household expenses. NCDs and poverty are strongly correlated. Thus, the initiatives taken to eliminate poverty in low-income countries are expected to be hindered by a rapid increase in NCDs, which increases health costs. People who are susceptible to dangerous items like smoking or unhealthy eating habits, together with restricted access to health care, become sick and die sooner than people with better societal status. NCD-related medical expenses quickly deplete household finances in low-resource environments. Large numbers of people are pushed into poverty every year, and growth is stifled by the high expenses of NCDs, which include treatment, which is frequently expensive and time-consuming, along with income loss (Bloom et al. 2011).

Diet and Lifestyle From the last few decades, it has been reported that NCDs have overtaken infectious and parasitic diseases as the primary cause of death (Martorell et al. 2000). The evolution of human diseases across time may be attributed to changes in dietary habits and lifestyle choices. NCDs are clearly linked to high consumption of sugar-loaded drinks and processed meat along with other unhealthy lifestyle habits, which includes smoking, lack of exercise, and increased body mass index (BMI) (Bloom et al. 2011). "Dietary transition" refers to availability, processing, dietary consumption, and modifications in energy expenditure (Drewnowski and Popkin 1997). In most countries, animal-based diets, processed foods high in fat and sugar, and refined carbs have supplanted cereal in places where traditional cuisine is healthier, natural, and higher in fiber. Consequently, there have been quick changes in diet and a sharp rise in NCDs in countries with low or middle incomes (Popkin 2015). Concurrent increases in food intake and declines in physical activity lead to NCDs. The primary variables that might induce or worsen metabolic dysfunctioning, which results in reproductive abnormalities such as PCOS, are diet and lifestyle choices (Diamanti-Kandarakis et al. 2006). A static lifestyle combined with bad eating habits can exacerbate insulin resistance and obesity, as well as increase PCOS's metabolic and reproductive symptoms.

Classification of Risk Factors for NCDs

Risk factors for NCDs are categorized into modifiable and non-modifiable factors (Fig. 1).

- *Modifiable Factors*: Modifiable risk factors are traits that people or societies can alter to enhance health outcomes. The four main ones that WHO usually mentions when discussing NCDs are unhealthy eating, physical inactivity, excessive alcohol consumption, and tobacco use (Bloom et al. 2014).

  The factors that are modifiable can be further categorized as follows (Piovani et al. 2022):

  (i) Biological or metabolic (such as hyperglycemia, hyperlipidemia, elevated blood pressure, and obesity).

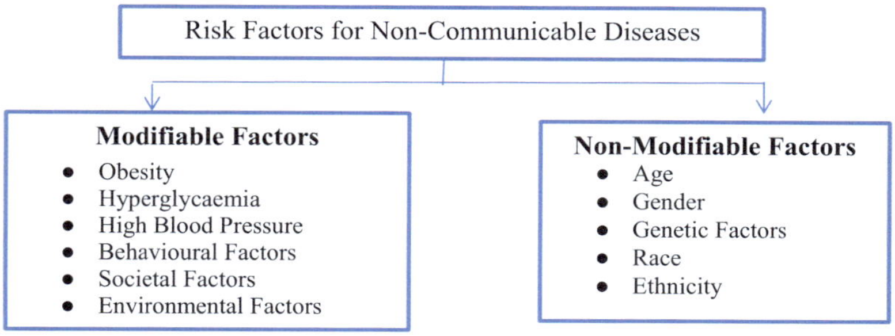

Fig. 1 Classification of risk factors for NCDs

(ii) Behavioral (Drug abuse, sedentary lifestyle, and poor nutrition).
(iii) Societal (poverty, low expenditure on health sector, and lack of accessibility to health care).
(iv) Lack of interaction between socioeconomic and environmental parameters (sunlight, air pollution, and climate change).

- *Non-modifiable factor:* Factors like age, gender, race, genetics, and ethnicity are examples of non-modifiable risk factors that cannot be changed. Their influence and partial determination of the efficacy of numerous preventive and treatment measures make them significant factors even though they cannot be the major objectives of treatments (Bloom et al. 2014).

Main NCD Diseases

- *Cardiovascular Diseases (CVDs):* CVD is a cluster of disorders that involve blood vessels, including hypertension and conditions related to carotid, cerebral and peripheral circulation besides heart conditions, such as ischemic heart disease, congenital heart disease, cerebrovascular disease, coronary heart disease, stroke, rheumatic heart disease, and peripheral arterial disease (Benjamin et al. 2018). Among NCDs, CVDs are the leading cause of the global burden of disease and cause more deaths globally annually than both cancer and chronic respiratory illnesses put together (Blundell and Hine 2019). Men experience greater rates of CVDs than women, even though both sexes are equally affected. In developed nations, cardiovascular diseases remain the root cause of death for women (Budreviciute et al. 2020). These diseases, especially heart attacks and strokes, are thought to claim the lives of 17 million people annually throughout the world. CVDs were once only found in industrialized nations; however, they are currently forming or spreading quickly to developing nations. It is true that in 1998, developing nations accounted for 86% of the Disability Adjusted Life Years (DALYs) triggered by CVDs, and in 1999, cardiovascular diseases caused a third of all fatalities worldwide, with 78% occurring in low- and middle-income countries. CVDs will become the prominent cause of death in developing nations undergoing socioeconomic transition due to changes in lifestyle brought about by urbanization and industrialization. Alcohol and tobacco use, sedentary lifestyles, and unhealthy eating habits are among the other threats for CVDs. Unfortunately, the harm that these risk factors inflict has an impact on developing countries' rising life expectancy (Nishida et al. 2004; Reddy 2002). Due to the high cost and duration of treatment for CVDs, low- and middle-income nations frequently reallocate their social resources to healthcare. This implies that risk factors are more common in lower socioeconomic categories; henceforth, the diseases and mortality are also higher in these groups (Boutayeb and Boutayeb 2005).

- *Diabetes:* The rising incidence and prevalence of diabetes have drawn attention from all around the world. This disease poses a serious risk to one's life in addition to being chronic. Moreover, it can also cause various complicated disorders like kidney failure, heart disease, foot ulcers, and ophthalmological disorders that might even cause blindness (Budreviciute et al. 2020). Hyperglycemia is a complication of the two primary forms of diabetes, i.e., Type 1 diabetes that is characterized by inadequate insulin production by beta cells of pancreatic, and Type 2 Diabetes that is characterized by improper insulin response by body cells (Bellou et al. 2018). Some other forms of diabetes include Type-3 diabetes, which is linked with Alzheimer's disease, causing neurons to be insensitive to insulin (De La Monte and Wands 2008) and Gestational diabetes mellitus, which influences pregnant women who are glucose intolerant (Gilmartin et al. 2008). Although there is a genetic component to diabetes, a number of lifestyle variables, including obesity, high sugar intake, and inactivity, can greatly accelerate the disease's progression. Nevertheless, changes in lifestyle are able to prevent diabetes and its long-term effects and can even reverse their condition. A "healthy dietary pattern" encompasses wide range of eating habits and dietary components like less consumption of alcohol, decreased intake of sweetened drinks, and processed meat (Esposito et al. 2014).
- *Cancer:* Cancer is considered as second leading cause of death and the primary public health concern. There are several factors linked with cancer that are common with other major NCDs. Three types of carcinogens, which are identified as the main causes of cancer: (i) Physical carcinogens (ultraviolet & ionizing radiation); (ii) Chemical carcinogens; (iii) Biological carcinogens (hormonal and genetic factors and various infections like bacterial, viral or parasitic). It is believed that smoking tobacco is the primary cause of cancer followed by poor nutrition (Nishida et al. 2004). Moreover, the predominant malignancies such as postmenopausal breast, colon, endometrial, Kidney and esophageal cancers are consistent to body weight and physical inactivity (Willett 1995). The main cause of lung cancer is smoking, and heavy smokers are more at risk (Tomatis et al. 1990). Furthermore, inadequate intakes of vegetables, fruits, and associated nutrients were found in these patients. Thus, reducing the consumption of smoking and increasing the consumption of fruits and vegetables can help to decline the chance of lung cancer. Moreover, physical inactivity and dietary practices are associated to breast cancer, which is most common and recurrent cancer in the world. During childhood and adolescence, dietary consumption

appears to have an impact on excess adiposity and hormonal processes, which appear to play a main role in the progression of breast cancer (Key et al. 2001). Therefore, the risk of breast cancer can be decreased by maintaining an ideal weight throughout life. There is also a considerable association between diet and colorectal cancer. Low intakes of high-fat diets, vegetables, fruits, dietary fiber, vitamins, and minerals have been linked with a high chance of colon cancer (Norat et al. 2002). Reduced intake of processed meat can lower the risk of developing colorectal cancer. Stomach cancer was once considered as the leading cause of death in developed countries. Vitamin C consumption and eating patterns are linked to it (Budreviciute et al. 2020). Infection with *Helicobacter pylori* is the majorly known risk factor for stomach cancer and is regarded as a category I carcinogen (Wroblewski et al. 2010). Three times fewer cancers in industrialized than in developing nations are brought on by infections. Avoiding infection is crucial for preventing cancer. This can be achieved by consumption of well-cooked food, drinking pure water, minimizing the excessive use of antibiotics to avoid antibiotic-resistance, and intake of vitamins as per the dietary guidelines (Mikkelsen et al. 2019). Lifestyle modification can reduce risk of cancer as the numerous studies have demonstrated clear correlation between lifestyle and cancer risk. Gender, age, smoking, alcohol, history of family illness, and diet are the primary risk factors for cancer (Key et al. 2001; Norat et al. 2002). A person's behavior can be changed to prevent cancer through proper nutrition, exercise, managing weight, obesity, quit smoking; prevention of unsafe sex, oncogenic viruses, harmful radiations, and alcohol consumption (Stein and Colditz 2004). Globally, there were more than 7 million cancer-related deaths. Developing nations accounted for 53% of incidence and 56% of mortality. The incidence and fatality rate rose by 2.4% year between 1990 and 2000. The total number of cancer cases in the developed countries rose by 29% from 2000 to 2020, while as in developing countries 73% increase was anticipated (mainly due to an decrease in the young population, modernization, and change in diet) (Boutayeb and Boutayeb 2005).

- *Chronic respiratory diseases (CRDs):* CRD is a broader term in itself as it encompasses various other diseases pertaining to lungs and respiratory pathway. Age being a major factor in the majority of CRD morbidity and death. Asthma, pulmonary hypertension, chronic obstructive pulmonary disease (COPD), sleep apnea, and allergies are few examples of CRD. In developing countries, the majority of CRD deaths are caused by asthma and COPD. The two main risk factors for CRDs are environmental and genetic, with environmental factors predominating. These include socioeconomic variables, air pollution; smoke and other passive smoke exposures (Leynaert et al. 2012). While not entirely avoidable, CRDs are somewhat reversible (Bellou et al. 2018). Hence, CRDs can be successfully prevented by leading healthy lifestyle throughout infancy; eluding respiratory infections, occupational and environmental toxins. The indoor and outdoor contaminants can be prevented by the use of natural gas, filtration, and ventilation (Schulze et al. 2017). A significant burden on global health systems is posed by chronic respiratory illnesses. When it comes to diagnosing and treating chronic, non-communicable respiratory conditions like asthma and COPD, most developing nations lack standard operating procedures. People who live in these nations will die before turning forty because they are illiterate and impoverished, and they have very little or no access to healthcare. Their population shares are as follows: 45% in Sub-Saharan Africa, and South-East Asia, 34% in the Arab world, and 15% in Latin America (Ait-Khaled et al. 2001; Boutayeb and Boutayeb 2005).

Economic Burden of the Four Major Non-communicable Diseases

Non-communicable diseases pose serious threat to economic growth and development in addition to human health. NCDs, being one of the primary causes of death, accounts 63% of mortalities universally. Currently, developing nation's accounts 80% of these deaths. The loss of life and the resulting disabilities also threaten the sector's ability to compete globally as most of the people who die from these diseases are in their productive years of life. Economists are more concerned with NCDs as these diseases have long-term macro-economic effects on capital accumulation, GDP, and labor supply showing severe effects in developing countries, even though research on the subject is still in its early stages (Abegunde et al. 2007; Bloom et al. 2014). There is a paucity of comprehensive documentation regarding the estimated economic cost of the four main NCDs, namely, CVDs, cancer, diabetes, and chronic respiratory conditions. Evidence from several nations reported that NCDs causes a significant financial burden that goes far beyond the health sector. In reality, executives and decision-makers in the global business community have identified NCDs as a substantial danger to economic loss, ranking them with other serious hazards including under-investment, lack of infrastructure, fiscal crises, and unemployment, as per the annual report of World Economic Forum. The problem of NCDs and their anticipated progression in the ensuing years are concerns shared by the business and health communities. To estimate this burden, three distinct methods were used; all of them produce intimidatingly enormous numbers but none of the conclusion. As per sickness approach, the overall cost of five different diseases mentioned below is as follows:

Cancer-projected cost US$ 290 billion in 2010 and expected cost US$ 458 billion in 2030.
Cardiovascular disease- projected to cost $863 billion in 2010 and expected cost $1.04 trillion in 2030.
Diabetes—projected cost US$ 500 billion in 2010 and expected cost US$ 745 billion by 2030.
Mental illness—projected cost up from US$ 2.5 trillion in 2010 and US$ 6.0 trillion by 2030.
COPD- Projected cost US$ 2.1 trillion in 2010 and expected US$ 4.8 trillion in 2030 (Bloom et al. 2014)

National-Level Economic Burden

Currently, the most common NCDs in India are cancer, diabetes, chronic respiratory diseases, and CVDs. As per the report of International Diabetes Federation (IDF), 33% of adults in India have diabetes but are not properly diagnosed, which makes it difficult to manage the disease. Type 2 diabetes being prevalent disease, affected nearly 63 million people in India in 2012, making it the "diabetes capital of the world" (Bloom et al. 2014). Furthermore, according to estimates from the IDF, 33% of adults in India who have diabetes do not have a diagnosis, which makes it difficult to properly manage the condition. The two associated demographic phenomena, the growth of the older population and the increase in global population, are expected to raise the burden of NCDs globally (Bloom et al. 2011). In many countries like India, 5.3% of the population are of the age of 65 and above and is expected to reach to 9.2% by 2035.

The burden of disease in India and globally will be significantly impacted by the growing older population, as they are the most susceptible to diseases (Bloom et al. 2011; Williams and Krakauer 2012). It is especially crucial to recognize the possible risks to this economic change in developing nations like India, where the burden of NCDs is on rise, while poverty is being reduced through fast economic transformation (Bloom et al. 2014). Table 1 presents the economic burden trend in India from 2012 to the present, together with the expected future economic loss.

Management of Risk Factors

NCDs pose the biggest threat to health systems in today's society. In order to manage NCDs, risk factor management is crucial. Numerous approaches are needed for the management of NCDs from various angles and at various levels, including the national and individual levels.

Therefore, it is concluded that best strategy for managing NCDs is to be focused on maintaining healthy lifestyle at individual level (Budreviciute et al. 2020).

Table 1 Economic burden of NCDs in India from 2012 to 2023* (Bloom et al. 2014)

NCD category	Economic loss between 2012 and 2030 (in trillions of 2010 dollars)
CVDs	2.17
Chronic respiratory diseases	0.98
Cancer	0.25
Diabetes	0.15
Total NCDs, excluding mental health conditions	3.55
Mental health conditions	1.03
Overall total	4.58

Lifestyle Intervention for Prevention of Non-communicable Diseases

Preventive medicinal approach (lifestyle modification) helps us to deal with various diseases pertaining to poor lifestyle approach. It encourages the people to improve their quality of life by changing sedentary life style including balanced diet, physical activity, stress management, proper sleep, and nurturing social bonds. Stress reduction techniques like meditation and yoga helps to reduce stress and improve overall mental health (Khoury et al. 2015). Chronic illnesses like obesity, diabetes, metabolic syndrome, and cardiovascular disease are the targets of lifestyle management. This field has the potential to greatly lower rates of morbidity and death and the cost burden related to managing chronic diseases because it aims to treat the underlying causes of many chronic diseases (Sallis et al. 2016; Lin et al. 2010). Lifestyle modification can help to prevent invasive and costly procedure such as laboratory investigation, medicinal cost, and surgeries (Livingston et al. 2021). In addition, a variety of people can receive lifestyle interventions because they are frequently inexpensive or even free (Lee et al. 2012). It is increasingly important for medical professionals to incorporate lifestyle medicine into their clinical practices as chronic diseases continue to rise in prevalence. Integrating evidence-based lifestyle interventions into clinical practice involves treating patients with the goal of improving their health and controlling chronic conditions. This method recognizes the significance of lifestyle elements in the development and advancement of many health disorders, including physical exercise, stress management, sleep, nutrition, and social relationships. Healthcare practitioners must take an extensive strategy that takes into account each patient's specific requirement, choices, and goals in order to successfully incorporate lifestyle medicine into clinical practice (Bell et al. 2002). They also need to be very knowledgeable about the most recent research on lifestyle interventions, and how it affects health outcomes (Maizes et al. 2009). In general, lifestyle medicine is a workable approach to lessening the financial load and hardship of chronic diseases. Lifestyle modifications offer unique alternative method for the prevention and treatment of several chronic diseases.

Dietary Interventions

Prevention, treatment, and regression of NCDs linked to obesity are greatly improved by preventive nutrition (PN), a significant field of nutrition study. The reach of PN is continuously changing. In actuality, it considers not just the nutritional characteristics in and of themselves but also the existence, both qualitatively and quantitatively, of natural bioactive substances and their potential health benefits when included in a balanced diet (Pinelli et al. 2003; Mastrogiovanni et al. 2019). Some basic dietary approaches for the prevention or management of non-communicable diseases are as follows.

Mediterranean Diet Since the 1950s, several studies have examined the Mediterranean Diet (MD) as a reference nutritional model for the prevention of NCDs. The Mediterranean diet was recognized by UNESCO in 2010 as a part of humanity's intangible cultural heritage (Di Daniele et al. 2017). Seven Countries Study of Cardiovascular Diseases (SCSCD) was the first research to examine the potential advantages of PN. It is recognized for being the first to characterize the "Mediterranean Diet" and to have established its protective effect against CHD (Keys et al. 2017). Moderate consumption of white meat (poultry), eggs, legumes, and dairy products), low consumption of saturated fats, red meat, and red wine are characteristics of this nutritional model. It has its roots in a variety of Mediterranean basin cultures and countries. High intakes of fruits and vegetables, seeds, cereals, extra virgin olive oil (unsaturated fats), and fish (Keys et al. 2017). Diet being a commonest risk factor for most of the NCDs; therefore, better options like evidence-based nutrition need to be adopted so as to avoid non-communicable diseases (Billingsley et al. 2018).

- *Unsaturated fats:* Unsaturated fats such as omega-3 alpha-linolenic and omega-6 linoleic acids play a significant role in the prevention and reduction of CVDS (Billingsley et al. 2018).
- *Vitamins:* The Majority of the substances needed for normal body function are synthesized by human body; nonetheless, food is the primary source of vital nutrients. The classes of important organic compounds such as co-factors or vitamins (e.g., B, C, A, D, E, and K) are needed in trace amounts for the majority of enzymes to function. The impact of low or absent vitamin levels can be intense on health. For an aging society, it seems to be extremely important to have attention on the requirement of meeting adequate dietary intakes of vital nutrients (Bruins et al. 2019). Another study found that vitamin E treatment considerably enhance liver function and decrease liver fat in people with non-alcoholic fatty liver disease (Sanyal et al. 2010).
- *Fish intake:* Daily intake of 40–60 grams of fish has shown to be optimal for reducing the risk of coronary heart disease mortality by about 50% in high-risk populations. Another study that included data from 36 different nations found a link between eating fish and a lower risk of deaths from CVD and other causes. Regretfully, very little fish is consumed, even in certain nations like those in North Africa that are well-known for having abundant fish stocks (Boutayeb and Boutayeb 2005).
- *Fruits and vegetables:* It is advised to consume 400–500 grams of fresh fruits and vegetables daily to lower the risk of heart disease and hypertension. Unfortunately, developing nations are again influenced by western way of living (Boutayeb and Boutayeb 2005).
- *Fasting and keto-diet:* The number of dietary restriction (DR) practices, including fasting and caloric restriction (CR), has been studied extensively as lifestyle modifications that improve health and can influence various disease risk variables (Varady and Hellerstein 2007). However, one of the biggest problems with diet limitation in today's world is the abundance of food availability. Food is always

present in wealthy, contemporary nations, and restricting one's diet calls for a great deal of self-control. Nonetheless, it is commonly known that ketogenic diet reduces appetite (Paoli et al. 2015). As a result, adopting a ketogenic diet can be a suitable starting point for reaping the benefits of fasting and CR because it significantly declines the experience of hunger. It has been demonstrated that adopting a ketogenic diet is an effective strategy for restoring the proper ratio and control of the hunger and satiety hormones, ghrelin and leptin, and for reducing appetite (Sumithran et al. 2011). The Keto diet is essentially very low in carbohydrates, moderate in protein, and high in fat. Breakdown of fatty acids by the liver provides ketone bodies, which are released into blood stream and used as a source of energy by the tissues. The body's fat reserves can last for several days following exercise, in contrast to the high-carb diets' limited energy availability that depletes after just 1.5 hours of exercise.

- *Regulation of hunger in keto-diet (KD):* Normally, glucose levels have an impact on hunger levels; however, following the initial phase of ketogenic adaption, energy levels have a greater influence. It is a frequent experience that during ketogenic adaption, the levels of ghrelin and leptin eventually return to normal and the overall urge for food gradually decreases. As a result, stable energy and satiety levels make eating only twice a day usual. To maintain adequate quantities of micro-nutrients like vitamins and minerals, it is essential to consume organ meats such liver, kidney, heart, and bone marrow (Volek and Phinney 2011). Numerous scientific studies demonstrated that KD is an effective regimen capable of addressing all four of the metabolic risk factors responsible for NCDs and MetS that were discussed earlier (Sakharoff 2019). One of the main uses of KD is the treatment of insulin resistance, which is the main abnormality and characteristic of PCOS and MetS (Wirth 1995). Insulin resistance is linked to reduced glucose metabolism in relation to high-carb diets. Therefore, reduced carbohydrate metabolism, which is a hallmark of insulin resistance, can be effectively controlled by dietary carbohydrate restriction combined with a KD regimen. Numerous other investigations found that the degree of oxidative stress and Reactive Oxygen Species (ROS) formation for diets high in carbs was correlated with or proportionate to rise in blood sugar and blood insulin levels (Mohanty et al. 2000). But when utilized as fuel in the mitochondria ketones prevent dysregulated ROS generation with less oxidative stress than glucose molecules (Maalouf et al. 2007). Ketone bodies also protect mitochondria and reduce blood levels of oxidative stress and free radicals in the brain (Kim et al. 2007).

Physical Activity

It is referred as any activity of the body involving skeletal muscles that requires energy expenditure (WHO 2017; Saqib et al. 2020). Furthermore, WHO indicates that physical activity at the community level is the primary indicator of health (Sadrollahi et al. 2016), and physical inactivity attributes 3.2 million deaths around the world each year. Sedentary lifestyles are the primary cause of this increase in mortality (WHO 2014). Regular physical activity lowers the risk of several chronic diseases such as Type 2 Diabetes, CVDs and cancer (Warburton et al. 2006). Energy balance and weight management are fundamentally influenced by physical exercise. Frequent physical activity provides many advantages for psycho-physical health as per WHO. Physical activity and NCDs are correlated and help in their prevention. Obesity is also contributing to the global spread of chronic diseases, which is mostly caused by physical inactivity. Physical activity has a major impact in preventing non-communicable diseases at any age. Any physical activity that causes your body to burn calories or use energy is good for your health. People in any age range benefit from physical activity because it enhances mental health, boosts self-esteem, improves sleep, increases energy, and lowers the risk of chronic illnesses (Leonard and Robertson 1992). Regular physical activity declines the risk of various diseases like CVD, stroke, hypertension, cancer, and osteoporosis. Additionally, exercise raises energy levels, which enhances overall well-being and reduces stress. Within this framework, the WHO has initiated a worldwide action plan to reduce sedentary behavior and physical inactivity by 10% in 2025 and 15% in 2030. Therefore, physical activity is considered as the most important global goal for prevention, diagnosis, and treatment of non-communicable diseases. Being physically active is now synonymous with leading a healthy life. WHO recommends that children and adolescents between the age group of 5–17 years must engage in physical activity for at least 60 minutes in a day, while as adults (18–64 years) must engage for at least 150 minutes per week (Saqib et al. 2020).

Stress Management

It has been demonstrated that the majority of diseases from the twentieth century are strongly correlated with psychosocial factors like stress, anxiety, or depression. According to estimates, stress is a factor in 75% of medical symptoms. Therefore, encouraging people to adopt health-promoting activities is one strategy to help minimize stress and save healthcare expenditures. The goal of stress management programs should be to assist patients in achieving optimal well-being and leading richer, more fulfilling, tension-free lives. A substantial decrease in HbA1C (0.5%) was linked to stress management training in one research (Surwit et al. 2002). Long-term individual treatment has been used in several researches that demonstrate the therapeutic benefit of stress management (Surwit et al. 2002). The Physician can assist patients in recognizing and evaluating their stress and can then provide strategies for managing it. However, this needs the assistance of a licensed psychologist or counsellor. Exercise, dietary modifications, yoga, meditation, relaxation techniques, stress management classes, counselling, and, when necessary, medication are just a few examples of coping mechanisms.

Support Groups to Prevent NCDs

Collaboration among people whether local or global level can contribute to strategies for preventing NCDs. Because NCDs directly contribute to the nation's declining income rate, NCD prevention is crucial. Large-scale productivity loss results from both the incapacity to work and the frequent lack of economic risks.

- The risk factor management approach, which tackles issues at the individual, societal, national, and international levels, is the foundation of the management plan to avoid NCDs. Initiatives like management awareness, multi-sectoral associations, resource allocation, and innovations are all part of it.
- The key element for the prevention strategy is individual's lifestyle management, focusing on innovative approaches and initiatives that can support in raising risk factor management awareness, national health policy decision-making, and creation of global health strategy. New methods for preventing NCDs must be developed, and leadership is essential to the change management process (Budreviciute et al. 2020).
- WHO and United Nations should collaborate to develop global policies and strategies with an aim of lowering the risk of NCDs (WHO 2020). Monitoring NCDs and assessing their progress on regional, national, and international levels is essential. Such agencies have the ability to provide grants to different research projects and to promote cooperation between academic institutions and health organizations.
- Research centers and institutes can play a leading role in the NCD prevention through their research projects and programs. NCDs risk is directly impacted by agricultural and food biotechnology research (Pullar et al. 2018). Through their research projects and initiatives, research centers and institutes can make a substantial contribution to society's efforts to avoid NCDs.
- The creation of diagnostic instruments has made it possible to quickly identify high-sensitivity biomarkers for NCDs, which can aid in early detection of diseases and facilitate quick treatment options (Damiati et al. 2017).
- It is necessary to motivate people to adopt healthy lifestyle to improve quality of health and to obtain an effective response for NCD prevention and control.
- Various interventions are used to encourage healthy lifestyle choices in public health management. Multi-sectoral measures are one of the most important interventions carried out to prevent NCDs in India. Some of the multi-sectoral initiatives being implemented in India are: National Cancer Control Programme, National Trauma Control Programme, National Programme for Control of Blindness, National Mental Health Programme, National Tobacco Control Programme and National Programme for the control of Diabetes, stroke and CVDs (Arora et al. 2011).

Conclusion

In developed countries, there has been a significant drop in total fatalities associated with CVDs, whereas developing countries have experienced a considerable increase in these fatalities (Roth et al. 2017; Rarau et al. 2019). An effective strategy to avert the development and transmission of diseases serves as the main obstacle contributing to the rising number of patients with NCDs in low- and middle-income regions. Not merely using coping mechanisms from developed nations, each nation must create its own management plan. Numerous effective models that account for low-cost methods of NCD prevention, diagnosis, and treatment need to be validated. For instance, Kenya has devised an affordable method for early diagnosis of hypertension and diabetes. To create a safe and healthy world, governments and international organizations should educate individuals about their surroundings and health. Innovations play a crucial role in effectively addressing the increasing problem of NCDs, often focusing on lifestyle initiatives that promote smoking cessation, nutritious eating, and healthier life choices. Lifestyle diseases are by their very nature multifactorial. Pharmacological focused treatments frequently involve combination of various chemicals that address several aspects of the illness, resulting in unfavorable medication interactions that can be harmful to health in and of them, causing iatrogenic comorbidities. In contemporary healthcare, iatrogenic effects represent a significant problem. Macary and Daniel claimed third most common cause of death in the United States is medical care-related comorbidities (Makary and Daniel 2016). Various approaches must be developed that will address the underlying causes of lifestyle diseases rather than just their symptoms, and these approaches need to be implemented. Metabolic techniques or methods, which are multifactorial in nature, are the opposite of pharmacological and invasive approaches (Sakharoff 2019). Rather than focusing on specific symptoms, these methods normalize the disease's origins by addressing the metabolic risk factors at its foundation. Most commonly, these methods are noninvasive and non-pharmacological. The focus of these non-pharmacological metabolic methods is on healing by prevalent drug-free metabolism optimization as opposed to total reliance on pharmaceuticals and invasive therapies.

References

Abegunde DO, Mathers CD, Adam T et al (2007) The burden and costs of chronic diseases in low-income and middle-income countries. Lancet 370:1929–1938

Ait-Khaled N, Enarson D, Bousquet J (2001) Chronic respiratory diseases in developing countries: the burden and strategies for prevention and management. Bull World Health Organ 79:971–979

Alexander CM, Landsman PB, Grundy SM (2006) Metabolic syndrome and hyperglycemia: congruence and divergence. Am J Cardiol 98:982–985

Arora M, Chauhan K, John S et al (2011) Multi-sectoral action for addressing social determinants of non-communicable diseases and mainstreaming health promotion in national health programmes in India. Indian J Commun Med 36:S43–S49

Barker DJ (1998) In utero programming of chronic disease. Clin Sci (1979) 95:115–128
Barker DJ, Osmond C (1986) Infant mortality, childhood nutrition, and ischaemic heart disease in England and Wales. Lancet 327:1077–1081
Bell IR, Caspi O, Schwartz GE et al (2002) Integrative medicine and systemic outcomes research: issues in the emergence of a new model for primary health care. Arch Intern Med 162:133–140
Bellou V, Belbasis L, Tzoulaki I et al (2018) Risk factors for type 2 diabetes mellitus: an exposure-wide umbrella review of meta-analyses. PLoS One 13:e0194127
Benjamin EJ, Virani SS, Callaway CW et al (2018) Heart disease and stroke statistics-2018 update: a report from the American Heart Association. Circulation 137:e67–e492
Billingsley HE, Carbone S, Lavie CJ (2018) Dietary fats and chronic non communicable diseases. Nutrition 10:1385
Bloom DE, Cafiero E, Jane-Llopis E et al (2011) The global economic burden of non-communicable diseases (No. 8712). In: The Global Economic Burden of Noncommunicable Diseases. Geneva: World Economic Forum. Available via www.weforum.org/EconomicsOfNCD. Accessed on 12 March 2024
Bloom DE, Cafiero-Fonseca ET, Candeias V et al (2014) Economics of non-communicable diseases in India: The costs and returns on investment of interventions to promote healthy living and prevent, treat, and manage NCDs. World Economic Forum, Harvard School of Public Health. Available via http://www.weforum.org/issues/healthy-living. Accessed on 18 March 2024
Blundell HJ, Hine P (2019) Non-communicable diseases: ditch the label and recapture public awareness. Int Health 11:5–6
Boutayeb A, Boutayeb S (2005) The burden of non-communicable diseases in developing countries. Int J Equity Health 4:1–8
Bruins MJ, Van Dael P, Eggersdorfer M (2019) The role of nutrients in reducing the risk for noncommunicable diseases during aging. Nutrition 11:85
Budreviciute A, Damiati S, Sabir DK et al (2020) Management and prevention strategies for non-communicable diseases (NCDs) and their risk factors. Front Public Health 8:574111
Chandrasekaran S, Sagili H (2018) Metabolic syndrome in women with polycystic ovary syndrome. Obstet Gynaecol 20:245–252
Damiati S, Kupcu S, Peacock M et al (2017) Acoustic and hybrid 3D-printed electrochemical biosensors for the real-time immunodetection of liver cancer cells (HepG2). Biosens Bioelectron 94:500–506
De la Monte SM, Wands JR (2008) Alzheimer's disease is type 3 diabetes-evidence reviewed. J Diabetes Sci Technol 2:1101–1113
Di Daniele N, Noce A, Vidiri MF et al (2017) Impact of Mediterranean diet on metabolic syndrome, cancer and longevity. Oncotarget 8:8947
Diamanti-Kandarakis E, Kandarakis H, Legro RS (2006) The role of genes and environment in the etiology of PCOS. Endocrine 30:19–26
Drewnowski A, Popkin BM (1997) The nutrition transition: new trends in the global diet. Nutr Rev 55:31–43
Esmailnasab N, Moradi G, Delaveri A (2012) Risk factors of non-communicable diseases and metabolic syndrome. Iran J Public Health 41:77
Esposito K, Chiodini P, Maiorino MI et al (2014) Which diet for prevention of type 2 diabetes? A meta-analysis of prospective studies. Endocrine 47:107–116
Gilmartin AB, Ural SH, Repke JT (2008) Gestational diabetes mellitus. Rev. Obstet Gynecol 1: 129–134
Key TJ, Verkasalo PK, Banks E (2001) Epidemiology of breast cancer. Lancet Oncol 2:133–140
Keys A, Menotti A, Karvonen MJ et al (2017) The diet and 15-year death rate in the seven countries study. Am J Epidemiol 185(11):903
Khoury B, Sharma M, Rush SE et al (2015) Mindfulness-based stress reduction for healthy individuals: a meta-analysis. J Psychosom Res 78:519–528

Kim DY, Davis LM, Sullivan PG (2007) Ketone bodies are protective against oxidative stress in neocortical neurons. J Neurochem 101:1316–1326

Lee IM, Shiroma EJ, Lobelo F et al (2012) Effect of physical inactivity on major non-communicable diseases worldwide: an analysis of burden of disease and life expectancy. Lancet 380:219–229

Leonard WR, Robertson ML (1992) Nutritional requirements and human evolution: a bioenergetics model. Am J Hum Biol 4:179–195

Leynaert B, Sunyer J, Garcia-Esteban R et al (2012) Gender differences in prevalence, diagnosis and incidence of allergic and non-allergic asthma: a population-based cohort. Thorax 67:625–631

Lin JS, O'Connor E, Whitlock EP (2010) Behavioral counseling to promote physical activity and a healthful diet to prevent cardiovascular disease in adults: a systematic review for the US Preventive Services Task Force. Ann Intern Med 153:736–750

Livingston KA, Freeman KJ, Friedman SM et al (2021) Lifestyle medicine and economics: a proposal for research priorities informed by a case series of disease reversal. Int J Environ Res Public Health 18:11364

Maalouf M, Sullivan PG, Davis L (2007) Ketones inhibit mitochondrial production of reactive oxygen species production following glutamate excitotoxicity by increasing NADH oxidation. Neuroscience 145:256–264

Maizes V, Rakel D, Niemiec C (2009) Integrative medicine and patient-centered care. Explore (NY) 5:277–289

Makary MA, Daniel M (2016) Medical error-the third leading cause of death in the US. BMJ Open 3:353

Martorell R, Khan LK, Hughes ML et al (2000) Overweight and obesity in preschool children from developing countries. Int J Obes Relat Metab Disord 24:959–967

Mastrogiovanni F, Mukhopadhya A, Lacetera N et al (2019) Anti-inflammatory effects of pomegranate peel extracts on in vitro human intestinal caco-2 cells and ex vivo porcine colonic tissue explants. Nutrients 11:548

McCullough AJ (2011) Epidemiology of the metabolic syndrome in the USA. J Dig Dis 12:333–340

Mikkelsen B, Williams J, Rakovac I et al (2019) Life course approach to prevention and control of non-communicable diseases. BMJ 364:257

Mohanty P, Hamouda W, Garg R et al (2000) Glucose challenge stimulates reactive oxygen species (ROS) generation by leucocytes. J Clin Endocrinol Metab 5:2970–2973

Nishida C, Uauy R, Kumanyika S et al (2004) The joint WHO/FAO expert consultation on diet, nutrition and the prevention of chronic diseases: process, product and policy implications. Public Health Nutr 7(1A):245–250

Norat T, Lukanova A, Ferrari P et al (2002) Meat consumption and colorectal cancer risk: dose-response meta-analysis of epidemiological studies. Int J Cancer 98:241–256

Paoli A, Bosco G, Camporesi EM et al (2015) Ketosis, ketogenic diet and food intake control: a complex relationship. Front Psychol 6:27

Pinelli P, Galardi C, Mulinacci N et al (2003) Minor polar compound and fatty acid analyses in monocultivar virgin olive oils from Tuscany. Food Chem 80:331–336

Piovani D, Nikolopoulos GK, Bonovas S (2022) Non-communicable diseases: the invisible epidemic. J Clin Med 11:5939

Popkin BM (2015) Nutrition transition and the global diabetes epidemic. Curr Diab Rep 15:1–8

Pullar J, Allen L, Townsend N (2018) The impact of poverty reduction and development interventions on non-communicable diseases and their behavioural risk factors in low and lower-middle income countries: a systematic review. PLoS One 13:e0193378

Rarau P, Pulford J, Gouda H (2019) Socio-economic status and behavioural and cardiovascular risk factors in Papua New Guinea: a cross-sectional survey. PLoS One 14:e0211068

Reddy KS (2002) Cardiovascular diseases in the developing countries: dimensions, determinants, dynamics and directions for public health action. Public Health Nutr 5:231–237

Roth GA, Johnson C, Abajobir A et al (2017) Global, regional, and national burden of cardiovascular diseases for 10 causes, 1990 to 2015. J Am Coll Cardiol 70:1–25
Sadrollahi A, Hosseinian M, Alavi NM (2016) Physical activity patterns in the elderly Kashan population. Iran Red Crescent Med J 18(6):e25008
Sakharoff M (2019) Buteyko breathing technique and ketogenic diet as potential hormetins in nonpharmacological metabolic approaches to health and longevity. In: Rattan SIS, Kyriazis M (eds) The science of hormesis in health and longevity. Elsevier Academic Press, pp 257–274
Sallis RE, Matuszak JM, Baggish AL et al (2016) Call to action on making physical activity assessment and prescription a medical standard of care. Curr Sports Med Rep 15:207–214
Sanchez-Garrido MA, Tena-Sempere M (2020) Metabolic dysfunction in polycystic ovary syndrome: pathogenic role of androgen excess and potential therapeutic strategies. Mol Metab 35: 100937
Sanyal AJ, Chalasani N, Kowdley KV et al (2010) Pioglitazone, vitamin E, or placebo for non-alcoholic steatohepatitis. N Engl J Med 362:1675–1685
Saqib ZA, Dai J, Menhas R (2020) Physical activity is a medicine for non-communicable diseases: a survey study regarding the perception of physical activity impact on health wellbeing. Risk Manag Healthc Policy 11:2949–2962
Schulze F, Gao X, Virzonis D et al (2017) Air quality effects on human health and approaches for its assessment through microfluidic chips. Genes 8:244
Stein CJ, Colditz GA (2004) Modifiable risk factors for cancer. Br J Cancer 90:299–303
Sumithran P, Prendergast LA, Delbridge E et al (2011) Long-term persistence of hormonal adaptations to weight loss. N Engl J Med 365:1597–1604
Surwit RS, Van Tilburg MA, Zucker N et al (2002) Stress management improves long-term glycemic control in type 2 diabetes. Diabetes Care 25:30–34
Tomatis L, Aitio A, Day NE et al (1990) Cancer: causes, occurrence and control. International Agency for Research on Cancer, Lyon
Varady KA, Hellerstein MK (2007) Alternate-day fasting and chronic disease prevention: a review of human and animal trials. Am J Clin Nutr 86:7–13
Volek JS, Phinney SD (2011) Low carbohydrate living. Beyond Obesity, New York
Warburton DE, Nicol CW, Bredin SS (2006) Health benefits of physical activity: the evidence. CMAJ 174:801–809
Willett WC (1995) Diet, nutrition, and avoidable cancer. Environ Health Perspect 103:165–170
Williams R, Krakauer R (2012) The challenge of non-communicable diseases and geriatric conditions. In: Beard JR, Biggs S, Bloom DE, Fried LP, Hogan P, Kalache A, Olshansky SJ (eds) Global population ageing: peril or promise. World Economic Forum, Geneva, pp 78–81
Wirth A (1995) Non-pharmacological therapy of metabolic syndrome. Herz 20:56–69
World Health Organization (2014) Diet and physical activity: A public health priority (Global Strategy on Diet, Physical Activity and Health. Available via https://www.who.int/publications/i/item/9241592222. Accessed on 12 March 2024
World Health Organization (2020) WHO package of essential non communicable (PEN) disease interventions for primary health care. Available via https://www.who.int/publications/i/item/who-package-of-essential-noncommunicable-(pen)-disease-interventions-for-primary-health-care. Accessed on 12 March 2024
World Health Organization (WHO) (2017) A practice guide to effective population-based food policy actions to promote healthy diets. World Health Organization, Regional Office for South-East Asia. Available via https://iris.who.int/handle/10665/312107. Accessed on 12 March 2024
Wroblewski LE, Peek RM Jr, Wilson KT (2010) Helicobacter pylori and gastric cancer: factors that modulate disease risk. Clin Microbiol Rev 23:713–739

Natural Products for the Management of Oral Cancer: A Lifestyle Problem and its Solution

Vikas Kumar, Komal Kumari, and Rizwanul Haque

Abstract

Lifestyle diseases pose a significant challenge in contemporary society, and their prevalence is steadily rising. Considerable progress can be achieved by implementing minor modifications in our lifestyle and dietary practices. This chapter delineates several natural products and compounds that exhibit efficacy in addressing diverse lifestyle diseases and particularly focused on oral cancer and written in a straightforward manner that will be highly advantageous for the general public, students, and new researchers. We have examined the mechanisms and signaling pathways that significantly contribute to oral cancer. Additionally, we have explored several natural compounds that effectively suppress these processes, which can be highly informative for students and new research endeavors.

Keywords

Alkaloids · Phenols · Inhibitors · Signaling pathway · Oral cancer

Lifestyle Diseases

Lifestyle diseases are a prominent public health issue, and our daily habits or behaviors mostly contribute to having them. They are also related to preventable lifestyle behavior risk factors such as tobacco use, harmful consumption of alcohol, unhealthy diets, and physical inactivity. These behaviors lead to the onset of

V. Kumar · K. Kumari · R. Haque (✉)
Department of Biotechnology, School of Earth, Biological and Environmental Science, Central University of South Bihar, Gaya, Bihar, India
e-mail: vikaskr@cusb.ac.in; komal@cusb.ac.in; rhaque@cub.ac.in

S. Raisuddin et al. (eds.), *Environmental Toxicants and Lifestyle Diseases*,
https://doi.org/10.1007/978-981-95-0575-3_24

noncommunicable diseases (NCDs), a chronic disease or condition that develops slowly and typically has long-lasting effects through combinations of genetic, physiological, behavioral, and environmental factors. Lifestyle diseases are often used interchangeably with chronic diseases. NCDs are chronic conditions that are not spread from one individual to another. These diseases represent a major threat to global health as the primary cause of death in the twenty-first century. The main types of NCDs include cardiovascular diseases, cancers, chronic respiratory diseases, and diabetes. These lifestyle diseases can be having several risk factors and that can be categorized in three major groups.

There are three main groups of risk factors for NCDs

1. Modifiable behavioral risk factors (Controllable Risk Factors)
2. Non-modifiable risk factors (Uncontrollable Risk Factors)
3. Metabolic risk factors

Controllable risk factors Increased NCDs risk by some types of behavioral risk factors like bad drinking, feeding stuff, or living in general and particularly harmful products such as tobacco, unhealthy eating habits; lack of a regular diet for the meal initializer day-lengthening hormones residents a messed up biological lock. Another vital cause of NCDs is the modern work environment, desk jobs, and stress. Controllable risk factors include

(i) Diet and body weight: This is of paramount importance because a healthy diet not only helps to prevent but also manage several lifestyle disorders that have become common place in modern times. Type 2, as well as the other type of lifestyle disease, including hypertension is greatly influenced by the way of eating and physical activity in which we have to make a mission statement on an important issue regarding our health lifestyles. Indeed, multiple studies found that diet and lifestyle were by far the biggest determinants of whether or not you would get one of these chronic killer diseases. A balance, nutritious diet combined with daily physical activity can produce real and potential benefits in preventing as well as managing associated complications of chronic diseases such as cardiovascular risks or diabetes.

(ii) Daily levels of physical activity: Physical inactivity causes NCDs to be increasingly prevalent. People who are less active tend to have a higher risk of all-cause mortality compared with those who engage in moderate-intensity physical activity. Moreover, being active lowers the risk of depression, high blood pressure, and stroke (Lee et al. 2012). In 2016, however, more than a quarter of all adults worldwide due to be singled out were not physically active enough: as per the WHO's suggestion that people do at least 150 min of moderate-intensity physical activity every week (Guthold et al. 2018). According to Guthold et al. (2018), despite global efforts to encourage the promotion of physical activity, reducing inactivity, there has not been any change in the prevalence of physical inactivity on a worldwide level for past 15 years.

(iii) Level of sun exposure: Sun exposure also plays a big role in how likely an individual is to develop specific lifestyle diseases. Excessive and prolonged sunlight exposure can increase the likelihood of skin cancer. Evidence supports solar radiation exposure as an important occupational risk, outdoor workers frequently underappreciate the risks and do not protect themselves accordingly. Solar radiation exposure can lead to severe health issues, including skin and eye cancers, cataracts, and macular degeneration, often overlooked by employers.

(iv) Smoking and alcohol abuse: Harmful use of alcohol is a leading cause of premature mortality and disability worldwide. It has been shown to lead to not only heart attacks, liver cancer, and various mental and behavioral disorders but also all other noncommunicable illnesses and communicable.

(v) Salt and sodium intake: Consuming a diet rich in salt raises the possibility of heart disease and stroke as well as blood pressure. The suggested daily intake of sodium is a maximum of 2 g of sodium or 5 g of salt in order to lower the risk. According to studies from 2010, people worldwide consume 9–12 g of salt on average every day, which is twice the daily recommended amount (Powles et al. 2013).

(vi) Tobacco use: Smoking and smokeless tobacco are two of the leading global risk factors for major NCD-related diseases and deaths. The toxic health effects of tobacco use itself, combined with adverse physiological effects of exposure to secondhand smoke for nonsmokers. Globally, in 2016, 34% of men and 6% of women aged 15 years or over used tobacco. High-income nations had the highest rates of female tobacco smoking.

Uncontrollable risk factors Some risk factors that increase our likelihood of getting a lifestyle disease we cannot control. These are non-modifiable risk factors. Uncontrollable risk factors include

(i) Age: As one grows older, the risk of contracting lifestyle diseases may usually increase. This is probably due to the gradual deterioration of physiological activities and the harmful influence of long-term risk factors. As people age, their bodies begin to break down, the process hastening the onset of chronic diseases.

(ii) Gender: Gender impacts lifestyle diseases at a wider scale. It increased the risk of obesity and diabetes, especially in women, who are more often threatened with cardiovascular events. This may be due to hormonal, metabolic as well as socio-cultural interplay contributing to gender disparity.

(iii) Ethnicity: The emergence of lifestyle diseases is also increasingly understood to be influenced by ethnicity. Ancestry, dietary heritage, and cultural practices of certain ethnic groups might relate to specific conditions.

(iv) Heredity: Lifestyle diseases are also affected by one's heredity, in simple words a person's genetic composition. Epidemiological studies agree that people with other close relatives who have the same disease, have a history of heart disease or diabetes, or point to cancers at an early age are at greater risk of developing such diseases.

Metabolic risk factors These mediators are culprits in agitating key disruptive changes in the metabolic systems that leads to an enhanced vulnerability to noncommunicable diseases (NCDs):

(i) Increased blood pressure: There is a lot of evidence supporting the link between having high blood pressure and being more likely to get lifestyle diseases. High blood pressure is among the major risk factors for heart disease and stroke.
(ii) Increased blood glucose level: High blood sugar levels over time can be harmful to the body regardless of whether they come from type 1 or type 2 diabetes. There is a higher risk of complications such as heart attack, stroke, kidney failure, lower limb amputation, blindness, and nerve damage in the heart, blood vessels, eyes, kidneys, and nerves.

There are many types of lifestyle diseases, but the following diseases are categorized as major lifestyle diseases:

1. *Cardiovascular diseases (CVDs)*: Cardiovascular diseases arise from gradual defects in blood vessels and the heart throughout the body. The lifestyle and diet of an individual really diminish or lessen the likelihood or threat of getting a cardiovascular disorder (high blood pressure, atherosclerosis, heart attack or stroke). Doctors determine heart diseases by using different kinds of tests, such as electrocardiography, ultrasonography, and angiographies. Significant progress has been made by researchers to understand genetic and molecular mechanism behind cardiovascular diseases development and progression (Laing et al. 2019). The data gathered reveal that genes in interaction with proteins can led to cardiovascular disease. Deregulation of energy production and signaling in the mitochondria disrupts metabolic function increase oxidative stress and inflammation, which are features of a multitude of chronic diseases, including cardiovascular pathophysiology (Khullar et al. 2017). In general, gene expression in cardiovascular diseases is influenced by epigenetic mechanisms (like DNA methylation and histone modifications) altering the activity of transcription factors, leading to dysregulation of both coding as well non-coding genes, which further reflects over phenotypic conditions like diabetic cardiomyopathy and heart failure (Khullar et al. 2017). Cardiovascular disease can be treated by many methods, including lifestyle changes, such as stopping smoking, controlling high blood pressure, and medications. Surgery removal of blockages in the coronary arteries or coronary bypass surgery is also used. Proper nutritious diet, healthy diet, no smoking, regular exercise, and good sleep all these have essential checks for CVDs.

2. *Diabetes*: Diabetes is a metabolic disease that affects the way your body uses food to produce energy and helps growth of your system. When the cells in body cannot use glucose, this increases sugar levels and condition known as diabetes. Abnormal insulin signaling pathways have impairment of sugar metabolism resulting in change in glucose homeostasis that is characteristic of diabetes most commonly hyperglycemia (Banday et al. 2020). Genetics plays a crucial role in both type 1 and type 2 diabetes, wherein heritable gene variants result in impairment of the pancreatic beta cell function or insulin sensitivity in target tissues (Patil et al. 2017). Epigenetic modifications, including DNA methylation, histone changes, and non-coding RNA regulation are now also recognized to play an important role in the pathogenesis of diabetes mellitus and its complications. The comprehension of the molecular events in association with diabetes and its sequelae may enhance diagnosis as well as prevent or therapeutic rate (Khullar et al. 2017). Metabolic conditions where the pancreas is unable to produce enough insulin or when it simply fails to respond appropriately, this condition will cause high blood sugar (Gilbert and Liu 2012). Glucagon secreted by pancreatic alpha cells increases blood glucose and dysfunction in it leads to diabetes (Onyango 2020). GLUT4 is a glucose transporter responsible for insulin-mediated glucose uptake in peripheral tissues including muscle and fat tissue (Govers 2014). Dysfunction in GLUT4 leads to diabetes. Insulin Receptor Signaling (IRS) proteins are critical mediators of insulin signaling and have major roles in glucose and lipid metabolism. Compromised IRS function can result in insulin resistance and the onset of type 2 diabetes (Boucher et al. 2014). PI3K is an important glyco-lipid metabolism modulator and innate insulin signaling enzyme (Huang et al. 2018). PI3K dysfunction results in diabetes.
3. *Chronic respiratory diseases*: Chronic respiratory diseases (CRDs) including asthma, chronic obstructive pulmonary disease (COPD), and cystic fibrosis are multifactorial disorders with complicated genes–environmental interactions. These diseases are accompanied by chronic inflammation, airway remodeling, and abnormal respiratory. Age and genetics were two risk factors that remained unmodifiable, thereby increasing the risk of CRDs. Cystic fibrosis is an inherited disorder resulting from mutations in the transmembrane conductance regulator gene, leading to the buildup of thick, viscous mucus in the lungs, and other organs, causing recurrent infections, inflammation, and pulmonary damage. The lung mucus in cystic fibrosis patients becomes more viscous by dehydration of solid phase. The viscous mucus provides a suitable habitat for microbial colonization that gives rise to chronic infection, primarily with *Pseudomonas aeruginosa* strains displaying phenotypic flexibility and ability to form biofilms, which are difficult to eliminate due to their adaptability (Beaume et al. 2015).
   COPD is a multifactorial disease characterized by persistent airflow limitation, inflammation, and structural changes in the airways and lungs, primarily due to genetic predisposition and cigarette smoke exposure (Brandsma et al. 2020). It has been demonstrated that the inflammatory component of COPD is characterized by an increased number of circulating alveolar macrophages, neutrophils, and T lymphocytes as for instance.

Asthma is a chronic airway inflammation disorder, and its clinical characteristics are hyper-responsiveness, reversible obstruction, and respiratory symptoms. The cause of autism is a mystery, and it appears to involve multiple genetic as well environmental factors.

Cystic fibrosis is caused by mutations of the conductance channel regulatory gene, called CFTR. Cytokines, chemokines, and adhesion molecules are proteins that contribute to chronic inflammation in respiratory diseases (Almughem et al. 2020). Chronic respiratory diseases are influenced by proteins involved in airway smooth muscle hypertrophy, goblet cell hyperplasia, and basement membrane thickening, which contribute to structural changes. The reduction in the production of antioxidants, such as nitric oxide, can potentially contribute to the development of respiratory diseases (Bezerra et al. 2023). Some of the significant comparisons are concerned with bacterial virulence factors, these allow opportunistic pathogens such as *Pseudomonas aeruginosa* to rapidly install and become a more passive responsive but orchestrated participant in airway niche during chronic respiratory disease. Matrix metalloproteinases (MMPs) are enzymes that degrade the extracellular matrix in chronic respiratory diseases leading to tissue destruction and aberrant remodeling. Nonetheless, it is also possible that the exaggerated production and altered glycosylation of mucin proteins lead to insufficient clearance promoting further respiratory illnesses (Bush 2019).

4. *Cancer*: Cancer results from uncontrolled cellular proliferation in human tissues. Cancer is the second leading cause of death, only after CVD. It is lethal because it can metastasize from one organ to another and strike different tissues. It is marked by the uncontrolled growth of abnormal cells, which have the characteristic of keeping on growing in the patient's body and showing frequent divisions, resulting in the formation of cancer. Cancer becomes an intricate disease on both genetic and molecular levels within cells, with different changes to cellular pathways. Cancer is caused by genetic mutations in critical genes, termed tumor suppressor genes and oncogenes (Chakravarthi et al. 2016). When a tumor suppressor gene undergoes mutation, it will be inactivated, leading to uncontrolled cell division and eventually giving birth to cancer (Shen et al. 2018). Most of the genes that are dysregulated in cancer have been thought to be due to genetic mutations and epigenetic changes, such as abnormal DNA methylation or alteration in histone modifications (Mehdi and Riazalhosseini 2017). From the cancer biology perspective, NF-κB is a master transcription factor that controls hundreds of genes involved in cell proliferation, survival, and inflammation. It acts as an activator–deactivator switch on pathway activation or inhibition. It can be deregulated in cancer, promoting tumor growth and progression (Markopoulos et al. 2018). Some oncoproteins such as RAS, MYC, and HER2 are pro-survival factors for cancer cells (Markopoulos et al. 2018). Tumor suppressor proteins such as p53, BRCA1, and PTEN inhibit the cell proliferation normal cases (Chen et al. 2020). The gene cascades that are regulated both in the upstream and downstream mostly represent histone-modifying enzymes, which might play an important role on cancer progression. Furthermore, signaling pathways that coordinate gene expression of immune response and inflammation-related

genes promote cancer development as well (Feller and Lemmer 2012). Additionally, numerous other proteins involved in cell cycle, apoptosis, angiogenesis, or metastasis might be essential for carcinogenic processes as their activation or inhibition. Individuals possessing specific genetic abnormalities may also be at risk of developing cancer. Exposure to radioactivity, viruses, diseases, UV light, and tobacco can damage the genes responsible for growth and development. Early detection and treatment of cancer may help increase a person's chances of surviving. There are multiple types of cancer particularly oral cancer comes under the lifestyle diseases. The oral cancer is viewed as a great public health concern with substantial morbidity and mortality in the entire population.

Oral Cancer

Oral Cancer (mouth cancer) develops in the oral cavity, which comprises the lips, tongue, cheeks, gums, floor or roof of your mouth, and hard palate. Oral cancer was previously known as squamous cell carcinoma because most cancers (approximately 90%) are histologically derived from the squamous cells and almost always occur in older patients, with a median age of diagnosis at about 64 years old or an incident rate that decreases sharply after 85. According to the WHO, oral squamous cell carcinoma (OSCC) is the most common malignant neoplasm of the lip, mouth, and oropharynx, representing approximately 90% of all diagnosed cases of head and neck cancer.

Types of Oral Cancer

Oral cancer is a common health problem, and the occurrence of oral cancer varies worldwide, with certain areas having a higher prevalence. Oral cancer manifests in diverse forms, such as tumors, ulcers, verrucous growths, and mixed forms. Health professionals may face difficulties in diagnosing early-stage lesions, such as in situ or microinvasive carcinomas. According to Feller and Lemmer (2012), there are many types of oral cancer, which are following:

1. Squamous cell carcinoma
2. Verrucous carcinoma
3. Adenocarcinoma
4. Mucoepidermoid carcinoma

Risk Factors of Oral Cancers

Due to the widespread prevalence, smoking tobacco and smokeless tobacco products, especially chewing tobacco and betel quid, are major risk factors for oral cancer, which has high incidence rates in South Asia. Long-term alcohol use raises the risk of oral cancer, and alcohol and tobacco use compound each other to increase this risk. Virus infection, particularly HPV-16 strain, has been connected to oral cancers, those affecting the oropharynx. These cancers typically affect younger people and differ from cancers linked to alcohol and tobacco use in terms of epidemiology and clinical features. An increased risk of oral cancer is linked to a diet low in fruits and vegetables, which are high in antioxidants and other protective compounds. On the other hand, diets heavy in processed foods and low in vital nutrients may be linked to the onset of cancer. Chronic sun exposure is strongly linked to lip cancer, specifically squamous cell carcinoma of the lip, especially in fair-skinned people who live in sunny climates.

Signaling Pathways Involved in Oral Cancer

One of the pathogenic factors in oral cancer is the tumor suppressor protein TP53, which can inhibit cell division and growth (Wahab et al. 2021). If the tumor suppressor gene TP53 is lost, cells would stop dying when they should and divide uncontrollably, a hallmark of cancer. There are several protein–protein interactions found in oral cancer pathogenesis. In non-cancerous cells, transforming growth factor-β (TGF-β) is a cytokine with multifunctional antagonistic roles as it acts both as an anti-tumorigenic cytokine inhibiting cell proliferation and activating of apoptosis. For instance, TGF-β can attack on tumor progression in cancerous cells inducing epithelial–mesenchymal transition, angiogenesis, and metastasis (Gu and Feng 2018). Epidermal growth factor receptor (EGFR) is a receptor tyrosine kinase, which is overexpressed or mutated in >90% of oral carcinoma (Chiang et al. 2020). It is also capable of promoting the cell growth, survival, and metastasis of oral cancer by activating EGFR-mediated signaling pathway. Nuclear factor-κB (NF-κB) is a transcription-factor family necessary for the activation of genes associated with inflammation, cell survival, and proliferation. This suggests that this pathway could be exploited as a therapeutic target in oral squamous cell carcinoma pathogenesis (Puar et al. 2018). The MAPK pathway is an important cellular machinery and has been shown to impede both at genetic as well as protein level making it progress in oral cancer (Braicu et al. 2019). PI3K/Akt signaling pathway is one of the most commonly deregulated pathways in oral cancer, possibly rendering these cancers to chemo-resistance and tumor spread. Wnt/β-catenin signaling pathway is an important protective effect on eggs, embryonic development, regulation of tissue function, and dysregulation of this pathway has been implicated in oral cancer (Shiah et al. 2016). The Notch signaling pathway regulates cell fate determination, proliferation, and differentiation. These provide insights into the involvement of this pathway that is mis regulated during or after oral carcinogenesis and might offer a potential therapeutic approach (Lee et al. 2016).

Prevention of Lifestyle Diseases

Lifestyle diseases are the primary cause of increased morbidity and mortality rates and the largest barrier to global development. Preventing these diseases involves controlling or reducing risk factors associated with them. Many diseases, including NCDs, can be prevented by changing lifestyle or behavioral factors. Implementing inexpensive strategies within communities can minimize major upstream modifiable risk factors associated with lifestyle diseases. Public health strategies, such as health promotion and prevention, are cost-effective when compared to treating noncommunicable disorders. Stronger health systems, multi-stakeholder participation, and leadership from governments and healthcare providers are needed for effective lifestyle disease intervention.

To reduce the risk of CVD, it is recommended to reduce fat and salt intake, maintain a healthy weight, avoid smoking, exercise regularly, and check cholesterol and blood pressure levels annually. Overweight individuals have higher CVD rates, and those with a family history should stop smoking. Regular exercise strengthens the cardiovascular system and reduces stress. Lifestyle changes like healthy diet, exercise, and weight loss are also essential. Stress and negative emotions can lead to high blood pressure, making conditions prime for CVD.

There are many modifiable risk factors that can help prevent cancer. The first is not smoking, as it causes one-third of cancer deaths in America. The second is reducing fat intake and eating more vegetables, as high saturated fats are linked to colon and rectum cancers. The third is controlling weight and engaging in physical activity, as obesity is a contributing factor. The fourth is making doctor appointments to discuss cancer treatments, preventions, and risk factors. Individuals can choose to live well away from these risks.

Diabetes is a genetic disease, affecting specific ethnic groups. Lifestyle changes like diet, exercise, and weight management can help manage the condition. Type 1 diabetes cannot be prevented, while type-2 diabetes can be minimized by maintaining a healthy weight, using balanced food, dieting, avoiding tobacco, and reducing stress levels.

Chronic respiratory diseases can be effectively prevented and managed through various strategies. Quitting smoking, reducing pollution exposure, and protecting workers from hazardous substances and dust are key. Screening for diseases, early intervention, and effective management can improve patient outcomes. Promoting physical activity, diet, and lifestyle changes can also help prevent and control these diseases.

Management of Lifestyle Diseases Using Natural Products

Natural Products

Natural products derived from various resources, such as plants, microorganisms, and marine organisms, have been favorably received by the recent advancements in discovering anticancer drugs (Nerkar et al. 2021). They have been shown to share some of the physiological properties of drugs and retain molecular attributes found in biological systems. Hence, these may be particularly helpful in the framework of subsequent development and optimization by methods that orient towards synthetic organic chemistry. More than 60% of the currently used cancer drugs are either natural products or based on a natural product lead compound.

A few of the important classes of natural products are as follows:

1. Natural products of plant origin
2. Bacterial natural products
3. Natural products from marine

Natural compounds of plant origin Biological molecules found in plants have been extensively studied for their anticancer potential. It was proved that these plants are used in traditional medicine and have possible application for the treatment of different diseases (Asma et al. 2022). The antioxidant activity of many plants derived natural products makes them critical for the prevention and treatment of cancer. These properties act as a free radical production inhibitor and reduce the spread of oxidative reactions that play an important role in cancer progression (Elansary 2014). This is the effect of antioxidants as well as anticancer properties like reduction in tumor size, apoptosis, and metastasis inhibition by plant-derived natural products (Asma et al. 2022). There are many types of plant-derived natural products with anticancer properties such as alkaloids, terpenoids, flavonoids, and phenolic compounds.

Plant-derived natural products have potential as anti-cancer drugs, but challenges like poor bioavailability, selectivity, and structural modification are needed. Currently, many plant-derived products and their synthetic analogs are in clinical trials or approved for cancer therapy. Alkaloids and terpenoids, like paclitaxel and docetaxel, are already in clinical use, while polyphenols are under investigation for potential anticancer molecules (Seca and Pinto 2018). Flavonoids and phenolics from medicinal plants, rich in chemopreventive and anti-tumor activities, are promising candidates for development as anticancer drugs due to their existing activity needing systematic development (Perveen and Al-Taweel 2017).

Microbial derived natural products Microbial natural products have been regarded as a precious source of structurally diverse and biologically active molecules. This class of compounds contributed remarkably to the development of many therapeutic agents for a wide array of diseases. Derived from bacteria, fungi, Actinomycetes and their natural secondary metabolites with wide structural diversity and varied biological activity. Many of these compounds have in vitro or in vivo efficacy as antimicrobial, anti-cancer, and/or anti-inflammatory drugs. Different types of natural products are obtained from microbes like phenolics, terpenes, bacteriocins, peptides, enzymes, natural polymers, fatty acids, and organic acids.

Microbial natural products have a broad spectrum of biological activities, so these compounds are crucial molecules for the treatment and management of many diseases. The anticancer activity of microbial natural products has been pioneered by the discovery and development of bleomycin, anthracyclines as clinically approved chemotherapeutics. Besides, their anti-inflammatory and immunomodulatory attributes have shown them to be beneficial in treating autoimmune as well as chronic inflammatory diseases.

Natural products from marine Bio-marine-based natural molecules are an attractive source for pharmaceutical, cosmetic, and nutraceutical applications due to the huge biodiversity present in marine environments. Marine organisms, including macroalgae (seaweeds), invertebrates, and microbes, operate with a vast array of metabolic pathways that allow them to produce secondary metabolites used as chemical defense or for communication, signaling ecological roles. For a variety of reasons, marine habitats have been increasingly recognized as an untapped resource for novel bioactive molecules. These compounds have well-identified favorable biological activities such as antimicrobial, anti-inflammatory, antioxidative, and anticancer properties (Barzkar et al. 2019). Based on structural and chemical properties, there are a lot of types for marine natural products such as terpenes, steroids, alkaloids, phenolic compounds, peptides, saccharides, lipids, proteins, enzymes, and halogenated compounds.

Therapeutic Benefits of Natural Products Against Oral Cancer

Plant-derived natural products, e.g., taxanes, vinca alkaloids (VAs), camptothecins, and anthracyclines have widespread applications in cancer therapy. Through harnessing the power of bacteria and fungi, these produce potent anticancer agents, such as dactinomycin or bleomycin, which are microbially sourced natural compounds. Some marine-based natural compounds, especially those from sponges, tunicates, soft corals, and algae, showed anticancer activities. Recent studies have shown that biologically compatible natural products derived from plants, marine organisms, and endophytic fungi are believed to be active in the treatment and prevention of oral cancer with less toxicity and desired biological functions. The new research has identified that some plant compounds, like polyphenols, terpenoids, and alkaloids, have significant role in fighting against oral cancer. These agents are proven to inhibit cellular growth, induce apoptosis (programmed cell death), and prevent neovascularization. Powerful research methods have been created to enable detailed study of the molecular pathways producing these effects, while validating some of them for potential clinical application.

Therapeutic benefits of alkaloids Alkaloids are a large class of naturally occurring organic compounds known primarily for their presence in plants and almost always having nitrogen with some basic properties. Secondary metabolites are found in different plants and exhibit a variety of pharmacological effects, including anticancer activity (Tilaoui et al. 2021). Studies have shown that alkaloids have antimutagenic and anti-cancer effects in oral cancer. These cytotoxic effects lead to apoptosis, inhibit cell growth, and activate signal transduction pathways. However, the current research on alkaloids' therapeutic potential is in its early stages, and more research is needed to fully understand their mechanisms and clinical utility. Natural plant-derived alkaloids have potential applications in oral cancer therapy, as they induce apoptosis and improve the potency of conventional cancer therapies when combined with them. These investigations have opened new horizons for the medicinal potential of alkaloids, and it is hoped that future drugs and treatment strategies based on alkaloid molecules may prove effective in controlling oral carcinogenesis.

Therapeutic effects of steroids Conventional cancer therapies like chemotherapy cause collateral damage to normal cells, leading to severe side effects like mucositis and stomatitis. mTOR inhibitors, developed to manage these issues, have been linked to oral toxicity. However, evidence suggests that stomatitis complications can be effectively treated with corticosteroids. However, treatment with corticosteroids (i.e., either topically, injected into the lesion, or taken orally, depending on various other factors) causes almost all patients to experience relief from their pain. This case series highlights the possible use of corticosteroids to ameliorate or prevent toxicity that might otherwise limit the dose of new cancer treatments and thereby improve efficacy.

Therapeutic benefits of phenolics Conventional modalities such as chemotherapy and radiation are generally characterized by anemial efficacy along with long term, serious side effects (Perveen and Al-Taweel 2017). In this regard, natural phenolic compounds have emerged as an interesting approach that can be used to take advantage of therapeutic effects on the one hand and minimize side effects (Kim et al. 2023). Phenolics account for a large part of the antioxidants in herbs, and plant materials or fruit waste-derived phenolic extracts have higher potential anticancer effects (Zhang et al. 2022). Phenolic compounds regulate cell pathways, affecting cancer progression. They activate or inactivate death receptors, mitochondrial pathways, inhibit cell proliferation, and suppress transcriptional factors in oral cavity cancer management (Kim et al. 2023). The ability to be absorbed by the body and even reach specific sites using nanotechnology is considered a great advancement in this field for phenolic compounds (Xie et al. 2016). A plethora of natural phenolic compounds has been identified for potential therapeutic intervention against oral cancer, demonstrating numerous anticancer mechanisms. Studies in this field may encourage the development of safer and more effective treatment options for people suffering from oral cancer. Natural phenolic compounds are promising alternatives to other traditional methods.

Therapeutic benefits of flavonoids Flavonoids are plant-derived polyphenolic compounds with well-documented antioxidant, anti-inflammatory, and anticancer attributes supported by numerous studies (Abotaleb et al. 2019). There have been several epidemiological studies that have shown a strong association between intake of flavonoids and decreased incidence of various types of cancer, such as oral cancer (Zhao et al. 2019). Zamani et al. demonstrated that flavonoids show specific toxicity to oral cancer cells and have little effect on normal oral cells (Zamani et al. 2016).

Various mechanisms account for the anti-cancer activities of flavonoids, and these include apoptosis (cell death), cell cycle arrest, signaling pathways that regulate proliferation, and angiogenesis. The results of such studies underscore the relevance of this kind of interaction by demonstrating that acylation facilitates alterations to existing biological effects imparted on flavonoids (Fernando et al. 2016). However, the clinical applications of flavonoids are severely restricted due to their low-water solubility and stability. Flavonoids hold promise as naturally occurring agents for the chemoprevention or chemotherapy of oral cancer. Nevertheless, further research is needed to optimize their therapeutic efficacy and develop successful strategies for implementation in clinical practice.

Natural Inhibitor for Signaling Pathways Involved in Oral Cancer Progression

Natural Inhibitor of PI3K/Akt/mTOR Pathway

The PI3K/AKT/mTOR signaling pathway is an extremely important cellular signal transduction, which plays a substantial role in the progression and development of oral cancer (Porta et al. 2014). This pathway is responsible for integrating internal and external cues to control fundamental cellular processes like cell metabolism, growth, proliferation, and survival. Self-regulation of the PI3K/AKT/mTOR signaling pathway is a frequent trait in many cancers, among others oral cancer (Rascio et al. 2021). This pathway can be activated abnormally by various genetic and epigenetics alterations, including somatic mutations in oncogenes or tumor suppressor genes. Activation of this pathway causes significant dysregulation in the control of cellular proliferation and viability, increased growth, capacity to metastasize elsewhere within or distant from the body. The PI3K/AKT/mTOR pathway is a whole system, in which Phosphatidylinositol 3-kinase (PI3K) activates the conversion of phosphoinositide to second messenger, phosphatidylinositol-3,4,5-trisphosphate (PIP), which then leads to activation of serine/threonine kinase AKT. Among the targets of AKT activation is mTOR, which phosphorylates several downstream effectors important for cell growth and metabolism (Hassan et al. 2013). There is also an extensively researched PI3K/AKT/mTOR pathway dysregulation in oral cancer (Rascio et al. 2021). The PI3K/AKT/mTOR signaling pathway is emerging as an attractive target for novel anticancer drug design and therapeutic management in oral cancer.

Interestingly, many natural compounds and phytochemicals have been identified to inhibit mTOR pathway thus possessing therapeutic potential against cancer. These compounds are listed in Table 1.

Table 1 List of natural inhibitors for PI3K/AKT/mTOR pathway

Natural compounds	Sources
Curcumin	*Curcuma longa*
Resveratrol	*Vitis vinifera*
Sulforaphane	Cruciferous vegetables
Platycodin-D	*Platycodon grandiflorum*
Trilinolein	*Panax notoginseng*
Formononetin	*Astragalus membranaceus*
Ginsenoside 20	*Panax ginseng*
Icaritin	*Epimedium brevicornum*
Asparanin A	*Asparagus officinalis L.*
Licoricidin	*Glycyrrhiza uralensis*
Morusin	*Morus australis*
Mollugin	*Rubia cordifolia L*
Gambogenic acid	*Garcinia hanburyi*
Geniposide	*Gardenia jasminoides*
Linalool	*Lavandula officinalis*
Oroxin B	*Oroxylum indicum*
Chelerythrine	*Toddalia asiatica*
Prodiginosin	*Serratia marcescens*
Diosgenin	*Dioscorea nipponica*
Eriocalyxin B	*Isodom eriocalyx*

Key Points of PI3K/AKT/mTOR Pathway

PI3K signaling pathway

- Crucial in managing cellular processes e.g., growth, proliferation, survival via activation of mTOR/Akt. Its disruption will be causing to many types of cancer including oral cancer.

AKT signaling pathway

- Altered in many types of cancer, including oral cancer.
- Targeting the AKT pathway has been found to be a potential therapeutic strategy in oral cancer management.

mTOR signaling pathway

- Regulatory in cell proliferation, growth, and metabolism
- Inhibition of mTOR signaling leads to further dysregulation, affecting drug therapies targeting mTOR.
- PTEN, a tumor suppressor of the PI3K/Akt/mTOR pathway commonly mutated in many cancer types.

Natural Inhibitor of Wnt/β-Catenin Signaling Pathway

Wnt/β-Catenin signaling pathway serves as an important mediator to control distinct cellular activities such as cell proliferation, differentiation, and migration. Dysregulation of this pathway may lead to oral cancer genesis and progression (Yang et al. 2016). In the absence of Wnt ligands, β-Catenin is constantly targeted for ubiquitin-dependent proteolysis. Upon Wnt binding, the destruction complex is destabilized allowing β-Catenin to accumulate in cytoplasm leading to its nuclear translocation. In the nucleus, β-Catenin interacts with TCF/LEF transcription factors and activates genes leading to proliferation, survival, and migration (Zhang et al. 2020). The Wnt/β-Catenin pathway has shown extensive activation in various oral cancer types such as OSCC, tongue, and gingival cancers (Serafino et al. 2017). Over-expression of Wnt ligands, leading to competitively inhibit the action of antagonists, has been reported in oral cancer (Serafino et al. 2017). The Wnt/β-Catenin pathway has been intensely studied as a potential cancer therapeutic by using oral cancers. A wide range of small molecules, as well as natural compounds, have been investigated in preclinical and clinical studies for inhibiting the activity of this pathway. Because of the potential important role for Wnt/β-Catenin signaling in cancer, numerous natural inhibitors have been synthesized or discovered to control this pathway. It is for this reason that these inhibitors are important to molecular therapy, given the good performance may be used as therapeutic agents, offering a more specific and safer alternative of treatment than traditional chemotherapies (Serafino et al. 2017).

Key Points of Wnt/β-Catenin Signaling Pathway

Wnt

- Wnt1, Wnt3a, and Wnt10b are specific ligands that bind the Frizzled receptors, as well as LRP5/6 molecules.

APCC, Axin, and GSK3β

- These are examples of destruction complex, one such complex is the β-Catenin destruction complex that phosphorylates β-Catenin resulting in its degradation.

β-Catenin

- It is a crucial component of the pathway and moves to the nucleus. In the nucleus, it forms a complex with TCF/LEF transcription factors. This complex activates target genes that are involved in cell proliferation and survival.

Many natural compounds act as inhibitor for Wnt/β-catenin signaling pathway, which are listed in Table 2.

Table 2 List of natural inhibitor of Wnt/β-catenin signaling pathway

Natural compounds	Sources
Piperine	*Piper nigrum*
Costunolide	*Saussurea lappa*
Trichosanthin	*Trichosanthes kirilowii*
Geniposide	*Gardenia jasminoides*
Tetrandrine	*Stephania tetrandra*
Ukrain	*Chelidonium majus*
Baohuoside-I	*Periploca sepium*
Eriocalyxin B	*Isodom eriocalyx*

Natural Inhibitor of MAPK/ERK Signaling Pathway

The MAPK/ERK signaling pathway is indispensable for the development and progression of oral cancer. The evolutionarily conserved signaling cascade is triggered by various extracellular agents including growth factors, cytokines, and other ligands. It regulates a spectrum of cellular functions, including proliferation, differentiation, and survival. The dysregulation of the MAPK/ERK pathway is commonly seen in oral cancer. This dysregulation leads to unbridled cellular proliferation, avoidance of apoptosis and the ability for metastasis. By contrast, the pathway is almost always constitutively activated downstream by overexpression or mutation of upstream receptors such as the epidermal growth factor receptor. It is a pattern commonly seen in head and neck squamous cell carcinomas (Gurbi et al. 2023). ERK1/2 kinases phosphorylate a variety of substrates in order to coordinate the cell response to mitogenic signals by mechanisms that have been described elsewhere. Over-activation of the MAPK/ERK pathway has also been related to abnormal growth, migration, and metastasis in oral cancer (Gurbi et al. 2023). Furthermore, the pathway has also been connected with enabling resistance to chemotherapy and immune escape in several types of cancer including oral (Salaroglio et al. 2019). Therefore, intervention of MAPK/ERK signaling pathway can provide a promising therapeutic strategy for oral cancer. Selective inhibitors targeting pivotal kinases of the pathway, BRAF, MEK, and ERK are being developed, with a number already evaluated in preclinical studies showing promising results.

Key Points of MAPK/ERK Signaling Pathway

MAPK

- Mitogen-activated protein kinases (MAPK) are serine/threonine-specific proteins that respond to extracellular stimuli and control various cellular activities, such as gene expression and mitosis.

ERK

- Members of the MAPK family, ERK1 and ERK2 are essential for tumor cell growth drivers that drive overexpression in tumors.

(continued)

Table 3 List of natural inhibitor of MAPK/ERK signaling pathway

Natural compounds	Sources
Emetine	*Cephaelis ipecacuanha*
Artemisinin	*Artemisia annua*
Actein	*Cimicifuga foetida*
Baicalin	*Scutellaria baicalensis*
Baicalein	*Scutellaria baicalensis*
Scopoletin	*Nicotiana glacua*
Chelerythrine	*Toddalia asiatica*
Dioscin	*Dioscorea nipponica*
Geranin	*Geranium thumbergii*
Dihydroartemisinin	*Artemisia annua*
Jaceosidin	*Artemisia princeps*
Honokiol	*Magnolia officinalis*

Activation of MAPK/ERK pathway

- Activation often occurs in response to growth stimuli, but requires upstream signaling events (receptor dimerization and autophosphorylation). This results in activation of small GTPases (e.g., Ras), MAPK kinase kinases (e.g., Raf), and subsequently, MAPK kinases themselves.

Downstream effect

- On activation ERK1/2 kinases phosphorylates variety of substrates like transcription factors, signaling effector, and cytoskeletal proteins. This coordinates with cellular response to mitogenic stimuli and tumor progression.

Natural inhibitors for MAPK/ERK signaling pathway are shown in Table 3.

Natural Inhibitor of Notch Signaling Pathway

The implication of this pathway has been extensively studied in various forms of cancer, including oral cancer and is significantly associated with the progression (Li et al. 2021). Notch signaling is activated by the binding of a Notch receptor with one of the five canonical ligands. The binding of Notch to Delta initiates an extracellular cleavage event that releases the intracellular domain of Notch transcriptional regulator. This domain then translocates to the nucleus and is a transcriptional co-regulator that dictates expression of distinct target genes (Gao et al. 2017). Research on the pathway of Notch-1 signaling in oral cancer has also been reported (Lee et al. 2016). In addition, Notch-1 is activated to enhance the occurrence and facilitation of epithelial–mesenchymal transition (EMT), a crucial tumorigenic/

Table 4 List of natural inhibitors of Notch signaling pathway

Natural compounds	Sources
Oridonin	*Rabdosia rubescens*
Curcumin	*Curcuma longa*
Resveratrol	*Vitis vinifera*
Withaferin A	*Withania somnifera*
Gambogenic acid	*Garcinia hanburyi*

metastatic process by regulating expression levels of specific genes involved in EMT (Li et al. 2021). Moreover, Notch-1 signaling pathway also played a role in maintaining and restoring oral cancer stem cells to strengthen the malignant inherent characteristics of this disease (Chen et al. 2019). The Involvement of Notch Signaling in Oral Cancer is Double-Edged Sword: Have Both Tumor-Promoting (Cancer) and Anti-Tumor Effects (Suppression).

Key Points of Notch Signaling Pathway

Notch receptor

- The mammalian cells contain four receptors: Notch 1–4 Binding of these specific ligands to their respective receptors present on the cell membrane makes them active.

Delta ligand

- Notch signaling is turned on when the transmembrane receptor binds to its (also transmembrane) Delta ligand.

Proteolytic cleavage

- Upon ligand binding, the Notch receptor undergoes proteolytic cleavage that releases the intracellular domain NICD.

NICD

- The C-terminal fragment of Notch released from the plasma membrane translocates to the nucleus as a transcription factor, regulating target genes for cell fate determination, differentiation, proliferation, and apoptosis.

Natural inhibitors for Notch signaling pathway are listed in Table 4.

Conclusions

Chronic diseases present a growing public health challenge, which is either due to or influenced by modifiable risk factors, namely poor dietary habits, physical inactivity, and unhealthy behaviors such as smoking. Adopting healthy habit a good diet, regular physical activity, and not smoking can radically reduce the risk of these chronic diseases. A healthy diet high in fruit, vegetables, whole grains, and lean protein can be associated with reduced risk of heart disease, some cancer types, and other chronic diseases. Risk of lifestyle diseases is also likely linked to diets heavy in processed foods, red meat, and added sugars. Taking regular exercise can help prevent and control conditions such as heart disease, type 2 diabetes, and some cancers. Oxidative stress is highlighted as one of the most promising factors to explain the efficacy of natural health products in preventing diseases and promoting health status, leading them to stand out for maintaining health conditions, especially with regard to chronic disease risk. Strategy including proper diet, regular exercise, and natural health supplements for the prevention and cure of lifestyle diseases is imperative. Lifestyle diseases are of different types and may not be curable, but they can at best be managed with approaches like improvements in our lifestyle, environment, and diet. Oral cancer is a major public health problem, especially affecting the lower socio-economic group. Prevention includes changes in socioeconomic status, minimizing consumption of tobacco and alcohol, and the promotion of healthy lifestyle. Early identification is crucial because most cases of esophageal cancer have already spread beyond cure by the time they are found, requiring elaborate, costly, and usually futile treatments. The present review proposes that both natural products and physical activity can significantly contribute to the prevention as well as treatment of oral cancer. A diet full of naturally occurring substances like fruits and vegetables that are so rich in antioxidants can reduce the risk of oral cancers. Physical activity in general has been shown to be chemopreventive in cancers, including oral cancer, through reducing inflammation, improving immune function and controlling hormone levels. What is needed is a comprehensive approach to combine smoking and drinking moderation with healthier lives, better diets, and more tests for early detection of certain cancers. These initiatives could be made more effective by complementing them with natural products and exercise as part of a holistic prevention and treatment strategy. Signal transduction pathways represent new targets for treatment, and their mechanism in the carcinogenesis of oral cancer has become an important topic for recent studies. Phytochemicals and mycotoxins have been attracted many attention as new leading compounds of potential therapeutic agents because of their biological activities diversity with few side effects. These molecules modulate the signal transduction pathways in cancer, thereby leading to suppression of tumor growth, invasion, and metastasis. Taken together, signal transduction pathways must be studied, and natural products should be developed as potential therapeutic interventions to improve patient care in the era of oral cancer research.

References

Abotaleb M, Samuel SM, Varghese E et al (2019) Flavonoids in cancer and apoptosis. Cancers (Basel) 11(1):28

Almughem FA, Aldossary AM, Tawfik EA et al (2020) Cystic fibrosis: overview of the current development trends and innovative therapeutic strategies. Pharmaceutics 12(7):1–29

Asma ST, Acaroz U, Imre K et al (2022) Natural products/bioactive compounds as a source of anticancer drugs. Cancers (Basel) 14(24):6203

Banday MZ, Sameer AS, Nissar S (2020) Pathophysiology of diabetes: an overview. Avicenna J Med 10(4):174–188

Barzkar N, Jahromi ST, Poorsaheli HB (2019) Metabolites from marine microorganisms, micro, and macroalgae: immense scope for pharmacology. Mar Drugs 17(8):464
Beaume M, Kohler T, Fontana T et al (2015) Metabolic pathways of Pseudomonas aeruginosa involved in competition with respiratory bacterial pathogens. Front Microbiol 6:321
Bezerra FS, Lanzetti M, Nesi RT et al (2023) Oxidative stress and inflammation in acute and chronic lung injuries. Antioxidants 12(3):548
Boucher J, Kleinridders A, Ronald Kahn C (2014) Insulin receptor signaling in normal and insulin-resistant states. Cold Spring Harb Perspect Biol 6(1):9191
Braicu C, Buse M, Busuioc C et al (2019) A comprehensive review on MAPK: a promising therapeutic target in cancer. Cancers (Basel) 11(10):1618
Brandsma CA, Vanden Berge M, Hackett TL et al (2020) Recent advances in chronic obstructive pulmonary disease pathogenesis: from disease mechanisms to precision medicine. J Pathol 250(5):624–635
Bush A (2019) Pathophysiological mechanisms of asthma. Front Pediatr 7:68
Chakravarthi BVSK, Nepal S, Varambally S (2016) Genomic and epigenomic alterations in cancer. Am J Pathol 186(7):1724–1735
Chen Z, Zuo X, Pu L et al (2019) Hypomethylation-mediated activation of cancer/testis antigen KK-LC-1 facilitates hepatocellular carcinoma progression through activating the Notch1/Hes1 signalling. Cell Prolif 52(3):12581
Chen L, Liu S, Tao Y (2020) Regulating tumor suppressor genes: post-translational modifications. Signal Transduct Target Ther 5(1):90
Chiang KH, Shieh JM, Shen CJ et al (2020) Epidermal growth factor-induced COX-2 regulates metastasis of head and neck squamous cell carcinoma through upregulation of angiopoietin-like 4. Cancer Sci 11(6):2004–2015
Elansary HO (2014) Natural antioxidants and their role against human cancer. J Plant Biochem Physiol 2:2
Feller L, Lemmer J (2012) Oral squamous cell carcinoma: epidemiology, clinical presentation and treatment. J Cancer Ther 03(04):263–268
Fernando W, Coombs MRP, Hoskin DW et al (2016) Docosahexaenoic acid-acylated phloridzin, a novel polyphenol fatty acid ester derivative, is cytotoxic to breast cancer cells. Carcinogenesis 37(10):1004–1013
Gao J, Long B, Wang Z (2017) Role of notch signaling pathway in pancreatic cancer. Am J Cancer Res 7(2):173–186
Gilbert ER, Liu D (2012) Epigenetics the missing link to understanding β-cell dysfunction in the pathogenesis of type 2 diabetes. Epigenetics 7(8):841–852
Govers R (2014) Cellular regulation of glucose uptake by glucose transporter GLUT4. Adv Clin Chem 66:173–240
Gu S, Feng XH (2018) TGF-β signaling in cancer. Acta Biochim Biophys Sin Shanghai 50(10): 941–949
Gurbi B, Brauswetter D, Pénzes K et al (2023) MEK is a potential indirect target in subtypes of head and neck cancers. Int J Mol Sci 24(3):2782
Guthold R, Stevens GA, Riley LM et al (2018) Worldwide trends in insufficient physical activity from 2001 to 2016: a pooled analysis of 358 population-based surveys with 1.9 million participants. Lancet Glob Health 6(10):1077–1086
Hassan B, Akcakanat A, Holder AM et al (2013) Targeting the PI3-kinase/Akt/mTOR signaling pathway. Surg Oncol Clin N Am 22(4):641–664
Huang X, Liu G, Guo J et al (2018) The PI3K/AKT pathway in obesity and type 2 diabetes. Int J Biol Sci 14(11):1483–1496
Khullar M, Cheema BS, Raut SK (2017) Emerging evidence of epigenetic modifications in vascular complication of diabetes. Front Endocrinol (Lausanne) 8:237
Kim KH, Ki MR, Min KH (2023) Advanced delivery system of polyphenols for effective cancer prevention and therapy. Antioxidants 12(5):1048

Laing N, Kraus SM, Shaboodien G et al (2019) An overview of the genetic basis of cardiovascular disease. S Afr Med J 109(6):364–370

Lee IM, Shiroma EJ, Lobelo F et al (2012) Effect of physical inactivity on major non-communicable diseases worldwide: an analysis of burden of disease and life expectancy. Lancet 380(9838):219–229

Lee SH, Do SI, Lee HJ (2016) Notch1 signaling contributes to stemness in head and neck squamous cell carcinoma. Lab Investig 96(5):508–516

Li P, Chen Y, Peng Y et al (2021) Notch-1 signaling promotes reattachment of suspended cancer cells by cdc42-dependent microtentacles formation. Cancer Sci 112(12):4894–4908

Markopoulos GS, Roupakia E, Tokamani M et al (2018) Roles of NF-κB signaling in the regulation of miRNAs impacting on inflammation in cancer. Biomedicine 6(2):40

Mehdi A, Riazalhosseini Y (2017) Epigenome aberrations: emerging driving factors of the clear cell renal cell carcinoma. Int J Mol Sci 18(8):1774

Nerkar AG, Chakraborthy GS, Ukirde RD (2021) Anti-cancer agents from natural sources: a review. Curr Trends Pharm Pharm Chem 3(4):38–46

Onyango AN (2020) Mechanisms of the regulation and dysregulation of glucagon secretion. Oxidative Med Cell Longev 2020:3089139

Patil PD, Mahajan UB, Patil KR et al (2017) Past and current perspective on new therapeutic targets for type-II diabetes. Drug Des Devel Ther 11:1567–1583

Perveen S, Al-Taweel AM (2017) Phenolic compounds from the natural sources and their cytotoxicity. In: Soto-Hernandez M, Palma-Tenango M, Garcia-Mateos MR (eds) Phenolic compounds – natural sources, importance and applications. IntechOpen, London

Porta C, Paglino C, Mosca A (2014) Targeting PI3K/Akt/mTOR signaling in cancer. Front Oncol 4: 64

Powles J, Fahimi S, Micha R et al (2013) Global, regional and national sodium intakes in 1990 and 2010: a systematic analysis of 24 h urinary sodium excretion and dietary surveys worldwide. BMJ Open 3:3733

Puar YR, Shanmugam MK, Fan L (2018) Evidence for the involvement of the master transcription factor NF-κB in cancer initiation and progression. Biomedicine 6(3):82

Rascio F, Spadaccino F, Rocchetti MT et al (2021) The pathogenic role of PI3K/AKT pathway in cancer onset and drug resistance: an updated review. Cancers (Basel) 13(16):3949

Salaroglio IC, Mungo E, Gazzano E (2019) ERK is a pivotal player of chemo-immune-resistance in cancer. Int J Mol Sci 20(10):2505

Seca AML, Pinto DCGA (2018) Plant secondary metabolites as anticancer agents: successes in clinical trials and therapeutic application. Int J Mol Sci 19(1):263

Serafino A, Sferrazza G, Colini BA et al (2017) Developing drugs that target the Wnt pathway: recent approaches in cancer and neurodegenerative diseases. Expert Opin Drug Discov 12(2): 169–186

Shen L, Shi Q, Wang W (2018) Double agents: genes with both oncogenic and tumor-suppressor functions. Oncogenesis 7(3):25

Shiah SG, Shieh YS, Chang JY (2016) The role of Wnt signaling in squamous cell carcinoma. J Dent Res 95(2):129–134

Tilaoui M, Ait Mouse H, Zyad A (2021) Update and new insights on future cancer drug candidates from plant-based alkaloids. Front Pharmacol 12:719694

Wahab KF, Salamah AY, Ali A (2021) Protein-protein interaction analysis through network topology (oral cancer). J Healthc Eng 2021:6623904

Xie J, Yang Z, Zhou C (2016) Nanotechnology for the delivery of phytochemicals in cancer therapy. Biotechnol Adv 34(4):343–353

Yang K, Wang X, Zhang H et al (2016) The evolving roles of canonical WNT signaling in stem cells and tumorigenesis: implications in targeted cancer therapies. Lab Investig 96(2):116–136

Zamani SS, Hossieni M, Etebari M et al (2016) Pharmacokinetics of calycopterin and xanthmicrol, two polymethoxylated hydroxyflavones with anti-angiogenic activities from Dracocephalum kotschyi Bioss. Daru 24(1):1–10

Zhang X, Wang L, Qu Y (2020) Targeting the β-catenin signaling for cancer therapy. Pharmacol Res 160:104794
Zhang Z, Li X, Sang S et al (2022) Polyphenols as plant-based nutraceuticals: health effects, encapsulation, nano-delivery, and application. Foods 11(15):2189
Zhao J, Yang J, Xie Y (2019) Improvement strategies for the oral bioavailability of poorly water-soluble flavonoids: an overview. Int J Pharm 570:118642

Natural Products as Promising Drug Candidates for Management of Lifestyle Diseases

Hussein Sakr and Srinivasa Rao Sirasanagandla

Abstract

Lifestyle factors could contribute to the risk of developing cardiovascular diseases and liver diseases. Evidence from experimental and clinical studies has shown that exposure to various heavy metals and endocrine-disrupting chemicals increases the risk of these lifestyle diseases. Humans are exposed to these pollutants through various routes. Increased oxidative stress, inflammation, apoptosis, altered autophagy regulation, hormonal activity, and estrogen mimicking nature are thought to be the possible mechanisms of harmful effects associated with environmental pollutants such as plastic pollutants, particulate matter, and heavy metals. In Indian and Chinese traditional medicine, various natural products are mentioned in treating these lifestyle diseases. Researchers from all over the world showed tremendous efforts in exploring the scientific evidence behind the beneficial effects of various natural products. In this chapter, the beneficial effects of natural products and/or natural compounds such as curcumin, ginseng, *Ginkgo biloba*, *Ganoderma lucidum*, *Gynostemma pentaphyllum*, flax seeds, silymarin, anthocyanins, and garlic chemical constituents against cardiovascular diseases and liver diseases have been discussed. More studies dealing with the molecular mechanism and drug delivery through nanoformulation should be undertaken for drug design and better clinical use. Clinical trails should

H. Sakr
Department of Physiology, College of Medicine and Health Sciences, Sultan Qaboos University, Muscat, Al-Khoudh, Oman
e-mail: hsakr@squ.edu.om

S. R. Sirasanagandla (✉)
Department of Human and Clinical Anatomy, College of Medicine and Health Sciences, Sultan Qaboos University, Muscat, Al-Khoudh, Oman
e-mail: srinivasa@squ.edu.om

S. Raisuddin et al. (eds.), *Environmental Toxicants and Lifestyle Diseases*,
https://doi.org/10.1007/978-981-95-0575-3_25

be performed based on the preclinical evidence for therapeutic use of these natural compounds.

Keywords

Cardiovascular disease · Metabolic syndrome · Traditional medicine · Natural compounds · Environmental toxins · Lifestyle diseases

Cardiovascular Disorders and Natural Products

Cardiovascular diseases (CVDs) are well known to cause death in both developed and developing nations. Hypertension is the leading risk factor for CVDs like ischemic heart disease and heart failure. Primary hypertension is a subtype of hypertension that increases with aging, environmental pollutants, and hereditary susceptibility (Mills et al. 2020). The incidence of hypertension is rising globally as a result of population aging, greater exposure to lifestyle risk factors such as inadequate diets (high salt and low potassium consumption), and a lack of physical exercise. Over the preceding two decades, the prevalence of hypertension decreased somewhat in high-income countries (HICs) but increased significantly in low- and middle-income countries (LMICs) (Mills et al. 2016). Food-based dietary nutrients function as antioxidants in scavenging free radicals to reduce oxidative stress (Rao 2002). Natural products can effectively ameliorate the CVD by reducing the oxidative stress (Rao 2002).

Particulate matter (PM) and gaseous pollutants, namely, ozone, carbon monoxide, nitrogen oxides, and sulfur dioxide are the main constituents of ambient air pollution. Hazardous air pollutants include benzene, acrolein, polycyclic aromatic hydrocarbons, volatile organic compounds (VOCs), and others are also present in air pollution. The life expectancy in the world is lowered by 2.9 years when exposed to ambient air pollution (Lelieveld et al. 2020), and other diseases (Lelieveld et al. 2023). PM is the main factor contributing to bad outcomes and cardiovascular risk among the many components of ambient air pollution. PM is a complicated combination of particles with varying sizes that come from a wide range of sources. Inorganic materials, heavy metals, water, and other organic compounds from the incomplete combustion of petroleum fuels are commonly found in particles.

Unregulated ultrafine PM (PM <0.1 μm in diameter) has been shown to be associated with CVDs though less research is conducted in this field. The use of fossil fuels, industrial operations, and the production of electricity are the main sources of PM2.5. Around 50% of PM come from vehicle emissions (Heydari et al. 2020). Exposure to M2.5 has been linked to atherosclerosis, coagulation and thrombosis, hypertension, endothelial dysfunction, and arrhythmias (Basith et al. 2022). Short-term exposures (hours to days) to high concentrations of PM2.5 are linked to acute outcomes, such myocardial infarction, perhaps because they produce abrupt plaque rupture (Bevan et al. 2021).

On the other hand, atherosclerosis is associated with exposures to environmental pollutants (Sirasanagandla et al. 2022). Simultaneously, epidemiological relationships implying oxidative stress-induced vascular endothelium damage and inflammation are supported by animal models and in vitro investigations (Bevan et al. 2021). The production of reactive oxygen species in tissues triggers inflammation by activating the key transcriptional regulator of inflammation, NFκB (Kampfrath et al. 2011). Furthermore, PM2.5 promotes arterial hypertension by activating the mouse sympathetic nervous system (Ying et al. 2014) and raising calcium sensitivity through Rho/ROCK signaling (Sun et al. 2008). Moreover, PM2.5 has been linked to atherosclerotic cardiovascular disease by disrupting circadian rhythms, which results in altered lipid metabolism and sleep disruptions (Ribble et al. 2023). Additionally, wildfires are becoming a major global source of PM2.5 exposure and a danger to cardiovascular health (Hadley et al. 2022). Globally, wildfires have increased in size and frequency as a result of climate change. Over the past 20 years, daily exposure to wildfire smoke has grown in 77% of nations worldwide, with China and India seeing the biggest increases (Watts et al. 2019). The link between PM2.5 from wildfire smoke and 675,000 yearly fatalities worldwide (Hadley et al. 2022) highlights the threat that wildfire smoke presents to human health everywhere. In addition to gaseous pollutants and entrained soils, wildfire smoke also contains PM that is largely different from ambient PM in that it is made up of sugars, resins, waxes, and inorganic salts, as well as products from incomplete combustion of cellulose and lignin present in wood (Naeher et al. 2007).

Resveratrol

Resveratrol (RES) is present in various naturally available plant products. It is abundant in red wine and grapes. RES has demonstrated to have various pharmacologic qualities, including actions that are cardioprotective, neuroprotective, anti-inflammatory, antioxidant, and anticarcinogenic (Sakr et al. 2015). The capacity of RES to upregulate eNOS (Wallerath et al. 2002), which supports nitric oxide-mediated vasodilation, is one of its cardioprotective mechanisms. RES treatment has been shown to prevent the pathogenesis of atherosclerosis (Li and Förstermann 2000). Additionally, in vitro study demonstrated that RES potentiates the action of HMG-CoA reductase drugs that is used in the treatment of hypercholesterolemia by increasing the expression of LDL-cholesterol receptors on the cell membrane of hepatocytes (Yashiro et al. 2012). In previous study, RES decreased the circulating triglycerides and LDL-cholesterol (Göçmen et al. 2011). RES inhibited the atherosclerosis process through inhibition of the migration of smooth muscles to the intima of the blood vessels (Lin et al. 2014). It has been shown that RES acts in the early stages of the atherogenic process. Further, it inhibits the activation of the NF-κB pathway, which in turn reduces the production of VCAM-1 and ICAM-1 (Deng et al. 2011) via blocking the expression of NADPH oxidase 1, and the synthesis of MCP-1 in macrophages exposed to LPS, RES may also prevent the generation of foam cells via acting on the Akt and FoxO3a pathways (Park et al. 2009). It alters the miRNAs that are associated with pro- or anti-inflammatory properties (such as miR-663) (Latruffe et al. 2015). Moreover, RES exerted anti-inflammatory effects by decreasing the pro-inflammatory cytokines (Latruffe et al. 2015).

A meta-analysis of seven clinical trials revealed that the clinical trials either had no effect at all or had positive effects on the lipid profile (Sahebkar 2013). For instance, individuals with type 2 diabetes who received therapy with 250–1000 mg RES/day had decreased LDL cholesterol (Bhatt et al. 2012). Similarly, administration of RES-reduced plasma triglycerides in healthy adult smokers (500 mg/day) and obese, healthy males (150 mg/day) (Bo et al. 2013). RES resulted in a 20% drop in oxidized LDL and a 4.5% decrease in LDL-cholesterol in individuals at high CV risk receiving statin medication for primary prevention (Tomé-Carneiro et al. 2012).

Hypertension is known to cause CVDs. RES has demonstrated anti-hypertensive benefits in the animal models of hypertension (Zordoky et al. 2015). Remarkably, in animal models linking hypertension to insulin resistance, very low doses of RES markedly reduced blood pressure (Li and Förstermann 2000), which may indicate that RES might be more effective in individuals with metabolic syndrome or diabetes. The antihypertensive effect of RES is related to its antioxidant action through the energy metabolism regulator AMPK, SIRT-1, Nrf2, and other endothelium-dependent pathways (Li and Förstermann 2000). This leads to vasodilation through enhanced availability of •NO in connection with elevated eNOS expression and activity (Gordish and Beierwaltes 2014). RES also blocks the AII-mediated smooth muscle contraction, leading to its relaxation and vasodilation through AMPK activation (Cao et al. 2014).

RES is known to prevent myocardial infarction. Its protective effects are by preventing platelet aggregation, preparing muscle tissue for ischemia-reperfusion injury (IRI), and maybe promoting tissue regeneration in the location of infarction (Raj et al. 2014). Preclinical research revealed that pretreatment with RES protected against the harmful consequences of cardiac reperfusion following IRI, particularly by lowering arrhythmia and infarct size. The effectiveness of these pretreatments raises the possibility that RES might prepare the heart to fend off IR damage (Shalwala et al. 2014). RES is known to produce preconditioning of the myocardium to combat the IR by altering Akt and Bcl-2 (Das et al. 2005).

RES treatment can prevent the cardiac hypertrophy and improve the cardiac function (Gu et al. 2014). Effects of RES treatment subsequent to heart failure induction, which may have more significance in medical settings, are yet to be fully elluciated. For example, RES was administered 4 weeks following surgery at a period when left ventricular ejection fraction (LVEF) was dramatically decreased. In a study of MI in mice, RES treatment for 2 weeks increased the LVEF (Kanamori et al. 2013). Given that RES has positive effects on inflammation, diabetes, dyslipidemia, hypertension, and obesity, it may be a promising pharmaceutical strategy to treat metabolic syndrome, which raises the risk of developing CVD.

Curcumin

The main active ingredient in turmeric root, curcumin, has long been utilized in traditional medicine to treat cardiovascular diseases. Curcumin has pleiotropic effects and attenuates several factors that lead to an elevated risk for CVDs in obesity and aging, according to newer mechanistic, animal, and human research (Cox et al. 2022). Curcumin's potential atheroprotective effects have been investigated using a variety of animal models of atherosclerosis, such as mice lacking either Ldlr or Apoe, or both, as well as atherogenic Western diets. Turmeric extract has been shown in early experiments to reduce the development of atherosclerosis in rabbits fed an atherogenic diet (Quiles et al. 2002). Similarly, in atherosclerosis animal model, curcumin treatment was seen to decrease the size of atherosclerotic plaques (Coban et al. 2012). The anti-atherogenic properties of curcumin may stem from its capacity to reduce elevated plasma cholesterol levels and lipid peroxidation, two essential elements in the onset of atherosclerosis. Additionally, curcumin may modify the ratio of LDL to HDL-cholesterol to a more advantageous, anti-atherogenic effect (Majeed et al. 2021).

Ginseng

Two British physicians who were studying Chinese herbal medicines in the early 1900s, F. Porter Smith and G.A. Stuart, were the first to employ ginseng in Western medicine (Smith and Stuart 1973). Ginseng has been utilized to treat hypertension. Ginseng improves vascular function, which leads to its hypotensive effects. It is interesting to note that ginsenosides promote vasorelaxation in a variety of vessels, including the cerebral arteries of monkeys (Toda et al. 2001), the rat aortas (Kim et al. 1999), and murine coronary arteries (Pan et al. 2013). While ginsenoside Rg3 activates eNOS, ginseng can boost eNOS expression and NO production (Lee et al. 2016a). Vasorelaxation mediated by NO is induced by Korean red ginseng, which improves vascular tone. According to Shin et al. (2013a), these effects are mediated via the augmentation of eNOS dimer formation, the rise in NO production, and the suppression of arginase activity. Additionally, it has been demonstrated that the *Panax ginseng* G115 extract inhibits angiotensin I-induced contractions of the bovine mesenteric arteries as well as ACE activity in human umbilical vein endothelial cells (HUVECs) (Persson et al. 2006). The anti-oxidant, anti-inflammatory, anti-hyperlipidemic, and Ca2+ channel-regulating actions of ginseng (Shin et al. 2013b) are further CVD-management attributes.

In hypertensive rats, ginseng can lower adrenal catecholamine output, which aids in vasorelaxation (Jang et al. 2011). Ginseng has been linked to hypertension in several studies (Jang et al. 2011). Indeed, ginseng could have effects that are reliant on biphasic concentrations. Ginseng raises blood pressure at low dosages while lowering blood pressure at greater concentrations (Jang et al. 2011). Ginseng improves lipid profiles, which reduces the risk of atherosclerosis. More than 30 years have passed since ginseng's advantageous effects on lipid metabolism were first reported (Qureshi et al. 1983). Red ginseng supplementation improves lipid profiles in both people and rats by raising total plasma levels of HDL-C and decreasing total plasma levels of triglycerides, cholesterol, LDL-C, free fatty acids (FFA), and platelet adhesiveness (Deng et al. 2017). By influencing the expression of genes involved in cholesterol metabolism, black ginseng can reduce hypercholesterolemia in rats (Saba et al. 2016).

Ginkgo biloba

Ginkgo biloba is referred as the maidenhair tree. It is known as a "living fossil" (Hori et al. 2012) for long many years survival without any alterations. The main components of ginkgo biloba that contribute to its pharmacological and therapeutic activities are terpenoids and flavonoids (ginkgo-flavone glycosides) (Lacour et al. 1991). These components of ginkgo biloba are widely recognized for their anti-inflammatory and antioxidant properties. Ginkgo biloba's anti-inflammatory and antioxidant properties are helpful in treating a wide range of illnesses, including those affecting the central nervous system, lungs, and heart.

Ginko biloba extract (GBE) with vasodilator and antihypertensive effects can be effusive in cardio-protection (Perez-Vizcaino et al. 2009). Other activities include ACE inhibition (Mansour et al. 2011), cholinergic pathway activation, endothelial cell health promotion (Mesquita et al. 2017), and altering lipid profile (Huang et al. 2018). GBE can reduce the proliferation of VSMCs through its anti-inflammatory and anti-atherothrombotic effects. GBE can also control the inflammatory response in blood vessels by regulating the MAPKs and the activity of the ROS-producing enzyme nicotinamide adenine dinucleotide phosphate. Consequently, in smooth muscle cells of aorta, MAPKs inhibit the production of toll-like-receptor-4 (Lin et al. 2007). Additionally, in human coronary smooth muscle cells generated by oxidized LDL and 4-hydroxynonenal, GBE can reduce the formation of MMP-1 (Akiba et al. 2007). In the same paradigm, Ginkgolide B, a component of GBE, reduced endothelial dysfunction by preventing the development of vascular cell adhesion molecule-1 (VCAM-1), intercellular adhesion molecule-1 (ICAM-1), and monocyte chemotactic protein-1 (MCP-1) in oxidized LDL-induced HUVECs. Furthermore, in mouse RAW264.7 macrophages stimulated with oxidized LDL, Ginkgolide B therapy decreased the production of many inflammatory cytokines (Feng et al. 2018).

Vascular aging is the main factor involved in the risk of CVDs development, including CHD and hypertension, and is frequently associated with low-grade inflammation, and stiffness of blood arteries (Lakatta and Levy 2003). In a study, GBE treatment reduced arterial stiffness and enhanced endothelial health in mesenteric arterioles (Cuong et al. 2019). In another study, GBE increased vascular elasticity in these elderly mesenteric arterioles by decreasing the intermediate collagen fiber layer, curving the inner elastic membrane, and narrowing the EC gap. Additionally, Akt/FoxO3a pathway alterations were observed to prevent the vascular dysfunction (Cuong et al. 2019).

Ganoderma lucidum

Ganoderma lucidum, commonly referred to as "lingzhi" or "reishi," is a kind of fungus. Its parts are utilized to create several commercial *G. lucidum* products with therapeutic properties. Commercial versions of *G. lucidum* include tea, nutritional supplements, and powders. For over 2000 years, the therapeutic usage of *G. lucidum* has been very prevalent in Asian countries. The mushroom's spores, mycelia, and fruiting bodies may all have their bioactive components extracted using ethanol or hot water (Heleno et al. 2012). *G. lucidum* bioactive substances exhibit various biological properties (Sanodiya et al. 2009). Triterpenoids possess hepatoprotective, hypocholesterolemic, anti-tumor, anti-histaminic, and anti-angiogenic properties. Triterpenoids from Ganoderma lucidum also block complement and exhibit anti-platelet aggregation properties. It should be mentioned that the only source of ganoderic acids, which are triterpene fatty acids, is *G. lucidum*. Ganoderic acids A, B, and C exhibit hypoglycemic effects among the 200 bioactive chemicals found in Ganoderma lucidum extracts (Hikino et al. 1985), whereas other ganoderic acids are found to have hypotensive effects (Morigiwa et al. 1986).

Experimental evidence clearly showed that antioxidants are crucial in preventing atherosclerosis. *G. lucidum* with known antioxidant properties has been shown to protect Caenorhabditis elegans, the model organism, from oxidative stress caused by heavy metals and paraquat through the mTOR/S6K signaling pathway and food restriction route, respectively (Cuong et al. 2019). According to Shi et al. (2002) *G. lucidum* has been shown to shield human lymphocyte DNA from oxidative damage brought on by hydrogen peroxide. Additionally, it was discovered that *G. lucidum* controls Nrf2 expression, which controls the expression of antioxidant genes including HO-1, GST, and NQO-1. In C2C12 myoblasts, *G. lucidum* ethanol extract improved Nrf2 phosphorylation, which increased HO-1 (Lee et al. 2016b). *G. lucidum* yielded three peptides known as ACE inhibitory peptides: QLVP, QDVL, and QLDL. These peptides have the ability to inhibit ACE activity and are utilized in the treatment of hypertension. Through its interactions with ACE's Gln242 and Lys472, QLVP can block ACE. In HUVECs, QLVP also decreased the mRNA and protein production of the vasoconstrictor peptide endothelin-1 and increased the phosphorylation of eNOS mediated by angiotensin 1 (Wu et al. 2019). In a mouse model of cardiac reperfusion injury, extracts from *G. lucidum* polysaccharide enhanced with selenium have protective properties against oxidative damage. Se-GLP supplementation markedly decreased the inflammation and improved the oxidative status (Shi et al. 2010).

Gynostemma pentaphyllum

Jiaogulan or *Gynostemma pentaphyllum* is an herbaceous climbing vine. It is found in subtropical China, Japan, Myanmar, and India. It originated in South China and later distributed to other parts (Li et al. 2016). It grows near rivers and in the shade of forests that surround the Yangtze River basin and the southern regions of China

(Shu-kun 1995). Drinks, snacks, face cleansers, and bath oils all contain it as a dietary supplement (Li et al. 2016). The plant exhibits a high degree of population variation and a low amount of genetic diversity (Zhang et al. 2019). Extracts from *G. pentaphyllum* have been shown by phytochemical research to include flavonoids, polysaccharides, amino acids, and gypenoside saponins (Yan et al. 2013). It is known to possess various biological effects such as anti-inflammatory, anti-cancer, antimicrobial, antioxidant, anti-inflammatory, antilipidemic, neuroprotective, and anti-obesity effects (Park et al. 2014). It has been used to treat hypertension and hepatitis (Lin et al. 2000). Dmarane-type triterpene saponins, also known as gypenosides or gynosaponins, are significant among the several types of bioactive chemicals (Kim and Han 2011).

In Wistar rats, the anti-atherosclerotic properties of a combination of *G. pentaphyllum* and *Fermentum rubrum* Hongqu gypenosides were studied. The anti-atherosclerotic potential of *G. pentaphyllum* was highlighted by the research findings, which showed that the HG combination exhibited anti-atherosclerotic effects that were superior to statin, simvastatin, and therapy (Gou et al. 2018). Furthermore, in male Sprague-Dawley rats with fatty liver disease, HG reduced the blood levels of anti-inflammatory cytokines and restored antioxidant defense components; therefore, mitigating oxidative stress indicators. The HG combination in this investigation has shown athero-protective properties (Gou et al. 2018).

Liver Diseases and Natural Products

Lifestyle factors could contribute to development of multiple diseases. Alcohol consumption in excess amounts would contribute to the development of alcohol-associated liver disease (ALD). It is one of the major causes of death worldwide. In 2019, among liver-related deaths, 45.6% of men were due to alcohol consumption, while among women liver-related deaths 39.0% were alcohol-related (Thursz et al. 2019). ALD is characterized by hepatic steatosis, hepatocyte inflammation, fibrosis, and cirrhosis. Chronic alcoholism can cause genetic polymorphisms of enzymes that are involved in the metabolism of alcohol. This could be one of the possible factors in the progression of fatty liver and further consequences of liver pathology (Tian et al. 2010). However, the specific genes that are linked to liver pathology in ALD are increased yet to be fully elucidated. The alcohol metabolism affects cell survival by increasing the ROS formation, alterations in immune responses, and gut microbiome are also postulated in the progression and pathogenesis of ALD (Nagy et al. 2016).

Nonalcoholic fatty liver disease (NAFLD) is a highly prevalent disease in various parts of the world (Younossi et al. 2016). Hepatic steatosis without other secondary causes is the characteristic feature of this disease. Nonalcoholic steatohepatitis is the subtype of NAFLD in which steatosis is accompanied by inflammation, fibrosis, and hepatocyte injury. NASH is clinically a silent disease that can lead to severe form of cirrhosis as well as end-stage liver disease (Chalasani et al. 2018). NAFLD is known to increase the risk of developing other chronic diseases such as diabetes mellitus and dyslipidemia, kidney failure, atherosclerosis, and heart failure. The estimated prevalence of NAFLD is around 25% all over the world (Younossi et al. 2016). However, a great variability in different regions was reported. The highest prevalence of NAFLD was reported in the Middle Eastern and South American populations, while the lowest rates were in the African population. Every year, the prevalence of NAFLD is estimated to increase by 3.6 million new cases (Estes et al. 2018). Men were more at risk of developing both NAFLD and NASH than women (Pan and Fallon 2014). Furthermore, racial and ethnic variations were also reported to influence the prevalence of NAFLD and NASH diseases. For example, NAFLD is reported to be more prevalent in the Hispanic group than in the black population (Pan and Fallon 2014). The two-hit hypothesis explains the complex pathogenesis of NAFLD. Firstly, the hepatic TG accumulation followed by inflammatory responses increased oxidative stress and dysfunction of mitochondria (Dowman et al. 2010).

Environmental Toxins Induced Liver Diseases

Humans are exposed to various heavy metals through plastic usage, atmosphere, industrial pollution, and agricultural pesticides. Most heavy metals such as arsenic, chromium, mercury, cadmium, and lead are known to deposit in the different tissues of the body and affect the function of different body systems, leading to serious health problems (He et al. 2005). Evidence from animal and human studies revealed that these metals alter lipid metabolism increase oxidative stress and cause hepatic steatosis and inflammation (Xie et al. 2023). Organochlorine insecticides and chlordane metabolites can cause the NAFLD (Yang and Park 2018). Other pollutants including polycyclic aromatic hydrocarbons, bisphenol A and its analogues, perfuoroalkyl Acids, dioxins, atrazine, persistent organic pollutants, polychlorinated biphenyls, and poly aromatic hydrocarbons were also known to be associated with the development of NAFLD (Sirasanagandla et al. 2023). Figure 1 represents the mechanisms involved in the pathogenesis of NAFLD.

Flax Seeds (*Linum usitatissimum*)

Flax flowering herbs belong to the family Lineaceae. Its small reddish brown color flat seeds have a nutty taste and are traditionally used as a functional food. It is also called linseed and is used for industrial applications (Rubilar et al. 2010). It is rich in ω-3 α-linolenic acid and dietary fiber. It has potential antioxidant and estrogen-mimicking properties. It has been traditionally used as a functional food (Kajla et al. 2015). Pre-clinical and clinical established its health benefits against various diseases including cancer, atherosclerosis, osteoporosis, metabolic syndrome, and neurological disorders (Kajla et al. 2015). In a randomized controlled trial on 50 NAFLD patients, flax seed supplementation along with lifestyle modification for 4 months significantly improved the BMI, serological parameters, and liver function tests and reduced liver fibrosis without any side effects (Yari et al. 2016). Treatment with

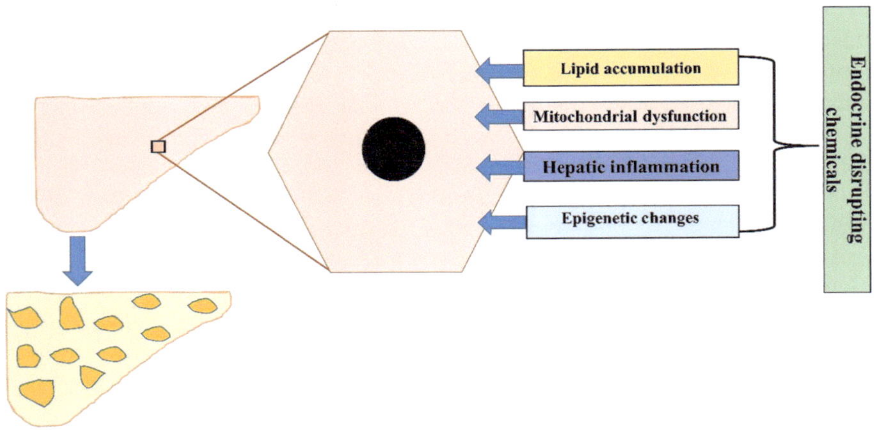

Fig. 1 Schematic diagram showing the effect of environmental factors on the hepatic steatosis or fatty liver

hesperidin and flaxseed for 12 weeks significantly improved the symptoms in NAFLD patients (Yari et al. 2021). Flax seeds (20 g/day) supplementation along with a low-energy diet for 12 weeks in NAFLD patients improved the fatty liver grade and reduced the weight as well as IL-6 when compared to control (Rezaei et al. 2020). Flax seed dietary supplementation significantly ameliorated alcoholic liver disease. Anti-inflammation and improving the gut microbiota were the main mechanism for these effects (Zhang et al. 2017). Flax seed powder treatment in a high-fat diet-fed mice model of non-alcoholic fatty liver improved the symptoms by modulating the gut microbiota as well as bile acids metabolism by regulating the FXR/TGR5 signaling pathways (Yang et al. 2023).

Resveratrol

Resveratrol is structurally 3,4′,5-trihydroxystilbene available naturally in various sources including berries, grapes, and peanuts. Its potential health benefits are extensively studied in both preclinical and clinical studies. This compound has potential antioxidant, anti-inflammatory, neuro and cardioprotective, and anti-carcinogenic properties (Meng et al. 2023). Its therapeutic properties against NAFLD and alcoholic liver diseases were reported in both preclinical and clinical studies. In clinical trial studies, RES supplementation along with a balanced diet and recommended physical activity significantly improved the symptoms in NAFLD patients (Faghihzadeh et al. 2015). In another clinical trial, RES supplementation in 60 NAFLD patients, significantly improved the symptoms (Chen et al. 2015). The RES micronized formulation treatment significantly prevented liver damage and triglyceride accumulation in NAFLD patients. The beneficial effects of RES were mediated through the activation of AMPK and SIRT1 (Theodotou et al. 2019). In an experimental study, RES treatment ameliorated liver alcohol-induced liver injury by regulating the HIF-1α protein expression and liver mitochondrial oxidative stress (Ma et al. 2017). In a high-fat diet-induced NAFLD mouse model, RES treatment improved liver steatosis and insulin resistance by altering the gut microbiota (Du et al. 2021). RES treatment in the NAFLD model improves lipid metabolism by altering the PKA/AMPK/PPARα signaling pathway (Huang et al. 2020).

Curcumin

Curcumin and curcuminoids are active principles of turmeric root, which belongs to the Zingiberaceae (Zinger) family. Tumeric has a commonly grown crop in most Asian countries and Latin America. It is commonly used as a spice in food to add flavor and color in these countries. Turmeric is a largely produced exported product of India. It is traditionally well known to possess anti-diabetic, anti-aging, anti-cancer, anti-inflammatory, antioxidant, and neuroprotective properties. Most of the important biological activities are through its anti-inflammatory properties. In a clinical trial, curcumin treatment at a dose of 1000 mg in 102 NAFLD patients for 8 weeks significantly reduced the levels of liver enzymes and cholesterol (Panahi et al. 2017). In another clinical trial, curcumin treatment in 80 NAFLD patients at a dose of 500 mg/day for 8 weeks improved lipid profile and reduced blood glucose and BMI (Rahmani et al. 2016). In meta-analysis, it was reported that curcumin supplementation improved metabolic markers and reduced body weight and body mass in NAFLD patients (Jalali et al. 2020). In mice, curcumin treatment for 6 weeks attenuated chronic alcohol-induced liver injury by preventing oxidative stress (Rong et al. 2012). In another clinical trial, curcuminoids supplementation improved the inflammatory cytokines and showed anti-steatotic effects (Saberi-Karimian et al. 2020). Curcumin treatment significantly ameliorated the liver injury in a model of alcohol-induced liver injury by regulating the mitochondrial dysfunction and preventing the endoplasmic reticulum stress (Wang et al. 2019).

Silymarin/Silybin

Silybum marianum, is referred as milk thistle, is traditionally used in the treatment of liver diseases. Silymarin is one of the most standard extracts derived from milk thistle. It contains flavonolignans and flavonoids (Federico et al. 2017). Silymarin extract contains one of the important compounds called Silybin, which constitutes around 50% of the total extract. It has several biological properties (Federico et al. 2017). Milk thistle edible Mediterranean annual herb is found in Asia, some parts of Europe, Africa, North and South Americas, and Australia. Silymarin is frequently used to treat cancer, osteoarthritis, cardiovascular diseases, and eye ailments (Akhtar et al. 2023). Therapeutic effects of both silymarin and/or silybin are extensively studied in various liver diseases due to their potential anti-inflammatory, antioxidant, as well as antifibrotic effects (Federico et al. 2017). Experimental studies demonstrated that the main mechanisms of ameliorative effects of silymarin/silybin against NAFLD progression are mediated by reducing insulin resistance, hepatocyte triglyceride accumulation, and ROS formation preventing the mitochondrial dysfunction, alternations in inflammatory responses, and attenuating the apoptosis and fibrogenesis (Federico et al. 2017).

In a recent systematic review, silymarin has been shown to prevent liver damage and hepatic steatosis by regulating the energy metabolism of NAFLD patients (Li et al. 2024). A clinical trial in 30 NAFLD patient's supplementation with two 2 pills of Realsil complex containing silybin along with phosphatidylcholine and vitamin E significantly reduced anthropometric parameters, triglycerides, total cholesterol, and fatty liver index. In another clinical trial, silybin treatment significantly improved serological parameters and liver enzymes in NAFLD patients (Federico et al. 2006). Both silymarin and vitamin E along with a hypocaloric diet significantly ameliorated the non-invasive NAFLD index (Aller et al. 2015). Treatment with Eurosil 85(®)-based nutraceutical containing both silymarin + vitamin E for 3 months improved the symptoms in NAFLD patients (Sorrentino et al. 2015).

Anthocyanins

Anthocyanins are structurally flavonoid compounds naturally found in various flowers, vegetables and fruits. Among various sources, berry fruits and strawberries are known to be the richest source of these compounds. They possess various biological activities including antioxidant, anticancer, cardio-protection and hepatoprotection (Sun et al. 2018). It has been demonstrated that blueberry anthocyanins can effectively prevent liver fibrosis by altering the hepatic stellate cell's epigenetics (Zhan et al. 2016) and regulating autophagy flux (Du et al. 2023). Anthocyanins prevent fatty liver by altering the lipid metabolism (Valenti et al. 2013). In a recent study, a high amount of anthocyanin supplementation been reported to reduce the risk of NAFLD. However, pronounced results were observed in non-diabetic and non-hypertensive patients (Xiang et al. 2023). The molecular mechanism behind the anthocyanins was demonstrated to prevent NAFLD-associated liver damage by reducing oxidative stress, lipogenesis inhibition via altering the Srebp1c levels, and lipolysis enhancement via PPARα regulation (Valenti et al. 2013). In another research, anthocyanins improve NAFLD symptoms by altering the gut microbiome and preventing dyslipidemia (Nakano et al. 2020).

Garlic

Garlic also scientifically called *Allium sativum* is widespread all around the world. It has been used traditionally as a home remedy for treating diseases in all parts of the world (Suleria et al. 2015). It has a powerful aroma with a pungent taste. It is known to contain trace elements mainly organic sulfur, dietary fibres, proteins, and various phytochemical compounds including phenols and steroids (Adaki et al. 2014). The main compounds include disulfide, diallyl trisulfide, g-glutamyl S-allylcysteine, and S-allylmercaptocysteine. Garlic is known to possess antioxidant, antihypertensive, cardioprotective, anti-atherosclerotic, and anticancer properties. It is also frequently used to lower cholesterol (Adaki et al. 2014). Experimental evidence explored the beneficial effects of garlic in attenuating NAFLD. A recent systematic review revealed that garlic can effectively regulate NAFLD and related associated complications (Pourreza et al. 2022). The main mechanisms involved in these effects include reducing inflammation, oxidative stress, blood glucose, and body weight and altering lipid metabolism (Pourreza et al. 2022). A recent meta-analysis conducted on four RCTs demonstrated that garlic supplementation in 212 patients significantly reduced the liver enzymes and lowered the body weight, total cholesterol, and LDL levels in chronic liver disease patients (Xiaohui et al. 2023). A research group found that garlic-derived exosome can attenuate NAFLD complications through macrophage-hepatocyte crosstalk and by regulating the 6-phosphofructo-2-kinase/fructose-2, 6-biphosphatase 3 (*PFKFB3*) expression (Liu et al. 2023).

Conclusion

Lifestyle diseases, including, MetS, NAFLD, and CVDs, are one of the major health issues currently in both developing and developed countries. Humans are exposed to environmental pollutants through plastic usage, atmosphere, industrial pollution, and agricultural pesticides. Evidence from experimental and clinical trials has been shown that natural products can mitigate the lifestyle diseases. However, further research is need to fully understand the molecular mechanisms involved in the potential beneficial effects of these natural products. Clinical trials should be conducted to explore the bioavailability and drug delivery of these natural compounds to optimize the safety and future drug design.

References

Adaki S, Adaki R, Shah K et al (2014) Garlic: review of literature. Indian J Cancer 51:577–581

Akhtar MN, Saeed R, Saeed F et al (2023) Silymarin: a review on paving the way towards promising pharmacological agent. Int J Food Prop 26:2256–2272

Akiba S, Yamaguchi H, Kumazawa S et al (2007) Suppression of oxidized LDL-induced PDGF receptor β activation by *Ginkgo biloba* extract reduces MMP-1 production in coronary smooth muscle cells. J Atheroscler Thromb 14:219–225

Aller R, Izaola O, Gomez S et al (2015) Effect of silymarin plus vitamin E in patients with non-alcoholic fatty liver disease. A randomized clinical pilot study. Eur Rev Med Pharmacol Sci 19:3118–3124

Basith S, Manavalan B, Shin TH et al (2022) The impact of fine particulate matter 2.5 on the cardiovascular system: a review of the invisible killer. Nanomaterials (Basel) 12:2656

Bevan GH, Al-Kindi SG, Brook RD et al (2021) Ambient air pollution and atherosclerosis: insights into dose, time, and mechanisms. Arterioscler Thromb Vasc Biol 41:628–637

Bhatt JK, Thomas S, Nanjan MJ (2012) Resveratrol supplementation improves glycemic control in type 2 diabetes mellitus. Nutr Res 32:537–541

Bo S, Ciccone G, Castiglione A et al (2013) Anti-inflammatory and antioxidant effects of resveratrol in healthy smokers a randomized, double-blind, placebo-controlled, cross-over trial. Curr Med Chem 20:1323–1331

Cao X, Luo T, Luo X et al (2014) Resveratrol prevents AngII-induced hypertension via AMPK activation and RhoA/ROCK suppression in mice. Hypertens Res 37:803–810

Chalasani N, Younossi Z, Lavine JE et al (2018) The diagnosis and management of nonalcoholic fatty liver disease: practice guidance from the American Association for the Study of Liver Diseases. Hepatology 67:328–357

Chen S, Zhao X, Ran L et al (2015) Resveratrol improves insulin resistance, glucose and lipid metabolism in patients with non-alcoholic fatty liver disease: a randomized controlled trial. Dig Liver Dis 47:226–232

Coban D, Milenkovic D, Chanet A et al (2012) Dietary curcumin inhibits atherosclerosis by affecting the expression of genes involved in leukocyte adhesion and transendothelial migration. Mol Nutr Food Res 56:1270–1281

Cox FF, Misiou A, Vierkant A et al (2022) Protective effects of curcumin in cardiovascular diseases-impact on oxidative stress and mitochondria. Cells 11:342

Cuong VT, Chen W, Shi J et al (2019) The anti-oxidation and anti-aging effects of *Ganoderma lucidum* in *Caenorhabditis elegans*. Exp Gerontol 117:99–105

Das S, Cordis GA, Maulik N et al (2005) Pharmacological preconditioning with resveratrol: role of CREB-dependent Bcl-2 signaling via adenosine A3 receptor activation. Am J Physiol Heart Circ Physiol 288:2

Deng YH, Alex D, Huang HQ et al (2011) Inhibition of TNF-α-mediated endothelial cell-monocyte cell adhesion and adhesion molecules expression by the resveratrol derivative, trans-3,5,4′-trimethoxystilbene. Phytother Res 25:451–457

Deng J, Liu Y, Duan Z et al (2017) Protopanaxadiol and protopanaxatriol-type saponins ameliorate glucose and lipid metabolism in type 2 diabetes mellitus in high-fat diet/streptozocin-induced mice. Front Pharmacol 8:506

Dowman JK, Tomlinson J, Newsome P (2010) Pathogenesis of non-alcoholic fatty liver disease. QJM 103:71–83

Du F, Huang R, Lin D et al (2021) Resveratrol improves liver steatosis and insulin resistance in non-alcoholic fatty liver disease in association with the gut microbiota. Front Microbiol 12: 611323

Du J, Liu L, Fan H et al (2023) Anthocyanins improve liver fibrosis in mice by regulating the autophagic flux level of hepatic stellate cells by mmu_circ_0000623. Food Sci Nutr 11:3002–3018

Estes C, Razavi H, Loomba R et al (2018) Modeling the epidemic of nonalcoholic fatty liver disease demonstrates an exponential increase in burden of disease. Hepatology 67:123–133

Faghihzadeh F, Adibi P, Hekmatdoost A (2015) The effects of resveratrol supplementation on cardiovascular risk factors in patients with non-alcoholic fatty liver disease: a randomised, double-blind, placebo-controlled study. Br J Nutr 114:796–803

Federico A, Trappoliere M, Tuccillo C et al (2006) A new silybin-vitamin E-phospholipid complex improves insulin resistance and liver damage in patients with non-alcoholic fatty liver disease: preliminary observations. Gut 55:901–902

Federico A, Dallio M, Loguercio C (2017) Silymarin/silybin and chronic liver disease: a marriage of many years. Molecules 22:191

Feng Z, Yang X, Zhang L et al (2018) Ginkgolide B ameliorates oxidized low-density lipoprotein-induced endothelial dysfunction via modulating Lectin-like ox-LDL-receptor-1 and NADPH oxidase 4 expression and inflammatory cascades. Phytother Res 32:2417–2427

Göçmen AY, Burgucu D, Gümüşlü S (2011) Effect of resveratrol on platelet activation in hypercholesterolemic rats: CD40-CD40L system as a potential target. Appl Physiol Nutr Metab 36:323–330

Gordish KL, Beierwaltes WH (2014) Resveratrol induces acute endothelium-dependent renal vasodilation mediated through nitric oxide and reactive oxygen species scavenging. Am J Physiol Renal Physiol 306:15

Gou SH, Liu BJ, Han XF et al (2018) Anti-atherosclerotic effect of *Fermentum rubrum* and *Gynostemma pentaphyllum* mixture in high-fat emulsion-and vitamin D3-induced atherosclerotic rats. J Chin Med Assoc 81:398–408

Gu XS, Wang ZB, Ye Z et al (2014) Resveratrol, an activator of SIRT1, upregulates AMPK and improves cardiac function in heart failure. Genet Mol Res 13:323–335

Hadley MB, Henderson SB, Brauer M et al (2022) Protecting cardiovascular health from wildfire smoke. Circulation 146:788–801

He ZL, Yang XE, Stoffella PJ (2005) Trace elements in agroecosystems and impacts on the environment. J Trace Elem Med Biol 19:125–140

Heleno SA, Barros L, Martins A et al (2012) Fruiting body, spores and in vitro produced mycelium of *Ganoderma lucidum* from Northeast Portugal: a comparative study of the antioxidant potential of phenolic and polysaccharidic extracts. Food Res Int 46:135–140

Heydari S, Tainio M, Woodcock J et al (2020) Estimating traffic contribution to particulate matter concentration in urban areas using a multilevel Bayesian meta-regression approach. Environ Int 141:105800

Hikino H, Konno C, Mirin Y et al (1985) Isolation and hypoglycemic activity of ganoderans A and B, glycans of *Ganoderma lucidum* fruit bodies. Planta Med 51:339–340

Hori T, Ridge RW, Tulecke W, Tredici PD, Tremouillaux-Guiller J, Tobe H (eds) (2012) *Ginkgo biloba* a global treasure: from biology to medicine. Springer, Tokyo

Huang WC, Chen YL, Liu HC et al (2018) Ginkgolide C reduced oleic acid-induced lipid accumulation in HepG2 cells. Saudi Pharm J 26:1178–1184

Huang Y, Lang H, Chen K et al (2020) Resveratrol protects against nonalcoholic fatty liver disease by improving lipid metabolism and redox homeostasis via the PPARα pathway. Appl Physiol Nutr Metab 45:227–239

Jalali M, Mahmoodi M, Mosallanezhad Z et al (2020) The effects of curcumin supplementation on liver function, metabolic profile and body composition in patients with non-alcoholic fatty liver disease: a systematic review and meta-analysis of randomized controlled trials. Complement Ther Med 48:102283

Jang SJ, Lim HJ, Lim DY (2011) Inhibitory effects of total ginseng saponin on catecholamine secretion from the perfused adrenal medulla of SHRs. J Ginseng Res 35:176–190

Kajla P, Sharma A, Sood DR (2015) Flaxseed-a potential functional food source. J Food Sci Technol 52:1857–1871

Kampfrath T, Maiseyeu A, Ying Z et al (2011) Chronic fine particulate matter exposure induces systemic vascular dysfunction via NADPH oxidase and TLR4 pathways. Circ Res 108:716–726

Kanamori H, Takemura G, Goto K et al (2013) Resveratrol reverses remodeling in hearts with large, old myocardial infarctions through enhanced autophagy-activating AMP kinase pathway. Am J Pathol 182:701–713

Kim JH, Han YN (2011) Dammarane-type saponins from *Gynostemma pentaphyllum*. Phytochemistry 72:1453–1459

Kim ND, Kang SY, Kim MJ et al (1999) The ginsenoside Rg3 evokes endothelium-independent relaxation in rat aortic rings: role of K^+ channels. Eur J Pharmacol 367:51–57

Lacour M, Ez-Zaher L, Raymond J (1991) Plasticity mechanisms in vestibular compensation in the cat are improved by an extract of *Ginkgo biloba* (EGb 761). Pharmacol Biochem Behav 40:367–379

Lakatta EG, Levy DJC (2003) Arterial and cardiac aging: major shareholders in cardiovascular disease enterprises: part II: the aging heart in health: links to heart disease. Circulation 107:346–354

Latruffe N, Lançon A, Frazzi R et al (2015) Exploring new ways of regulation by resveratrol involving miRNAs, with emphasis on inflammation. Ann N Y Acad Sci 1:97–106

Lee KH, Bae IY, Park SI et al (2016a) Antihypertensive effect of Korean red ginseng by enrichment of ginsenoside Rg3 and arginine–fructose. J Ginseng Res 40:237–244

Lee YH, Kim JH, Song CH et al (2016b) Ethanol extract of *Ganoderma lucidum* augments cellular anti-oxidant defense through activation of Nrf2/HO-1. J Pharmacopuncture 19:59

Lelieveld J, Pozzer A, Pöschl U et al (2020) Loss of life expectancy from air pollution compared to other risk factors: a worldwide perspective. Cardiovasc Res 116:1910–1917

Lelieveld J, Haines A, Burnett R et al (2023) Air pollution deaths attributable to fossil fuels: observational and modelling study. BMJ 383:e077784

Li H, Förstermann U (2000) Nitric oxide in the pathogenesis of vascular disease. J Pathol 190:244–254

Li Y, Lin W, Huang J et al (2016) Anti-cancer effects of *Gynostemma pentaphyllum* (thunb.) makino (jiaogulan). Chin Med 11:1–16

Li S, Duan F, Li S et al (2024) Administration of silymarin in NAFLD/NASH: a systematic review and meta-analysis. Ann Hepatol 29:101174

Lin CC, Huang PC, Lin JM (2000) Antioxidant and hepatoprotective effects of *Anoectochilus formosanus* and *Gynostemma pentaphyllum*. Am J Chin Med 28:87–96

Lin FY, Chen YH, Chen YL et al (2007) *Ginkgo biloba* extract inhibits endotoxin-induced human aortic smooth muscle cell proliferation via suppression of toll-like receptor 4 expression and NADPH oxidase activation. J Agric Food Chem 55:1977–1984

Lin YC, Chen LH, Varadharajan T et al (2014) Resveratrol inhibits glucose-induced migration of vascular smooth muscle cells mediated by focal adhesion kinase. Mol Nutr Food Res 58:1389–1401

Liu J, Li W, Bian Y et al (2023) Garlic-derived exosomes regulate PFKFB3 expression to relieve liver dysfunction in high-fat diet-fed mice via macrophage-hepatocyte crosstalk. Phytomedicine 112:154679

Ma Z, Zhang Y, Li Q et al (2017) Resveratrol improves alcoholic fatty liver disease by downregulating HIF-1α expression and mitochondrial ROS production. PLoS One 12: e0183426

Majeed ML, Ghafil FA, Fatima G et al (2021) Anti-atherosclerotic and anti-inflammatory effects of curcumin on hypercholesterolemic male rabbits. Indian J Clin Biochem 36:74–80

Mansour SM, Bahgat AK, El-Khatib AS et al (2011) *Ginkgo biloba* extract (EGb 761) normalizes hypertension in 2K, 1C hypertensive rats: role of antioxidant mechanisms, ACE inhibiting activity and improvement of endothelial dysfunction. Phytomedicine 18:641–647

Meng Q, Li J, Wang C et al (2023) Biological function of resveratrol and its application in animal production: a review. J Anim Sci Biotechnol 14:25

Mesquita TR, De Jesus IC, Dos Santos JF et al (2017) Cardioprotective action of *Ginkgo biloba* extract against sustained β-adrenergic stimulation occurs via activation of M2/NO pathway. Front Pharmacol 8:220

Mills KT, Bundy JD, Kelly TN et al (2016) Global disparities of hypertension prevalence and control: a systematic analysis of population-based studies from 90 countries. Circulation 134: 441–450

Mills KT, Stefanescu A, He J (2020) The global epidemiology of hypertension. Nat Rev Nephrol 16:223–237

Morigiwa A, Kitabatake K, Fujimoto Y et al (1986) Angiotensin converting enzyme-inhibitory triterpenes from *Ganoderma lucidim*. Chem Pharm Bull (Tokyo) 34:3025–3028

Naeher LP, Brauer M, Lipsett M et al (2007) Woodsmoke health effects: a review. Inhal Toxicol 19: 67–106

Nagy LE, Ding WX, Cresci G et al (2016) Linking pathogenic mechanisms of alcoholic liver disease with clinical phenotypes. Gastroenterology 150:1756–1768

Nakano H, Wu S, Sakao K et al (2020) Bilberry anthocyanins ameliorate NAFLD by improving dyslipidemia and gut microbiome dysbiosis. Nutrients 12:3252

Pan JJ, Fallon MB (2014) Gender and racial differences in nonalcoholic fatty liver disease. World J Hepatol 6:274–283

Pan SY, Zhou SF, Gao SH et al (2013) New perspectives on how to discover drugs from herbal medicines: CAM's outstanding contribution to modern therapeutics. Evid Based Complement Alternat Med 2013:627375

Panahi Y, Kianpour P, Mohtashami R et al (2017) Efficacy and safety of phytosomal curcumin in non-alcoholic fatty liver disease: a randomized controlled trial. Drug Res 67:244–251

Park DW, Baek K, Kim JR et al (2009) Resveratrol inhibits foam cell formation via NADPH oxidase 1-mediated reactive oxygen species and monocyte chemotactic protein-1. Exp Mol Med 41:171–179

Park SH, Huh TL, Kim SY et al (2014) Antiobesity effect of *Gynostemma pentaphyllum* extract (actiponin): a randomized, double-blind, placebo-controlled trial. Obesity (Silver Spring) 22: 63–71

Perez-Vizcaino F, Duarte J, Jimenez R et al (2009) Antihypertensive effects of the flavonoid quercetin. Pharmacol Rep 61:67–75

Persson IA, Dong L, Persson K (2006) Effect of Panax ginseng extract (G115) on angiotensin-converting enzyme (ACE) activity and nitric oxide (NO) production. J Ethnopharmacol 105: 321–325

Pourreza S, Azar PS, Sanaie S et al (2022) Therapeutic effects and mechanisms of action of garlic (allium sativum) on nonalcoholic fatty liver disease: a comprehensive systematic literature review. Evid Based Complement Alternat Med 2022:6960211

Quiles JL, Mesa MD, Ramírez-Tortosa CL et al (2002) *Curcuma longa* extract supplementation reduces oxidative stress and attenuates aortic fatty streak development in rabbits. Arterioscler Thromb Vasc Biol 22:1225–1231

Qureshi A, Abuirmeileh N, Din Z et al (1983) Suppression of cholesterogenesis and reduction of LDL cholesterol by dietary ginseng and its fractions in chicken liver. Atherosclerosis 48:81–94

Rahmani S, Asgary S, Askari G et al (2016) Treatment of non-alcoholic fatty liver disease with curcumin: a randomized placebo controlled trial. Phytother Res 30:1540–1548

Raj P, Louis XL, Thandapilly SJ et al (2014) Potential of resveratrol in the treatment of heart failure. Life Sci 95:63–71

Rao AV (2002) Lycopene, tomatoes, and the prevention of coronary heart disease. Exp Biol Med 227:908–913

Rezaei S, Sasani MR, Akhlaghi M et al (2020) Flaxseed oil in the context of a weight loss programme ameliorates fatty liver grade in patients with non-alcoholic fatty liver disease: a randomised double-blind controlled trial. Br J Nutr 123:994–1002

Ribble A, Hellmann J, Conklin DJ et al (2023) Fine particulate matter (PM2.5)-induced pulmonary oxidative stress contributes to increases in glucose intolerance and insulin resistance in a mouse model of circadian dyssynchrony. Sci Total Environ 877:162934

Rong S, Zhao Y, Bao W et al (2012) Curcumin prevents chronic alcohol-induced liver disease involving decreasing ROS generation and enhancing antioxidative capacity. Phytomedicine 19: 545–550

Rubilar M, Gutiérrez C, Verdugo M et al (2010) Flaxseed as a source of functional ingredients. J Soil Sci Plant Nutr 10:373–377

Saba E, Jeon BR, Jeong DH et al (2016) Black ginseng extract ameliorates hypercholesterolemia in rats. J Ginseng Res 40:160–168

Saberi-Karimian M, Keshvari M, Ghayour-Mobarhan M et al (2020) Effects of curcuminoids on inflammatory status in patients with non-alcoholic fatty liver disease: a randomized controlled trial. Complement Ther Med 49:102322

Sahebkar A (2013) Effects of resveratrol supplementation on plasma lipids: a systematic review and meta-analysis of randomized controlled trials. Nutr Rev 71:822–835

Sakr HF, Abbas A, Elsamanoudy A et al (2015) Effect of fluoxetine and resveratrol on testicular functions and oxidative stress in a rat model of chronic mild stress-induced depression. J Physiol Pharmacol 66:515–527

Sanodiya BS, Thakur GS, Baghel RK et al (2009) *Ganoderma lucidum*: a potent pharmacological macrofungus. Curr Pharm Biotechnol 10:717–742

Shalwala M, Zhu SG, Das A et al (2014) Sirtuin 1 (SIRT1) activation mediates sildenafil induced delayed cardioprotection against ischemia-reperfusion injury in mice. PLoS One 9:e86977
Shi YL, James AE, Benzie IF et al (2002) Mushroom-derived preparations in the prevention of H2O2-induced oxidative damage to cellular DNA. Teratog Carcinog Mutagen 22:103–111
Shi WL, Han H, Chen GZ et al (2010) Extraction, characterization of the polysaccharide extracts from Se-enriched *G. lucidum* (Se-GLP) and its inhibition against oxidative damage in ischemic reperfusion mice. Carbohydr Polym 80:774–778
Shin W, Yoon J, Oh GT et al (2013a) Korean red ginseng inhibits arginase and contributes to endothelium dependent vasorelaxation through endothelial nitric oxide synthase coupling. J Ginseng Res 37:64
Shin YM, Jung HJ, Choi WY et al (2013b) Antioxidative, anti-inflammatory, and matrix metalloproteinase inhibitory activities of 20 (S)-ginsenoside Rg3 in cultured mammalian cell lines. Mol Biol Rep 40:269–279
Shu-kun C (1995) A classificatory system and geographical distribution of the genus Gynostemma BL. (Cucurbitaceae). J Syst Evol 33:403–410
Sirasanagandla SR, Al-Huseini I, Sofin RG et al (2022) Perinatal exposure to bisphenol A and developmental programming of the cardiovascular changes in the offspring. Curr Med Chem 29:4235–4250
Sirasanagandla SR, Al-Mushaiqri M, Al Ghafri F et al (2023) Protective effects of perinatal resveratrol on bisphenol A exposure-induced cardiovascular alterations and hepatic steatosis in adult offspring mice: a histopathological study. Appl Sci 13:13163
Smith FP, Stuart GA (1973) Chinese medicinal herbs: Georgetown Press, San Francisco
Smith FP, Stuart GA (2003) Chinese medicinal herbs: a modern edition of a classic sixteenth century manual. Dover Publications, Mineola
Sorrentino G, Crispino P, Coppola D et al (2015) Efficacy of lifestyle changes in subjects with non-alcoholic liver steatosis and metabolic syndrome may be improved with an antioxidant nutraceutical: a controlled clinical study. Drugs R D 15:21–25
Suleria HAR, Butt MS, Khalid N et al (2015) Garlic (*Allium sativum*): diet based therapy of 21st century–a review. Asian Pac J Trop Dis 5:271–278
Sun Q, Yue P, Ying Z et al (2008) Air pollution exposure potentiates hypertension through reactive oxygen species-mediated activation of Rho/ROCK. Arterioscler Thromb Vasc Biol 28:1760–1766
Sun J, Wu Y, Long C et al (2018) Anthocyanins isolated from blueberry ameliorates CCl4 induced liver fibrosis by modulation of oxidative stress, inflammation and stellate cell activation in mice. Food Chem Toxicol 120:491–499
Theodotou M, Fokianos K, Moniatis D et al (2019) Effect of resveratrol on non-alcoholic fatty liver disease. Exp Ther Med 18:559–565
Thursz M, Kamath PS, Mathurin P et al (2019) Alcohol-related liver disease: areas of consensus, unmet needs and opportunities for further study. J Hepatol 70:521–530
Tian C, Stokowski RP, Kershenobich D et al (2010) Variant in PNPLA3 is associated with alcoholic liver disease. Nat Genet 42:21–23
Toda N, Ayajiki K, Fujioka H et al (2001) Ginsenoside potentiates NO-mediated neurogenic vasodilatation of monkey cerebral arteries. J Ethnopharmacol 76:109–113
Tomé-Carneiro J, Gonzálvez M, Larrosa M et al (2012) Consumption of a grape extract supplement containing resveratrol decreases oxidized LDL and ApoB in patients undergoing primary prevention of cardiovascular disease: a triple-blind, 6-month follow-up, placebo-controlled, randomized trial. Mol Nutr Food Res 56:810–821
Valenti L, Riso P, Mazzocchi A et al (2013) Dietary anthocyanins as nutritional therapy for nonalcoholic fatty liver disease. Oxidative Med Cell Longev 2013:145421
Wallerath T, Deckert G, Ternes T et al (2002) Resveratrol, a polyphenolic phytoalexin present in red wine, enhances expression and activity of endothelial nitric oxide synthase. Circulation 106: 1652–1658

Wang B, Gao X, Liu B et al (2019) Protective effects of curcumin against chronic alcohol-induced liver injury in mice through modulating mitochondrial dysfunction and inhibiting endoplasmic reticulum stress. Food Nutr Res 63:e3567

Watts N, Amann M, Arnell N et al (2019) The 2019 report of the lancet countdown on health and climate change: ensuring that the health of a child born today is not defined by a changing climate. Lancet 394:1836–1878

Wu Q, Li Y, Peng K et al (2019) Isolation and characterization of three antihypertension peptides from the mycelia of *Ganoderma lucidum* (Agaricomycetes). J Agric Food Chem 67:8149–8159

Xiang S, Li Y, Li Y et al (2023) Higher anthocyanin intake is associated with a lower risk of non-alcoholic fatty liver disease in the United States adult population. Front Nutr 10:1265507

Xiaohui L, Jinqi L, Xiaofang X et al (2023) Garlic supplementation for the treatment of chronic liver disease: a meta-analysis of randomized controlled trials. Afr Health Sci 23:409–415

Xie Z, Aimuzi R, Si M et al (2023) Associations of metal mixtures with metabolic-associated fatty liver disease and non-alcoholic fatty liver disease: NHANES 2003–2018. Front Public Health 11:1133194

Yan W, Niu Y, Lv J et al (2013) Characterization of a heteropolysaccharide isolated from diploid Gynostemma pentaphyllum Makino. Carbohydr Polym 92:2111–2117

Yang JS, Park Y (2018) Insecticide exposure and development of nonalcoholic fatty liver disease. J Agric Food Chem 66:10132–10138

Yang C, Yang L, Yang Y et al (2023) Effects of flaxseed powder in improving non-alcoholic fatty liver by regulating gut microbiota-bile acids metabolic pathway through FXR/TGR5 mediating. Biomed Pharmacother 163:114864

Yari Z, Rahimlou M, Eslamparast T et al (2016) Flaxseed supplementation in non-alcoholic fatty liver disease: a pilot randomized, open labeled, controlled study. Int J Food Sci Nutr 67:461–469

Yari Z, Cheraghpour M, Alavian SM et al (2021) The efficacy of flaxseed and hesperidin on non-alcoholic fatty liver disease: an open-labeled randomized controlled trial. Eur J Clin Nutr 75:99–111

Yashiro T, Nanmoku M, Shimizu M et al (2012) Resveratrol increases the expression and activity of the low density lipoprotein receptor in hepatocytes by the proteolytic activation of the sterol regulatory element-binding proteins. Atherosclerosis 220:369–374

Ying Z, Xu X, Bai Y et al (2014) Long-term exposure to concentrated ambient PM2.5 increases mouse blood pressure through abnormal activation of the sympathetic nervous system: a role for hypothalamic inflammation. Environ Health Perspect 122:79–86

Younossi ZM, Koenig AB, Abdelatif D et al (2016) Global epidemiology of nonalcoholic fatty liver disease-meta-analytic assessment of prevalence, incidence, and outcomes. Hepatology 64:73–84

Zhan W, Liao X, Yu L et al (2016) Effects of blueberries on migration, invasion, proliferation, the cell cycle and apoptosis in hepatocellular carcinoma cells. Biomed Rep 5:579–584

Zhang X, Wang H, Yin P et al (2017) Flaxseed oil ameliorates alcoholic liver disease via anti-inflammation and modulating gut microbiota in mice. Lipids Health Dis 16:1–10

Zhang X, Su H, Yang J et al (2019) Population genetic structure, migration, and polyploidy origin of a medicinal species *Gynostemma pentaphyllum* (Cucurbitaceae). Ecol Evol 9:11145–11170

Zordoky BN, Robertson IM, Dyck JR (2015) Preclinical and clinical evidence for the role of resveratrol in the treatment of cardiovascular diseases. Biochim Biophys Acta 6:1

Diet and Nutrition in Disease Prevention and Maintenance of Healthy Lifestyle

Srijit Das and Norsham Juliana

Abstract

Diet and nutrition play a crucial role in maintaining good health. Many of the non-communicable diseases are influenced by faulty diets. The non-communicable diseases include cardiovascular diseases, cancer, diabetes, and respiratory diseases. To make matters worse, if there is an association between an unhealthy diet and environmental pollution, it increases morbidity and mortality. This chapter discusses the role of diet in disease prevention with an emphasis on foods, which can prove to be beneficial to the human body. The active chemical compounds present in the food are discussed for better understanding. Educating the public regarding the importance of intake of healthy food and a better lifestyle may help combat various non-communicable life-threatening diseases.

Keywords

Diet · Nutrition · Food · Antioxidants · Health · Non-communicable disease

S. Das (✉)
Faculty of Medicine and Health Sciences, Sultan Qaboos University, Muscat, Oman
e-mail: s.das@squ.edu.om

N. Juliana
Faculty of Medicine and Health Sciences, Universiti Sains Islam Malaysia, Nilai, Negeri Sembilan, Malaysia
e-mail: njuliana@usim.edu.my

S. Raisuddin et al. (eds.), *Environmental Toxicants and Lifestyle Diseases*,
https://doi.org/10.1007/978-981-95-0575-3_26

Introduction

Non-communicable diseases are major health concerns in many developing countries. Improper dietary practices, lifestyle, lack of physical activity, substance abuse, and addiction constitute different risk factors that are responsible for various non-communicable diseases, which include diabetes, coronary artery diseases, stroke, osteoporosis, and cancer (Fig. 1). There has been a rise in fast and packaged food consumption worldwide. The nutritional component in these foods may be less. In addition, harmful constituents such as monosodium glutamate, dyes, and preservatives such as nitrites, sugar, and flavorings may be detrimental to health. In many developed countries, the increase in cardiometabolic deaths has been attributed to suboptimal food habits. Hence, it is important to educate and increase the awareness of the common people regarding healthy food habits to prevent various diseases.

Diseases Related to Faulty Diet and Nutrition

The food and beverages that we consume in our diet are necessary for maintaining health and well-being. A diet that is balanced and nutritious is necessary for maintaining good health and prevent any disease. Whereas, a diet is deficient in essential nutrients or is too high in harmful substances, it may cause several health

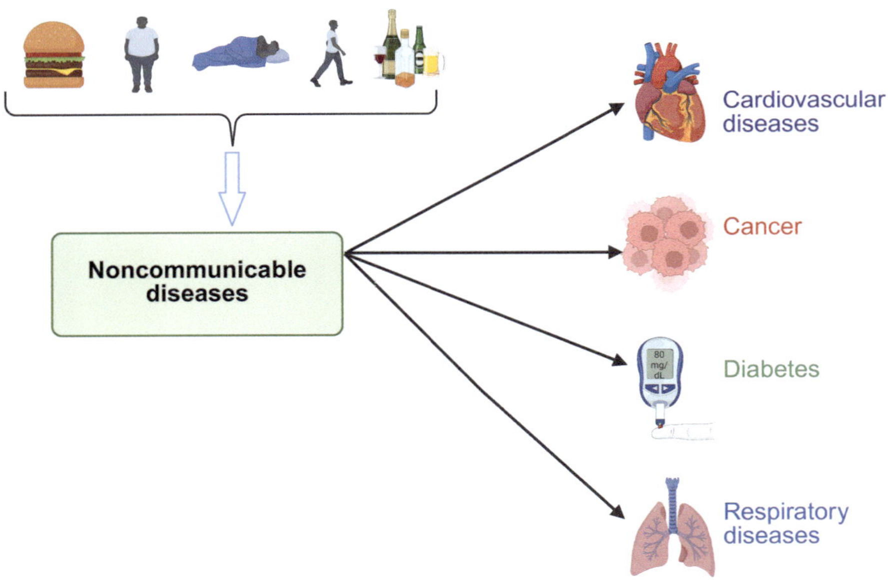

Fig. 1 Few non-communicable diseases related to unhealthy diet

issues. Hence, all individuals should understand the relationship between diet and diseases for both prevention and management.

Generally, poor health is not attributed to a single nutritional component. Instead, diet-related diseases involve interactions between various factors such as environmental, behavioral, biological, social, and last but not least the inherited genetic factor. The interplay between diet and different risk factors that are responsible for the progression of diseases makes it difficult to assess the specific contribution of diet to poor health. In an optimal diet, the goal is to ensure that energy intake is sufficient for tissue maintenance, repair, and growth. The macronutrients, carbohydrates, proteins, and fats along with essential minerals necessary for maintaining health should be obtained by consuming a variety of nutritious foods.

Non-communicable diseases have been linked to chronic conditions resulting from faulty dietary intake. Type-2 diabetes mellitus, obesity, metabolic syndrome, chronic kidney disease (CKD), cardiovascular disease, cancer, and chronic respiratory diseases are a few of the NCDs. Based on recent studies, the incidence is projected to increase due to an increase in the age of the population, urbanization, and changes occurring in lifestyle. In countries that are industrialized and have high income, numerous studies demonstrated a direct association of socioeconomic factors and health status, showing that NCDs disproportionately affect people of lower socio-economic status more when compared to those belonging to the higher socio-economic status (Williams et al. 2018). The post-COVID-19 pandemic period has exacerbated the situation, with rising reports indicating an increased incidence of NCDs. This surge is largely attributed to unhealthy lifestyle changes adopted during the pandemic, including increased sitting and screen time, reduced physical activity, disrupted sleep patterns, and poor dietary habits (Xu et al. 2024a).

Type 2 Diabetes Mellitus

T2DM is a chronic disorder of the endocrine system, characterized by increased glucose levels, insulin resistance, and a disturbance in insulin secretion. The primary etiology includes various factors related to genetics and environment, with diet having a vital role in the disease progression. Over the past three decades, T2DM cases have increased by more than two times. In 2010, approximately 285 million people suffered from diabetes, and this was estimated to increase to 439 million by 2030, which amounts to 7.7% of the adult population who are aged between 20 and 79 years (Tiwari and Kumar 2018). Additionally, the incidence of the disease and the mortality cases arising out of T2DM also increased more than two times from 1990 to 2019. These trends underscore the growing global burden of T2DM (Ye et al. 2023).

The pathophysiology of T2DM is related to various factors that are important for the disease progression. Insulin resistance occurs when the peripheral tissues, which include muscle and adipose tissue become less responsive to insulin. This reduces sensitivity by the main energy user and storage leads to impaired glucose uptake, causing elevated glucose levels. Besides that, defective insulin secretion happens when the pancreatic β-cells fail to secrete sufficient insulin in response to glucose. Initially, the cells may compensate by secreting more insulin, but in the long run, their function declines, hence exacerbating hyperglycemia (Eapen 2019). The liver may also contribute to hyperglycemia by producing excess glucose due to impaired suppression by insulin. The elevated hepatic glucose output will further raise the blood sugar levels, compounding the problem of high blood glucose. As a result of the loss of β-cells function, there is inadequate production of insulin. The deterioration results in insufficient production of insulin, which leads to problems in maintaining the normal level of glucose in the blood. Ultimately resulting in uncontrolled hyperglycemia (Saxena et al. 2024).

Impaired glucose tolerance (IGT) is a condition that exhibits a lesser degree of hyperglycemia and has an intermediate stage in the process of development of non-insulin-dependent diabetes mellitus, thereby carrying an increased risk of progression to T2DM. Between one and three-quarters of individuals with IGT develop diabetes within a decade of diagnosis, with annual progression rates ranging from 1% to 10% (Rasmussen et al. 2016). If the progression of IGT is checked, the incidence and complications of the disease could certainly be reduced or delayed. Risk factors influencing the rate of progression from IGT to diabetes include age, obesity, hyperinsulinemia, insulin resistance, and a faulty diet high in refined sugars and unhealthy fats (Pramono et al. 2023).

Obesity

Obesity refers to a condition in which there is excessive body fat accumulation. As the condition leads to a significant risk for numerous health conditions, it is classified as a complex multifactorial disease. The primary cause is a chronic energy imbalance between intake and expenditure. Genetic variants, the obesogenic environment of high-calorie food and a sedentary lifestyle together with psychosocial factors, with diet being a crucial component, are the factors that lead to the imbalance. Since 1980, the worldwide prevalence of obesity has increased two times, with one-third of individuals now being overweight or obese (Alsannan et al. 2024). The increase is observed across all sexes and age groups. Interestingly, the prevalence of obese or overweight children in high-income countries, specifically the United States, Norway, France, Sweden, Australia, Japan, and Denmark has shown a decrease or plateaued over the past two decades (NCD Risk Factor Collaboration (NCD-RisC) 2017). On the contrary, obesity rates are rising in low- and middle-income countries, which mainly include urban areas. The rate of overweight African children below the age of 5 years has increased by 24% since the year 2000, and approximately half of Asian children in a similar age group were shown to be overweight or obese as of 2019 (Wariri et al. 2021).

Body mass index (BMI) has been the universally accepted key measure used by the World Health Organization (WHO) to define and diagnose obesity. Adults who have a BMI of 25.0–29.9 are considered as "overweight," while a BMI of 30.0 or higher is classified as "obese." Obesity is further categorized into three subclasses;

class I (BMI 30.0–34.9), class II (BMI 35.0–39.9), and class III (BMI $\geq$ 40.0). Nevertheless, BMI can vary significantly due to factors such as sex, ethnicity, and age. Another important parameter is the deposition of excess fat around the region of the abdomen (abdominal obesity) has been widely reported to be linked to higher health risks (Xu et al. 2022). A common practice in assessing abdominal obesity is by measuring the waist circumference. However, its measurement guidelines differ among organizations like WHO, the International Diabetes Federation (IDF), and the American Heart Association (AHA). There is still no universal standard for defining abdominal obesity (Issa et al. 2017; Lin and Li 2021).

Hypertension

Hypertension refers to high blood pressure, which is an important risk factor for development of the cardiovascular diseases. Its development and progression are influenced by a combination of genetic predisposition, lifestyle factors, and dietary habits. The disease is closely related to a chronic practice of poor nutritional intake including excess sodium intake, insufficient potassium, alcohol consumption, and excessive calorie intake that leads to obesity, which also contributes to the disease. Besides that, chronic stress and genetic factors also play a significant role with certain populations being more predisposed to developing hypertension than the rest. It is a well-accepted fact that diet has a crucial role in the development and management of hypertension. High sodium intake present in processed and fast foods is strongly linked to increased blood pressure. Sodium contributes to water retention and blood pressure. Conversely, diets low in potassium, which is considered as an essential mineral that helps balance the sodium levels in the body, may also contribute to high blood pressure. Potassium-rich foods as found in fruits and vegetables, help in the mitigation of the effects of sodium and support normal blood pressure levels. The Dietary Approaches to Stop Hypertension (DASH) diet has extensively been studied, and it shows good results in reducing blood pressure. This diet underlines the importance of intake of fruits, vegetables, whole grains, and low-fat dairy products and less intake of saturated fats, cholesterol, and sodium. The DASH diet's effectiveness is attributed to its high content of potassium, magnesium, calcium, and fiber, all of which contribute to better control of blood pressure (Kim et al. 2024). Reducing sodium intake is a cornerstone of hypertension management. According to the WHO recommendation, one should consume less than 5 g of salt (about one teaspoon) per day to decrease the risk of developing hypertension. Increasing potassium intake by consuming a diet rich in fruits, vegetables, legumes, and nuts is also an effective strategy for managing hypertension (Xu et al. 2024a, b). Additionally, maintenance of a healthy weight, engagement in regular physical activities, control of alcohol consumption, and managing stress are important modifications in the lifestyle that can help control blood pressure (Sadiq 2023).

Cardiovascular Diseases (CVD)

Unhealthy eating habits stand out as the main lifestyle factor that causes an increase in the risk of deaths related to CVD. Numerous studies conclusively showed that what is on our plate could be linked to the cause and progression of CVD. More than two-thirds of deaths related to CVD were linked to poor dietary intake. More than six million deaths are associated with the high intake of processed food, sugary drinks, trans and saturated fats, and products with additional salt and sugar. On the other hand, increasing the intake of omega-3 fatty acids from seafood to approximately 200–300 mg, 200–430 g of fruits or vegetables, 16–25 g of nuts, and 100–150 g of whole grains is recommended to maintain a healthy heart (Bowen et al. 2016). All the recommendations were provided by the Global Burden of Disease Study 2017, which was based on the study conducted in 195 countries between 1990 and 2017 (Safiri et al. 2020).

There are 11 risk factors linked to mortality due to ischemic heart disease, and these include diet, high blood pressure, elevated serum low-density lipoprotein (LDL) cholesterol, high level of plasma glucose, tobacco consumption, increased BMI, air pollution, decreased physical activity, impaired kidney function, exposure to lead, and alcohol use (Lui et al. 2021). Five out of the 11 are all related to poor eating habits; thus, it was estimated that the proportion of deaths could be prevented by eliminating each risk factor. Considering that all other factors stayed constant, the study observed that 69.2% of ischemic heart disease deaths occurring all over the world could be prevented with healthier diets (Lui et al. 2021). Additionally, 54.4% of these deaths could be avoided by maintaining systolic blood pressure at 110–115 mmHg, and 41.9% could be prevented by assuring serum LDL cholesterol levels at 0.7–1.3 mmol/L (Niklas et al. 2019). While approximately 25.5% of deaths could be avoided by maintaining serum fasting plasma glucose at 4.8–5.4 mmol/L. High BMI was the fifth leading contributor to CVD deaths in women and sixth in men. Hence, 18.3% of deaths in women could be prevented if BMI was maintained at 20–25 kg/m^2 (Wang et al. 2023). All the studies conducted recommend that CVD is largely preventable by maintaining a healthy lifestyle that comprises healthy choices of food and eating habits.

Osteological Disorders

Bone health can be significantly influenced by poor eating habits and inadequate energy intake. The complex relationship between nutrition and bone health highlights the critical role of a balanced diet in maintaining skeletal integrity and preventing bone-related diseases like osteoporosis, rickets, and osteomalacia. Part of the important nutrients in maintaining bone health are protein, calcium, vitamin D, and micronutrients specifically magnesium, phosphorus, and vitamin K. Protein provides the structural matrix for bone, and it is required for production of collagen, which gives tensile strength to the bones. Low protein intake leads to a reduction in the bone formation and repair, leading to increased bone fragility and higher chances of fractures. Chronic protein deficiency may contribute to osteoporosis, particularly in elderly populations (Kędzia et al. 2023). Using the WHO definition of osteoporosis, approximately 6.3% of men and 21.2% of women who are aged above 50 years are affected. Calculation based on world population implies that an estimated 500 million individuals, including both men and women around the globe are affected with osteoporosis. Therefore, the numbers significantly represent the health burden across all regions of the world (Global Burden Disease Fracture Collaborators 2021).

Another important nutrient, calcium, is the fundamental mineral for bone formation and maintenance. Deficiency of calcium intake in diet is considered a widespread problem globally, with approximately 50% of the individuals in the entire world having inadequate access to dietary calcium. While individuals in low- and middle-income countries (LMICs) are at more risk of low calcium intake, many individuals in high-income countries (HICs) also fall short of necessary recommendations. Interestingly, some LMICs with lower calcium intakes paradoxically show lower rates of osteoporotic fracture compared to HICs, although data on this are sparse. Customized calcium intake recommendations are based on dietary factors, health behaviors, and risk profiles (Shlisky et al. 2022). Vitamin D is an important nutrient, which is needed for calcium absorption in the gut that helps in the maintenance of proper serum calcium and phosphate levels for normal bone mineralization. Insufficient levels of vitamin D impair calcium absorption, leading to hypocalcemia and subsequent bone demineralization. This deficiency can cause osteomalacia in adults and rickets in children, resulting in soft and weakened bones. Micronutrients such as magnesium, phosphorus, and vitamin K are also required for maintaining bone health. Magnesium is involved in bone structure and the regulation of calcium metabolism. Phosphorus is the key component of hydroxyapatite, the mineral complex that gives bones their rigidity. Vitamin K is crucial for the synthesis of osteocalcin, a protein involved in bone mineralization. Deficiencies in these micronutrients may lead to impaired bone health and increased susceptibility to osteological disorders.

The rise in obesity prevalence worldwide also affected bone health. Excessive energy intake leading to obesity can have a dual impact on bone health. While the increased body weight stimulates bone formation, it also places additional stress on the skeletal system, potentially leading to joint degeneration and osteoarthritis. Besides that, obesity is associated with chronic low-grade inflammation, which can have a detrimental effect on bone metabolism and thereby lead to higher chances of developing osteoporosis. High sodium intake leads to raised calcium in the excreted urine, reducing calcium availability for bone maintenance and thereby lead to an increased risk of osteoporosis. Excessive consumption of sugary food and beverages is responsible for poor overall diet quality, displacing nutrient-dense foods essential for bone health. High sugar intake is also linked to a rise in inflammatory markers, which have harmful effect on bone metabolism.

Cancer

Cancer is another chronic disease that has higher morbidity and mortality in the entire world. While genetic factors have a crucial role in cancer development, lifestyle choices, particularly dietary habits, are the significant modifiable risk factors. However, establishing the precise relationships between the modifiable risk factors with cancer remains a challenge. Diets high in processed meats, refined sugars, and unhealthy fats are linked to chronic inflammation and oxidative stress. These conditions may lead to DNA damage and subsequently, mutations that initiate cancer. For example, high-fat diets are associated with an increase in the level of free radicals and a decrease in antioxidant defenses, resulting in oxidative damage to cellular components. A high-sugar diet and refined carbohydrates that cause blood glucose and insulin levels to increase are responsible for an increased risk of stomach, breast, and colorectal cancers. High insulin levels can stimulate cell division and enhance the growth and metastasis of the cancer cells. Consuming a diet high in total fat, saturated fatty acids (SFAs), and trans-fatty acids (TFAs) may disrupt hormonal regulation, due to oxidative stress and inflammation, and cause a change in growth factor signaling and lipid metabolism. Hence, several factors were reported to contribute to carcinogenesis. A few dietary components contain carcinogens or form carcinogenic compounds during cooking. Processed meats, for instance, contain nitrates and nitrites that can form carcinogenic nitrosamines. Cooking methods that involve high-temperature methods, such as grilling or frying, produce heterocyclic amines (HCAs) and polycyclic aromatic hydrocarbons (PAHs), and these are associated with a greater risk of colorectal and other cancers.

To reduce cancer risk, it is important to limit foods that boost the level of insulin, such as those high in sugar and refined carbs, and instead focus on a healthy, balanced diet rich in lean protein, healthy fats, fruits, vegetables, and whole grains. Dietary interventions like calorie restriction, fasting-mimicking diets, and ketogenic diets have shown promise in modulating tumor metabolism and helping in the efficacy of cancer treatments, but still then, more research is needed.

Immunodeficiency Conditions

Nutrition has a vital role in the maintenance of the immune system. Malnutrition leads to immunodeficiency states. According to researchers, protein-energy malnutrition was reported to be associated with disturbance of cell-mediated immunity, function of the phagocytes, complement system, secretory immunoglobulin A antibody, and the production of cytokines (Chandra 1997). Several micronutrients such as iron, folate, vitamins A, C, D, and B12, zinc, copper, magnesium, and selenium influence the immune system of the body.

The T cells coordinate and control the immune responses. The T and the B cells, which are the constituents of the adaptive immune response, can identify the antigens from the different pathogens, and thereby produce antibodies to target and destroy the pathogen. Vitamin A helps in adaptive immunity. Supplementation of vitamin A in trials, which were conducted in pediatric age groups, showed a beneficial effect in increasing T-cells, particularly of the CD4 subpopulation (Villamor and Fawzi 2005). Spinach, carrots, and leafy vegetables contain abundant vitamin A. Recently, there has been a decrease in the consumption of green leafy vegetables. Lack of vitamin A leads to night blindness, skin disorders, and gastrointestinal disturbances.

Citrus fruits are rich in Vitamin C. Vitamin C deficiency leads to scurvy and bleeding gums, while vitamin D deficiency causes bone deformities and rickets in children. Vitamin E deficiency although rare, leads to muscle weakness, and malabsorption. Vitamins C, D, and E increase the natural killer cells activity. Through T cells, vitamins C, D, and E play a role in adaptive immunity.

Iron is present in red meat, eggs, beans, legumes, and spinach. Iron deficiency leads to anemia. Interestingly, recent research reports reveal that a lower level of iron in the blood was responsible for the inhibition of immune response to vaccines because T lymphocytes need iron for their metabolism (Frost et al. 2021). Iron has an important role in cytokine production and action, DNA synthesis, and phagocytosis (Haryanto et al. 2015). Citrus fruits and green leafy vegetables contain folic acid. Folate deficiency leads to anemia. Folate maintains innate immunity and increases the production of NK cells (Haryanto et al. 2015).

Seaweeds, dairy products, fish, iodized salts, and eggs contain iodine. The iodine deficiency state leads to hypothyroidism.

Zinc is found in red meat and poultry products. Zinc is responsible for both innate and acquired immunity, skin integrity, and modulation of cytokine release and affects virus replication (Haryanto et al. 2015). Recently, during the COVID-19 outbreak, supplementation of zinc was found to be effective. Meat products, shellfish, nuts, mushrooms, seafish, and sweet potatoes contain copper. The micronutrient is a part of the Cu/Zn-superoxide dismutase, which scavenges the superoxide radicals. Reports show that interleukin 2 is decreased in a copper deficiency state and there is a decrease in T cell proliferation, affecting the immune system (Percival 1998).

Magnesium is present in green leafy vegetables, whole grains, seeds, bananas, and whole wheat. A decrease in magnesium levels causes nausea, muscle weakness cramps, and fatigue. Magnesium levels influence the T cell function. Magnesium has anti-inflammatory action.

Seafood, poultry products, broccoli, meat, and Brazil nuts are sources of selenium. Selenium was reported to influence both innate and acquired immunity and acted as an antioxidant agent, which checked oxidative stress (Haryanto et al. 2015). The count of CD4 increased, and the progression of HIV-1 was reported to be suppressed by selenium (Haryanto et al. 2015).

Dental Problems

A poor diet may also cause several dental problems. Dental caries increases with age. Dental caries has been reported to affect ≤80% of the global population with approximately one-fourth of the US adult population suffering from such (Moynihan 2016). In the United States, dental caries is one of the most prevalent non-communicable chronic diseases (Moynihan 2016). Dental caries occurs as a dynamic process involving susceptible tooth surfaces, and the most common pathogen is the *Streptococcus mutans* bacteria (Tungare and Paranjpe 2023). Unhealthy foods rich in sugar are responsible for the cause of dental caries.

Dietary food contains monosaccharides (glucose, galactose, and fructose) and disaccharides (sucrose, maltose, lactose). Sucrose is the most common dietary sugar. The naturally occurring sugars in fruits, grains, and vegetables may not be harmful because of the fiber content, water, and polyphenolic compounds present in them (Tungare and Paranjpe 2023). Other than the intrinsic natural sugars, they are classified by WHO as free sugars. Free sugar includes all monosaccharides and disaccharides, which are added to the manufactured food for cooking, and those sugars that naturally occur in honey, syrups, and different fruit juices (WHO 2015). The increased intake of carbonated sweetened beverages was also responsible for an increase in the incidence of T2DM. The main culprit is the free sugar, which needs to be avoided.

Foods that contain starch include bread, pasta, potatoes, and rice. Raw starch has a low cariogenic potential but cooked starch accounts for a third and a half as cariogenic as sucrose (Moynihan 2005). The same study found that sucrose and starch combined were possibly more cariogenic than sugar (Moynihan 2005).

Each time sugar is consumed, there is an increase in demineralization of the enamel (Moynihan and Petersen 2004). It was reported that when sugar was consumed with meals up to four times per day, it was less associated with caries development while increased sugar consumption between meals was linked to a higher incidence of dental caries (Gustafsson et al. 1954). Research reports also stated the "sugar-caries relationship" to be dependent on oral hygiene, fluoride exposure, and socio-economic status (Moynihan and Petersen 2004).

Mental Health Problems

Diet is important for various disturbances in the nervous and respiratory systems. Meat consumption increases high-quality protein and saturated and *trans*-fatty acids. An earlier research study showed that higher healthy sleep scores were achieved with increased intake of vegetables, fruit, and fish but decreased with consumption of unprocessed red meat and processed meat (Hepsomali and Groeger 2021). Another earlier study showed negative associations between habitual meat intake (processed and unprocessed red meat) and the duration and quality of sleep (Lana et al. 2019). A high-protein diet causes substantial changes in the plasma tryptophan ratio, and it has an effect on the concentration of tryptophan and serotonin synthesis in brain

(Wurtman et al. 2003). The changes affect the synthesis of melatonin, serotonin, and dopamine, which causes a reduction of the synthesis of brain sleep inductors, thereby deteriorating the sleep parameters (Wurtman et al. 2003).

An interesting study was conducted on fish against meat consumption on sleep. The study had an experimental group that consumed 300 g of fatty fish three times per week for 6 months, while the control group also consumed an equal amount of meat (Hansen et al. 2014). Results revealed that the consumption of fish had a positive effect on sleep and daily functioning (Hansen et al. 2014).

Consumption of processed meats, which are rich in saturated- and trans-fatty acids, leads to changes in the gut microbiota that could result in increased circulation of serum biomarkers associated with inflammation and obesity (Chai et al. 2017). As per the 2016 International Agency for Research on Cancer Working Group, carcinogenicity was strong for red meat consumption but was moderate for processed meat consumption (Bouvard et al. 2015). Many harmful carcinogenic products such as heterocyclic aromatic amines and polycyclic aromatic hydrocarbons are formed because of meat processing. Hence, there is a need to curb the excess intake of meat products.

Latest Concept of Food and Gut Microbiota

The gut microbiome, also known as gut microbiota, includes the complete microbial community housed in the mammalian gastrointestinal tract, with the majority located in the colon. Five major phyla of microbiota are found in the human gut, and they are *Firmicutes*, *Bacteroidetes*, *Actinobacteria*, *Proteobacteria*, and *Verrucomicrobe*, with the two dominating phyla, Firmicutes and Bacteroidetes, making up to 90% of the gut microbiota (Arumugam et al. 2011). The gut microbiota is important for maintaining good health. Dysbiosis of gut microbiota leads to various diseases such as diabetes, obesity, and various gastrointestinal diseases. Recently, considerable interest was shown in targeting the gut microbiota through proper diet and nutritional interventions with an aim to promote gut health or even consider it as an additional therapy for the treatment of inflammatory bowel disease (Wastyk et al. 2021). Carbohydrates, proteins, and lipids are the main constituents of food components that modulate the gut epithelial barrier and the mucosal immune system (Wastyk et al. 2021). The gut microbiome interacts with the gut epithelia and the mucosal immune system to control and regulate homeostasis, which is vital for maintaining good gut health (Wastyk et al. 2021). If there is any imbalance in the gut health, the incidence of metabolic disorders also increases. According to research studies, a typical Western diet can lead to decreased diversity and gut microbiome dysbiosis, thereby leading to a leaky gut and chronic inflammation, which is commonly seen in inflammatory bowel disease (Wastyk et al. 2021). Figure 2 is a schematic representation of the events occurring due to gut microbiota.

Dietary fiber may be the most important dietary influencer of the gut microbiota (Chassaing et al. 2017). Dietary soluble fiber is considered soluble when it is readily fermentable by bacteria found in the gastrointestinal tract of mice and humans,

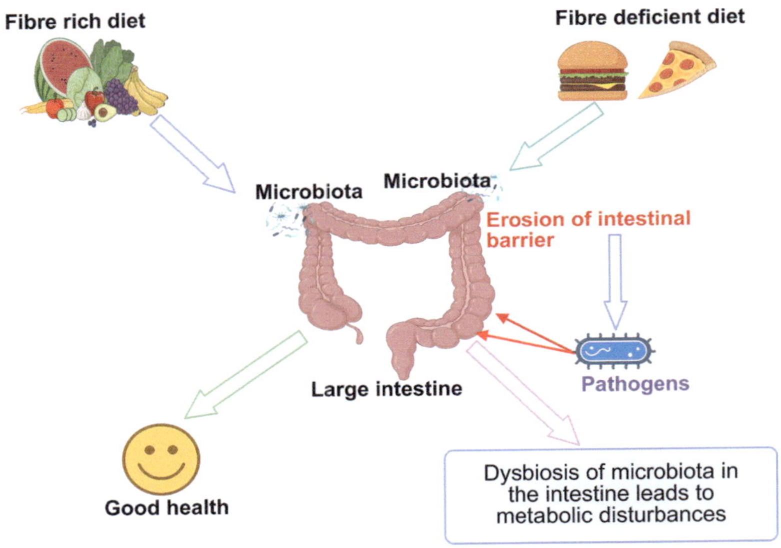

Fig. 2 Schematic diagram to show the role of diet on gut microbiota

whereas the insoluble, includes cellulose that is resistant to breakdown (Chassaing et al. 2017). Research studies reported that a deficiency of fiber results in microbiota consuming the mucus layer, which results in the erosion of the intestinal barrier, making it more vulnerable to any attack by invasive pathogens (Desai et al. 2016).

Environmental Pollutants and Nutritional Aspects

Air pollution is responsible for causing various diseases involving the cardiovascular and respiratory systems of the body (Fig. 3). These diseases include ischemic heart disease, stroke, chronic obstructive pulmonary disease, asthma, and lung cancer. Although the WHO specifies air quality guidelines, it is reported that every 9 out of 10 individuals breathe toxic environmental pollutants that are present in the air (Tiotiu et al. 2020).

Asthma is a chronic inflammatory disease that involves the respiratory system. The prevalence of asthma was reported to be between 1% and 18% of the population in various countries, and the majority of asthma cases are due to traffic-related causes (GINA Reports 2022). The harmful particles include nitrogen oxides, particulate matter, carbon monoxide, hydrocarbons, and carbon dioxide.

Consumption of fruits and vegetables could reduce the risk of developing asthma. Flavonoids are present in fruits and vegetables. A higher intake of flavonoids reduced the incidence of asthma. Flavonoids such as quercetin, kaempferol, myricetin, hesperetin, and naringenin were reported to decrease the mortality and morbidity of ischemic heart disease (Knekt et al. 2002). Apple is rich in quercetin.

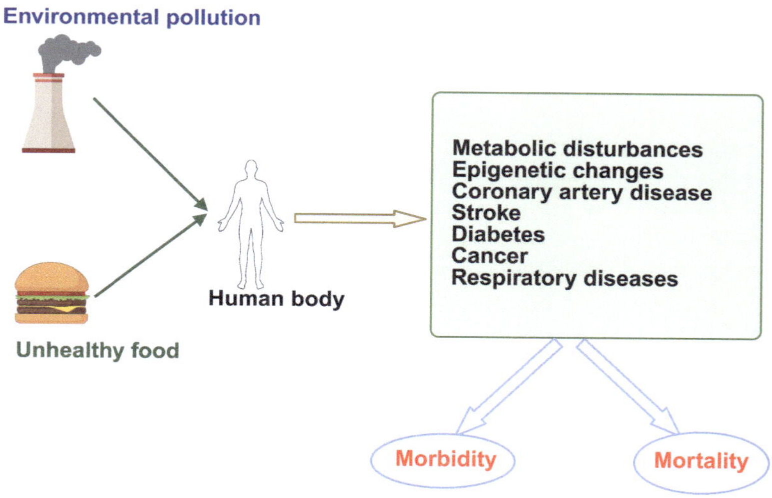

Fig. 3 Schematic diagram showing the association of unhealthy diet with exposure to environmental pollution leading to various diseases

According to earlier studies, apple consumption gave the prediction for the development of ischemic heart disease, lung cancer, type 2 diabetes, and asthma (Knekt et al. 2002). Flavonoids possess potent antioxidant properties, which can counteract oxidative stress. Flavonoids can act on cytokines, which trigger inflammatory conditions. Flavonoids are responsible for the inhibition of regulatory enzymes or transcription factors involved in controlling mediators, which are associated with inflammation (Maleki et al. 2019). Research studies showed a greater beneficial effect on asthma with the consumption of apples and oranges (Maleki et al. 2019).

Exposure to long-term air pollution may cause atherosclerosis, ischemic heart disease, and endothelial dysfunction. Particulate matter (PM) is considered to be an important risk factor for the cause of hypertension and diabetes (Kim et al. 2020). Due to environmental pollution, there may be oxidative stress in the body. Oxidative stress may be responsible for the increase in the blood pressure, disturbances in endothelial function, and cardiac disturbances. In this regard, a study conducted on subjects aged 50–71 years in six states of the United States observed that the Mediterranean diet reduced cardiovascular disease mortality, due to prolonged exposure to air pollutants (Lim et al. 2019).

Conclusion

It is important to reduce exposure to environmental toxins. Proper control measures need to be formulated to curb unnecessary exposure. Unhealthy food habits associated with exposure to environmental toxins may predispose to metabolic and epigenetic changes, coronary artery disease, diabetes, stroke, cancer, and respiratory diseases. An immuno-compromised state in any individual makes him or her more vulnerable and more prone to the development of various disorders and diseases. Proper dietary choices and nutrition are necessary to curb the oxidative stress affecting various systems and organs of the body. Foods rich in antioxidants and anti-inflammatory properties should be consumed for better health and to reduce vulnerability to environmental toxins. Healthy nutrition and better lifestyle methods should be followed and inculcated at an early stage. Long-term planning and health policies should aim to reduce the morbidity and mortality, which arise due to poor nutrition associated with environmental pollution.

References

Alsannan B, Alharmi J, Alrahal F et al (2024) Prevalence and quality of life among overweight and obese women with different severity and types of urinary incontinence. Med Princ Pract 33:47–55

Arumugam M, Raes J, Pelletier E et al (2011) Enterotypes of the human gut microbiome. Nature 473:174–180

Bouvard V, Loomis D, Guyton KZ et al (2015) Carcinogenicity of consumption of red and processed meat. Lancet Oncol 16:1599–1600

Bowen KJ, Harris WS, Kris-Etherton PM (2016) Omega-3 fatty acids and cardiovascular disease: are there benefits? Curr Treat Options Cardiovasc Med 18:69

Chai W, Morimoto Y, Cooney RV et al (2017) Dietary red and processed meat intake and markers of adiposity and inflammation: the multiethnic cohort study. J Am Coll Nutr 36:378–385

Chandra RK (1997) Nutrition and the immune system: an introduction. Am J Clin Nutr 66:460S–463S

Chassaing B, Vijay-Kumar M, Gewirtz AT (2017) How diet can impact gut microbiota to promote or endanger health. Curr Opin Gastroenterol 33:417–421

Desai MS, Seekatz AM, Koropatkin NM et al (2016) A dietary fiber-deprived gut microbiota degrades the colonic mucus barrier and enhances pathogen susceptibility. Cell 167:1339–1353.e21

Eapen JT (2019) Insulin resistance: is it a feedback mechanism or a therapeutic target in type 2 diabetes? Int J Res Med Sci 7:3202

Frost JN, Tan TK, Abbas M et al (2021) Hepcidin-mediated hypoferremia disrupts immune responses to vaccination and infection. Med 2:164–179.e12

GINA Reports (2022) Global initiative for asthma-GINA. Available via https://ginasthma.org/gina-reports/. Accessed on 27 July 2024

Global Burden Disease Fracture Collaborators (2021) Global, regional, and national burden of bone fractures in 204 countries and territories, 1990-2019: a systematic analysis from the global burden of disease study 2019. Lancet Healthy Longev 2:e580–e592

Gustafsson BE, Quensel CE, Lanke LS et al (1954) The Vipeholm dental caries study; the effect of different levels of carbohydrate intake on caries activity in 436 individuals observed for five years. Acta Odontol Scand 11:232–264

Hansen AL, Dahl L, Olson G et al (2014) Fish consumption, sleep, daily functioning, and heart rate variability. J Clin Sleep Med 10:567–575

Haryanto B, Suksmasari T, Wintergerst E et al (2015) Multivitamin supplementation supports immune function and ameliorates conditions triggered by reduced air quality. Vitam Miner 4:2–15

Hepsomali P, Groeger JA (2021) Diet, sleep, and mental health: insights from the UK biobank study. Nutrients 13:2573

Issa S, Ibrahim SMS, Musawwi AAA (2017) A study on abdominal obesity at Basra university staffs. Clin Med Res 6:69
Kędzia G, Woźniak M, Samborski W, Grygiel-Górniak B (2023) Impact of dietary protein on osteoporosis development. Nutrients 15:4581
Kim JB, Prunicki M, Haddad F et al (2020) Cumulative lifetime burden of cardiovascular disease from early exposure to air pollution. J Am Heart Assoc 9:e014944
Kim B, Yu MY, Shin J (2024) Effect of low sodium and high potassium diet on lowering blood pressure and cardiovascular events. Clin Hypertens 30:2
Knekt P, Kumpulainen J, Järvinen R et al (2002) Flavonoid intake and risk of chronic diseases. Am J Clin Nutr 76:560–568
Lana A, Struijk EA, Arias-Fernandez L et al (2019) Habitual meat consumption and changes in sleep duration and quality in older adults. Aging Dis 10:267–277
Lim CC, Hayes RB, Ahn J et al (2019) Mediterranean diet and the association between air pollution and cardiovascular disease mortality risk. Circulation 139:1766–1775
Lin X, Li H (2021) Obesity: epidemiology, pathophysiology, and therapeutics. Front Endocrinol (Lausanne) 12:706978
Lui M, Safiri S, Mereke A et al (2021) Burden of ischemic heart disease in central Asian countries, 1990-2017. Int J Cardiol Heart Vasc 33:100726
Maleki SJ, Crespo JF, Cabanillas B (2019) Anti-inflammatory effects of flavonoids. Food Chem 299:125124
Moynihan PJ (2005) The role of diet and nutrition in the etiology and prevention of oral diseases. Bull World Health Organ 83:694–699
Moynihan P (2016) Sugars and dental caries: evidence for setting a recommended threshold for intake. Adv Nutr 7:149–156
Moynihan P, Petersen PE (2004) Diet, nutrition and the prevention of dental diseases. Public Health Nutr 7:201–226
NCD Risk Factor Collaboration (NCD-RisC) (2017) Worldwide trends in body-mass index, underweight, overweight, and obesity from 1975 to 2016: a pooled analysis of 2416 population-based measurement studies in 128·9 million children, adolescents, and adults. Lancet 390:2627–2642
Niklas A, Marcinkowska J, Kozela M et al (2019) Blood pressure and cholesterol control in patients with hypertension and hypercholesterolemia: the results from the Polish multicenter national health survey WOBASZ II. Pol Arch Intern Med 129:864–873
Percival SS (1998) Copper and immunity. Am J Clin Nutr 67:1064S–1068S
Pramono A, Fitranti D, Nugroho K et al (2023) The association between unhealthy food consumption and impaired glucose metabolism among adults with overweight or obesity: a cross-sectional analysis of the Indonesian population. J Obes 2023:2885769
Rasmussen S, Johansen NB, Witte DR et al (2016) Incidence of register-based diabetes 10 years after a stepwise diabetes screening programme: the ADDITION-Denmark study. Diabetologia 59:989–997
Sadiq IZ (2023) Lifestyle medicine as a modality for prevention and management of chronic diseases. J Taibah Univ Med Sci 18:1115–1117
Safiri S, Kolahi AA, Hoy D et al (2020) Global, regional, and national burden of neck pain in the general population, 1990-2017: systematic analysis of the global burden of disease study 2017. BMJ 368:m791
Saxena S, Saxena R, Singh AP et al (2024) Metabolic reasons of diabetes mellitus: an update. Int J Clin Biochem Res 11:8–11
Shlisky J, Mandlik R, Askari S et al (2022) Calcium deficiency worldwide: prevalence of inadequate intakes and associated health outcomes. Ann N Y Acad Sci 1512:10–28
Tiotiu AI, Novakova P, Nedeva D et al (2020) Impact of air pollution on asthma outcomes. Int J Environ Res Public Health 17:6212
Tiwari K, Kumar D (2018) Recent methods for diagnosis of diabetes mellitus. Himal J Health Sci 3: 58–63

Tungare S, Paranjpe AG (2023) Diet and nutrition to prevent dental problems. In: StatPearls [Internet]. StatPearls Publishing, Treasure Island. Available via https://www.ncbi.nlm.nih.gov/books/NBK534248/. Accessed on 26 July 2024

Villamor E, Fawzi W (2005) Effects of vitamin a supplementation on immune responses and correlation with clinical outcomes. Clin Microbiol Rev 18:446–464

Wang Q, Yang X, Zhang H et al (2023) Temporal trends in mortality of tuberculosis attributable to high fasting plasma glucose in China from 1990 to 2019: a joinpoint regression and age-period-cohort analysis. Front Public Health 11:1225931

Wariri O, Nkereuwem E, Erondu NA et al (2021) A scorecard of progress towards measles elimination in 15 west African countries, 2001-19: a retrospective, multicountry analysis of national immunisation coverage and surveillance data. Lancet Glob Health 9:e280–e290

Wastyk HC, Fragiadakis GK, Perelman D et al (2021) Gut-microbiota-targeted diets modulate human immune status. Cell 184:4137–4153.e14

WHO (2015) Sugars intake for adults and children. WHO, Geneva. Available via https://pubmed.ncbi.nlm.nih.gov/25905159/. Accessed on 26 July 2024

Williams J, Allen L, Wickramasinghe K et al (2018) A systematic review of associations between non-communicable diseases and socioeconomic status within low- and lower-middle-income countries. J Glob Health 8:020409

Wurtman RJ, Wurtman JJ, Regan MM et al (2003) Effects of normal meals rich in carbohydrates or proteins on plasma tryptophan and tyrosine ratios. Am J Clin Nutr 77:128–132

Xu F, Earp JE, Adami A et al (2022) The sex and race/ethnicity-specific relationships of abdominal fat distribution and anthropometric indices in US adults. Int J Environ Res Public Health 19:15521

Xu X, Shi Z, Zhou L et al (2024a) Impact of COVID-19 on risks and deaths of non-communicable diseases in the Western Pacific region. Lancet Reg Health West Pac 43:100795

Xu X, Zeng L, Jha V et al (2024b) Potassium-enriched salt substitutes: a review of recommendations in clinical management guidelines. Hypertension 81:400–414

Ye J, Wu Y, Yang S et al (2023) The global, regional and national burden of type 2 diabetes mellitus in the past, present and future: a systematic analysis of the global burden of disease study 2019. Front Endocrinol 14:1192629

MIX
Papier aus verantwortungsvollen Quellen
Paper from responsible sources
FSC® C105338

If you have any concerns about our products,
you can contact us on
ProductSafety@springernature.com

In case Publisher is established outside the EU,
the EU authorized representative is:
Springer Nature Customer Service Center GmbH
Europaplatz 3, 69115 Heidelberg, Germany

Printed by Libri Plureos GmbH
in Hamburg, Germany